W0245768

Krause · Jäger (Eds.)
High Performance Computing in Science and Engineering '01

Springer-Verlag Berlin Heidelberg GmbH

Egon Krause · Willi Jäger

Editors

High Performance Computing in Science and Engineering '01

Transactions of the High Performance Computing Center
Stuttgart (HLRS) 2001

With 252 Figures, 81 in Color and 45 Tables

 Springer

Editors

Egon Krause
Aerodynamisches Institut der RWTH Aachen
Wuellnerstraße zw. 5 u. 7
52062 Aachen, Germany
e-mail: ek@aia.rwth-aachen.de

Willi Jäger
Institut für Wissenschaftliches Rechnen
Universität Heidelberg
Im Neuenheimer Feld 368
69120 Heidelberg, Germany
e-mail: jaeger@iwr.uni-heidelberg.de

Library of Congress Cataloging-in-Publication Data applied for

Die Deutsche Bibliothek - CIP-Einheitsaufnahme
High performance computing in science and engineering ... : transactions of
the High Performance Computing Center Stuttgart (HLRS) .... - 1998 (1999)-.
- Berlin ; Heidelberg ; New York ; Barcelona ; Hong Kong ; London ; Milan ;
Paris ; Tokyo : Springer, 1999
    Erscheint jährl. - Bibliographische Deskription nach 2001 (2002)
2001 . - (2002)
    ISBN 978-3-642-62719-4   ISBN 978-3-642-56034-7 (eBook)
    DOI 10.1007/978-3-642-56034-7

Front cover figure: Numerical Simulation of Mixing and Combustion
in Supersonic Flows (Institute of Combustion Technology DLR Stuttgart)

Mathematics Subject Classification (2000): 65Cxx, 65C99, 68U20

ISBN 978-3-642-62719-4

This work is subject to copyright. All rights are reserved, whether the whole or part of the material
is concerned, specifically the rights of translation, reprinting, reuse of illustrations, recitation, broad-
casting, reproduction on microfilm or in any other way, and storage in data banks. Duplication of
this publication or parts thereof is permitted only under the provisions of the German Copyright Law
of September 9, 1965, in its current version, and permission for use must always be obtained from
Springer-Verlag. Violations are liable for prosecution under the German Copyright Law.

http://www.springer.de
© Springer-Verlag Berlin Heidelberg 2002
Originally published by Springer-Verlag Berlin Heidelberg 2002
Softcover reprint of the hardcover 1st edition 2002

The use of general descriptive names, registered names, trademarks, etc. in this publication does not
imply, even in the absence of a specific statement, that such names are exempt from the relevant pro-
tective laws and regulations and therefore free for general use.

Typeset by the authors. Edited by Kurt Mattes, Heidelberg.
Cover design: *design & production* GmbH, Heidelberg

SPIN: 10853340   46/3142LK - 5 4 3 2 1 0 – Printed on acid-free paper

Preface

Prof. Dr. Egon Krause

Aerodynamisches Institut
RWTH Aachen
Wüllnerstr. zw. 5 u. 7, D-52062 Aachen

Prof. Dr. Willi Jäger

Interdisziplinäres Zentrum für Wissenschaftliches Rechnen
Universität Heidelberg
Im Neuenheimer Feld 368, D-69120 Heidelberg

Promoting computational science in research and technology is a main aim of the activity in high-performance scientific computing. In order to achieve this goal advanced techniques in computer technologies and computational methods have to be developed, tested and integrated into new working tools to solve the challenging problems in modeling and simulating complex processes in the various fields of science and technology.

This series of publications provides an annual survey of the progress in scientific computing performed at the High Performance Computing Center Stuttgart (HLRS) and the Scientific Supercomputing Center Karlsruhe (SSC) and makes the most important results available to the scientific public. This fourth volume covers the period beginning October 2000 to the present and contains 44 selected contributions to computational fluid dynamics, physics, chemistry and engineering, to reactive flows, structural dynamics and other fields of computational sciences, presented at the Fourth Results and Review Workshop on High Performance Computing in Science and Engineering, held at the HLRS, October 8th and 9th, 2001.

According to the rules of access to the supercomputer facilities of the HLRS and SSC the Steering Committee is responsible for the scientific control of the program. The main emphasis is put on the computational part, however, also the general scientific quality was checked by specialists in the evaluation procedure. Here, the support of WiR, a network of competence in computational sciences in Baden-Württemberg, by now has become indispensable and is absolutely essential for almost all activities. The investigations reported here were selected by members of the Steering Committee out of a total number of projects of 100, processed on the machines of the HLRS, and another 20 projects on the IBM SP complex of the SSC. They were presented in lectures or poster sessions at the workshop. Altogether, scientists from 44 German universities and research centers participated in the investigations.

The articles published in this volume were subjected to an internal review. The authors were encouraged to emphasize computational techniques used in solving the problems investigated. The importance of the newly computed results for the specific disciplines, as interesting they may be from a scientific point of view, were not in the focus here.

Following the main goal of the workshop and also of the publication, communication and discussion of computational methods and tools were stirred. Time and again experience has shown, that mathematical and computational methods can be carried over from one field to another; tools developed for solving specific problems can often be applied in other fields, sometimes seeming to be rather far away from the original one. Interdisciplinary contacts and exchange of experience and ideas have become very important, especially in the new discipline of high-performance scientific computing. For example the following topics arise in many investigations and are of general interest: Implementation of artificial boundary conditions, grid generation techniques, methods of discretization, iterative techniques for solving the resulting generally large systems of algebraic equations, and, last but not least, visualization techniques. Visualization is of outmost importance for investigations of multidimensional processes. High-dimensional problems with multiple independent or dependent variables are arising in many applications and are a computational challenge of growing importance.

At the HLRS about 37 percent of the machine capacity available on the SX4, the newly installed SX5, and on the T3E were allocated for flow simulations, ranging over a wide spectrum of problems, from investigations of aerodynamic flows, of laminar-turbulent transitional and turbulent flows to simulations of flow processes in chemical engineering, and of flows of reacting gases. It is interesting to note that 25 percent of all flow simulations were carried out on the SX5, and only some 5 percent on the T3E. The reason for this choice seems to be the large demand in memory in these simulations. In physics the trends are reversed: About 45 percent of the total machine capacity was used to simulate physical processes, including solid state physics; of these only some 2,5 percent were carried out on the SX5, and over 40 percent on the T3E. In contrast to fluid dynamics and physics, machine capacity reserved for simulation of chemical processes was only some 7 percent, for problems in structural dynamics about 4 percent. The rest, a small fraction of the available machine time, was used to simulate processes in life sciences and computer science. In Karlsruhe at the SSC similar trends are observed: On the SP-SMP, for example, 50 percent of the machine capacity were used for flow simulations, 26 percent for problems in physics, 13 percent for problems in chemistry, and 8 percents for problems in computer science.

These figures show that the HLRS and the SSC machines are preferably used in fluid dynamics and physics. As in previous years, these two disciplines represent the majority of the user community, followed by chemists in the third place. Corresponding to this distribution this volume contains 16 contributions covering problems in fluid dynamics, including reacting flows, 17 concerned with problems in physics, including solid state physics, 4 with problems in chemistry, 4 with computer science problems, and 3 with problems in structural dynamics.

Experience gained since the establishment of the HLRS in Baden-Württemberg shows that the network WiR is of outmost importance for using high-technology tools efficiently. Scientists specialized in computational sciences and experts in different fields are needed not only for the planning and evaluation necessary to manage high-performance computing centers but also for developing and providing new methods and tools and for education on all levels as well. WiR was formed several years ago by leading scientists and their teams at the Universities of Freiburg, Heidelberg, Karlsruhe, Stuttgart and Tübingen, now including altogether some 25 chairs and also scientists from industry. The members represent the scientific disciplines aerodynamics, applied mathematics, astrophysics, chemistry, combustion technology, computer science, hydraulics, hydrodynamics, numerical analysis, optimization, solid mechanics and turbo machinery. Presently WiR is strongly engaged in setting up several long-term programs for scientific computing and computational science. One such initiative is aimed at strengthening research in massively parallel computing. Some of these results are included in this volume in the chapter on computer science.

The articles published in this volume clearly demonstrate in a convincing manner the progress made in high performance computing at the HLRS and the SSC during the past years. Thus, it is not only shown that the investments in these activities were justified, but also promising perspectives for the future development are provided. Most of the problems solved today, could not have been tackled only a few years ago. Continued improvements of algorithms and substantial increase in computational speed and enlargement of storage capacity made this success possible. However, considering the demands of the users, the hardware situation of the HLRS and the rapid progress in hardware developments, it became clear that the capacity and performance of the HLRS machines need urgent improvement to further be ready for international competitiveness.

In order to be able to satisfy these demands in the future, the HLRS investigated the trends of development in the major fields served in November 2000 by a specially assigned committee of experts with the aim to provide recommendations for future installations. Such an investigation appeared to be of outmost necessity at that time, as the vector computers, the machines that so far had carried a large portion of the workload of the HLRS, seem to loose ground, while clusters of microprocessors are presently being used in increasing number. Even more important, it was recognized, that the Accelerated Strategic Computing Initiative (ASCI) in the USA is presently dictating the speed of progress with systems of a peak performance of about 10 TFLOPS already built, and others aiming at 30 TFLOPS in the planning. In Japan the Earth Simulator project aims at a peak performance of 40 TFLOPS. The recommendation committee reached the conclusion, that the HLRS will have to deliberate design, installation, and use of a follow-on system that can provide at least a sustained level of performance of 3 TFLOPS, implying a peak performance between 10 and 15 TFLOPS, if international competitiveness is to be maintained.

Although highest performances can presently only be reached with parallel computing machines, it must be realized, that it is massive parallelism. With several thousands of processors working in parallel the complexity of computing is substantially increased. The key issues to be addressed are then the integration of hard- and software into a truly working tool. One way to reach this goal might be a central global file system at the core of a configuration that integrates pre-processing, simulation and postprocessing. Also new ways will have to be found to ease parallelization for the user, who will have to adapt his algorithms to the architecture of the machine to be installed.

We gratefully acknowledge the continued support of the Land Baden-Württemberg in promoting and supporting high-performance computing on an internationally competitive basis. Grateful acknowledgment is also due to the Deutsche Forschungsgemeinschaft (DFG); many projects processed on the machines of the HLRS and the SSC could not have been carried out without the support of the DFG. Also, the increasing activities of the WiR, strengthening scientific investigations on a large scale in the State of Baden-Württemberg are gratefully acknowledged. Finally, we thank the Springer Verlag for publishing this volume and thus helping to position the local activities into an international frame. We hope that this series of publications is contributing to the global promotion of high performance scientific computing.

Stuttgart, October 2001   W. Jäger
                                                                E. Krause

Contents

Computational Fluid Dynamics

Physics

Prof. Dr. Hanns Ruder, Dr. Roland Speith

Institut für Astronomie und Astrophysik, Abteilung Theoretische Astrophysik
Universität Tübingen, Auf der Morgenstelle 10, D-72076 Tübingen

In the last decade, the methods of scientific computing and the usage of supercomputers have become a common and successful tool of research in all branches of physics. This is demonstrated by the wide range of physical applications that is covered by the following articles, which present a selection of the currently running projects with general physical background at the supercomputing centres in Stuttgart and Karlsruhe.

The power of high performance computing in research can be seen by the fact that the bigger part of the presented work consists of long-term projects that successfully have continued and progressed over several years. A particular example is the IMD-project (Schaaf et al.) where a massively parallel molecular dynamics package has been developed. This article reports on recent investigations performed with this general purpose program, especially on simulations of dislocations in quasicrystals. An important point in this work is the appropriate visualisation of the three-dimensional results, which is a non-trivial problem.

Another project which has been carried out at the supercomputing centre in Stuttgart for some years now is the calculation of the dynamics of convection and dynamos in rotating spherical shells (Grote, Busse, Simitev). The objective of this project is the numerical study of the process of magnetic field generation by geophysical flows in rotating fluid shells like the liquid iron core of the Earth. In the report of this year, especially coherent structures in turbulent convection are investigated.

A third example for a successfully ongoing work is the modelling of the propagation of seismic waves in the Earth's upper mantle (Ryberg et al.). In the presented paper, the application of synthetic seismograms as a tool for the interpretation of observed seismograms is discussed. Other ongoing projects that are to mention are N-body simulations to investigate the dynamics of star clusters and stellar systems with a massive central black hole (Hemsendorf et al.), or the computation of eigenvalues and eigenfunctions of the hyperbolic 3D-space (Aurich) which are related to certain cosmological models as well as to quantum chaos.

Also several new projects started during the last year. Three examples of them are presented as well: The three-dimensional simulation of astrophysical jets propagating through inhomogeneous media (Thiele, Camenzind), the modelling of nano-pores and their phase transitions (Nielaba), and the determination of properties of quantum chromo dynamics at finite temperatures by comparing simulations with imaginary and with real chemical potentials (Hart, Laine, Philipsen).

Simulation of Dislocations in Icosahedral Quasicrystals with IMD

Gunther Schaaf, Franz Gähler, Christopher Kohler, Ulrich Koschella, Nicoletta Resta, Johannes Roth, Christoph Rudhart, and Hans-Rainer Trebin

Institut für Theoretische und Angewandte Physik, Universität Stuttgart, Pfaffenwaldring 57, 70550 Stuttgart, Germany

Abstract. We report on recent investigations performed with IMD (ITAP Molecular Dynamics), a general purpose program for classical molecular dynamics simulations on workstations and massively parallel supercomputers. Especially the simulations of dislocations in icosahedral quasicrystals are described. The quasiperiodic structure leads to new interesting properties. The visualization of a dislocation is much more complicated than in periodic crystals and is presented in detail. An overview of the software used is also provided.

1 Introduction

IMD (ITAP Molecular Dynamics) [1] is a program package for classical molecular dynamics simulations [2]. It is designed to run efficiently both on single processor workstations and massively parallel supercomputers. IMD supports a large number of different thermodynamic ensembles, among which is the microcanonical or NVE ensemble, the canonical ensemble (NVT) and the NPT ensemble where the pressure instead of the volume of the sample is fixed. IMD offers several options for the study of all kinds of mechanical properties. These enable various deformations of the sample like shear deformations, or loading the sample for the study of cracks. For further details we refer to the IMD home page [3], where more information on the most recent version can be found.

Several forms of atomic interactions are supported, which are suitable for different types of materials. The simplest interactions are given by central pair potentials. Such two-body potentials have mainly been used for quasicrystals, where more refined potentials are still lacking. For an adequate description of most materials, it is necessary to use more complicated potentials, which include also three-body or many-body terms. For covalently bound materials, like ceramics and semiconductors, it is essential to include also three-body terms in the interactions, which depend on the angle between adjacent bonds. Such interactions are provided in the form of Stillinger-Weber potentials [4], and Tersoff potentials [5]. Adequate interactions for metals must take into account the delocalized conduction electrons and the consequences of the Pauli principle. This can be achieved with the Embedded Atom Method (EAM) [6], which has been implemented in IMD. In EAM potentials, in addition to the pair force a cohesive force term is calculated for each atom, which depends on the local electron density into which the atom is "embedded". This density is calculated from contributions of neighbouring atoms.

The parallelization method and the parallel performance of the pair interactions has been described in [1]. More details on the implementation and the parallelization of the different many-body interactions can be found in [7]. Besides the implementation and testing of new types of interaction, IMD has mainly been used in production runs for the study of the physical properties of quasicrystals.

2   Simulation of dislocations in icosahedral quasicrystals

The structure of a periodic crystal can be described as a periodic arrangement of copies of a single structural unit, the unit cell of the crystal. For quasicrystals, which are quasiperiodic rather than periodic, a similar description requires at least two such structural units. Geometrically, these units can be described as *tiles* decorated with atoms, which are forming a quasiperiodic tiling.

Like in ordinary crystals the plasticity of quasicrystals is mainly due to the generation and motion of dislocations. This has been proved in several experimental investigations [8]. While quasicrystals are brittle at room temperature there is a ductile high-temperature regime starting at about 70-80% of the melting temperature.

Theoretical work considering the deformation process qualitatively has not helped much in understanding the microscopic properties. Therefore MD simulations of shear deformations at various temperatures have been performed on a three-dimensional icosahedral quasicrystalline model system: the 3D-Penrose-Tiling [9] decorated with two types of atoms according to a structure model by Henley and Elser [10]. This atomic structure is stable if the atoms interact via simple pair potentials like the *Lennard-Jones potential*

$$V(r_{ij}) = \varepsilon_{ij}((\frac{\sigma_{ij}}{r_{ij}})^{12} - 2(\frac{\sigma_{ij}}{r_{ij}})^6) \tag{1}$$

where i and j denote the two atom types [11]. The structure was relaxed and a shear deformation was applied at various temperatures.

2.1   Method

In former simulations [12] a dislocation had been put into a quasicrystalline sample of 1,504,080 atoms by applying a displacement field of Peierls-Nabarro type

$$u_x(x) = -\frac{b}{2\pi} \arctan \frac{x}{\zeta} \tag{2}$$

where b is the length of the Burgers vector **b**. This Burgers vector was obtained by an energy consideration according to [13]: the sample was cut along an easy glide plane (a plane with a large separation between the atoms) and the two halves were shifted rigidly. The mismatch energy was calculated, and shift directions corresponding to its minima were chosen as candidates for Burgers vectors.

The sample was relaxed by a microconvergence method [14] corresponding to an undercooling of the sample. The dislocation could be stabilized in this way. Then a shear deformation of 2.1% along the direction of the Burgers vector was applied. The dislocation moved through the sample creating a stacking fault in its wake. Both simulations required partly fixed and partly free boundaries, which causes large inhomogeneities in the distribution of system properties like temperature and pressure.

Therefore, in a second series of simulations we decided to apply the shear deformation in a continuous way, using a non-equilibrium MD (NEMD) algorithm. NEMD had previously been used for the simulation of transport processes, especially for the computation of the shear viscosity of fluids. In that case a Couette flow is modeled by the application of a linear velocity field. The shear viscosity is the ratio of the pressure and the applied velocity gradient. This has several advantages over the linear response method where the time correlation functions are integrated: it is less expensive in computer time and avoids the poor signal-to-noise ratio of the latter. As NEMD algorithms require a modification of the equations of motion they are often called *synthetic algorithms*.

The formal analogy between a Couette flow and a shear deformation can be used to formulate a Hamiltonian of a dynamics explicitly containing the shear deformation. Such a deformation adds energy to the system that can be expressed in terms of the infinitesimal deformation $\nabla_\alpha u_\beta$ where u_β is the displacement

$$\delta E = \sum_{\alpha,\beta} \sigma_{\alpha\beta} \delta(\nabla_\alpha u_\beta). \tag{3}$$

The rate of dissipated energy is then

$$\dot{E} = \sigma_{\alpha\beta}\delta(\dot{\nabla}_\alpha u_\beta) = \frac{d}{dt}\left[\sum_i (q_\alpha^{(i)} p_\beta^{(i)})\nabla_\alpha u_\beta\right] \tag{4}$$

where $q_\alpha^{(i)}$ and $p_\alpha^{(i)}$ are the position and momentum coordinates of the i-th particle. With H_0 being the Hamiltonian of the unperturbed system the following ansatz is chosen

$$H_{Doll} = H_0 + \sum_i (q_\alpha^{(i)} p_\beta^{(i)})\nabla_\alpha u_\beta \tag{5}$$

yielding the equations of motion

$$\dot{q}_\alpha^{(i)} = \frac{p_\alpha^{(i)}}{m_i} + q_\beta^{(i)} \nabla_\alpha u_\beta, \quad \dot{p}_\alpha^{(i)} = f_\alpha^{(i)} - q_\beta^{(i)} \nabla_\alpha u_\beta \qquad (6)$$

The fantasy name *Doll's tensor* shall stress that we deal with a synthetic algorithm here. The dissipation is the same as the one from linear response theory. It even does not change if the velocity gradient is transposed in the momentum equation:

$$\dot{p}_\alpha^{(i)} = f_\alpha^{(i)} - q_\beta^{(i)} \nabla_\beta u_\alpha. \qquad (7)$$

Contrary to the Doll's tensor equations, these equations of motion reduce to the Hamiltonian equations if the velocity field is constant. Their name *SLLOD equations* shall remind of the transposition.

The SLLOD-ensemble must be applied together with periodic boundary conditions in all directions. As energy is permanently added to the system, a thermostat has to be used. In our case a Nosé-Hoover-Thermostat [15] was applied, where a heat bath is coupled to the system, which effectively rescales the momenta of the particles. A critical parameter is the mass of the thermostat, which must be chosen appropriately.

2.2   Visualization

A dislocation is a line defect that can be created by an incomplete slip of two crystal halves along a *glide plane*. The dislocation line is the boundary between the displaced and undisplaced parts of the crystal. If the dislocation *glides* — moves within its glide plane — the slipped region is enlarged. In periodic crystals the atoms within a narrow region around the dislocation line — the *dislocation core* — are displaced from their ideal positions, but far away from the dislocation line the atoms are again in perfect order, provided the Burgers vector of the dislocation is a primitive translation of the crystal lattice. The same is true for the atoms in the wake of a gliding dislocation. This is a direct consequence of the translational periodicity of the crystal lattice. Quasicrystals lack such a translational periodicity, so that every dislocation leaves behind a *stacking fault*, a plane of atomic misalignment.

In ordinary crystals every atom of a given type has the same local environment, so that all these atoms have the same potential energy. Therefore, dislocations in ordinary crystals can easily be identified by plotting only the atoms having an excessive potential energy. They form narrow "tubes" indicating the position of the dislocation line. The situation is more complicated in a quasicrystal. In our model, for example, the atoms of one type are subject to potential energies varying over a range from -24 to -15 LJ units, while the excess energy of an atom in the dislocation core is only a few LJ units. We have implemented the following visualization methods in IMD:

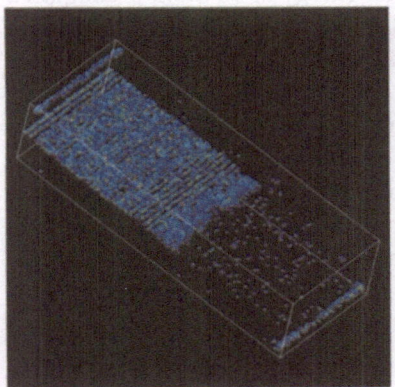

Fig. 1. Visualization of the dislocation line and the stacking fault. Only atoms with a large potential energy compared to the initial value are plotted.

Fig. 2. Volume visualization of the distribution of the kinetic energy. Like in Fig. 1 the energy increases from blue to red colours.

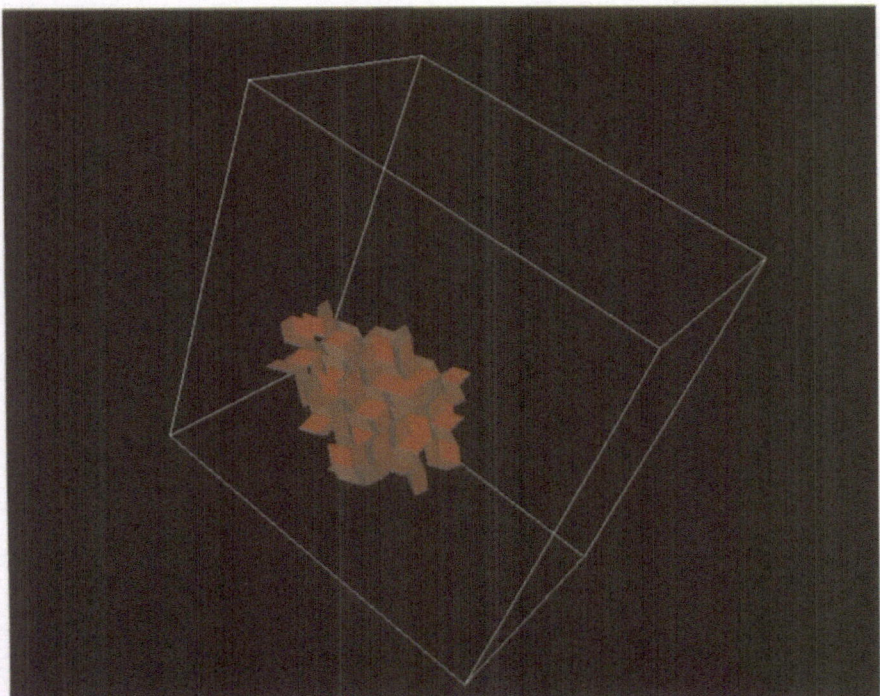

Fig. 3. Visualization of the boundary surface between defective and undefective regions. Rhombohedra with distorted bonds form the defective region.

- *Potential energy difference.* The potential energy of an atom is compared to its energy in the initial state not containing the defect. If the difference exceeds a threshold the atom is plotted (Fig. 1). The dislocation core and the stacking fault cannot be separated by this method.
- *Kinetic energy.* The kinetic energy is sampled into an array of boxes. We obtain pictures like Fig. 2 which do not allow for a precise determination of the current position of the dislocation.
- *Retiling of the sample.* The initial rhombohedral tiles and their edges — in the following referred to as *bonds* — are stored in a list. For a given snapshot of the simulation the rhombohedra with strongly distorted bonds are deleted from the list. In general, each rhombohedron face belongs to two rhombohedra except those whose rhombohedra have been deleted. By plotting the faces where one rhombohedron is deleted, but the other is not, we obtain a surface separating the defective and the undefective region (Fig. 3). This allows for a precise determination of the current position of the dislocation.

2.3   Results

Simulations at three different temperatures corresponding to 20, 50, 66% of the melting temperature have been performed. The stress-strain curves (Fig. 4) show a behaviour corresponding to Hooke's law. At deformations

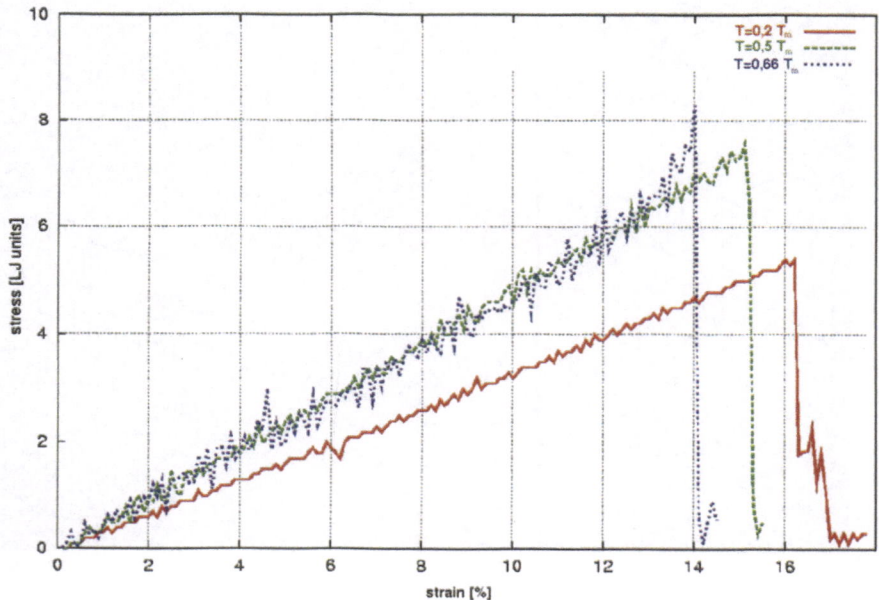

Fig. 4. Stress-strain curve for various temperatures.

between 14 and 17% there is a yield drop indicating the onset of plastic deformation. The critical strain decreases with increasing temperature while both the critical stress and the shear modulus increase. This unphysical behaviour is a consequence of the NVT ensemble, where the volume is kept constant. The hydrostatic pressure of the sample is larger for the high temperature simulations. Simulations with an NPT ensemble provide a solution to this problem, but the resulting volume fluctuations are very difficult to control.

Due to the periodic boundaries in all directions, a dislocation dipole has nucleated enclosing a stacking fault (Fig. 3). The two dislocations move away from each other, enlarging the stacking fault. The dislocation line bulges out during its motion. This is due to pinning at *clusters*, atomic sites of high symmetry [16]. A detailed analysis of the dislocation motion is in progress.

3   Further applications: fracture and shock wave simulations

IMD has also been used for the study of other mechanical properties of quasicrystals. Propagation of mode-I cracks have been studied in two-dimensional decagonal ordered and randomized model quasicrystals in dependence of temperature. For low temperatures and perfect ordered systems the crack is propagating by dislocation emission. Due to the quasiperiodicity, dislocations, which are emitted in a zig-zag fashion, are followed by phason walls. Along these walls the interfacial energy is lowered, and the material is opening, causing a rough cleavage surface. At intermediate temperatures this dislocation emission process still works, but with much longer dislocation paths. At more elevated temperatures the formation of voids in front of the crack was observed.

The phason-phonon coupling constant of a two-dimensional quasicrystalline model system was determined using a decomposition of its elastic energy into symmetry invariant terms. By applying a phononic strain to the samples at fixed phasonic strain, parabolas were obtained whose vertices are displaced from zero. From this displacement the coupling constant can be calculated. By varying the choice of the phasonic strain the complete set of quasicrystalline elastic constants could be determined.

Finally, shock waves have been simulated in icosahedral quasicrystals and structurally similar crystals [17]. The behaviour of these materials is much more complicated than that of simple crystals, where stacking faults are generated, but the structure itself remains undestroyed. In the materials studied here the structure is increasingly fragmented with increasing strength of the shock waves, until the structure finally turns completely amorphous at very high shock wave intensity.

4 Software for Visualization

For the visualization of volume data, the package volimd [18] is available. It can be used to render energy distributions, for example. A new version of the volume visualization module is currently under development in cooperation with J. Schulze-Döbold of SFB 382 [19,20].

Even though IMD provides socket connections for online visualization, to which volimd can connect, this approach has turned out to be not very useful, at least in three dimensions. The reason is that elaborate post processing is necessary to filter the *interesting* information from the enormous amount of *available* information.

Our main interest lies in the determination of strongly localized defects of the size of a few atoms in a bulk of several million atoms. If the volume visualization is refined to the atomic level the data files soon become intractably large. If the defect atoms are singled out by one of the methods described in section 2.2, the volume picture is still too diffuse to be useful.

Currently the most practicable way to figure out what happens in the simulation box is to postprocess the data after the simulation by methods like those described in section 2.2, and then to display the results by programs which have been developed for molecular modelling and for the representation of polyhedra.

Some of the programs found useful are be listed below. Most of them are freely available, and all of them provide output routines which permit to create picture files.

- Atom positions can efficiently be visualized using the program rasmol available at www.OpenRasmol.org.
- A very useful program to visualize polyhedra and the tiles of quasicrystals is geomview, available at www.geomview.org. Input for geomview can easily be created by qhull, which has originally been developed to compute convex hulls and Voronoi decompositions. It is available at www.geom.umn.edu/software/download/qhull.html.
- VMRL scenes have been created to display vector fields, which can be rendered by any VRML viewer, like the OpenInventor based ivview, or the older vrweb (available at www.iicm.edu/vrweb). More about VRML can be found at www.vrml.org.

More insight into physical processes happening during simulation can be achieved if the dimension "time" is taken into account. Often complicated phenomena are understood only if the simulation run is presented as a computer movie.

If embedded into shell-scripts, rasmol, geomview or even gnuplot can be used to generate sequences of pictures. With montage from the ImageMagick package pictures may be combined to display several aspects of the simulations at the same time. The sequence of pictures can then be transformed into computer movies with the SGI mediaconvert or dmconvert programs. For high-quality movies the Stuttgart computing center RUS provides support through its Videoservice.

A number of computer movies generated from IMD simulations can be found on the IMD home page [3].

References

1. Stadler, J., Mikulla, R., Trebin, H.-R.: IMD: A Software package for molecular dynamics studies on parallel computers, Int. J. Mod. Phys. C **8** (1997) 1131.
2. Allen, M. P., Tildesley D. J.: Computer simulation of liquids. Oxford Science Publications, Oxford (1987).
3. http://www.itap.physik.uni-stuttgart.de/~imd
4. Stillinger, F. H., and Weber, T. A., Phys. Rev. B **31** (1985) 5262.
5. Tersoff, J., Phys. Rev. B **39** (1989) 5566.
6. Daw, M. S., and Baskes, M. I.:, Embedded-atom method: Derivation and application to impurities, surfaces, and other defects in metals, Phys. Rev. B **29** (1984) 6443.
7. Bitzek, E., Gähler, F., Hahn, J., Kohler, C., Krdzalic, G., Roth, J., Rudhart, C., Schaaf, G., Stadler, J., Trebin, H.-R.: IMD: A Software package for molecular dynamics studies on parallel computers, Int. J. Mod. Phys. C **8** (1997) 1131.
8. Wollgarten, M., Beyss, M., Urban, K., Liebertz, H., Köster, U.: Shock waves in quasicrystals, Mat. Sci. Eng. A **71** (1993) 549.
9. Kramer, P., and Neri, R.: On periodic and non-periodic space-fillings of E^m obtained by projection, Acta Crystallographica **40** (1984) 580.
10. Henley, C. L., Elser, V.: Quasicrystal structure of $(Al, Zn)_{49}Mg_{32}$, Phil. Mag. B **53** (1986) 59.
11. Roth, J., Schilling, R., Trebin, H.-R.: Stability of monoatomic and diatomic quasicrystals and the influence of noise, Phys. Rev. B **41** (1990) 2735.
12. Schaaf, G., Mikulla, R., Roth, J., Trebin, H.-R.: Numerical simulation of dislocation motion in three-dimensional icosahedral quasicrystals, Phil. Mag. A, **80** (2000), 1657.
13. Vitek, V.: Theory of the core structures of dislocations in body-centered-cubic metals, Cryst. Latt. Def., **5** (1974), 1.
14. Beeler, J. R.: Radiation Effects Computer Simulations, North Holland Amsterdam, New York, Oxford (1983).
15. Nosé, S.: Molecular dynamics at constant temperature and pressure, Computer Simulation in Materials Science. Kluwer Academic Publishers, Dordrecht (1991).
16. Schaaf, G., Mikulla, R., Roth, J., Trebin, H.-R.: Numerical simulation of dislocation motion in an icosahedral quasicrystals, Mat. Sci. Eng. A **294-296** (2000), 799.
17. Roth, J.: Shock waves in quasicrystals, Mat. Sci. Eng. A **294-296**, 799 (2000), 753.
18. http://www.uni-stuttgart.de/RUSuser/vis/People/roland/vol.html
19. http://www.hlrs.de/people/schulze_doebold/virvo/index.htm
20. Schulze-Döbold, J.: Three Dimensional Computer Graphics: Time Efficient Display of Surfaces of Revolution, Computer Science Master's Thesis, University of Massachusetts Dartmouth, (1998).

Buoyancy Driven Convection in Rotating Spherical Shells and Its Dynamo Action

E. Grote, F.H. Busse, and R. Simitev

Institute of Physics, University of Bayreuth, D-95440 Bayreuth

Abstract. Scientific results based on computations carried out at the Stuttgart Supercomputing Center are presented. Coherent structures in turbulent convection are found and the form of magnetic fields generated by the dynamo action has been determined as a function of the parameters of the problem. The saturation of magnetic energy in dependence on the Rayleigh number is studied in a particular case.

1 Introduction

The availability of sufficiently large computer capacities has attracted a number of groups to the numerical study of the process of magnetic field generation by geophysically realistic flows in rotating spherical fluid shells which are expected to represent the liquid iron core of the Earth. There is general agreement that the dynamo process for the generation of the geomagnetic field can indeed be understood on the basis of computer simulations even if the parameter range of the Earth's core is still a bit out of reach. The dynamo computations of Glatzmaier and Roberts (1995a,b), Kuang and Bloxham (1997), Busse et al. (1998), Olson et al. (1999), Christensen et al. (1998, 1999), Grote et al. (1999, 2000a,b) and others have reached a rotation dominated regime where considerable similarities between numerically simulated models and the observed properties of geomagnetism can be expected. After some of the basic aspects have been understood secondary influences such as the distribution of buoyancy or the effect of laterally inhomogeneous boundary conditions must be investigated in order to understand more special properties of convection driven dynamos. In this way it will be possible to identify some of the conditions affecting the geodynamo which are not accessible through observational means.

In the following we shall first describe briefly in section 2 the basic equations and the numerical scheme for their integration in time. In section 3 the implementation of the numerical methods on a supercomputer will be discussed. Section 4 is devoted to the description of the computational results for convection in rapidly rotating spherical shells without magnetic fields. Based on an understanding of the coherent structures of turbulent convection, the dynamo action and the reaction of the Lorentz force can be understood more readily. These two latter topics will be discussed in sections 5 and 6, respectively. In section 7 some results on the influence of lateral variations of the temperature at the outer boundary of the spherical shell will be described. The paper ends with concluding remarks in section 8.

2   Mathematical Formulation of the Problem and Numerical Method

For the description of finite amplitude convection in rotating spherical shells and its dynamo action we follow the standard formulation used in earlier work by the authors (Busse et al., 1998; Grote et al., 1999, 2000b). But we assume that a more general static state exists with the temperature distribution $T_S = T_0 - \beta d^2 r^2/2 + \Delta T \eta r^{-1}(1-\eta)^{-2}$ where η denotes the ratio of inner to outer radius of the shell and d is its thickness. ΔT is the temperature difference between the boundaries in the special case without heat sources, $\beta = 0$. In this way rather general forms of the basic temperature profile can be realized and even a partially stably stratified region can be obtained as is often discussed as a possibility for the outer core of the Earth. The gravity field is given by $\boldsymbol{g} = -\gamma d \boldsymbol{r}$ where $\boldsymbol{r}$ is the position vector with respect to the center of the sphere and r is its length measured in units of d. In addition to the length d, the time d^2/ν, the temperature $\nu^2/\gamma \alpha d^4$ and the magnetic flux density $\nu(\mu \varrho)^{1/2}/d$ are used as scales for the dimensionless description of the problem where ν denotes the kinematic viscosity of the fluid, κ its thermal diffusivity, ϱ its density and μ is its magnetic permeability. The density is assumed to be constant except in the gravity term where its temperature dependence given by $\alpha \equiv (d\varrho/dT)/\varrho = $ const. is taken into account. Since the velocity field $\boldsymbol{u}$ as well as the magnetic flux density $\boldsymbol{B}$ are solenoidal vector fields, the general representation in terms of poloidal and toroidal components can be used,

$$\boldsymbol{u} = \nabla \times (\nabla v \times \boldsymbol{r}) + \nabla w \times \boldsymbol{r} \ , \tag{1a}$$

$$\boldsymbol{B} = \nabla \times (\nabla h \times \boldsymbol{r}) + \nabla g \times \boldsymbol{r} \ . \tag{1b}$$

By multiplying the $(\mathrm{curl})^2$ and the curl of the Navier-Stokes equations in the rotating system by $\boldsymbol{r}$ we obtain two equations for v and w

$$[(\nabla^2 - \partial_t)L_2 + \tau \partial_\varphi]\nabla^2 v + \tau Q w - L_2 \Theta = -\boldsymbol{r}\cdot\nabla \times [\nabla \times (\boldsymbol{u}\cdot\nabla\boldsymbol{u} - \boldsymbol{B}\cdot\nabla\boldsymbol{B})] \tag{2a}$$

$$[(\nabla^2 - \partial_t)L_2 + \tau \partial_\varphi]w - \tau Q v = \boldsymbol{r}\cdot\nabla \times (\boldsymbol{u}\cdot\nabla\boldsymbol{u} - \boldsymbol{B}\cdot\nabla\boldsymbol{B}) \tag{2b}$$

where ∂_t and ∂_φ denote the partial derivatives with respect to time t and with respect to the angle φ of a spherical system of coordinates r, θ, φ and where the operators L_2 and Q are defined by

$$L_2 \equiv -r^2\nabla^2 + \partial_r(r^2\partial_r)$$

$$Q \equiv r\cos\theta\nabla^2 - (L_2 + r\partial_r)(\cos\theta\partial_r - r^{-1}\sin\theta\partial_\theta)$$

The heat equation for the dimensionless deviation Θ from the static temperature distribution can be written in the form

$$\nabla^2\Theta + \left[R_i + R_e\eta r^{-3}(1-\eta)^{-2}\right]L_2v = P(\partial_t + \boldsymbol{u}\cdot\nabla)\Theta \qquad (2c)$$

and the equations for h and g are obtained through the multiplication of the equation of magnetic induction and of its curl by $\boldsymbol{r}$,

$$\nabla^2 L_2 h = P_m[\partial_t L_2 h - \boldsymbol{r}\cdot\nabla\times(\boldsymbol{u}\times\boldsymbol{B})] \qquad (2d)$$

$$\nabla^2 L_2 g = P_m[\partial_t L_2 g - \boldsymbol{r}\cdot\nabla\times(\nabla\times(\boldsymbol{u}\times\boldsymbol{B}))] \qquad (2e)$$

The Rayleigh numbers R_i and R_e, the Coriolis parameter τ, the Prandtl number P and the magnetic Prandtl number P_m are defined by

$$R_i = \frac{\alpha\gamma\beta d^6}{\nu\kappa}, \quad R_e = \frac{\alpha\gamma\Delta T d^4}{\nu\kappa}, \quad \tau = \frac{2\Omega d^2}{\nu}, \quad P = \frac{\nu}{\kappa}, \quad P_m = \frac{\nu}{\lambda} \qquad (3)$$

where λ is the magnetic diffusivity. We assume stress-free boundaries with fixed temperatures,

$$v = \partial_{rr}^2 v = \partial_r(w/r) = \Theta = 0 \quad \text{at} \quad r = r_i \equiv \eta/(1-\eta)$$

$$\text{and at} \quad r = r_o = (1-\eta)^{-1} \qquad (4a)$$

Throughout this paper the case $\eta = 0.4$ will be considered unless indicated otherwise. For the magnetic field electrically insulating boundaries are used such that the poloidal function h must be matched to the function $h^{(e)}$ which describes the potential fields outside the fluid shell

$$g = h - h^{(e)} = \partial_r(h - h^{(e)}) = 0 \qquad \text{at } r = r_i \text{ and } r = r_o. \qquad (4b)$$

The numerical integration of equations (2) together with boundary conditions (4) proceeds with the pseudo-spectral method as described by Tilgner and Busse (1997) which is based on an expansion of all dependent variables in spherical harmonics for the θ, φ-dependences, i.e.

$$v = \sum_{l,m} V_l^m(r,t)P_l^m(\cos\theta)\exp\{im\varphi\} \qquad (5)$$

and analogous expressions for the other variables, w, Θ, h and g. P_l^m denotes the associated Legendre functions. For the r-dependence expansions in Chebychev polynomials are used. For further details see also Busse et al. (1998).

For most of the computations to be reported in the following 33 collocation points in the radial direction and spherical harmonics up to the order 64 have been used. But in a few cases up to 65 collocation points and spherical harmonics up to the order 128 have been used.

3 Implementation of the Numerical Algorithm on the Stuttgart Supercomputer

The functions $V_l^m(r,t), W_l^m(r,t), T_l^m(r,t), H_l^m/r,t)$ and $G_l^m(r,t)$ corresponding to the fields v, w, Θ, h and g, respectively, are being stored in r, l, m-space where the radial dependence is discretized into a set of grid points. In order to allow fast transforms from normal space to Chebychev expansion and vice versa, the N collocation points are chosen to lie at $x_n = \cos\left(\pi \frac{n-1}{N-1}\right)$ where x is defined by $x \equiv 2r - (r_o + r_i)$. The dynamic equations are converted into a system of ODEs in time through the enforcement of the full equations at every collocation point. The decomposition is thus merely used to compute radial derivatives.

The main reason for this "pseudo-spectral" method is the fact that in a pure spectral method the computation of the nonlinear terms would be very expensive both in CPU time and in memory consumption. On the other hand, spectral methods provide a better convergence behavior than finite difference or finite element methods. Thus, the goal is to benefit from both techniques as much as possible. In our preference for the pseudo spectral method we feel confirmed by recent comparison done by Fornberg and Merrill (1997).

Time stepping is performed by a combination of an Adams-Bashforth second order scheme treating all the right hand sides of equations (2) explicitly, whereas the terms at the left hand sides are included in an implicit Crank-Nicolson step. At the beginning of a time integration an Euler step is used to start up the scheme.

In order to take advantage of the multiprocessor architecture the code is parallelized in azimuthal direction, i.e., all coefficients sharing a common index m are stored at the same processor. This means that for a typical calculation usually 64 processors are being used. All the communication between them is done with the help of the *Message Passing Interface* (MPI).

At the beginning of each time step all fields and their first and second derivatives are given in r, l, m-space. At this stage the spectral coefficients of v, w, Θ, g, h are stored such that coefficients with identical m are stored at the same processor.

The calculation of the fields v, w, Θ, g, h requires adding associated Legendre functions and performing a Fourier transform. The summation over l in (5) is implemented as matrix vector multiplications and obviously parallelizes over m. The summation over m is the Fourier transform which requires interprocessor communication. Before the actual execution of the FFT, data are redistributed such that individual processors contain all data with a given index l. The fast Fourier transform algorithm can then be executed locally and in parallel for separate l and r. The nonlinear terms are now easily obtained since they only involve multiplications of local data. The transformation back into the r, l, m-space is performed with a FFT followed by a Gauss quadrature. These are technically the same operations as for the first transformation.

At the end of the FFT the original data distribution is restored, i.e. all variables at a given m are collected in the storage of individual processors. The Gauss quadrature is again expressed in terms of matrix vector multiplications which run independently on all processors for different m.

Once the nonlinear terms have been obtained, they can be combined as required by the Adams-Bashforth scheme and added to the terms of the implicit time step involving the variables at the present moment in time only. To complete the time step, a set of N linear equations must be solved for every l, m. Boundary conditions are also included in this set of equations. The coefficients in these equations are independent of m and are collected in separate matrices which are inverted during initialization and multiplied with vectors containing the spectral coefficients of v, w, Θ, g, h during the actual time step. These multiplications separate again in m and involve only local data for each processor. The discretized equations are formulated such that the updated fields are obtained in the n, l, m-space where the radial derivatives can be conveniently computed. A fast cosine transform brings the variables back into the r, l, m-space ready for use in the next time step. For further details we refer to the analogous numerical treatment of the problem of non magnetic convection by Tilgner and Busse (1997).

In summary, the computational burden lies mostly in matrix-vector multiplications, followed by fast cosine and Fourier transforms. The matrix vector multiplications carried out at each processor are of course readily vectorized. However, with the resolution used so far, each vector is relatively short (usually 64 elements). Only the FFT needs to shuffle data between processors. Interprocessor communication contributes little to the CPU time expenditure. Table 1 shows the accumulated CPU time per time step for a calculation running on different numbers of processors. The speed-up is limited by load balancing. For instance, the number of matrix vector multiplications to be performed is not always a multiple of the number of processors in use. However, the performance of the spectral method evidenced in Table 1 compares well with speed-ups commonly achieved with grid methods, e.g. finite elements.

Table 1. Total CPU time per time step t_p in seconds used by N_p processors for a resolution of 33 Chebychev polynomials and 64×65 real coefficients in the angular decomposition for each of the five scalar fields. Ideally, t_p is independent of N_p.

N_p	16	32	64
t_p	16.6	22.0	22.5

4 Convection in Rotating Spherical Shells at Finite Amplitudes

The main difficulty faced by the convection flow as it evolves with increasing Rayleigh number is the fact that its columnar structure is not well suited to transport heat from and to the curved boundaries. Since the term describing the transport of heat is basically the same as the term describing the release of potential energy, it is clear that the amplitude of convection can not grow rapidly with R_i or R_e if the heat transport to the walls is impeded. It is thus not surprising that the amplitude of the steadily drifting convection columns tends to saturate with increasing R as has been demonstrated, for example, in figure 7 of Ardes et al. (1997). Convection modes with a strong time dependence tend to replace convection in the form of steadily drifting columns because of their ability to enhance the heat transport. Most common are vacillating convection columns which either oscillate coherently or incoherently in that a $m = 1$ azimuthal modulation sets in. Vacillating convection can be found in the rotating cylindrical annulus (Or and Busse, 1987) in the spherical case at infinite Prandtl number (Zhang, 1992), at $P = 1$ (Sun et al., 1993; Grote and Busse, 2001) or at low Prandtl numbers (Ardes et al., 1997). But there are also other ways in which bifurcations from the steadily drifting columns occur, for example through a subharmonic bifurcation as shown in figure 1a. Every second convection column is stretched in the azimuthal direction until the outer tip of the spiraling column snaps off. In the meantime the stretching of the neighboring columns has begun and the process is repeated in a time periodic fashion. If one defines the period of this process as the time in which the same pattern is obtained except for a rotation about the axis, one finds that the period decreases from 0.032 at $R_i = 3.1 \cdot 10^5$ to 0.021 at $R_i = 3.3 \cdot 10^5$. As R_i is increased the synchronisation of the process of figure 1a disappears and an $m = 1$ modulation is superimposed onto the $m = 4$ pattern as shown in figure 1b for the case $R_i = 3.45 \cdot 10^5$. At the same time an amplitude vacillation occurs which is indicated by the periodic strengthening and weakening of the convection columns. At the slightly higher Rayleigh number of $R_i = 3.5 \cdot 10^5$ a transition to a quasiperiodic time dependence has already occurred which is associated with a strengthening of the $m = 1$ modulation. With increasing R_i the convection pattern becomes increasingly chaotic as is evident from the time dependence of the energies of various components of motion shown in figure 2. The energy of the axisymmetric toroidal component of motion grows especially rapidly with R_i because of the strong differential rotation created by the Reynolds stresses of the spiralling columns. The kinetic energy can be separated into four different kinds,

$$E_p^m = \frac{1}{2}\langle |\,\nabla \times (\nabla \bar{v} \times \boldsymbol{r}\,|^2\rangle, \quad E_t^m = \frac{1}{2}\langle |\,\nabla \bar{w} \times \boldsymbol{r}\,|^2\rangle \tag{6a}$$

$$E_p^f = \frac{1}{2}\langle |\,\nabla \times (\nabla \breve{v} \times \boldsymbol{r}\,|^2\rangle, \quad E_t^f = \frac{1}{2}\langle |\,\nabla \breve{w} \times \boldsymbol{r}\,|^2\rangle \tag{6b}$$

Fig. 1. a) Sequence in time (uppermost row left to right, then second row right to left) of plots of streamlines ($r\frac{\partial v}{\partial \varphi}$ = const.) in the equatorial plane in the case $R_i = 3.2 \cdot 10^5, \tau = 1.5 \cdot 10^4, P = 0.5, R_e = 0$. The equidistant ($\Delta t = 0.005$) plots cover a period such that the last plots closely resembles the first plot except for a shift in azimuth. **b)** Same as a), except that $R_i = 3.45 \cdot 10^5$ and $\Delta t = 0.0075$ (third row left to right, then lowermost row right to left). As is evident from the similarity of the first and fifth plot, the sequence covers about a period of the oscillation (except for a shift in azimuth).

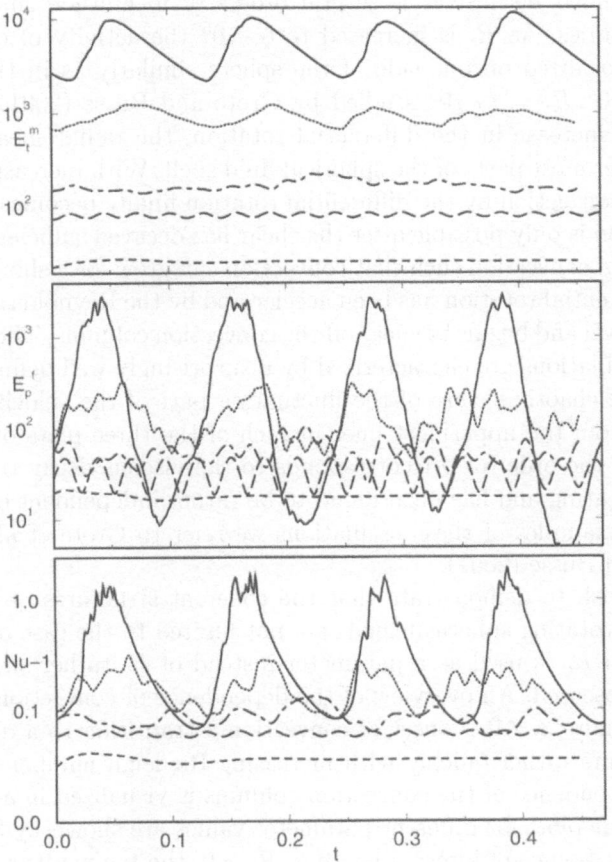

Fig. 2. Time series of the mean energy densities of the axisymmetric toroidal component (upper plot) and of the non-axisymmetric poloidal component of motion (middle plot) and of the Nusselt number at the inner boundary (lower plot) in the case $\tau = 1.5 \cdot 10^4, P = 0.5, R_e = 0$ for $R_i = 3 \cdot 10^5$ (short dash lines), $3.5 \cdot 10^5$ (long dash lines), $4 \cdot 10^5$ (dash dotted lines), $6 \cdot 10^5$ (dotted lines), and 10^6 (solid lines).

where the brackets $\langle \dots \rangle$ indicate the average over the spherical shell, $\bar{v}$ denote the axisymmetric component of v and $\check{v} \equiv v - \bar{v}$ indicates the non-axisymmetric component of v. Also presented in figure 2 are time series of the Nusselt number Nu which measures the convective heat transport and which is defined by

$$Nu = 1 - \frac{P}{r_i} \frac{\partial \bar{\bar{\Theta}}}{\partial r}\bigg|_{r=r_i} \tag{7}$$

where $\bar{\bar{\Theta}}$ indicates the average of Θ over the surface $r = $ const. The energy E_t^f has not been plotted because it closely parallels E_p^f even though it is about twice as large. The energy of the axisymmetric poloidal component has also not been included because it is several orders of magnitude smaller than the other energies. As R_i is increased to $5 \cdot 10^5$ the activity of convection tends to be localized on one side of the sphere similarly as in the case of $P = 1, \tau = 10^4, R_i = 7 \cdot 10^5$ studied by Grote and Busse (2001). Because of the strong increase in the differential rotation, the radial shear inhibits convection over most parts of the spherical fluid shell. With increasing R_i the inhibition of convection by the differential rotation finally becomes so strong that convection is only possible after the shear has decayed sufficiently in the near absence of convection such that convection can grow for a short moment until the differential rotation has been accelerated by the Reynolds stresses to its previous level and begins to shear off the convection columns. The resulting relaxation oscillations are characterized by a surprisingly well defined period in spite of the chaotic nature of the fluctuating part of the velocity field as can be seen from the uppermost lines in each of the three plots of figure 2. The period of the order of 0.1 corresponds to the viscous decay time of the differential rotating and has been found to be rather independent of P, R_i or τ. For other examples of these oscillations we refer to Grote et al. (2000b) and Grote and Busse (2001).

Here we wish to demonstrate that the coherent structures of turbulent convection in rotating spherical shells are not limited to the case of internal heating. When R_e is used as a parameter instead of R_i rather similar phenomena are observed. An overview of the dependence of convection on R_e is provided by figure 3. After onset of convection a transition to a quasi periodic form occurs rather quickly with increasing Rayleigh number. In figure 4 the time dependence of the convection columns is visualized in a sequence of plots. Similar plots for different parameter values are shown by Sun et al. (1993). In the absence of internal heating, $R_i = 0$, the temperature gradient near the inner boundary is especially strong and the amplitude of the spiralling convection columns decays more strongly with increasing r than in the case $R_e = 0$. For this reason the disruption of the convection columns and their subsequent reconnection as shown in the figure occurs more rapidly after onset. At $R_e = 8 \cdot 10^5$ convection flow has become chaotic as can be seen from the energy of the fluctuating poloidal component of motion. A visualization of the flow at a particular time can be gained from figure 5. The flow at $R_e = 10^6$ which is also shown in this figure, indicates the tendency towards localized convection although this feature does not seem to develop as dramatically as in the case exhibited in the paper of Grote and Busse (2001). At $R_e = 1.2 \cdot 10^6$ the relaxation oscillation begins to set in which becomes fully developed as we reach the highest Rayleigh number, $R_e = 1.7 \cdot 10^6$, of figure 3. The amplitude of the oscillation of the differential rotation is not quite as large as in the case of figure 2 and the period is slightly shorter. But the phenomenon itself does not depend on the detailed form of the basic temperature distribution.

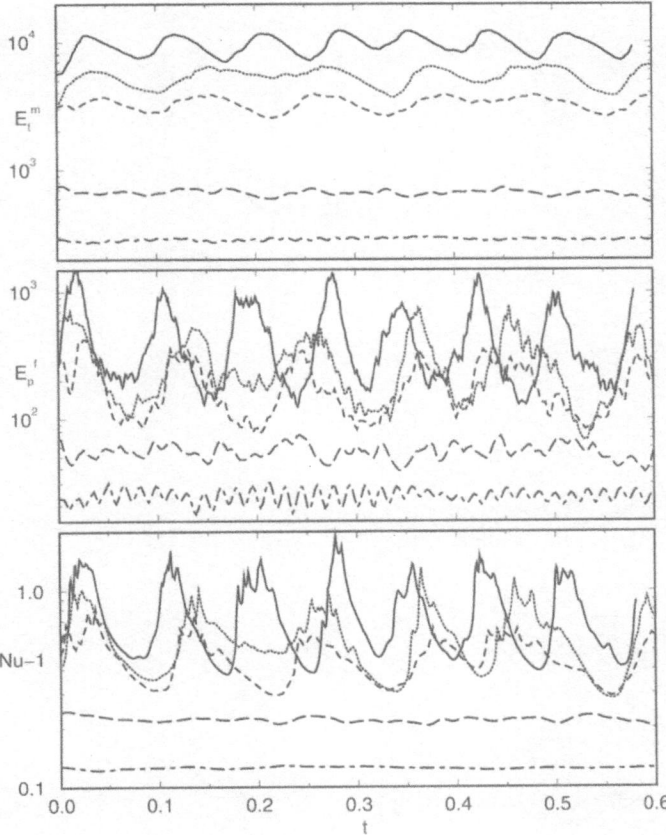

Fig. 3. Time series of the mean energy densities of the axisymmetric toroidal component (upper plot) and of the non-axisymmetric poloidal component of motion (middle plot) and of the Nusselt number (lower plot) in the case $\tau = 10^4, P = 1$, $R_i = 0$ for $R_e = 6 \cdot 10^5$ (dash dotted lines), $8 \cdot 10^5$ (long dash lines), $1.2 \cdot 10^6$ (short dash lines), $1.4 \cdot 10^6$ (dotted lines) and $1.7 \cdot 10^6$ (solid lines).

5   Convection Driven Dynamos

All convection flows realized in sufficiently rapidly rotating spherical shells appear to be capable of acting as dynamos if only the magnetic Prandtl number P_m is sufficiently large, such that a critical magnetic Reynolds number of the order 100 is exceeded even for small amplitudes of convection. A typical diagram is shown in figure 6 where the open symbols characterize the onset of dynamo action in the R_i-P_m-space for the Coriolis parameter $\tau = 10^4$ and the closed symbols do the same for $\tau = 3 \cdot 10^4$. In the latter case the values of R_i must be multiplied by the factor 10 as indicated on the right ordinate.

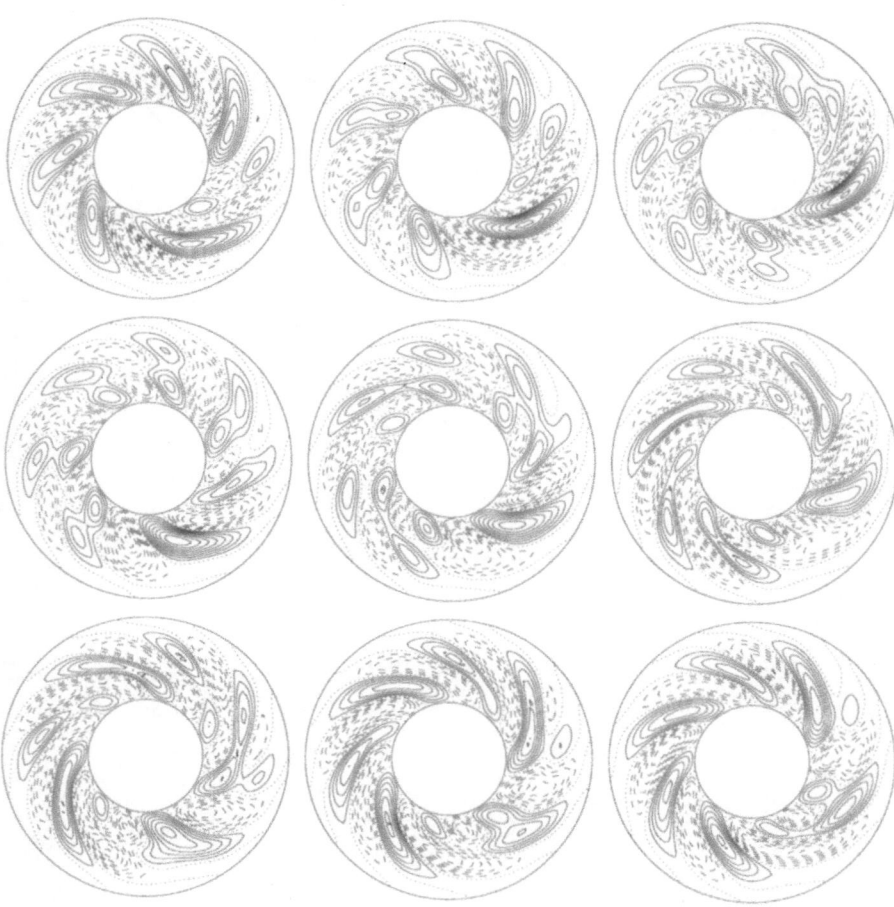

Fig. 4. Time sequence (left to right, then top to bottom) of equidistant ($\Delta t = 5 \cdot 10^{-3}$) plots of streamlines, $r\partial v/\partial\varphi = $ const. in the equatorial plane in the case $\tau = 10^4, R_e = 5.5 \cdot 10^5, R_i = 0, P = 1$. The kinetic energies exhibit oscillations with the period 0.018, similar to those shown in the case of $R_e = 6 \cdot 10^5$ in figure 6.

This figure is an extension of a similar one presented by Grote et al. (2000b). Analogous diagrams as a function of R_e instead of R_i have been obtained by Christensen et al. (1999). There are some typical differences in that dynamos other than dipolar ones are the exception in the cases considered by Christensen et al. (see also Kutzner and Christensen, 2000) while in the case of figure 6 dynamos of predominantly dipolar character are found only for relatively high values of P_m. They are replaced by hemispherical dynamos as P_m decreases and finally at low values of P_m quadrupolar dynamos are obtained. But the meaning of "high" and "low" values of P_m varies with τ in that the transition from dipolar to hemispherical dynamos shifts towards

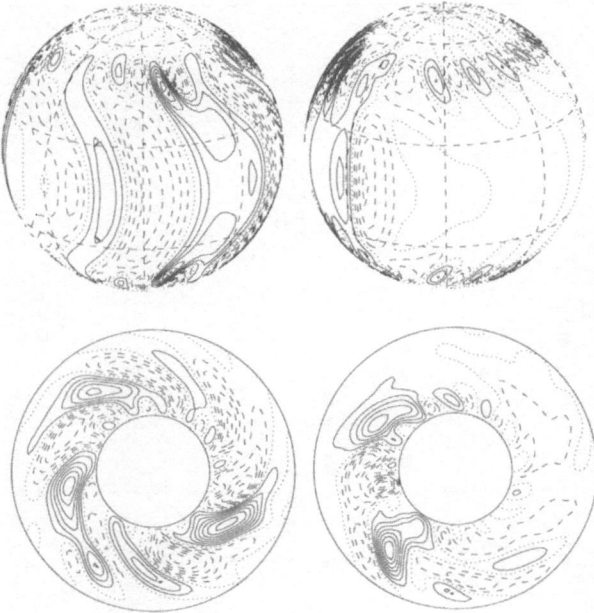

Fig. 5. Lines of constant radial component of the velocity on the mid surface of the shell, $r = (1 + \eta)/2(1 - \eta)$, (upper plots) and streamlines ($r\partial v/\partial \varphi = $ const.) in the equatorial plane for $R_e = 8 \cdot 10^5$ (left side) and $R_e = 10^6$ (right side) in the case $\tau = 10^4, P = 1, R_i = 0$.

lower P_m with increasing τ. In the case of $\tau = 3 \cdot 10^4$ it has not yet been possible to reach the regime of quadrupolar dynamos.

All hemispherical and quadrupolar dynamos exhibit an oscillatory character in that a dynamo wave propagates from the equator to the pole (or poles). An example is shown in figure 7. The emergence of magnetic flux with a new polarity appears to be initiated at the equator of the inner boundary as can best be seen in the plots of $\overline{B_\varphi}$ in the figure. The magnetic energy also changes in phase with the oscillation and reaches its maximum when the axisymmetric flux tubes with opposite signs of $\overline{B_\varphi}$ reach about equal amplitude. The perfect correlation between the amplitudes of spherical harmonics with $l - m = $ odd and with $l - m = $ even which characterizes an ideal hemispherical dynamo is nearly approached in the case $m = 0$ as can be seen from the upper plot of figure 8. In the case of the non-axisymmetric components the correlation is not as good, but is still quite remarkable.

A clear distinction of dynamos with different symmetry is usually only possible as long as convection occurs predominantly outside the tangent cylinder and still exhibits an approximate symmetry about the equator. Even in the case of hemispherical dynamos only minor deviation from this symmetry are caused by the action of the Lorentz force. As soon as the critical Rayleigh

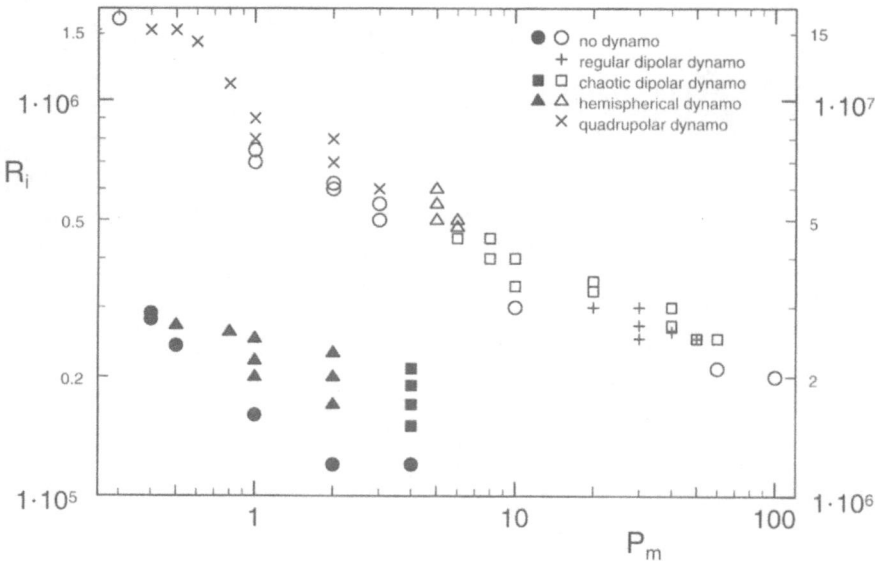

Fig. 6. Existence of convection dynamos of different types as a function of the Rayleigh number R_i and the magnetic Prandtl number P_m in the cases $\tau = 10^4$ (open symbols and crosses) and $\tau = 3 \cdot 10^4$ (filled symbols). In the latter case the values of R_i are given by the right ordinate. The Prandtl number is $P = 1$ in all cases.

number for onset of convection in the polar regions is exceeded, the influence of equatorial symmetry diminishes and mixtures of quadrupolar and dipolar components of the magnetic field are generated without the nearly perfect correlation that characterizes hemispherical dynamos. Usually the quadrupolar part of the magnetic field is still strongest outside the tangent cylinder while the dipolar components appear to be generated primarily in the polar regions. With increasing Rayleigh number convection grows more strongly in the polar regions than outside the tangent cylinder and as a consequence the magnetic field becomes more dipolar. In contrast to the dipolar dynamos indicated in figure 6 the high Rayleigh number dipolar fields still exhibit the oscillations which characterize the predominantly quadrupolar fields at lower values of R_i. An example of an oscillatory mainly dipolar dynamo is shown in figure 9.

6   Interaction of Magnetic Fields and Convection

The problem of convection in rotating systems in the presence of an imposed magnetic field has a long history. It can be formulated as a linear problem when the conditions for the onset of convection are of primary interest. Early

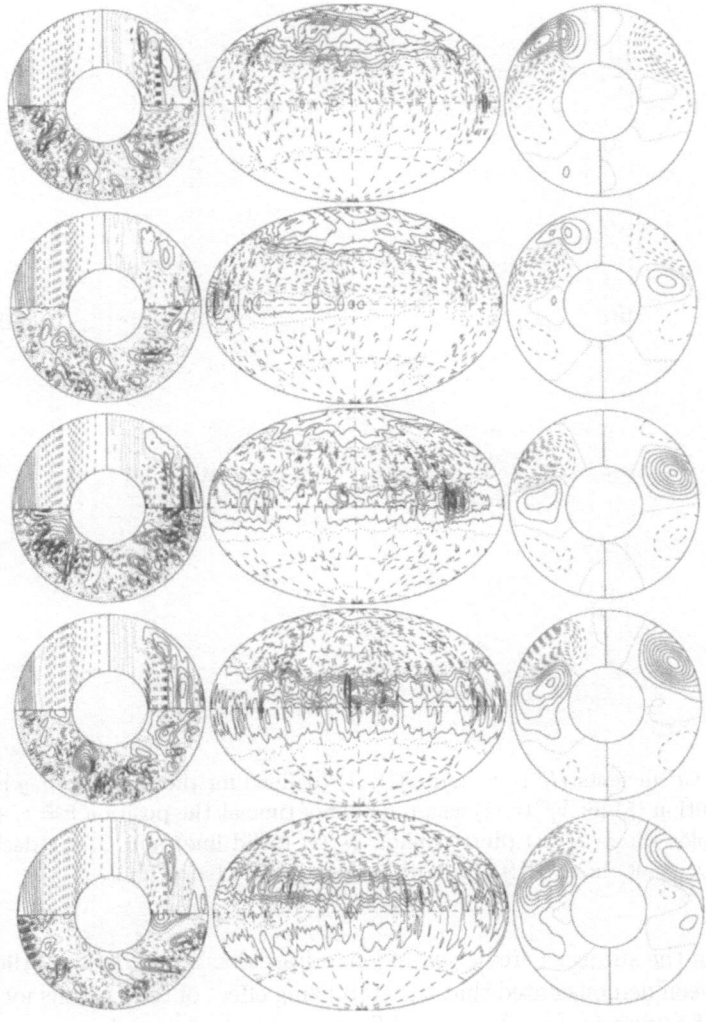

Fig. 7. Oscillating hemispherical dynamo in the case $R_i = 2.7 \cdot 10^6, \tau = 3 \cdot 10^4, P_m = 0.5, P = 1$. The plots represent a sequence in time (from top to bottom in equidistant steps with $\Delta t = 0.0012$) with the column displaying the velocity field with lines of constant $\overline{u_\varphi}$ in the upper left quarter, streamlines $r \sin\theta \partial \overline{v}/\partial\theta = $ const. in the upper right quarter and streamlines $r\partial v/\partial\varphi = $ const. in the lower half of each circle. The middle column shows of constant B_r at the outer surface, $r = (1-\eta)^{-1}$, and right column shows of constant $\overline{B_\varphi}$ in the left half and meridional field lines $r \sin\theta \partial \overline{h}/\partial\theta$ in the right half of each circle.

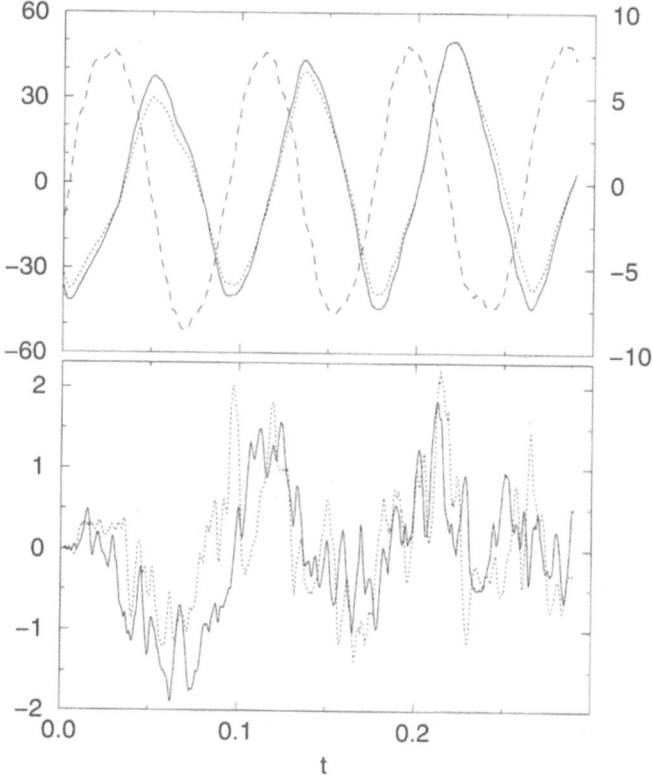

Fig. 8. Coefficients $H_l^m(r, t)$ and $G_l^m(r, t)$ (defined for the variables h, g in analogy to definition (5) for $V_l^m(r, t)$) as a function of time at the position $r = r_i + 0.5$. The upper plot shows $H_1^0(t)$ (dotted line), $H_2^0(t)$ (solid line) and $G_1^0(t)$ (dashed line). The lower plot shows $G_1^1(t)$ (dotted line) and $G_2^1(t)$ (solid line).

work on the subject is reviewed in Chandrasekhar's monograph (1961). Here it has been demonstrated that the stabilizing effect of the Coriolis force on the onset of convection in a horizontal fluid layer heated from below and rotating about a vertical axis can be partly released when a vertical magnetic field is imposed. Even an imposed horizontal field may lead to reduction of the critical Rayleigh number R_c for onset of convection if the rotation rate is high enough (Eltayeb, 1972). For a recent study of this problem which includes the possibility of oscillatory onset we refer to Roberts and Jones (2000). The optimal strength B_0 of the magnetic field for lowering R_c is usually given by an Elsasser number Λ of the order unity where Λ is defined by

$$\Lambda = \frac{B_0^2}{\Omega \varrho \mu \lambda} \tag{8}$$

Here Ω denotes the absolute value of the angular velocity of rotation.

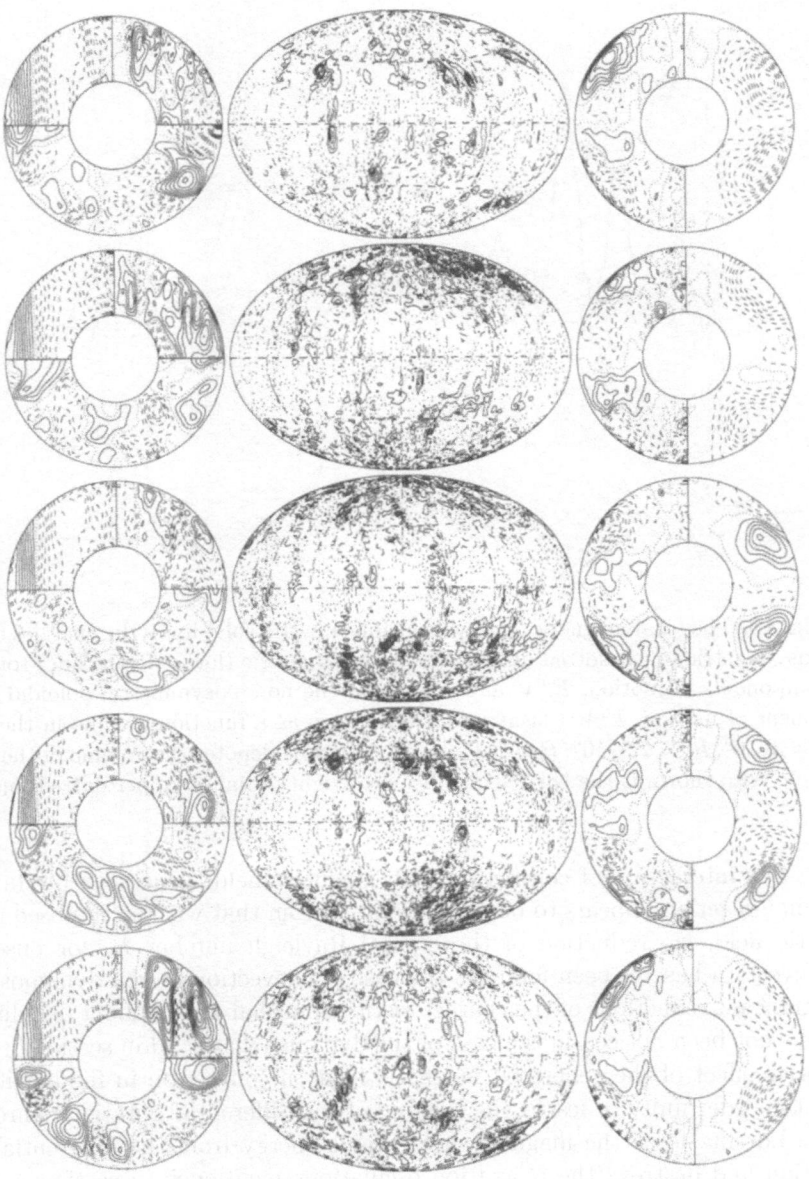

Fig. 9. Oscillating dipolar dynamo in the case $R_i = 1.4 \cdot 10^6, \tau = 5 \cdot 10^3, P_m = P = 1$. The same quantities as in figure 7 are plotted with the time step $\Delta t = 0.009$.

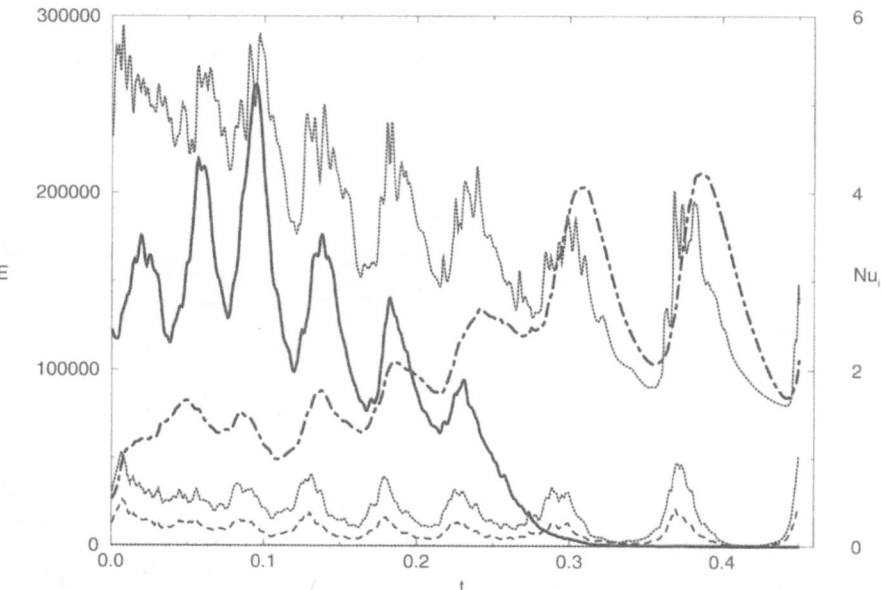

Fig. 10. The total magnetic energy (multiplied by 10, solid line), the energies of the axisymmetric (dash-dotted line) and non-axisymmetric (lower dotted line) toroidal components of motion, $E_t^m V$ and $E_t^f V$, and the non-axisymmetric poloidal component of motion, $E_p^f V$ (dashed line) are shown as a function of time in the case $\tau = 3 \cdot 10^4, R_i = 2.9 \cdot 10^6, P = 1, P_m = 0.4$, where V denotes the volume of the fluid shell. Also shown is the Nusselt number Nu (right ordinate, upper dotted line).

The interaction of convection with magnetic fields generated by its own dynamo action appears to be quite different from that with an imposed magnetic field. No reduction of the critical Rayleigh number R_c for onset of convection has yet been found in the case of convection driven dynamos and significant reductions of the characteristic azimuthal wavenumber of columns have not been noticed in the case of the dynamos discussed in section 5. The major effect of the generated magnetic field on convection in fluids with P of the order unity or less is the braking of the differential rotation. Through the Lorentz force the magnetic field drains energy from the differential rotation and destroys the relaxation oscillations mentioned in section 4. The radial extent of the convection columns is increased and the transport of heat is enhanced. These effects are clearly demonstrated in figure 10 where the relaxation oscillation with an increased energy of the differential rotation returns after the magnetic field has decayed as it often happens when the Rayleigh number is just below the value needed for sustained dynamo action.

The main question of the interaction between convection and the magnetic field generated by its dynamo action is the question of the mean amplitude of the magnetic field. With this in mind extensive computations have been

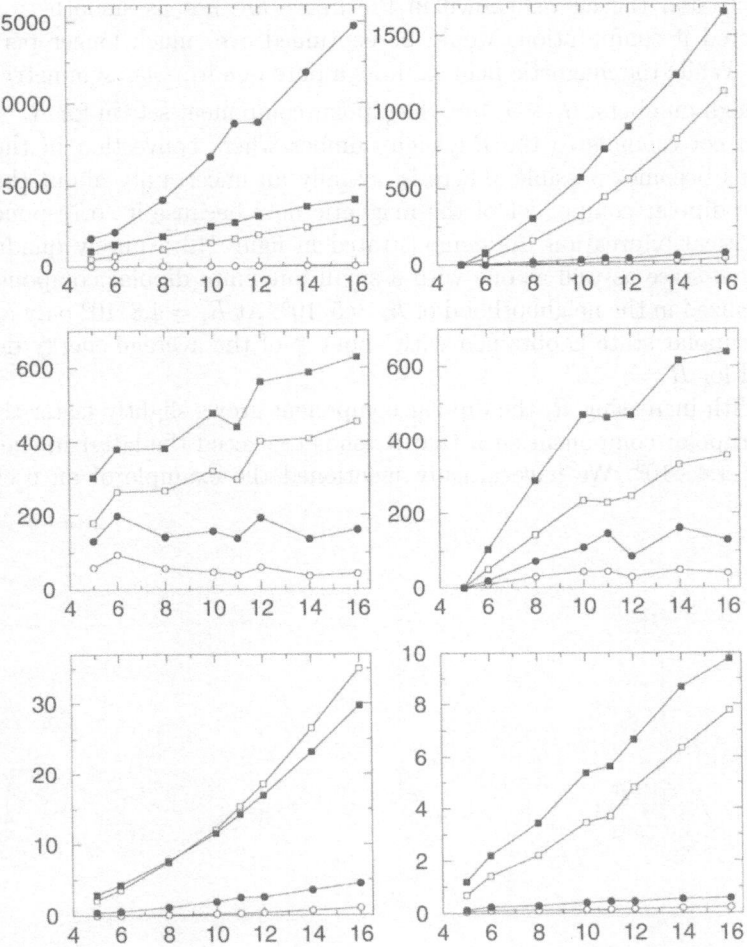

Fig. 11. Kinetic energy densities of symmetric (upper left) and antisymmetric (upper right) components, magnetic energy densities of quadrupolar (middle left) and dipolar (middle right) components, and viscous (lower left) and Ohmic (lower right) dissipation are plotted as a function of R_i for convection driven dynamos in the case $\tau = 5 \cdot 10^3, P = P_m = 1$. Filled (open) symbols indicate toroidal (poloidal) components of the energies and dissipations, circles (squares) indicate axisymmetric (non-axisymmetric) components. In the case of the dissipations the contributions have not been separated with respect to their equatorial symmetry. The values of R_i at the abszissa should be multiplied by 10^5. The scales of the ordinates in the two lower plots must also be multiplied by the factor 10^5.

performed to produce the dependences on R_i of the energies of the various components of magnetic and velocity fields shown in figure 11. Although the energies are obtained from averages over several Ohmic or viscous decay times, the influence of the statistical fluctuations cannot be eliminated entirely and the curves shown in the figure are not as smooth as can be expected if computations would be continued over much longer periods in time. While the magnetic field exhibits a pure quadrupolar symmetry at low Rayleigh numbers, $R_i \overset{<}{\sim} 5 \cdot 10^5$, the dipolar component sets in for $R_i \overset{>}{\sim} 5 \cdot 10^5$ which corresponds to the Rayleigh number where convection in the polar regions becomes possible. There is actually an uncertainty about the onset of the dipolar component of the magnetic field because it corresponds to a subcritical bifurcation. As demonstrated in figure 12 a purely quadrupolar dynamo state as well as one with a small but finite dipolar component can be realized in the neighborhood of $R_i = 5 \cdot 10^5$. At $R_i = 4.8 \cdot 10^5$ only a purely quadrupolar state is obtained with about $\frac{3}{4}$ of the average energy densities found for $R = 5 \cdot 10^5$.

With increasing R_i the dipolar component grows slightly faster than the quadrupolar component such that it tends to exceed the latter in energy for $R_i \geq 1.4 \cdot 10^6$. We have already mentioned the example of an oscillating

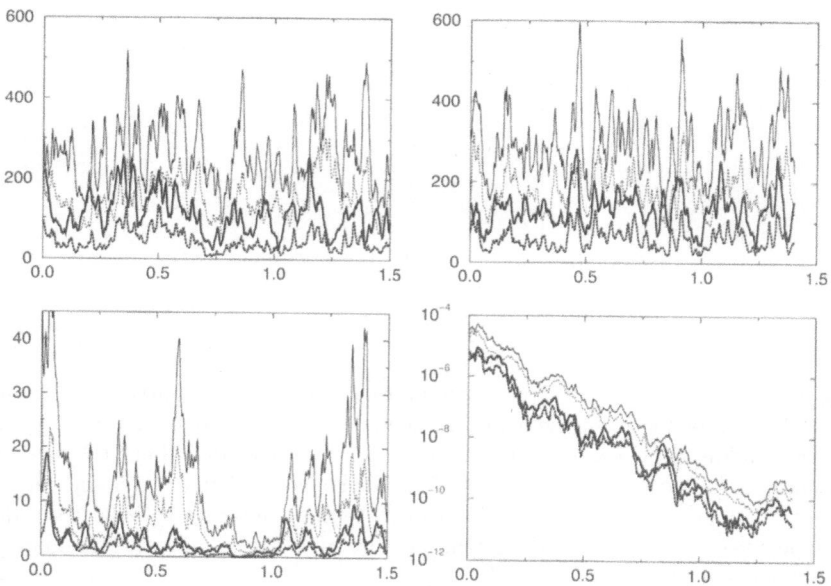

Fig. 12. Energies of quadrupolar (dipolar) components of the magnetic field are shown as a function of time in upper (lower) plots for $\tau = 5 \cdot 10^3, R = 5 \cdot 10^5, P = P_m = 1$. The left side corresponds to a dynamo of mixed parity, while the right side presents a purely quadrupolar dynamo. The toroidal energy densities M_t^m, M_t^f (definitions are analogous to those given by expressions (7)) and poloidal energy densities M_p^m, M_p^f are indicated by solid and dashed lines, respectively. Thick (thin) lines indicate energy densities of axisymmetric (non-axisymmetric) components.

predominantly dipolar magnetic field as shown in figure 9. The axisymmetric components of both, the dipolar and the quadrupolar parts of the magnetic field, clearly saturate with increasing R_i. They actually tend to decay with a further increase of R_i because flux expulsion from the convection eddies becomes a dominant process. Even the fluctuating components are affected by this process which limits their energies and leads to a highly filamentary structure of the magnetic field. This latter property is also evident from the Ohmic dissipation which does not exhibit the saturation of the magnetic energies, but instead parallels the growth of viscous dissipation albeit at a lower level.

Since the Elsasser number can be written in the form

$$\Lambda = 2MP_m\tau^{-1} \tag{9}$$

where M denotes the average density of the total magnetic field, we conclude from figure 11 that the value $\Lambda = 1$ is approached by the saturating magnetic field. While such a value makes sense on the basis of considerations discussed above, the open question remains, why this value is not reached at lower values of the Rayleigh number. The fact that the Ohmic dissipation amounts to a certain fraction of the viscous dissipation could be a more typical feature of the interaction between magnetic field and convection in the parameter regime that has been investigated so far. Further explorations of the parameter dependences of convection driven dynamos are clearly needed.

7   Effects of Lateral Variation of the Boundary Conditions

There is increasing evidence that the geodynamo is sensitive to the inhomogeneities at the core-mantle boundary. Owing to descending oceanic crust and plumes rising from the lowermost mantle the temperature boundary conditions are undoubtedly inhomogeneous. According to paleomagnetic evidence the preferred path of the virtual magnetic pole during reversals appears to go along the cold section of the lowermost mantle. Another influence of mantle convection manifests itself in the property that the frequency of reversals increases with a decrease in mantle convection activity. It is thus of considerable interest to investigate the influence of lateral variations of the temperature boundary conditions on convection driven dynamos.

In the ongoing computational work we are analyzing the case of the boundary condition

$$\Theta = AP_2^2(\cos\theta)\cos 2\varphi \quad \text{at } r = r_0 \tag{10}$$

which is imposed at the outer boundary instead of (4a). This particular form is motivated by the observation that the data from seismic tomography of the lower mantle indicate that the spherical harmonic (10) provides the major

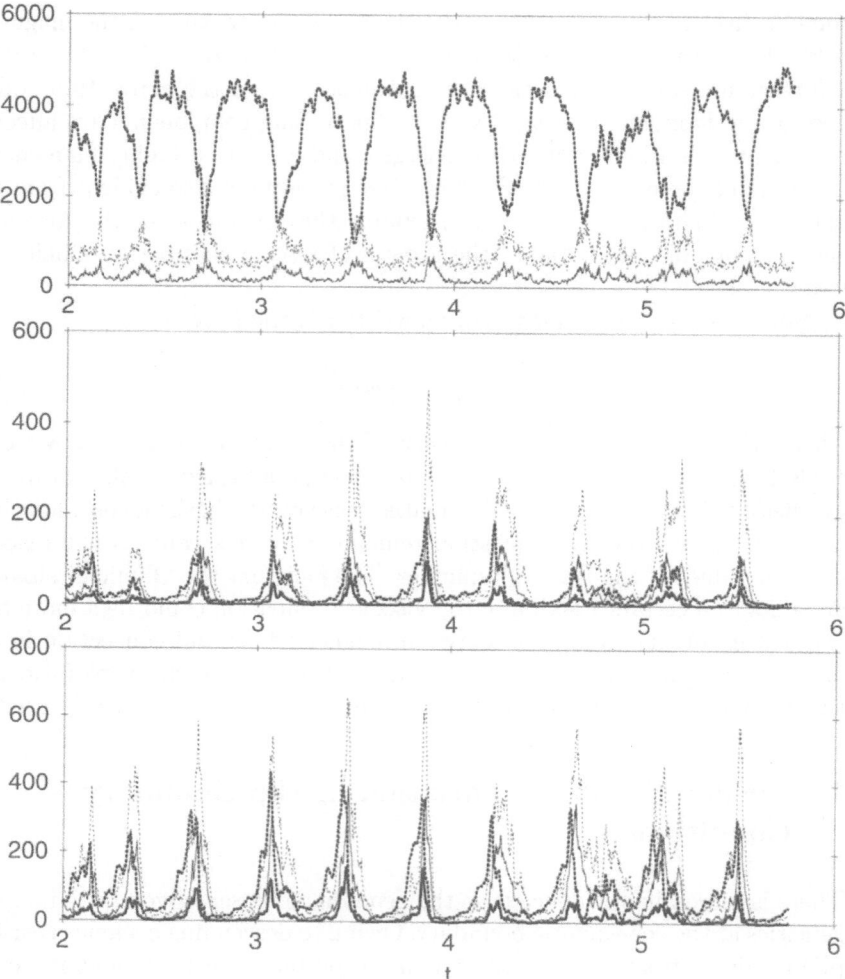

Fig. 13. Energy densities E_t^m (thick dotted line), E_t^f (thin dotted line) and E_p^f are shown in the upper plot for the case $\tau = 1.5 \cdot 10^4, R = 8.5 \cdot 10^5, P = 1, P_m = 3, A = -156$. Corresponding energy densities of the dipolar and quadrupolar components of the magnetic field are shown in the middle and the lower plot. Thick [thin] solid (dotted) lines correspond to axisymmetric [nonaxisymmetric] poloidal (toroidal) components of the magnetic fields.

contribution to the lateral inhomogeneity. Preliminary results have shown that the new boundary condition (10) does indeed lead to qualitative changes in the dynamics of convection and its dynamo action. Here we demonstrate this aspect through the results of figure 13 where relaxation oscillations of a new kind are shown which occur in the presence of the dynamo generated magnetic field in contrast to the relaxation oscillations of convection discussed in section 4. It is also remarkable that a strong dipolar component of the magnetic field is found for $P_m = 1$ while purely quadrupolar dynamos are realized for this value of P_m in the absence of lateral inhomogeneity.

8   Concluding Remarks

The dynamo studies discussed in this article have been motivated by geophysical applications. But there are still some difficulties which must be overcome before a realistic description of the geodynamo can be reached. It has been difficult, for instance, to reach a magnetic Prandtl number much less than unity. It may not be necessary to attain the value 10^{-6} based on the molecular diffusivities expected to characterize the liquid outer core of the Earth. But the effects of small scale turbulence are unlikely to raise the magnetic Prandtl number much beyond 10^{-2}. Christensen et al. (1999) have suggested that the critical value of P_m for dynamo action decreases like $750\tau^{-3/4}$ with increasing τ. We are less optimistic in this respect since our computations suggest a much weaker dependence.

A common property of most computational dynamos is that Ohmic dissipation is either less or not far above viscous dissipation. With decreasing Prandtl number the ratio of Ohmic to viscous dissipation usually increases as can be seen from figure 7b of Grote et al. (2000b). It thus seems prudent to proceed in the direction of lower P in order to increase that ratio, especially since low values of P are typical for liquid metals like those in planetary cores. On the other hand, the effective Prandtl number must be large in the Earth's core since large ratios between magnetic and kinetic energies can be expected only in the case of large P (Glatzmaier and Roberts, 1995; Busse et al., 1998). This latter property can be used for arguing that the concentration of light elements with its low diffusivity rather than the temperature provides the buoyancy for driving convection. A combination of both effects may even lead to new effects which have not yet been fully explored. These will be the subjects of future numerical simulations.

References

Ardes, M., Busse, F.H., and Wicht, J., Thermal Convection in Rotating Spherical Shells, *Phys. Earth Plan. Int.* **99**, 55-67 (1997)

Busse, F.H., Grote, E., and Tilgner, A., On convection driven dynamos in rotating spherical shells, *Studia geoph. et geod.* **42**, 211-223 (1998)

Chandrasekhar, S., "Hydrodynamic and Hydromagnetic Stability", Oxford, Clarendon Press (1961)

Christensen, U., Olson, P., and Glatzmaier, G.A., A dynamo model interpretation of geomagnetic field structures, *Geophys. Res. Lett.* **25**, 1565-1568 (1998)

Christensen, U., Olson, P., and Glatzmaier, G.A., Numerical Modeling of the Geodynamo: A Systematic Parameter Study, *Geophys. J. Int.* **138**, 393-409 (1999)

Eltayeb, I.A., Hydromagnetic convection in a rapidly rotating fluid layer, *Proc. Roy. Soc. Lond. A* **326**, 229-254 (1972)

Fornberg, B., and Merrill, D., Comparison of finite difference- and pseudospectral methods for convective flow over a sphere, *Geophys. Res. Lett.* **24**, 3245-3248 (1997)

Glatzmaier, G.A., and Roberts, P.H., A three-dimensional self-consistent computer simulation of a geomagnetic field reversal, *NATURE* **377**, 203-209 (1995a)

Glatzmaier, G.A., and Roberts, P.H., A three-dimensional convective dynamo solution with rotating and finitely conducting inner core and mantle, *Phys. Earth Plan. Int.* **91**, 63-75 (1995b)

Grote, E., and Busse, F.H., Dynamics of Convection and Dynamos in Rotating Spherical Fluid Shells, to be published in *Fluid Dyn. Res.* (2001)

Grote, E., Busse, F.H., and Tilgner, A., Convection driven quadrupolar dynamos in rotating spherical shells, *Phys. Rev. E* **60**, R5025-R5028 (1999)

Grote, E., Busse, F.H., and Tilgner, A., Effects of Hyperdiffusivities on Dynamo Simulations, *Geophys. Res. Lett.* **27**, 2001-2004 (2000a)

Grote, E., Busse, F.H., and Tilgner, A., Regular and Chaotic Spherical Dynamos, *Phys. Earth. Planet. Int.* **117**, 259-272 (2000b)

Kuang, W., and Bloxham, J., An Earth-like numerical dynamo model, *NATURE*, **389**, 371-374 (1997)

Kutzner, C., and Christensen, U., Effects of driving mechanisms in geodynamo models, *Geophys. Res. Lett.* **27**, 29-32 (2000)

Olson, P., Christensen, U., and Glatzmaier, G.A., Numerical modeling of the geodynamo: Mechanism of field generation and equilibration, *J. Geophys. Res.* **104**, 10383-10404 (1999)

Or, A.C., and Busse, F.H., Convection in a rotating cylindrical annulus. Part 2. Transitions to asymmetric and vacillating flows, *J. Fluid Mech.* **174**, 313-326 (1987)

Roberts, P.H., and Jones, C.A., The Onset of Magnetoconvection at Large Prandtl Number in a Rotating Layer I. Finite Magnetic Diffusion, *Geophys. Astrophys. Fluid Dyn.* **92**, 289-325 (2000)

Sun, Z.-P., Schubert, G., and Glatzmaier, G.A., Transitions to chaotic thermal convection in a rapidly rotating spherical fluid shell, *Geophys. Astrophys. Fluid Dyn.* **69**, 95-131 (1993)

Tilgner, A., and Busse, F.H., Finite amplitude convection in rotating spherical fluid shells, *J. Fluid Mech.* **332**, 359-376 (1997)

Zhang, K., Convection in a rapidly rotating spherical shell at infinite Prandtl number: transition to vacillating flows, *Phys. Earth Plan. Int.* **72**, 236-248 (1992)

Finite-Difference Simulations of Seismic Wavefields in Isotropic and Anisotropic Earth Models

Trond Ryberg[1], Georg Rümpker[1], Marc Tittgemeyer[2], and Friedemann Wenzel[3]

[1] GeoForschungsZentrum, Telegrafenberg, 14473 Potsdam, Germany
[2] Max-Planck Institute of Cognitive NeuroScience,
    Stephanstraße 1a, 04103 Leipzig, Germany
[3] Geophysikalisches Institut, Universität Karlsruhe,
    Hertzstraße 16, 76187 Karlsruhe, Germany

1 Introduction

The analysis of the propagation of elastic waves (seismic phases) plays an essential role in studying the structure, composition and evolution of Earth's interior. Elastic waves may be generated by earthquakes or artificial sources (explosives). Analysing the recorded wavefields yields information about the medium through which the waves have propagated. Information about the velocity structure within the Earth, provides important constraints on the mineralogical composition. In most cases, however, the direct derivation of a velocity model is not possible. The comparison of synthetic seismograms with observations can be used in the interpretation of the recorded seismic data and for the analysis of wave propagation effects. In this forward-modeling approach, an initial model of the velocity structure is modified until observed and calculated wavefields agree sufficiently well. We discuss applications of wavefield simulations in isotropic and anisotropic elastic media that have improved our knowledge about the elastic fine structure of Earth's interior. Our modeling is based on finite-difference schemes to solve the elastic wave equation. Efficient implementations of the corresponding code with further examples of seismic wavefield modeling are also given in [11].

2 The 2D and 3D isotropic wave equation: finite difference method and code parallelization

Seismic waves propagate in the Earth's interior according to the elastic wave equation. The wave equation is based on Newton's law (force balance) and Hooke's law (elastic material properties). It describes the temporal and spatial development of the field of motion (wave field) in an elastic solid body with respect to a given velocity model for both components, U and W:

$$\rho \frac{\partial^2 U}{\partial t^2} = \frac{\partial}{\partial x}\left[\lambda\left(\frac{\partial U}{\partial x} + \frac{\partial W}{\partial z}\right) + 2\mu \frac{\partial U}{\partial x} + \frac{\partial}{\partial z}\left(\mu \frac{\partial W}{\partial x} + \frac{\partial U}{\partial z}\right)\right], \quad (1)$$

$$\rho \frac{\partial^2 W}{\partial t^2} = \frac{\partial}{\partial z} \left[\lambda \left(\frac{\partial U}{\partial x} + \frac{\partial W}{\partial z} \right) + 2\mu \frac{\partial W}{\partial z} + \frac{\partial}{\partial x} \left(\mu \frac{\partial W}{\partial x} + \frac{\partial U}{\partial z} \right) \right]. \quad (2)$$

Here U and W represent the vertical and horizontal component of the wave field and μ and λ are the elastic constants, which describe elastic properties and can be combined functionally to give wave velocities of the model. Generally they are spatially varying functions.

To solve equations (1) and (2) numerically for a given model we have to resample the model on a grid, and replace the spatial and temporal partial derivatives by their finite difference equivalents. Replacing the partial derivatives in equation (1) by their second-order approximations ($O(\Delta t^2, \Delta x^2)$) equation (1) can be written as:

$$U_{i,j}^{n+1} = F(U_{i,j}^{n-1}, U_{i,j}^n, U_{i\pm1,j}^n, U_{i,j\pm1}^n, W_{i\pm1,j\pm1}^n) \quad (3)$$

here $U_{i,j}^n$ represents the vertical component for the n-th time step at the location i,j on the grid. Function F is a simple linear function of their arguments with coefficients which depend on the local values of λ and μ (i.e. the local velocities). Equation (3) gives the rule to calculate the n+1 time step of $U_{i,j}$ from its past values $U_{i,j}^n$ and $U_{i,j}^{n-1}$ and from the spatial neighbours $U_{i\pm1,j\pm1}^n$ and $W_{i\pm1,j\pm1}^n$. A similar scheme holds for $W_{i,j}^{n+1}$:

$$W_{i,j}^{n+1} = F'(W_{i,j}^{n-1}, W_{i,j}^n, W_{i\pm1,j}^n, W_{i,j\pm1}^n, U_{i\pm1,j\pm1}^n) \quad (4)$$

Expressions (3) and (4) can be efficiently implemented on a massive parallel computer system. To model wave propagation in three dimensions a straightforward extension of the wave equations (1) and (2) to the 3D case has been carried out. This leads to similar expressions for the "update" of the wave field for the U-component:

$$U_{i,j,k}^{n+1} = F(U_{i,j,k}^{n-1}, U_{i,j,k}^n, U_{i\pm1,j,k}^n, U_{i,j\pm1,k}^n, W_{i\pm1,j,k\pm1}^n, V_{i\pm1,j\pm1,k}^n). \quad (5)$$

Analog expressions exist for both horizontal components W and V.

Special care has to be taken for the model boundaries. Benchmark testing shows, that on massive parallel computer systems, the introduction of buffer zones, where the wave field is successively damped (non-physically, i.e. not based on the wave equation), is more efficient than the implementation of absorbing boundary conditions based on the one-way wave equation. In [11] we analyze the computational efficiency of the parallelization for different model sizes and numbers of CPUs used. Updating the wave field according to equations (1) and (2), in the 2D case, requires ~ 74 floating point operations per grid point (additions and multiplications). By chosing the appropriate number of CPUs a sustained performance of > 100 MFLOPS can be reached easily on a T3E-900/512. Caching problems can be avoided by proper allocation of memory. Best numerical performance is obtained when using the following HPF statements for memory allocation and CPU distribution. In the 2D case:

```
    REAL*4 A(0:8,0:35000,-1:5000)
$HPF PROCESSORS VEC(256)
$HPF DISTRIBUTE (*,BLOCK,*) ONTO VEC :: A
```

and in the 3D case:

```
    REAL*4 A(0:12,0:500,0:500,-1:500)
$HPF PROCESSORS ARR(16,16)
$HPF DISTRIBUTE (*,BLOCK,BLOCK,*) ONTO ARR :: A
```

In the 3D case, ~ 183 floating point operations per grid point are required; this results in a sustained performance of > 40 MFLOPS. The optimizations of the numerical kernel of the wave propagation simulation is carried out by trial and error. The strip-like distributions (*,BLOCK,*) and (*,BLOCK,BLOCK,*) is chosen because of the necessary special treatment of the upper model boundary (Earth's surface). Another problem occurred when reading in large models (file size >100 MBytes). Taking advantage of FORTRAN statements like:

```
extrinsic (hpf_serial) SUBROUTINE READCOLUMN
...
READ(2,REC=M)  (VALUEX(J),J=1,N)
RETURN
END
```

we significantly enhanced the file read procedure. Otherwise reading in files would result in a relatively poor performance compared to the actual simulation of the wave propagation.

To solve the elastic wave equation for typical 2D models of the size 2100×300 km at frequencies around 5 Hz a spatial and temporal discretization of 60 m and 0.0056 s is required. This results in a model of 35000 by 5000 grid points and an integration over ~ 50000 time steps. To compute the wave field for a single model, more than $6 \cdot 10^{14}$ floating point operations have to be carried out; the memory required is >25 GBytes. Model input and seismogram output increase the computational time by not more than 5%.

3   Propagation and blocking of the seismic phases travelling trough the Earth mantle

We tested the hypothesis of an upper mantle scattering waveguide by carrying out extensive numerical simulations for 2-dimensional Earth models with randomly distributed velocity fluctuations in the upper mantle [10]. The finite-difference modelling helped to reveal key features of wave propagation for this class of models. Most previous models of the Earth's upper mantle are

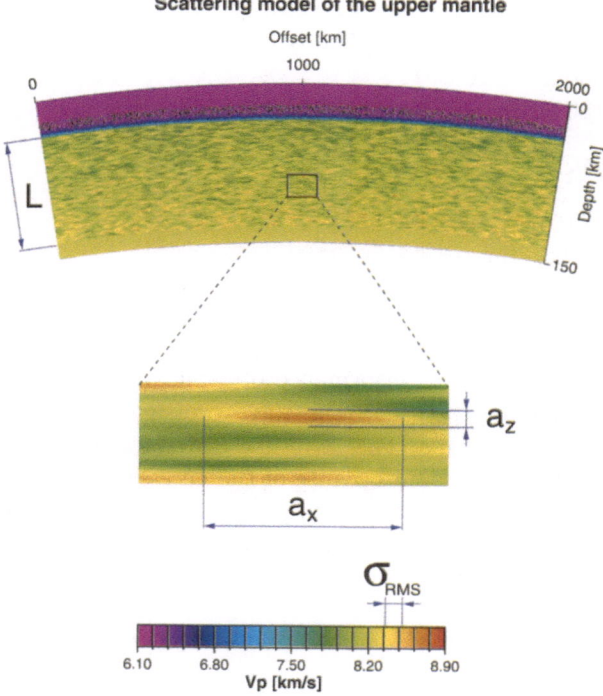

Scattering model of the upper mantle

Fig. 1. Long distance wave propagation of the teleseismic P_n and S_n is achieved by statistical fluctuations of elastic parameters in the uppermost mantle. Mantle heterogeneities have a Gaussian distribution with a horizontal and vertical correlation lengths of 20 km and 0.5 km, respectively. The standard deviation of the velocity fluctuations is 2 %. (Not to scale)

characterized by laterally homogeneous elastic structures. In contrast many seismological data sets show conclusive evidence for strong scattering. For instance, scattering within the uppermost mantle is prominently documented by the so-called high-frequency teleseismic P_n phase. This unusual high frequency seismic phase was observed at large distances (> 2000 km). Examples occur in the Russian PNE-program, and in other long range refraction experiments [14]. A waveguide, which is caused by random fluctuations of the mantle's elastic properties can explain the main features of this seismic phase [10]. This waveguide is characterized by random velocity fluctuations with average "length" a_x, "thickness" a_z and "magnitude" σ_{RMS} as can be seen in Figure 1. We focus on the statistical properties of the upper mantle's velocity fluctuations; these fluctuations act as scatterers. We find that randomly distributed velocity fluctuations in the uppermost mantle are characterized by a predominantly elongated shape with a horizontal and vertical correlation lengths of about 20 km and 0.5 km, respectively. Small but significant veloc-

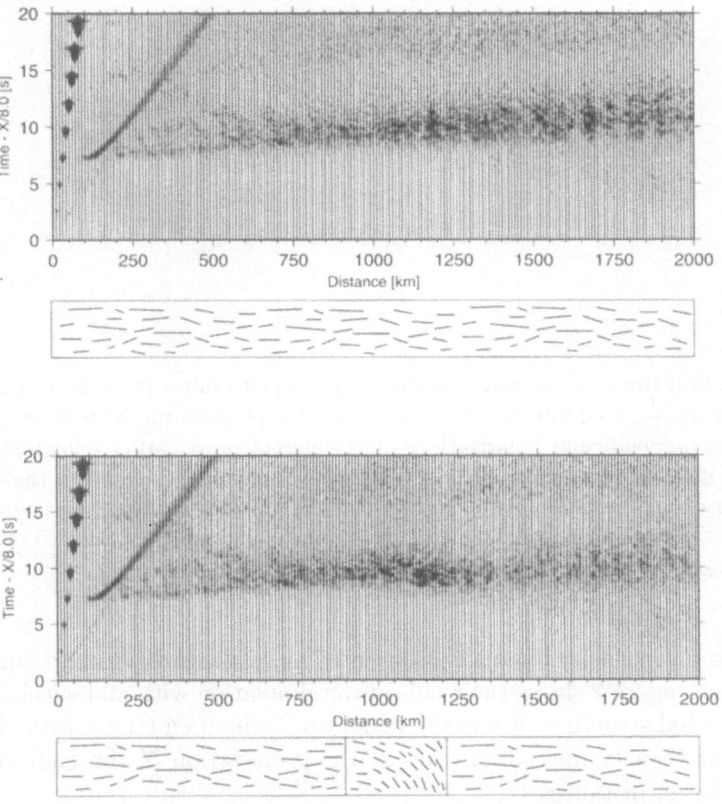

Fig. 2. Long distance propagation of the teleseismic P_n for a simple scattering model (top) and more complex model, simulating down-welling of the mantle (bottom). For the simple model (top) the P_n phase can be observed up to a distance of 2000 km from the source. For the complex model the propagation of the P_n phase is strongly attenuated.

ity fluctuations of 1.5 to 2 % located within a sufficiently thick heterogeneous upper mantle layer ($L \geq 100\,\text{km}$) are necessary to produce synthetic seismograms similar to the observations. Physically, the distinct properties of the velocity fluctuations in the upper mantle result in frequency-selective wave propagation within the heterogeneous zone, or a waveguide phenomenon.

To further test our hypothesis we studied the "blocking" (or truncation) of propagation of the P_n and S_n phases. It is well-known, that in stable regions of the Earth, such as continental shields and deep-ocean basins, the phases propagate very efficiently. However, it appears that the transmission of the phases is blocked by major suture zones such as boundaries at mid-ocean ridges, island arc structures, and subduction zones. This blocking could be related to modified upper mantle structures due to ongoing tectonic processes, such as up- and down-welling of mantle flow. In our simulations we tested

Shear-wave splitting

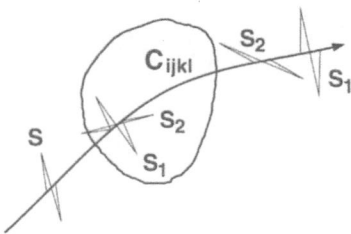

Fig. 3. Splitting of shear waves in anisotropic regions (after [4]). The incident shear wave S is separated into two phases, S_1 and S_2, propagating with different velocities and perpendicular polarizations. The different propagation velocities result in a time delay between both phases. This delay time, δt, is related to the extent of the anisotropic region and the strength of the anisotropy. The polarization direction of the fast shear wave, ϕ, can provide constraints on the preferred orientation (alignment) of anisotropic crystals in the medium.

models with velocity fluctuations oriented along these regional up- and down-wellings. Figure 2 shows the result of the simulation with and without such a flow-related structure. A mantle model that exhibits a region with vertically oriented velocity fluctuations blocks the propagation of the high-frequency P_n at larger distances.

4   Elastic anisotropy

Based on the implementations of FD codes for large and complex isotropic Earth models, we extended our research to anisotropic models. Elastic anisotropy and related effects on traveltimes and waveforms play an important role in both, exploration seismics and global seismology ([3]). So-called shear wave splitting, the most obvious anisotropic effect (similar to optical birefringence or double refraction), is in the focus of ongoing research. The anisotropic properties of the upper mantle are related to the crystal orientation of olivine (a highly anisotropic mantle mineral). The preferred orientation of olivine crystals over large spatial scales (compared to the wavelength of the seismic signal) is related to deformation and flow processes within the upper mantle. Studying the spatial distribution of the anisotropic properties of mantle materials thus enables us to extract important information about dynamic processes such as mantle convection and the subduction of oceanic lithosphere. In complex inhomogeneous regions, as in the case of high-frequency P_n propagation, the direct inversion of the wave field does not seem feasible. Therefore, forward modelling, i.e. comparison between numerical simulations of wave fields and observations at seismic stations, plays an important role in constraining structural and dynamic properties. Figure 3 shows wave propagation through an anisotropic region, described by the elastic tensor C_{ijkl}. As in the analog case of optical birefringence, the incident shear wave is separated into two phases propagating with different velocities and exhibiting perpendicular polarizations.

In recent years, the analysis of teleseismic shear wave splitting has become an important tool for the mapping of anisotropic structures of the upper mantle. Lateral variations of splitting parameters (fast polarization direction ϕ and delay time δt) over relatively short spatial scales are commonly observed and are usually interpreted in terms of similar variations of anisotropic properties. However, the relatively long-period shear waves may be sensitive to the anisotropic properties of more than one domain within the medium, thus requiring a more careful interpretation of the splitting parameters. We have used wave-field forward-modeling to improve these interpretations ([8]). The numerical solution of the anisotropic wave equation is described in the following section.

5   The anisotropic wave equation: theory and numerical formulation

Solving the anisotropic wave equation numerically is computationally demanding, however, it has the advantage that the synthetic seismograms are "complete" and include diffraction effects due to changes in material properties. Our numerical simulations of shear wave splitting are performed with respect to a 2D cartesian coordinate system. Thus, changes in material properties and wave field variations are restricted to the x_1-x_3 coordinate plane. This implies that all derivatives ∂_2 (or $\frac{\partial}{\partial x_2}$) vanish and the equation of motion in terms of the displacement u_i [1] may be written as

$$\rho u_{1,tt} = \sigma_{1,1} + \sigma_{5,3} ,$$
$$\rho u_{2,tt} = \sigma_{6,1} + \sigma_{4,3} ,$$
$$\rho u_{3,tt} = \sigma_{5,1} + \sigma_{3,3} ,$$

(6)

where σ_i is given by

$$\sigma_i = c_{i1}u_{1,1} + c_{i3}u_{1,3} + c_{i4}u_{2,3} + c_{i5}(u_{3,1} + u_{1,3}) + c_{i6}u_{2,1} .$$

(7)

The c_{ij} denote elastic constants in crystallographic notation. Equations (6)-(7) apply to a general anisotropic medium.

In the following, we assume anisotropic variations due to the preferred orientation of orthorhombic olivine, which can be characterized by nine independent elastic constants. Since our modelling will be limited to situations where the olivine a-axis is oriented horizontally (within the x_1-x_2 plane), the matrix representation of c_{ij} has the form

$$c_{ij} = \begin{pmatrix} c_{11} & \times & c_{13} & 0 & 0 & c_{16} \\ \times & \times & \times & 0 & 0 & \times \\ c_{13} & \times & c_{33} & 0 & 0 & c_{36} \\ 0 & 0 & 0 & c_{44} & c_{45} & 0 \\ 0 & 0 & 0 & c_{45} & c_{55} & 0 \\ c_{16} & \times & c_{36} & 0 & 0 & c_{66} \end{pmatrix} \tag{8}$$

where the symmetry of the elastic constants has been used. Matrix elements that do not contribute to (6) because of the 2D geometry are denoted by $\times$. Note that $c_{16} = c_{36} = c_{45} = 0$ whenever the a-axis is parallel to one of the coordinate axes.

The structure of (8) may be used in the numerical formulation of (6) to avoid the unnecessary evaluation of certain terms. The values for the remaining nine (non-zero) elastic constants can be defined initially at every point of the computational grid. A general orientation of the orthorhombic elastic tensor would require the definition of up to 15 non-zero elastic constants in the 2D case. For the numerical evaluation of (6), we extended the explicit second-order isotropic finite-difference formulation of equations (3) and (4) to the anisotropic case. To update the wave field, i.e. to calculate the wave field for the next time step, we need for the 2D case ~ 170 floating point operations per grid point (additions and multiplications). For a large number of CPUs (>10) a sustained performance of > 30 MFLOPS can be reached on a T3E-900/512.

Initially, the computational domain was chosen sufficiently large to avoid interference effects with spurious reflections from the grid boundaries. However, benchmark testing [11] has shown, that solving the wave equation on a massive-parallel computer system for large models (with respect to the wave length) can be done much faster by implementing buffer zones. Here, have used the same simple damping of the wave field within the buffer zone as used for the isotropic case. Numerical grid dispersion for the low-order approximation of the derivatives is overcome by using a finer grid discretization ($\Delta x_1 = \Delta x_3$) such that $\lambda/\Delta x > 50$ and $T/\Delta t > 100$ for the shortest wavelength λ.

To test the effectivity and correctness of our implementation we calculated the wave fields for several simple homogeneous models. Figure 4 shows the snapshots of the three components of the wave field due to a line source in homogeneous isotropic and anisotropic models. The latter exhibits hexagonal symmetry with vertical symmetry axis (parallel to z). As expected, the wave field of for the isotropic case, which was calculated using the anisotropic code, has a circular shape, indicating the isotropic character of the medium. The anisotropic case shows wavefronts which differ significantly from the isotropic case. In the angular range from $0°$ to $20°$ (taken from the vertical) several individual shear phases can be observed. The calculations have been compared with analytical solutions for this problem. Additionally, some code

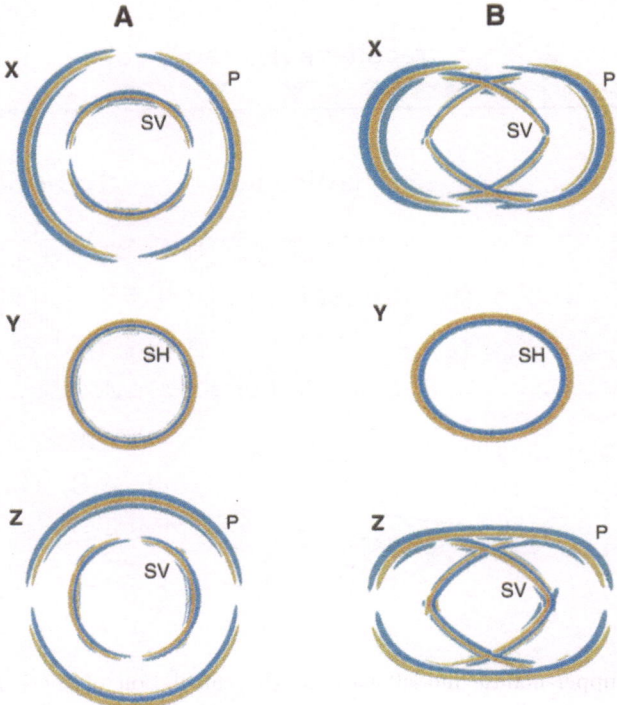

Fig. 4. (A) Snapshots of the wavefronts due to line source (along the y axis) within an homogeneous isotropic elastic medium. The source characteristics are somewhat artificial and are chosen to illustrate the wavefronts for the three components simultaneously. Both, P and S wave energy is radiated. The circular wavefronts propagate within the x-z plane and the three orthogonal displacement components are shown (as indicated by X, Y, Z). The fast wavefront corresponds to the P wave. The S wavefront exhibits displacement components within the $x - z$ plane, which are denoted SV. The corresponding displacement in y-direction is denoted SH. (B) Snapshots for the anisotropic case. The wavefronts for P, SV, and SH exhibit pronounced deformations due to the directional dependence of the wave speed. In addition, the SV wavefront is characterized by triplications which lead to multiple S-wave arrivals at certain positions within the medium.

optimization, mainly loop interchanging, has been carried out to improve the numerical performance.

6 Shear-wave splitting and laterally-varying anisotropy

In our modelling of anisotropic upper mantle structure, the region beneath the the seismic stations is characterized by a single layer of laterally-varying anisotropy which is embedded between two homogeneous isotropic regions (Figure 5). When setting-up this simple model, we were inspired by the study of [2] who estimated sensitivity ranges for shear waves for variations in the

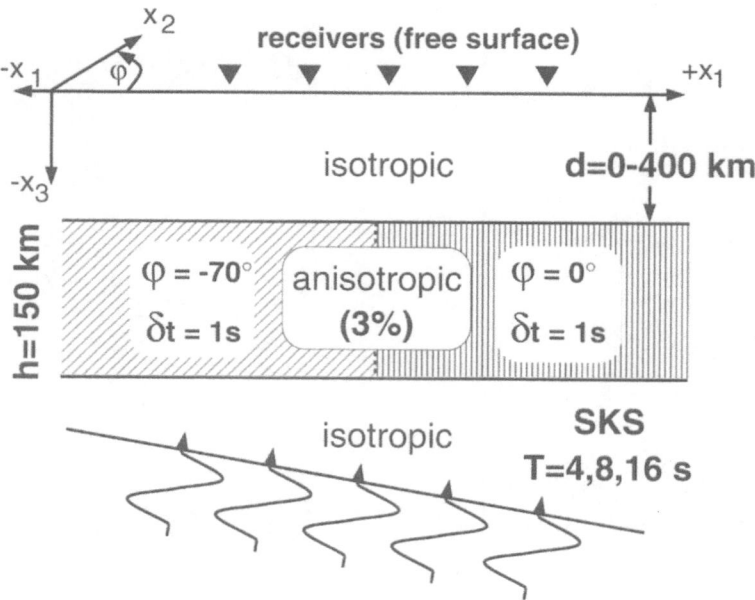

Fig. 5. 2-D upper-mantle model used for the calculation of the synthetic seismograms. The laterally-varying anisotropic layer is characterized by a horizontal olivine a-axis. For vertical propagation the shear waves accumulate a delay time δt of 1 s and the a-axis is parallel to the fast polarization direction ϕ. ϕ is the angular distance from the x_1 axis within the horizontal x_1-x_2 plane. In our modeling we investigated the effects of variations in depth of the anisotropic layer and period (wavelength) of the incident SKS wave.

depth of the anisotropy and for different dominant wavelengths of the incident wave.

In this paper we concentrate on teleseismic shear waves such as SKS phases, which travel through the Earth's core. Thus, we assume a plane horizontal SKS wavefront entering the region from below at small angles of incidence. The pulse shape of the incident wave is given by the derivative of a Gaussian. The initial polarization is taken at 45° between the x_1 and x_2 axes. The laterally-varying anisotropy within the layer is given by two homogeneous regions: the model exhibits horizontal a-axis orientations of -70° (relative to x_1, see Figure 5) for $x_1 < 0$ km and of 0° for $x_1 > 0$ km. The elastic constants for olivine ([6]) are scaled such that vertically propagating fast and slow shear waves accumulate a delay time (δt) of 1 s over the total thickness ($h = 150$ km) of the layer. This corresponds to about 3% of velocity anisotropy, which similar to values obtained from measurements on mantle rocks [7].

In Figure 6 we show a snapshot of the displacement field after a vertically incident shear wavefront has passed the anisotropic layer. The dominant pe-

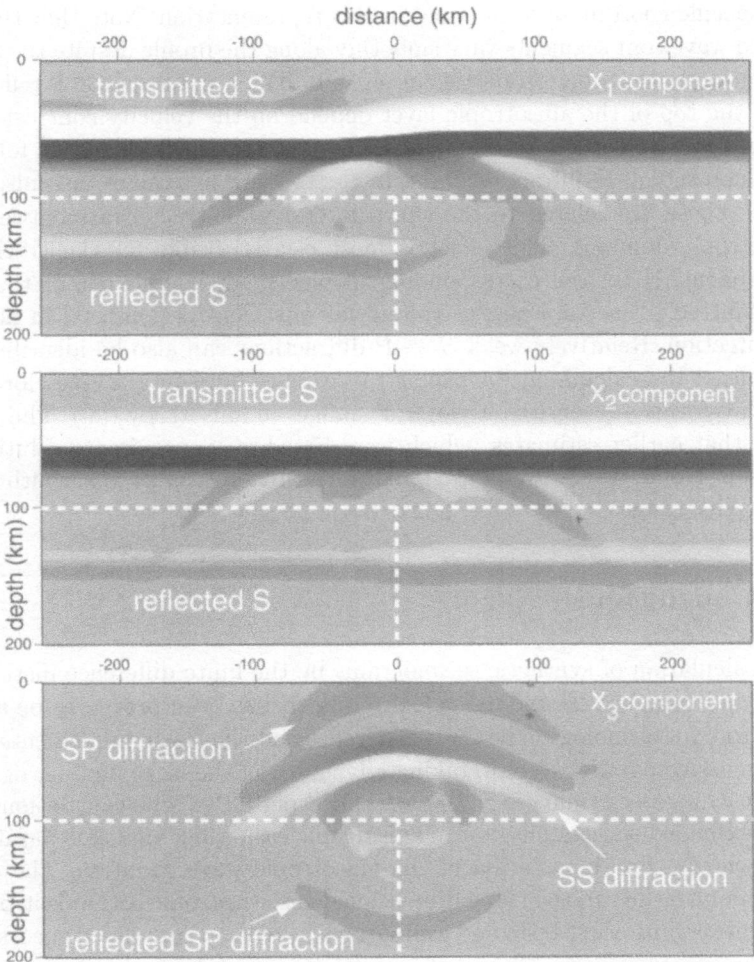

Fig. 6. Snapshots of the wavefield of an S wave transmitted through the laterally-varying model. The adjacent positive (white) and negative (black) regions that correspond to the main transmitted phase (x_1 and x_2 components). A nonlinear greyscale has been used to enhance weaker phases. The wavefield consists of the primary transmitted S phase, reflected and diffracted S phases, and SP conversions.

riod of the incident wave is $T = 8$ s, which corresponds to a wavelength of about 16 km in is this model. The main arrival, exhibiting x_1 and x_2 components, is the transmitted plane S phase. On the right-hand side of the model the polarization directions of the fast and slow shear waves are parallel to the x_1 and x_2 coordinate axes, respectively. In this region the incident wavefront separates into a fast and slow front with distinct non-zero displacement components. On the left-hand side the fast direction is oriented at $-70°$.

Therefore the fast and slow shear waves are not separated with respect to the specific coordinate frame used in this representation. Note that the individual wavefront segments vary smoothly along the profile despite the abrupt change in anisotropic properties at $x_1 = 0$. The amplitudes of S reflections from the top of the anisotropic layer depend on the velocity contrast at the anisotropic/isotropic boundary. In the central region, immediately following the main signal, S diffractions arise at the top and bottom of the anisotropic layer. These are related to the sharp lateral velocity contrast between the anisotropic domains. This effect is most clearly visible for the vertical x_3 component. Here, the corresponding displacement component of the main transmitted phase vanishes, as the initial wavefront is polarized in horizontal direction. Relatively weak $S \rightarrow P$ diffractions can also be identified. We have used the calculated wavefields to estimate horizontal ranges for which the waveforms are sensitive to lateral changes in anisotropy ([8]). The results show that earlier estimates, which were based on approximate solutions of the anisotropic wave equation, significantly underestimate the width of the sensitivity range.

7   Conclusions

The calculation of synthetic seismograms by the finite difference method for large and complex 2D and 3D velocity models has been proven to be an efficient tool in seismology. The finite difference code has been implemented on a T3E massive parallel system, taking advantage of the large number of CPUs and reaching performances > 100 MFLOPS per CPU. The calculations have been extended successfully to the anisotropic case using analogue parallelization schemes for the solution of the anisotropic wave equation. The extension of our results to the three-dimensional case (isotropic and anisotropic) is straightforward. First tests and evaluations have been completed successfully.

Acknowledgements. The numerical calculations of the synthetic seismograms were performed on the Cray T3E-900/512 of the High-Performance-Computing Center in Stuttgart. We are grateful for their continuous support. All figures were generated using the GMT software of [15].

References

1. Aki, K. E., and P. G. Richards, *Quantitative Seismology - Theory and Methods,* Freemann, San Francisco, 1980.
2. Alsina, D. and Snieder, R., Small-scale sublithospheric continental mantle deformation: constraints from SKS splitting observations, *Geophys. J. Int., 123,* 431–448, 1995.
3. Babuska, V., and M. Cara, *Seismic anisotropy in the Earth,* Kluwer Academic Publishers, Dordrecht, 1991.

4. Crampin, S, An introduction to wave propagation in anisotropic media, *Geophys. J. R. astr. Soc.*, *76*, 135–145, 1984.
5. Kelly, K. R., Ward, R. W., Treitel, S., and Alford, R. M., Synthetic seismograms: a finite-difference approach, *Geophysics*, *41*, 2–27, 1976.
6. Kumazawa, M., and O. L. Anderson, Elastic moduli, pressure derivatives, and temperature derivatives of single-crystal olivine and single-crystal forsterite, *J. Geophys. Res.*, *74*, 5.961–5.972, 1969.
7. Mainprice, D., and P. G. Silver, Interpretation of SKS-waves using samples from the subcontinental lithosphere, *Phys. Earth Planet. Inter.*, *78*, 257–280, 1993.
8. **Rümpker, G. and Ryberg, T., New "Fresnel-zone" estimates for shear-wave splitting observations from finite-difference modeling, *Geophys. Res. Lett.*, *27*, 2005–2008, 2000.**
9. **Ryberg, T. and Weber, M., Receiver function arrays - a reflection seismic approach, *Geophys. J. Int.*, *141*, 1–11, 2000.**
10. **Ryberg, T., Tittgemeyer, M., and Wenzel, F., Finite difference modelling of *P*-wave scattering in the upper mantle, *Geophys. J. Int.*, *141*, 787–801, 2000.**
11. **Ryberg, T., Tittgemeyer, M., and Wenzel, F., Finite difference modelling of elastic wave propagation in the Earth's uppermost mantle, in *High Performance Computing in Science and Engineering'99*, edited by E. Krause and W. Jäger, pp. 3–12, Springer, Berlin Heidelberg New York, 2000.**
12. **Ryberg, T., Tittgemeyer, M., and Wenzel, F., Finite difference modelling of seismic wave phenomena within the Earth's uppermost mantle, submitted to Transactions of the High Performance Computing Center Stuttgart (HLRS), 2000.**
13. Tittgemeyer, M., Wenzel, F., Fuchs, K., and Ryberg, T., Wave propagation in a multiple-scattering upper mantle - observations and modelling, *Geophys. J. Int.*, *127*, 492–502, 1996.
14. Tittgemeyer, M., Wenzel, F., Ryberg, T., Fuchs, K., Scales of heterogeneities in the continental crust and upper mantle. *Pure Appl. Geophys.*, *156*, 29–52, 1999.
15. Wessel, P. and Smith, W., Free software helps map and display data, *EOS Trans. Am. Geophys. Union*, *72*, 441, 445–446, 1991.

References high-lighted by bold face took extensive benefits from the project "Seismische Wellenfeldmodellierung" supported generously by the High-Performance Computer-Center in Stuttgart.

Collisional Dynamics of Black Holes, Star Clusters and Galactic Nuclei

Marc Hemsendorf,[1,2] Holger Baumgardt,[3] Christian Boily,[2]
Rainer Spurzem,[2] and Steinn Sigurdsson[4]

[1] Department of Physics & Astronomy, Rutgers University, 136 Frelinghuysen,
   Piscataway N.J. 08854-8019 U.S.A.
[2] Astronomisches Rechen-Institut, Mönchhofstrasse 12-14, Heidelberg D-69120
[3] Mathematics Department, The University of Edinburgh, James Clerk Maxwell
   Building, Kings Buildings, Mayfield Road, Edinburgh, EH9 3JZ, U.K.
[4] Department of Astronomy & Astrophysics, 525 Davey Lab., Penn State
   University, University Park PA 16802 U.S.A.

Abstract. We investigate with numerical N-body integrations the time-evolution
of stellar systems with centrally peaked density profiles harboring a massive black
hole (galactic nuclei) or with varied initial conditions (such as rotation) using direct-
integration integrators. The computer programmes *EUROSTAR* and *NBODY6++*
developped for use with mpi libraries maintain high accuracy over billions of integration timesteps, allowing a detailed analysis of the orbit of the black hole(s),
the precise diffusion of angular momentum between stars, and the ejection of stars
from a cluster orbiting a host galaxy. Below we consider each topic in turn.

1 Dissolution of star clusters

1.1 Theory

One of the most important aspects of the evolution of star clusters is their
slow dissolution due to internal (two-body relaxation, mass-loss of individual
cluster stars) and external (gravitational shocks from disc and bulge passages)
processes. This dissolution leads to an enrichment of the galactic environment
with cluster stars, and to a change in the properties of cluster systems in
galaxies.[63],[6]

Most studies of cluster dissolution so far have been done for the idealised
case of circular orbits. Here the tidal field is constant and gravitational shocks
do not exist, so star clusters dissolve due to internal processes alone. After the
massive stars have gone supernova, two-body relaxation is the most important
dissolution mechanism.

Two-body relaxation is caused by encounters between cluster stars. These
encounters lead to small changes in the energies of the stars and some stars
gain enough energy so that they can leave the cluster. The timescale on which
these energy changes happen is measured by the relaxation time [12], [53]:

$$t_{rh} = 0.138 \frac{\sqrt{N}\, r_h^{3/2}}{\sqrt{<m>}\,\sqrt{G}\,\ln(\gamma N)} \tag{1}$$

One expects that during each relaxation time a constant fraction of cluster stars is scattered above the critical energy necessary for escape and that the lifetimes of clusters are proportional to their relaxation times.

The scaling of the lifetimes was put to a test by the 'Collaborative Experiment' [22]. Here multi-mass clusters moving in circular orbits were studied. Contrary to the expectation, it was found that the lifetimes do not scale with the relaxation time. Despite many ideas, the reason for this scaling problem could not be identified.

1.2 Results

In order to clarify this scaling problem, we studied the dissolution of star clusters moving on circular orbits around a parent galaxy. The parent galaxy was modelled as a point mass. We restricted ourselves to single-mass clusters with no stellar evolution. Clusters containing between 128 and 16384 stars were studied, using the program NBODY6++ running with MPI link up on Cray T3E computers [54]. Figure 1 shows the results for the half-mass times, which are taken here as a measure for the lifetimes of the clusters.

The solid line shows a scaling proportional to the relaxation time, fitted to the result of the largest run. As one can see, it does not give a good description of the results. The reason for this discrepancy is that stars which have enough energy to escape need time until they leave the cluster. While they are still inside the cluster, potential escapers undergo further scatterings and some of them become bound again. This mechanism is especially important for clusters in external tidal fields, since the escape times of stars in such clusters can be very long. As Fukushige & Heggie [17] have shown, they can be of the same order as the cluster lifetime for stars with energies only slightly larger than the critical energy that is necessary for escape.

Taking the backscattering of potential escapers into account, and utilising the result of Fukushige & Heggie for the escape time of potential escapers, it can be shown [7] that the lifetimes should scale as: $T_{Diss} \propto t_{rh}^{3/4}$. This dependency is shown as a dotted line in Fig. 1. As can be seen, it gives a much better fit to the result of the N-body simulations. The remaining difference is due to the core-collapse of the clusters and the starting condition. Since the lifetimes increase much slower with the particle number, we expect that the lifetimes of globular clusters are a factor of 3 to 4 shorter than expected hitherto.

2 The Impact of Rotation on Cluster Dynamics

Star clusters are self-gravitating Newtonian systems of choice for a complex brew gravitational dynamics (e.g. [44] for a review). Observations of old,

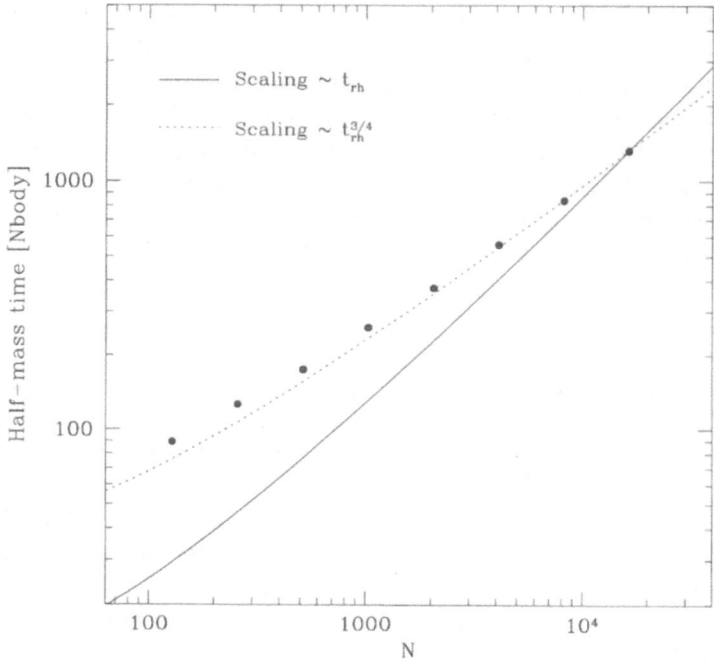

Fig. 1. Scaling of the half-mass times for clusters in an external tidal field. The dots show the results of the N-body simulations. They show a clear deviation from a scaling with the relaxation time (solid line). The dotted line shows the scaling obtained from taking the finite escape times of potential escapers into account. It gives a much better description of the results.

globular, stellar clusters have led to the formulation of spherically symmetric dynamical models of equilibria. The most successful and universally studied one-integral spherical models are the King profiles.king66 However, surveys of up to 100 Milky Way clusters have found small but significant departures from spherical symmetry [64] fits to their projected isophotes yield ellipticities $<\epsilon> \equiv 1- <a/b> \approx 0.07 \pm 0.01$. A study of 173 clusters in M31 found $<\epsilon> = 0.09 \pm 0.04$.[56] Observations of young clusters in the Large Magellanic Cloud revealed isophotal contours with ellipticities as large as $\epsilon = 0.3$ [15,28]. This raises the possibility that clusters are formed as strongly flattened structures which then evolve towards rounder configurations [16,10,60], and brings up important theoretical issues concerning processes which may drive this evolution.

Rotation stretches any stellar association along a preferred axis: observations of the clusters ω Centauri and M13 have shown that they are flattened by rotation [43,41,32]. Thus angular momentum, measured or possibly lost during evolution, offers a way to account for the morphology of clusters.

Yet to date there are few evolutionary models of clusters with initial angular momentum. We have started on a project to develop three-dimensional dynamical models of rotating star clusters. In this articles results for two n-body models of isolated clusters are presented. Previous modelling of rotating clusters is reviewed first.

2.1   Gas and Fokker-Planck Models of Rotating Clusters

Agekian [4] considered the effects of angular momentum diffusion on the equilibria of rotating fluid masses of uniform density. In his analysis, concentric spheroids rotating about their minor axis become rounder in time when the spheroids have initially an ellipticity $\epsilon = 1 - a/b \leq 0.735$, where a and b are the minor and major axes. Shapiro & Marchant [52] integrated the equations of motion for this fluid in the limit of adiabatic (slow) diffusion of momentum. Angular momentum losses are driven by mass elements moving in the direction of the stream leaving the system at a rate higher than those moving in the opposite direction. The energy required for escape comes from 'heat', attributed to two-body encounters. Thus, angular momentum losses are accrued over a local two-body relaxation timescale, t_{col}, which is inversely proportional to the mass density ($\propto 1/\rho_*$). In practice this hinders applications of the results to actual clusters, which show centrally peaked density profiles [44]. Nevertheless, the framework set by Agekian provides a start in linking rotating bodies and observed (non-rotating) globular clusters.

With zero rotation, the central region of a cluster evolves towards a cusp in density during what is known as the gravothermal catastrophe. Does rotation stop the formation of a cusp? Hachisu [20,21] discussed the time-evolution of self-gravitating cylindrical distributions of gas with angular momentum. He predicted a runaway collapse of the central region whenever angular momentum is expelled faster than a critical rate [29]. Hachisu dubbed this the 'gravogyro catastrophe', by analogy with the non-rotating case. These were until recently the only evolutionary models of rotating clusters. Two-dimensional orbit-averaged Fokker-Planck methods have now also been developed to address this issue. Following Goodman's [18] approach, Einsel & Spurzem [14] integrated the Fokker-Planck equation in energy-momentum space $[E, J_z]$. Their initial configurations are truncated King models with added bulk motion. This velocity field takes the form of a Maxwellian distribution such that the mean velocity scales in proportion to radius away from the centre, then drops off at large radii. Their adopted axisymmetric distribution function [32]

$$f(E, J_z) \propto \exp\left(-\beta\Omega_o J_z\right) \cdot \left[\exp\left(-\beta E\right) - 1\right] \qquad (2)$$

where β is the inverse square central velocity dispersion and Ω_o an angular velocity. The initial conditions are fixed by specifying the dimensionless parameters

$$\omega_o = \Omega_o / \sqrt{9G\rho_c/(4\pi)} \text{ and } W_o ,$$

i.e., the scales of angular momentum and gravitational potential, respectively. The latter is the King parameter.

In the 2D Fokker-Planck models core-collapse proceeds on a much shorter timescale than in the non-rotating case, confirming Hachisu's early intuition. However the central angular velocity does not increase at the high rates expected during the on-set of a gravo-gyro catastrophe; however near the end of core-collapse the central velocity dispersion bears the same relation to the central density as in the non-rotating self-similar collapse. This leaves open the question of what controls the final phase of evolution in these systems, i.e. whether or not rotation truly survives up to core-collapse. We chose to approach this problem using three-dimensional numerical integration.

Self-consistent n-body realisations of the distribution function were obtained from the equilibrium Fokker-Planck code FOPAX developed by Christian Einsel. The models are fully specified once values are assigned to (W_o, ω_o). Models with rotation have $\omega_o \neq 0$ and a range of flattening increasing with it. We observed for such models, save those with $\omega_o = 0.8$, that the central part remains quasi-spherical and in solid-body rotation. At larger radii, the models are all distinguished from one another.

Overall the fraction of kinetic energy invested in streaming motion ranges from 0% to 4%, 14% and 26% when $\omega_o = 0, 0.3, 0.5$ and 0.8. For comparisons, the cluster ω Centauri invests perhaps as much as 22 % of its kinetic energy in rotation [41].

2.2   N-body Simulations

The code NBODY6++ is an Aarseth-type integration code based on a Hermite expansion of the variables in time [3]. It has been ported to parallel architecture [55]; the calculations were performed on CRAY computers linked up with MPI library. The code treats particles as point-masses and stellar evolution options were switched off. The chain-regularisation algorithm for hierarchical stellar encounters as well as the standard 'KS' regularisation [48,3] ensures high-precision integration during close interactions. Only simulations with N = 5,000 equal-mass particles will be discussed. There are no external tides.

Figures 2 & 3 illustrate the time-evolution of the models. The central density, total angular momentum and mean and core radii are plotted as function of time in units of the two-body relaxation time t_{col} [44,11]. (Note: the lengths were normalised to their initial values.) The top panels show evolution for the case of $\omega_o = 0.5$, the bottom set for $\omega_o = 0.8$. Looking at these diagrams we find an evolution of the central density similar to the standard case with no rotation: the contraction of the central region leads to more close encounters and ejection, hence further contraction ensues, etc, until $t \simeq 5\ t_{col}$ when the density peaks sharply, indicating core-collapse: at the end of the simulations $r_c \approx 0.03$ and 0.015, respectively, for $\omega_o = 0.5$ and 0.8. At constant energy, core-contraction drives the expansion of the outer

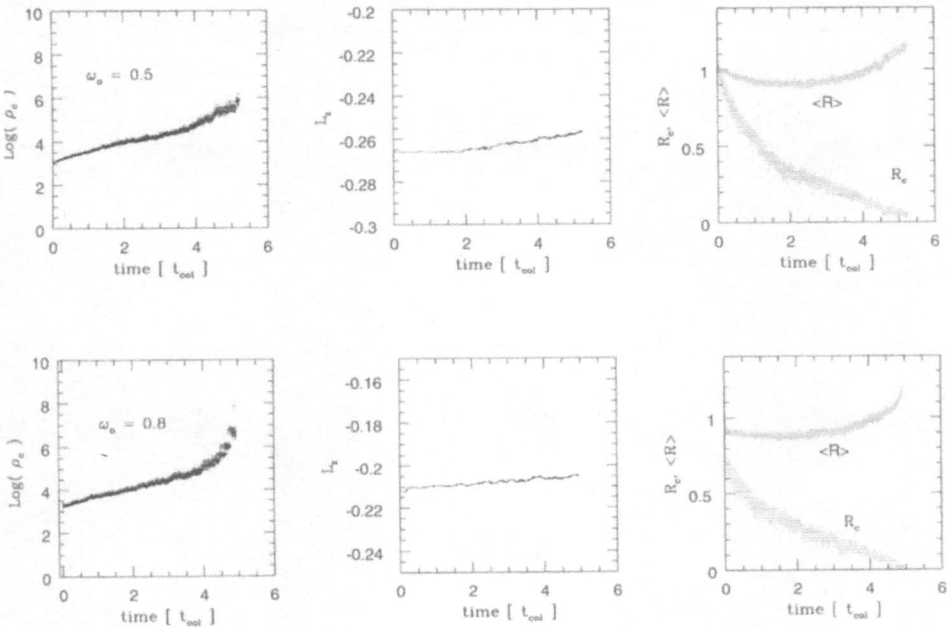

Fig. 2. Time-evolution of the central density (left-most panels), total z-angular momentum (middle) and the mean- and core-radii (right-hand panels) for runs with $W_o = 6.0$.

envelope and hence the mean radius expands rapidly at core-collapse. Note a subtle but noticeable difference between the two simulations, namely that the cluster with initially more rotation evolves faster; this is particularly visible in a comparison of radii at fixed time. A more convincing demonstration of the fast evolution of such clusters follows if we recall that in this unit of time, clusters without rotation reach core-collapse in around 12 t_{col}, which is more than twice as long.

To appreciate how many stars might be lost to galactic tides, were a tidal field present, we imagine the cluster orbiting the galaxy on a circular orbit. A tidal radius may then be defined from the initial configuration, by computing the radius r_t at which the mean density at time t equates the initial mean density: $r_t(t) = <r>[0] \times (M[t]/M[0])^{1/3}$. If we label as escapers all stars found outside $2\,r_t(t)$, we obtain an estimate of the number of stars likely to leave the cluster on a timescale short compared with t_{col}; the angular momentum they carry with them is deduced from summing up all the momenta of the stars left behind, and comparing with the initial value. Implementing this algorithm, we found the run with $\omega = 0.5$ (top panels) would have lost 3.7% of the initial angular momentum over the time of evolution, but only 1.0% (51:5000) of its mass. The second model, with more rotation, would have lost 0.96% (48:5000) of its mass, but only 1.4% of its total momentum to such escapers. This shows how the cluster redistributes angular momen-

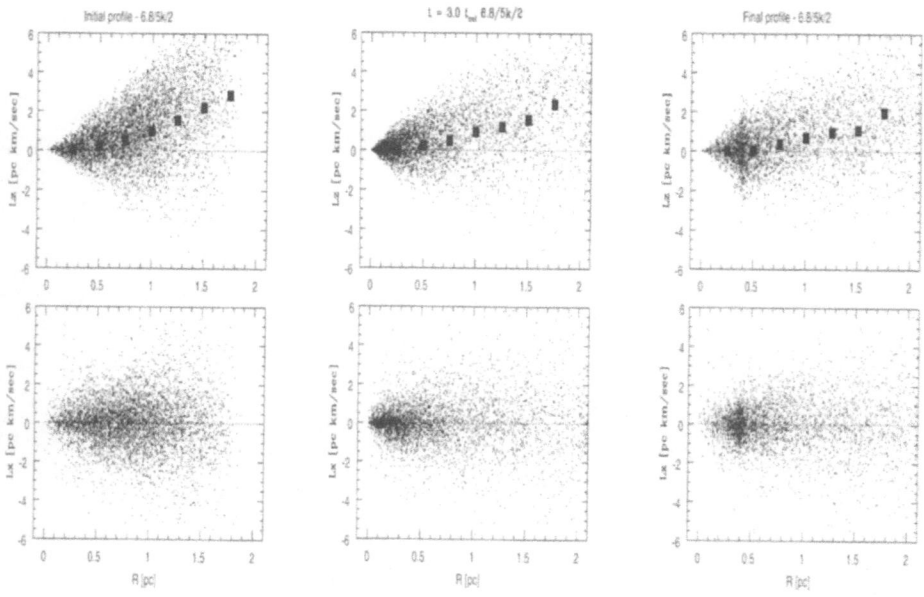

Fig. 3. Distribution of specific angular momentum components for three time-slots. See text for explanation of the black squares.

tum efficiently within itself, such that the core evolves towards core-collapse despite added rotational support. Evolution in the core is faster, the faster the core rotates initially, since the cluster is more compact which speeds up two-body effects.

Figure 3 graphs the specific angular momentum of the $\omega_o = 0.8$ cluster as a function of radius r for three different times. For comparison, two components are given: the z-axis component about which the cluster rotates; and the x-axis component. Dividing the cluster in ten concentric shells, we computed $\boldsymbol{L} = \boldsymbol{r} \times \boldsymbol{v}$ and summed up the momenta in each shell: the result is the series of black squares shown on the figure. Initially the (net) z-angular momentum increases steadily from the centre, outwards; the symmetry of the figure would make the sum over L_x's cancel out, and the black squares have been left out for this quantity. Evolution is monotonic, with the net z-momenta inside $r = 2\text{pc}$ decreasing, from which we deduce that an increasing fraction of the momentum is transferred to the volume $> 2\text{pc}$. Notice on Fig. 3 that the stars form a core around $r = 0.5\text{pc}$ in the final stage of the simulation (rightmost panels). It is not clear whether this is the result of an $m = 1$ (lopsided) instability, attributable to the dynamics of the system (from a d.f. point of view), or a case of core-wandering, likely due to the small number of particles inside $r = 0.5\text{pc}$ [58].

2.3   Conclusion

The faster evolution of clusters with rotation has been illustrated with two sample runs. The time to core-collapse we found from three-dimensional n-body simulations are in agreement with two-dimensional Fokker-Planck calculations [14]: the collapse time of 5.4 $t_{\rm col}$ obtained for the $\omega_o = 0.8$ agrees with the Fokker-Planck solution of 5.6 $t_{\rm col}$ for these parameters. This increases confidence in the results up to core collapse, obtained with two different algorithms.

3   Binary Black Holes in Galactic Centres

We follow the sinking of two massive black holes in a spherical stellar system. The massive particles become bound under the regime of dynamical friction. Once bound, the binary hardens by three body encounters with surrounding stars. Unlike other assumptions the massive system moves inside the core providing an enhanced supply of reaction partners for the hardening. These are the first results from simulations applying a hybrid "self consistent field" (SCF) and direct Aarseth N–body integrator (NBODY6), which synthesises the advantages of the direct force calculation with the efficiency of the field method. The code is aimed for use on parallel architectures and is therefore applicable for collisional N–body integrations with extraordinarily large particle numbers ($> 10^5$). It opens the perspective to simulate the dynamics of globular clusters with realistic collisional relaxation, as well as stellar systems surrounding a supermassive black hole in galactic nuclei.

The picture of galaxy evolution by hierarchical merging of galaxies seems to be well established [50]. On the other hand, there is evidence that larger galaxies harbour supermassive black hole in their centre [33]. It would therefore seem likely that two merging galaxies would each conceal a supermassive black hole. This brings up issues regarding the fate of these objects, such as the formation of a bound binary system and its effect on the dynamics of the merged galaxy, for example: if indeed these two black holes form a binary, on what timescale? Similarly, what would be the effect of surrounding stars on the hardening of this binary? On what timescale will both black holes coalesce?

These problems attract much attention in the astrophysical community [19]. Work on this has been done either by solving the perturbed two and three body problem in simplified models [61], [45] or by N-body simulations [42], [51], [37], [36]. As this shows, the modelling of binary black hole hardening turns out to be extremely challenging, algorithmically and computationally.

This work introduces some details of the problem of a sinking black hole binary. Furthermore it gives answers to the above questions from N-body simulations carried out with a new hybrid method for collisional stellar dynamics.

3.1   The sinking binary black hole problem

Following Begelman, Blandford & Rees [8], the central black holes of two galaxies will coalescence in the course of a galactic merger. During the early stages of the merger, the stellar component will form a nearly spherical system within the short timescale of violent relaxation. After that, the two supermassive black holes move through the stellar component with a velocity similar to the initial relative motion between the two galaxies. From this moment on, both massive bodies will feel dynamical friction from the surrounding stars. This friction leads the black holes to the newly-formed galactic centre, while the frictional force becomes more efficient with increasing density. Through this process, the black holes must inevitably 'find' each other and form a binary system [36].

After it forms, the bound binary hardens through dynamical friction. Assuming the centre of mass of the binary to be fixed at the centre of mass of the surrounding galaxy, dynamical friction must become less efficient with increasing binding energy of the binary. An encounter on a timescale longer than the period of the binary will harden it more efficiently, the shorter the period. Close encounters and three body interactions become more and more efficient at increasing the binding energy of the binary. The most efficient process for binary hardening in this stage are the stars which gain very large velocities in three body encounters with the black holes. Because this process can deplete stars from the surroundings of the binary, the hardening timescale may become very long. However, this will not true if the supply of stars undergoing close three-body encounters with the supermassive objects remains large. N-body simulations with high particle numbers can help set constraints on this problem [37], [51].

3.2   Results

To give an example of one of our models, we have followed the shrinking of two black holes of

$$M_\bullet := 1.00015 \cdot 10^7 \, M_\odot, \tag{3}$$

in a galactic nucleus of $M_{\text{tot}} = 10^9 \, M_\odot$, using 65.000 or in some case 128.000 stellar particles in the simulations. So the mean mass of a particle is $\bar{M} = 1.52588 \cdot 10^4 \, M_\odot$. This choice means that every stellar particle with mass M_* represents a compact star cluster with the order of 10^4 particles. The chosen mass for the black hole particle has approximately the same mass as the central black hole of M31 [33]. In a typical model, the initial distance between the black holes is 355.39 pc. They become bound after approximately 40 million years. The total simulated time is approximately 190 million years. At the end of the simulation the black holes' distance varies between 1 pc at the apo-center and 0.2 pc at the peri-center. The semi-major axis of the black hole binary at the time it becomes first bound is 21 pc. For dimensional analysis we can identify the time the binary becomes first bound with the time it has the first critical separation $a_h \approx GM/\sigma^2$ [49], where σ is the velocity dispersion of the stellar system outside of the cusp dominated by the black hole, and M is the total mass of the black hole binary. At this stage the dominant interaction process with the stellar system changes from dynamical friction to three-body encounters. The minimum separation of 0.2 pc at the end of our simulation is precise of the order of the second critical separation $a_{\mathrm{crit}} \approx 0.012\mathrm{pc} \cdot M_8^{0.8}$, where M_8 is the black hole mass in units of $10^8 M_\odot$. At a semi-major axis of the order of a_{crit} quick gravitational radiation merger sets in; a_{crit} is approximately 0.01 a_h [40], and since the semimajor axis of our black hole binary at the end of the simulation is still 0.6 pc, we are only a fector of three away from the gravitational radiation merger. This is closer than any previously published models of that kind, and similar to the models of Milosavljevic & Merritt [49], who use the same code for the central region than we.

The final eccentricity of the black hole binary in our model varies around 0.7. It depends on the initial eccentricity; we start with more eccentric orbit of the black hole than Milosavljevic & Merritt's models [49], whose black hole binary starts on a circular orbit, and they obtain smaller final eccentricities of only 0.3. This depends on the merging history of the nucleus and its density structure. The evolution of the orbital data of the black hole binary can be seen in Fig. 4. More details are published in Hemsendorf, Sigurdsson & Spurzem [23].

In total we have performed a number of (μp to now) ten different runs used to acquire a small statistical data basis. As one example we show the time evolution of the semi-major axis of the black hole binary in an ensemble averaged sense (error bars are intrinsic 1-σ scatter of the data). We also find a smaller than expected decrease of the motion of the black hole binary's centre of mass with increasing particle number. Despite of expectation that Brownian motion of the black hole binary should decrease with increasing N, we find that our results are consistent with no dependency on particle number. That would mean this effect is not a classical Brownian motion, but other effects, e.g. induced by the superelastic scatterings or collective interactions with the stellar system's core play a role. Our statistical data (compare Figs. 5 and 6 of Hemsendorf, Sigurdsson & Spurzem [23] for the available data on statistically averaged motion of the black hole binary) are not yet complete and reliable enough to give further conclusions here.

Figure 5 shows the motion of the centre of mass of the binary around the centre of mass of the stellar component. This motion prevents the binary from easily evacuating surrounding stars, which establishes an efficient hardening even at the late stages when dynamical friction becomes less important for this. Movies in MPEG format (1.6MB) of this process are available at

ftp://ftp.ari.uni-heidelberg.de/pub/staff/marc/MPEG/

simulation600.mpeg

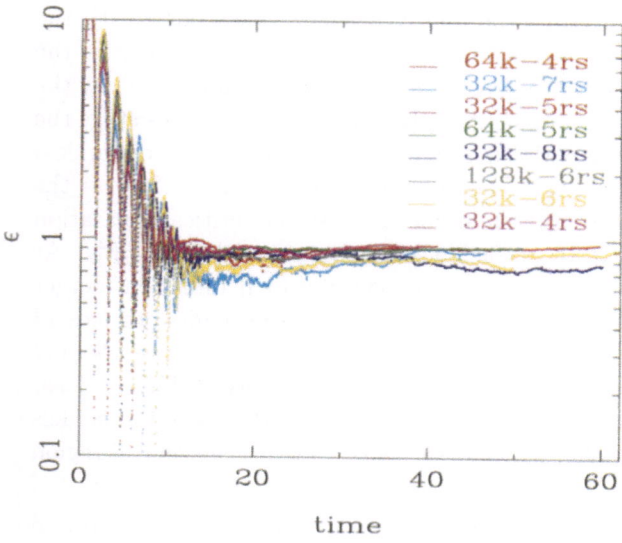

Fig. 4. Development of the orbit eccentricity of the black hole binary as a function of time in model time units (top panel) and its binding energy as a function of time, for a sequence of runs with different N, as indicated in the key. Runs denoted with s start with smaller initial eccentricity. Note, that if the binary is not yet bound there are formally values of $a > 0$ and $e > 1$. This means that in that phase the black holes are still not yet gravitationally bound to each other.

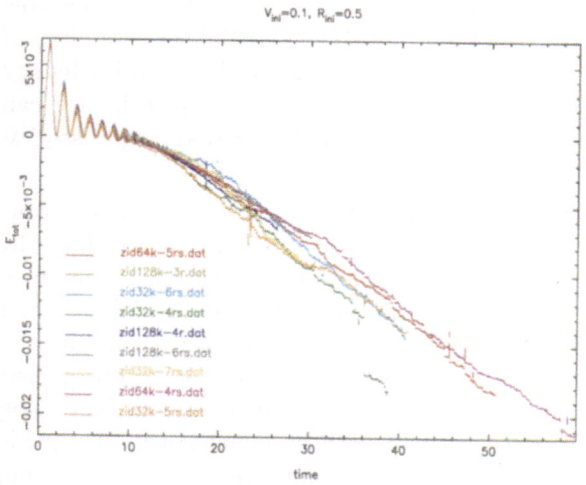

This movie (as well as simulations800.mpeg and simulations400.mpeg) illustrates in the first stages the standard shrinking of the binary black hole to the centre by dynamical friction, then, in the second stages, the feebback effects it has on the core as it moves through the nucleus, and the sometimes rather chaotic motion of the black hole binary, which in our interpretation is responsible for the relatively high eccentricity.

How our present results scale to the case of real particle number of galactic nuclei is the subject of future work. Within the next years and subject to appropriate computing equipment we will be able to follow the black hole binary into its gravitational radiation merger phase. We have not examined yet the question what happens if a third black hole comes in before the first binary merges. Conventional wisdom says (e.g. [62]) that slingshot ejections

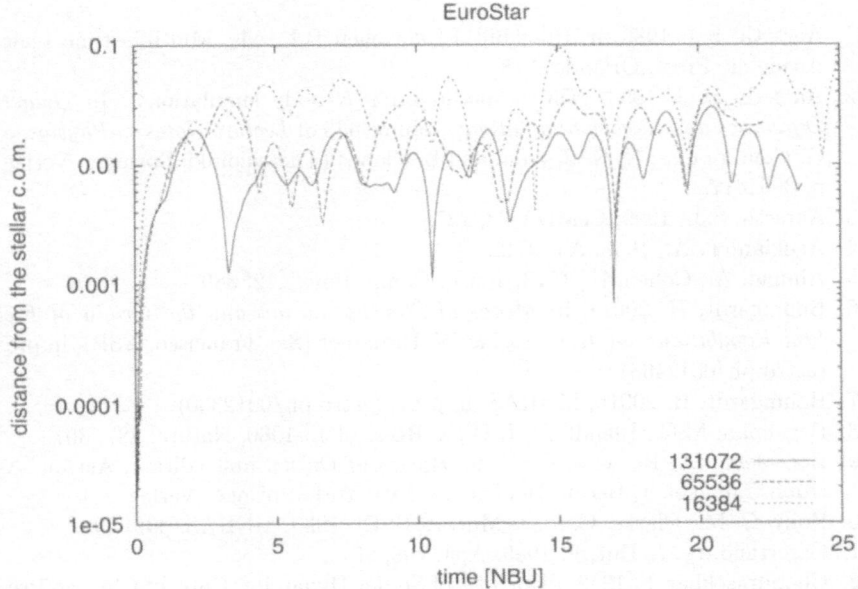

Fig. 5. The distance of the centre of mass of the black hole binary from the centre of mass of the stellar system as a function of simulated time in N-body time units.

would eject single or even binary black holes. Recent huge direct N-body models performed by Makino (personal communication) using a GRAPE-6 special purpose computer suggest however, that before that happens, there is a large chance that two black holes in the resonant three-body interaction come very close to each other (eccentricity 0.99) to merge quickly. We have, however, much more carefully than any other study examined the black hole motion in a self-consistent study with very large particle number.

EuroStar proved to be capable of integrating the binary black hole problem in galactic centres as a point mass system. The simulations introduced in this work are the first fully collisional simulations in this field, which could only be carried out on the up-to-date parallel computers accessible to the authors. The new code is going to be applied for simulating collisional dynamics for large N spherical systems including very massive particles.

Acknowledgements

The authors would like to thank S. Aarseth, D. Heggie, P. Kroupa, W. Sweatman, C. Theis, D. Merritt and G. Hensler for fruitful help and discussion. This project is funded by *Deutsche Forschungsgemeinschaft* (DFG) project Sp 345/9-1 and SFB 439. Technical help and computer resources awarded by *HLRS* in Stuttgart and *NIC* Jülich are gratefully acknowledged.

References

1. Aarseth, S.J. 1985, in Brackbill J,U., Cohen B.I., eds, Multiple time scales, Academic Press, Orlando, 378.
2. Aarseth, S. J. 1993. Direct methods for N-body simulations. In *Galactic Dynamics and N-body Simulations*, volume 433 of *Lecture Notes in Physics*, ed. G. Contopoulos, N. K. Spyrou, and L. Vlahos (Thessaloniki: Springer–Verlag), p. 365–417.
3. Aarseth, S.J., 1999, CeMDA 73, 127.
4. Agekian, T. A., 1958, AJ, 2, 22.
5. Ahmad, A., Cohen, L., 1973, Journ. Comp. Phys., 12, 389.
6. Baumgardt, H. 2001a, in *Modes of Star Formation and the Origin of Field Star Populations*, ed. E. Grebel & W. Brandner (San Francisco: ASP), in press (astro-ph/0012468)
7. Baumgardt, H. 2001b, MNRAS, in press (astro-ph/0012330)
8. Begelman, M.C., Blandford, R.D., & Rees, M.J., 1980, Nature, 287, 307.
9. Boccaletti, D., Pucacco, G. 1996. *Theory of Orbits*, first edition, Astron. Astroph. Lib. Vol. 1, Berlin, Heidelberg, New York: Springer Verlag.
10. Boily, C. M., Clarke, C. J., & Murray, S. D., 1999, MNRAS, 302, 399.
11. Casertano, R., & Hut, P. 1985, ApJ, 298, 80.
12. Chandrasekhar, S. 1942, Principles of Stellar Dynamics, Univ. of Chicago Press
13. Chandrasekhar, S., 1943, ApJ, 97, 255
14. Einsel, C., & Spurzem, R. 1999, MNRAS, 302, 81.
15. Elson, R. A. W., Fall, S. M., & Freeman, K. C. 1987, ApJ, 323, 54.
16. Frenk, C. S., & Fall, S. M. 1982, MNRAS, 199, 565.
17. Fukushige, T., Heggie, D. C., 2000, MNRAS 318, 753
18. Goodman, J., 1983, unpublished Ph.D. Thesis, Princeton University.
19. Gould, A., Rix, H.W., 2000, ApJ, 532, L29.
20. Hachisu, I. 1979, PASJ, 31, 523.
21. Hachisu, I. 1982, PASJ, 34, 313.
22. Heggie, D. C., Giersz, M., Spurzem, R., Takahashi, K., 1998, in *Highlights of Astronomy*, Vol. 11B, ed. J. Andersen, D. Reidel Publ. Comp. Dordrecht, 591
23. Hemsendorf, M., Sigurdsson, S., Spurzem, R., 2001, subm. ApJ, astro-ph/0103410
24. Hernquist, L., Ostriker, J.P., 1992, ApJ, 386, 375.
25. von Hoerner, S., 1960, Zs. f. Astroph., 50, 184.
26. von Hoerner, S., 1963, Zs. f. Astroph., 57, 47.
27. King, I. R. 1966, AJ, 71, 64.
28. Kontizas, E., et al., 1990, AJ, 100, 425.
29. Lagoute, C., Longaretti, P.-Y., 1996, A&A, 308, 441.
30. Lippert, T., Glaessner, U., Hoeber, H., Ritzenh"ofer, G., Schilling, K., Seyfried, A., 1996, Int'l J. Mod. Phys., 7, 485.
31. Lippert, T., Seyfried, A., Bode, A., Schilling, K., 1998, IEEE Transact. Par. Distr. Syst., 9, 1.
32. Lupton, R.H., Gunn, J.E., Griffin, R.F., 1987, AJ, 93, 1114.
33. Magorrian, J., et al., 1998, AJ, 115, 2285.
34. Makino, J., 1991a, ApJ,, 369, 200.
35. Makino, J., 1991b, PASJ, 43, 859.
36. Makino, J., 1997, ApJ, 478, 58.

37. Makino, J., Fukushige, T., Okumura, S.K., Ebisuzaki, T., 1993, PASJ, 45, 303.
38. Makino, J., Hut, P., 1988, ApJS, 68, 833.
39. Makino, J., Taiji, M., Ebisuzaki, T., Sugimoto, D., 1997, ApJ, 480, 432.
40. Merritt, D., 2001, ApJ, in press, astro-ph/0012264
41. Merritt, D., Meylan, G., Mayor, M, 1997, AJ, 114, 1074.
42. Merritt, D., Quinlan, G.D., 1998, ApJ, 498, 625.
43. Meylan, G., Mayor, M., 1986, A&A, 166, 122.
44. Meylan, G., Heggie, D. C., 1997, A&A Rev., 8, 1.
45. Mikkola, S., Valtonen, M.J., 1992, MNRAS, 259, 115.
46. Mikkola, S., 1997, CeMDA, 68, 87.
47. Mikkola, S., Aarseth, S. J., 1996, CeMDA, 64, 197.
48. Mikkola, S., Aarseth, S. J., 1998, NewA, 3, 309.
49. Milosavljevic, M., Merritt, D., 2001, subm. ApJ, astro-ph/0103350
50. Peebles, P.J.E. 1993, *Principles of physical cosmology*, Princeton : Princeton University Press
51. Quinlan, G.D., Hernquist, L., 1997, NewA, 2, 533.
52. Shapiro, S. L., Marchant, A. B., 1976, ApJ, 210, 757.
53. Spitzer, L. Jr., 1987, Dynamical Evolution of Globular Clusters, Princeton University Press, Princeton
54. Spurzem, R., Baumgardt, H. 2001, submitted to Monthly Notices of the Royal Astronomical Society
    (ftp://ftp.ari.uni-heidelberg.de/pub/staff/spurzem/edinpaper.ps.gz)
55. Spurzem, R., 1999, in Riffert, H., Werner, K. (eds), Computational Astrophysics, The Journal of Computational and Applied Mathematics (JCAM) 109, Elsevier Press, Amsterdam, 407.
56. Staneva, A., Spassova, N., Golev, V., 1996, A&AS, 116, 447.
57. Sugimoto, D., Chikada, Y., Makino, J., Ito, T., Ebisuzaki, T., Umemura, M., 1990, Nature, 345, 33.
58. Sweatman, W. L., 1993, MNRAS, 261, 497.
59. Sweatman, W. L., 1994, Journ. Comp. Phys., 111, 110.
60. Theis, C., Spurzem, R., 1999, A&A, 341, 361.
61. Valtonen, M.J., Mikkola, S., Heinämäki, P., Valtonen, H., 1994, ApJS, 95, 69.
62. Valtonen, M. J., 1996, MNRAS 278, 186
63. Vesperini, E. 2000, MNRAS, 318, 841
64. White, R. E., Shawl, S. J., 1987, ApJ, 317, 246.
65. Zhao, H.S., 1996, MNRAS, 278, 488.

A   Algorithmic and Computational Aspects

A.1   The Hermite Scheme

Assume a set of N particles with positions $r_i(t_0)$ and velocities $v_i(t_0)$ ($i = 1, \ldots, N$) is given at time $t = t_0$, and let us look at a selected test particle at $r = r_0 = r(t_0)$ and $v = v_0 = v(t_0)$. Note that here and in the following the index i for the test particle i and also occasionally the index 0 indicating the time t_0 will be dropped for brevity; sums over j are to be understood to include all j with $j \neq i$, since there should be no self-interaction. Accelerations $\boldsymbol{a}_0$ and their time derivatives $\dot{\boldsymbol{a}}_0$ are calculated explicitly:

$$\boldsymbol{a}_0 = \sum_j Gm_j \frac{\boldsymbol{R}_j}{R_j^3} \quad ; \quad \dot{\boldsymbol{a}}_0 = \sum_j Gm_j \left[\frac{\boldsymbol{V}_j}{R_j^3} - \frac{3(\boldsymbol{V}_j \cdot \boldsymbol{R}_j)\boldsymbol{R}_j}{R_j^5} \right] , \quad (4)$$

where $\boldsymbol{R}_j := \boldsymbol{r} - \boldsymbol{r}_j$, $\boldsymbol{V}_j := \boldsymbol{v} - \boldsymbol{v}_j$, $R_j := |\boldsymbol{R}_j|$, $V_j := |\boldsymbol{V}_j|$. By low order predictions,

$$\boldsymbol{x}_p(t) = \frac{1}{6}(t - t_0)^3 \dot{\boldsymbol{a}}_0 + \frac{1}{2}(t - t_0)^2 \boldsymbol{a}_0 + (t - t_0)\boldsymbol{v} + \boldsymbol{x} , \quad (5)$$

$$\boldsymbol{v}_p(t) = \frac{1}{2}(t - t_0)^2 \dot{\boldsymbol{a}}_0 + (t - t_0)\boldsymbol{a}_0 + \boldsymbol{v} , \quad (6)$$

new positions and velocities for all particles at $t > t_0$ are calculated and used to determine a new acceleration and its derivative directly according to Eq. 4 at $t = t_1$, denoted by $\boldsymbol{a}_1$ and $\dot{\boldsymbol{a}}_1$. On the other hand $\boldsymbol{a}_1$ and $\dot{\boldsymbol{a}}_1$ can also be obtained from a Taylor series using higher derivatives of $\boldsymbol{a}$ at $t = t_0$:

$$\boldsymbol{a}_1 = \frac{1}{6}(t - t_0)^3 \boldsymbol{a}_0^{(3)} + \frac{1}{2}(t - t_0)^2 \boldsymbol{a}_0^{(2)} + (t - t_0)\dot{\boldsymbol{a}}_0 + \boldsymbol{a}_0 , \quad (7)$$

$$\dot{\boldsymbol{a}}_1 = \frac{1}{2}(t - t_0)^2 \boldsymbol{a}_0^{(3)} + (t - t_0)\boldsymbol{a}_0^{(2)} + \dot{\boldsymbol{a}}_0 . \quad (8)$$

If $\boldsymbol{a}_1$ and $\dot{\boldsymbol{a}}_1$ is known from direct summation (from Eq. 4 using the predicted positions and velocities) one can invert the equations above to determine the unknown higher order derivatives of the acceleration at $t = t_0$ for the test particle:

$$\frac{1}{2}\boldsymbol{a}^{(2)} = -3\frac{\boldsymbol{a}_0 - \boldsymbol{a}_1}{(t - t_0)^2} - \frac{2\dot{\boldsymbol{a}}_0 + \dot{\boldsymbol{a}}_1}{(t - t_0)} \quad (9)$$

$$\frac{1}{6}\boldsymbol{a}^{(3)} = 2\frac{\boldsymbol{a}_0 - \boldsymbol{a}_1}{(t - t_0)^3} - \frac{\dot{\boldsymbol{a}}_0 + \dot{\boldsymbol{a}}_1}{(t - t_0)^2} , \quad (10)$$

This is the Hermite interpolation, which finally allows to correct positions and velocities at t_1 to high order from

$$\boldsymbol{x}(t) = \boldsymbol{x}_p(t) + \frac{1}{24}(t - t_0)^4 \boldsymbol{a}_0^{(2)} + \frac{1}{120}(t - t_0)^5 \boldsymbol{a}^{(3)} , \quad (11)$$

$$\boldsymbol{v}(t) = \boldsymbol{v}_p(t) + \frac{1}{6}(t - t_0)^3 \boldsymbol{a}_0^{(2)} + \frac{1}{24}(t - t_0)^4 \boldsymbol{a}_0^{(3)} . \quad (12)$$

Taking the time derivative of Eq. 12 it turns out that the error in the force calculation for this method is $\mathcal{O}(\Delta t^4)$, as opposed to the widely used leap-frog schemes, which have a force error of $\mathcal{O}(\Delta t^2)$. Additional errors induced by approximate potential calculations (particle mesh or TREE) create potentially even larger errors than that. However, it can be shown that the above Hermite method used for a real N-body integration sustains an error of $\mathcal{O}(\Delta t^4)$ for the entire calculation [34]. Many persons in the world know as Aarseth scheme (in particular the code version NBODY5 [1]) an integrator of the same order as the Hermite scheme, but using only accelerations on four time points instead of a and $\dot{a}$ on two time points. As is shown in [34], the Aarseth scheme is $\mathcal{O}(\Delta t^4)$ as well, but for the same number of time steps the absolute value of the energy error (not its slope) is clearly smaller in the Hermite scheme. This means that for a given energy error the Hermite scheme allows timesteps which are larger by some factor of order unity depending on the parameters of the system under study. The Hermite scheme has been commonly adopted during the past years, because it needs less memory, and allows slightly larger timesteps. More importantly, after the addition of a hierarchical (as opposed to individual) time step scheme it is well suited for parallelization on modern special and general purpose high performance computers [55]. The timestep scheme will be discussed now.

A.2   Choice of Timesteps – Parallelization

Aarseth [1] provides an empirical timestep criterion

$$\Delta t = \sqrt{\eta \frac{|a||a^{(2)}| + |\dot{a}|^2}{|\dot{a}||a^{(3)}| + |a^{(2)}|^2}} \ . \tag{13}$$

The error is governed by the choice of η, which in most practical applications is taken to be $\eta = 0.01 - 0.04$. It is instructive to compare this with the inverse square of the curvature κ of the curve $a(t)$ in coordinate space

$$\frac{1}{\kappa^2} = \frac{1 + |\dot{a}|^2}{|a^{(2)}|^2} \ . \tag{14}$$

Clearly under certain conditions the time step choice Eq. 13 becomes similar to choosing the timestep according to the curvature of the acceleration curve; since it was determined just empirically, however, it cannot generally be related to the curvature expression above. In [34] a different time step criterion has been suggested, which appears simpler and more straightforwardly defined, and couples the timestep to the difference between predicted and corrected coordinates. The standard Aarseth time step criterion Eq. 13 has been used in most N-body simulations so far (but compare the discussion in [59]).

Since the position of all field particles can be determined at any time by the low-order prediction Eq. 6, the time step of each particle (which determines the time at which the corrector Eq. 12 is applied) can be freely chosen according to the local requirements of the test particle; the additional error induced due to the use of only predicted data for the full N sums of Eq. 4 is negligibly small, for the benefit of not being forced to keep all particles in lockstep. Such an individual time step scheme is in particular for non-homogeneous systems very advantageous, as was quantitatively pointed out by [38]. Particles in the high density core of a star clusters need to be updated much more often than particles on orbits very far from the centre. They show that the gain in computational speed due to the individual time step scheme (as compared to a lockstep scheme where all particles share the minimum required time step) is of the order $N^{1/3}$ for homogeneous and N^1 for strongly spatially structured systems.

For the purpose of vectorization and parallelization it is better not to have the particles continuously distributed on a time axis. Consequently, [35] uses a hierarchical scheme, still on the basis of Eq. 13; but a change of the timestep is considered only if that equation yields a variation of Δt compared to the last step by more than a factor of 2 (increase or decrease). If this is the case a variation by 2 is applied only. Thus in model units all timesteps are selected from the set $\{2^{-i}|i = 0, ...i_{max}\}$ with $k = i_{max}$ determined by the condition that $\Delta t_{min} > 2^{-i_{max}}$ for the minimum timestep Δt_{min} determined from Eq. 13. For core collapse simulations of star clusters of a few ten thousand particles i_{max} goes up to about 20; empirically and theoretically [38] $\Delta t_{min} \propto N^{-1/3}$, so for large N i_{max} becomes larger, however, on the other hand, how large i_{max} grows for fixed N depends on the selected criteria for so–called KS regularisation of perturbed two–body motion (see below). The implementation of the block step scheme indeed uses an even stronger condition than the above described one, it is demanded that not only the time steps, but also the individual accumulated times of each particles are commensurate with the timestep itself. This ensures that for any particle i and any time $T_i = t_i + \delta t_i$ all particles with $\delta t_j < \delta t_i$ have for their own time $T_j = t_j + \delta t_j = T_i$, where the last equality is the non–trivial one. Such procedure is important for the parallelization of the algorithm. For example it has as a consequence that at the big time steps always huge groups of particles are due for correction, sometimes even all particles (at the largest steps). Such scheme allows an efficient parallelization of all operations necessary for calculation of a and $\dot{a}$ and for the update of particle positions and velocities (corrections). Special purpose computers have been built tailored to the Hermite codes, which are denoted as HARP ("Hermite Accelerator Pipeline") boards and stem from the bigger GRAPE–family [57,39]. Such HARP–boards have been made available also at some places outside Japan, including "Astronomisches Rechen–Institut" Heidelberg (for an application see e.g. [60]).

Another refinement of the Hermite or Aarseth "brute force" method is the two-time step scheme, denoted as neighbour or Ahmad-Cohen scheme [5]. For each particle a neighbour radius is defined, and a and $\dot{a}$ are computed due to neighbours and non-neighbours separately. Similar to the Hermite scheme the higher derivatives are computed separately for the neighbour force (irregular force) and non-neighbour force (regular force). Computing two timesteps, an irregular small Δt_{irr} and a regular large Δt_{reg}, from these two force components by Eq. 13 yields a timestep ratio of $\gamma := \Delta t_{reg}/\Delta t_{irr}$

being in a typical range of 5–20 for N of the order 10^3 to 10^4. The reason is that the regular force has much less fluctuations than the irregular force. The Ahmad-Cohen neighbour scheme is implemented in a self-regulated way, where at each regular time-step a new neighbour list is determined using a given neighbour radius r_{si} for each particle. If the neighbour number found is larger than the prescribed optimal neighbour number, the neighbour radius is increased or vice versa. In [1,38] more complicated algorithms to adjust the neighbour radius are described. On the contrary to [38], who find an optimal neighbour number of $N_{n,\text{opt}} \propto N^{3/4}$ we find that adopting a constant neighbour number of the order of $20 - 50$ is sufficient at least up to $N = 50000$. The reason is that by using special purpose machines or parallelization for parts of the code, an optimal neighbour number is not well defined, so the neighbour number can be selected according to accuracy and efficiency requirements [55]. After each regular time step the new neighbour list is communicated along with the new particle positions to all processors of the parallel machine, thus making it possible to do the irregular time step in parallel as well.

Using a two-time step or neighbour scheme again increases the computational speed of the entire integration by a factor of at least proportional to $N^{1/4}$ [34]. Both the regular and irregular timesteps are arranged in the hierarchical, commensurable way, and the total inherent parallelism in the resulting algorithm is discussed in detail in [55]. One can see that even for moderate particle numbers of 10^4 particles some 512 processors could be used efficiently. Sometimes there are only very few particles in the smallest steps to be integrated, which one might consider as being very prohibitive for parallelization. However, due to the large number of medium and large size blocks this effect is negligible for the overall performance [55]. By using more and more processors in the parallel execution one finds that the asymptotic scaling of the "brute force" N-body problem can be reduced effectively to an N scaling (Fig. 6). But in our present implementation the parallelization is done only according to parallel sections (do loops) in the code; there is no domain decomposition (distributing particles on the processor). Thus at the end of any timesteps new results have to be broadcast to all other processing units. A systolic algorithm is used for that which scales linearly in communication time with the number of processors. It is interesting to note an approach suggested by molecular dynamicists to use a new kind of hyper-systolic communication algorithm, which scales only by the square root of the processor number [30,31]. Presently we think that hyper-systolic algorithms can efficiently be used only if the sum over all particles for the acceleration and its time derivative (Eq. 4) should be directly parallelized. The number of interprocessor communications N_{comm} for the hyper-systolic algorithm is of the order $N\sqrt{n_{\text{PE}}}$; on the other hand our algorithm, which we would like to call here "parallel group execution algorithm" [55], has a scaling $N_{\text{comm}} \propto N^{2/3} n_{\text{PE}}$, because only subgroups of particles, whose size scales with $N^{2/3}$ have to be communicated across the processor network. In other

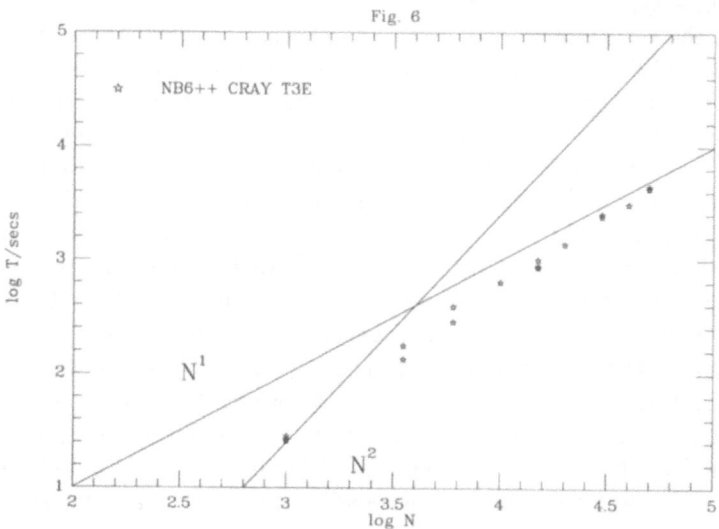

Fig. 6. CPU time needed for one N-body time unit as a function of particle number N using NBODY6++ on the CRAY T3E. The collection of data points includes runs with varying average neighbour number and processor/pipeline number, starting from 8 for low N up to 512 for the largest N, which are not individually discriminated in the figure.

words, asymptotically (above some critical particle number as a function of n_{PE}, the hyper-systolic algorithm should lose against the parallel group execution algorithm. However, these questions have not yet been examined in detail, for example what the critical N really are and which algorithm is more efficient for practically useful particle numbers of today. This is subject of present and future work.

If the two-body force between any pair of particles becomes dominant their (perturbed) relative motion is integrated in special regularized coordinates (taking into account perturbations from field particles), in which the singularity of the two-body motion is transformed into a slowly varying parameter (the binding energy) and does not occur in the integration variables. The rest of the N-body simulation generally regards the regularized pair as a compound particle located at the position and moving with the velocity of its centre of mass, except in the case when a perturber moves very close to a regularized pair (in such cases the pair is resolved). It was already discovered in the earliest published N-body simulations that the formation of close and eccentric binaries occurs as the rule rather than as an exception and that it was particularly difficult to accurately integrate them [25,26]. As a consequence two-body, three-body and chain regularizations were developed and implemented in order to accurately and efficiently integrate star clusters including all their close binaries, triples and hierarchical subsystems.

An excellent account of regularization, historically and scientifically, can be found in [46]. Most recent developments are the slow-down treatment of tight binaries [47] and a new method to gain accuracy and exact solutions in the unperturbed case using Stumpff functions [48].

For the binary black hole models of galactic nuclei described in the previous section a hybrid code was used, which embeds the direct N-body region in a larger system, where the potential is approximately computed by a serious evaluation. In Fig. 7 the speed-up of the combined code as a function of the number of nodes is displayed.

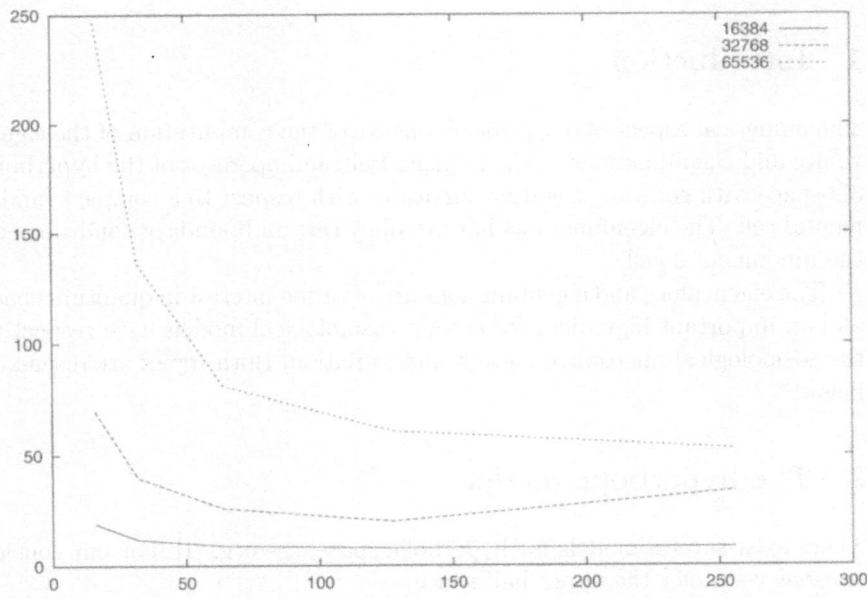

Fig. 7. Wall clock time for 200 *SCF* time steps with varying PE number on the *CRAY-T3E*, for the indicated particle numbers. particles, the lower for 16384.

The Computation of Highly Excited Hyperbolic 3D-Eigenmodes and Their Application to Quantum Chaos and Cosmology

Ralf Aurich

Abteilung Theoretische Physik
Universität Ulm
Albert-Einstein-Allee 11
89069 Ulm

1 Introduction

The numerical aspect of our project consists of the computation of the eigenvalues and eigenfunctions of the Laplace-Beltrami operator of the hyperbolic 3D-space with constant negative curvature with respect to a compact fundamental cell. The eigenfunctions have to obey certain boundary conditions on the fundamental cell.

The eigenvalues and eigenfunctions are of prime interest in quantum chaos and an important ingredient for certain cosmological models with respect to the cosmological microwave background radiation. Both topics are discussed below.

2 The hyperbolic model

There exist several models for hyperbolic space, see e. g. [1]. For our considerations we prefer the upper half space

$$\mathfrak{H}_3 = \{(x_1, x_2, x_3) \in \mathbb{R}^3 \mid x_3 > 0\} \ , \tag{1}$$

equipped with the Riemannian metric

$$ds^2 = \kappa \frac{dx_1^2 + dx_2^2 + dx_3^2}{x_3^2} = \kappa \gamma_{ij} dx^i dx^j \ . \tag{2}$$

The Gaussian curvature equals $-1/\kappa$ at every point of $\mathfrak{H}_3$; we set $\kappa = 1$.

The orbifold used in our project is obtained from a Kleinian group Γ which yields a pentahedron as a fundamental cell which in turn is symmetric along an intersection plane. The pentahedron is divided by this intersection plane into two equal tetrahedra. Thus the eigenmodes can be computed by desymmetrizing the pentahedron (for more details, see [2]). A group-symmetry consideration shows that the eigenmodes of the pentahedron obeying periodic boundary conditions decompose into two symmetry classes, one having

Dirichlet boundary conditions, i. e., $\psi = 0$, at the surface of the tetrahedron, and the other having Neumann boundary conditions, i. e., a vanishing normal derivative $\partial\psi/\partial n = 0$. Using the tetrahedron with Dirichlet and Neumann boundary conditions, respectively, facilitates the numerical computation of the eigenmodes. In the nomenclature of [3,4] and [5] this tetrahedron is called T_8. It has a volume vol$T_8 \simeq 0.3586534$ and is defined by the dihedral angles

$$\angle BC = \frac{\pi}{2}\,,\quad \angle CA = \frac{\pi}{3}\,,\quad \angle AB = \frac{\pi}{4}\,,$$

$$\angle DA = \frac{\pi}{2}\,,\quad \angle DB = \frac{\pi}{3}\,,\quad \angle DC = \frac{\pi}{5}\,,$$

where A, B, C and D are the four corner points. For the tetrahedron T_8 the first 749 eigenmodes corresponding to Dirichlet boundary conditions have been computed using the boundary element method as described in [2]. It is worthwhile to note that there are only nine compact tetrahedra in hyperbolic space and that T_8 is the only compact tetrahedron whose generating group is not arithmetic [4].

Turning now to the eigenmodes which arise from the stationary Schrödinger equation (in units $\hbar = 2m = 1$) and which also occur in the perturbation theory of the fluctuations in the cosmological background radiation, one has to consider the eigenvalue equation

$$-\Delta\psi(x) = E\psi(x)\,,\qquad \psi \in \mathrm{L}^2(\mathfrak{H}_3/\Gamma,\chi)\,. \tag{3}$$

Here Δ denotes the Laplace-Beltrami operator

$$\Delta = x_3^2\left(\frac{\partial^2}{\partial x_1^2} + \frac{\partial^2}{\partial x_2^2} + \frac{\partial^2}{\partial x_3^2}\right) - x_3\frac{\partial}{\partial x_3}\,. \tag{4}$$

$\mathrm{L}^2(\mathfrak{H}_3/\Gamma,\chi)$ is the space of all functions which are square integrable on the fundamental cell and Γ-automorphic, i. e., satisfy for all $g \in \Gamma$

$$\psi(g(x)) = \chi(g)\,\psi(x), \tag{5}$$

where χ is any one-dimensional unitary representation of Γ, a so-called *character*. For example take a reflection group – like we will do – generated by the reflections $g_1,\dots,g_n$ at faces of a hyperbolic polyhedron, and choose $\chi(g_i) = -1$ for all those generators, then one will obtain Dirichlet boundary conditions on the faces of the polyhedron. Choosing $\chi(g_i) = 1$ yields Neumann boundary conditions.

3   The method of computation

The eigenvalues and eigenfunctions are computed by using the boundary element method (BEM). Define the following differential operator $\mathcal{L}_x := -\Delta_x - E$, $x \in \mathfrak{H}_3$, then the eigenvalue equation (3) reads

$$\mathcal{L}_x\psi(x) = 0\,. \tag{6}$$

The free Green function given by

$$G(x, y; E) = \frac{1}{4\pi \sinh d(x, y)} \exp[i\, k\, d(x, y)], \qquad k = \sqrt{E - 1} \;, \qquad (7)$$

where $d(x, y)$ denotes the hyperbolic distance between x and $y \in \mathfrak{H}_3$, satisfies

$$\mathcal{L}_x \, G(x, y; E) \;=\; y_3^3 \, \delta(x - y) \;\;, \qquad (8)$$

with $\delta(x)$ being the three-dimensional Dirac delta distribution. Let Σ be the surface of the polyhedron, then a solution $\psi(x)$ of the eigenvalue equation (6) with Dirichlet boundary conditions can be represented as a surface integral for $x \notin \Sigma$

$$\psi(x) \;=\; \int_\Sigma d\Sigma_y \, G(x, y; E) \, \Phi(y) \;\;, \qquad (9)$$

as is verified by applying $\mathcal{L}_x$ on equation (9). The function $\Phi(y)$ with $y \in \Sigma$ has to be determined, such that $\psi(x)$ satisfies the boundary conditions for $x \in \Sigma$. To obtain a representation of $\Phi(y)$ the surface Σ is tessellated into triangles whose size is chosen to be of roughly a quarter de Broglie wavelength. On each triangle $\Phi(y)$ is approximated by a third-order polynomial $p(\xi, \eta) = c_{10}\xi^3 + c_9\eta^3 + c_8\xi^2\eta + c_7\xi\eta^2 + c_6\xi^2 + c_5\eta^2 + c_4\xi\eta + c_3\xi + c_2\eta + c_1$, where ξ and η are local coordinates on the unit triangle. The coefficients are expressed in terms of the function values Φ_k at ten points (ξ, η), three of which lie at the corners of the triangle, two on each of the edges and one in the interior of the triangle. This choice allows a continuous representation of $\Phi(y)$. Expanding Φ in terms of the above ansatz functions, a set of linear equations is obtained from (9) in terms of the expansion coefficients c_i which in turn are expressed by the function values Φ_k. The Dirichlet boundary conditions are imposed by setting $\psi(x) = 0$ for $x \in \Sigma$. The resulting system of equations $M_E \Phi = 0$, with the vector $\Phi = (\Phi_1, \Phi_2, \dots)$, has non-trivial solutions only if the determinant of the energy dependent matrix M_E vanishes, i.e., if E corresponds to an eigenvalue E_n. Since the matrix M_E is almost singular, a singular value decomposition is applied to M_E. The singular value decomposition is carried out by the LaPack routine "zgesvd". Then the zeros of the singular values betray the locations of the eigenvalues. To obtain the eigenvalues with respect to the two symmetry classes of the tetrahedron T_8 the Green function $G^\pm(x, y; E) = G(x, y; E) \pm G(x, \rho_\pi(y); E)$ is used, where $\rho_\pi(y)$ is the symmetry operation mapping the tetrahedron onto itself.

An analogous program computes the Neumann symmetry class which itself decomposes in the same parity classes as in the Dirichlet case. The Neumann-BEM program computes the eigenfunctions on the surface of the tetrahedron directly. An example of positive parity is shown in figure 1. This figure demonstrates that the polynomials of third-degree approximate the eigenfunction very well. One observes further that the structure is much finer in the lower part of the figure than in the upper part. This is an effect due to

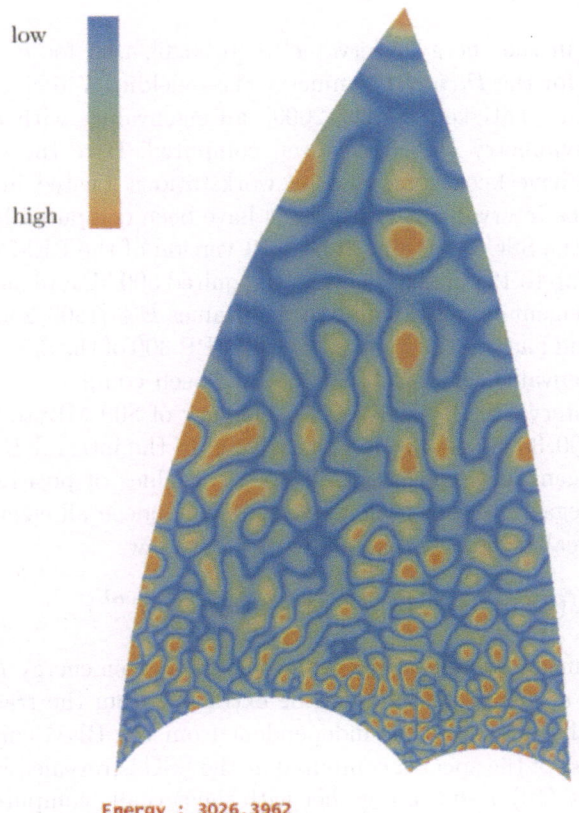

Energy : 3026.3962

Fig. 1. The intensity of an eigenfunction on two adjacent surfaces of the tetrahedron as approximated with two-dimensional polynomials of third degree as discussed in the text. This eigenfunction of positive parity has energy $E_n = 3026.3962$.

the metric (2) and had to be incorporated in the mesh which tessellates the surface. For this reason a new mesh generator had to be written since available standard software does not use a hyperbolic metric. From the eigenfunctions known on the surface, one can compute the eigenfunctions in the interior using a boundary integral analogous to (9).

Some animations of the eigenfunctions can be found at

http://www.uni-ulm.de/~raurich/Orbifold/tetraeder.html

The program can be parallelized using MPI with very little communication overhead since every node scans another energy interval. This version was used on the IBM RS/6000 SP.

4   The status of computation

All eigenvalues in the energy interval $E \in [0, 3026]$, i.e., for $k < 55$, have been computed for the *Dirichlet* symmetry class yielding 746 eigenmodes.

Since the last "Tätigkeitsbericht 2000" all eigenvalues with $k < 55$ for the *Neumann* symmetry class have been computed. Here the eigenmodes with $E < 1000$ have been computed on workstations located in Ulm. The eigenmodes in the interval $E \in [1000, 1500]$ have been computed by the IBM RS/6000 SP of the SSC Karlsruhe. The MPI version of the BEM program is employed using up to 128 nodes. These jobs required 500 MByte memory. The computation of eigenmodes belonging to eigenvalues $E \in [1500, 2500]$ requires more memory and has been carried out on the VPP 300 of the SSC Karlsruhe, whereas the eigenvalues $E \in [2500, 3026]$ have been computed on the VPP 5000. For the interval $E \in [1500, 2500]$ a memory of 800 MByte is required and for $E \in [2500, 3026]$ one needs 1700 MByte. In the interval $E \in [0, 3026]$ all Neumann eigenvalues consisting of 662 eigenvalues of positive and 656 eigenvalues of negative parity are computed. That indeed all eigenvalues are computed is revealed by a comparison with Weyl's law

$$\overline{\mathcal{N}}(E) \; = \; c_3 k^3 + c_2 k^2 + c_1 k + c_0 + O(e^{-\alpha k}) \quad , \tag{10}$$

which gives the mean number of eigenvalues below a given energy $E = k^2 + 1$. The coefficients c_i are known and can be extracted from the trace formula (see [2] for details) and are thus independent from the BEM computation. The completeness of the spectra computed at the SSC is revealed by figure 2. There Weyl's law (10) is shown together with the actually computed number $\mathcal{N}(E)$ in panels a) and b) for both parity classes with Neumann boundary conditions. Both curves agree so well that they are indistinguishable in the figure. In panels c) and d) the difference

$$\Delta\mathcal{N} \; := \; \mathcal{N}(E) - \overline{\mathcal{N}}(E) + \frac{1}{2}$$

is shown revealing the chaotic fluctuations. In these difference plots one observes the completeness of the spectra. If one eigenvalue would be missing, the mean of the curve would jump from that value on by one unit. Since this does not happen, both spectra are complete up to $E = 3026$. This is a severe requirement for a quantum chaological analysis.

The program utilizes the vector unit very well. For a typical job of the 800 MByte class, one gets a CPU System time 9414 msec, CPU User time 12382496 msec and Vector User time 11449273 msec, i.e., 93% of the computer time is carried out in the vector modus. An even slightly better behaviour is observed for a job of the 1700 MByte class. There one gets a CPU System time 10666 msec, CPU User time 17463040 msec and Vector User time 16520527 msec, i.e., 95% of the computer time is carried out in the vector modus. A typical job running on the VPP 5000 with 4832 Mbyte gives

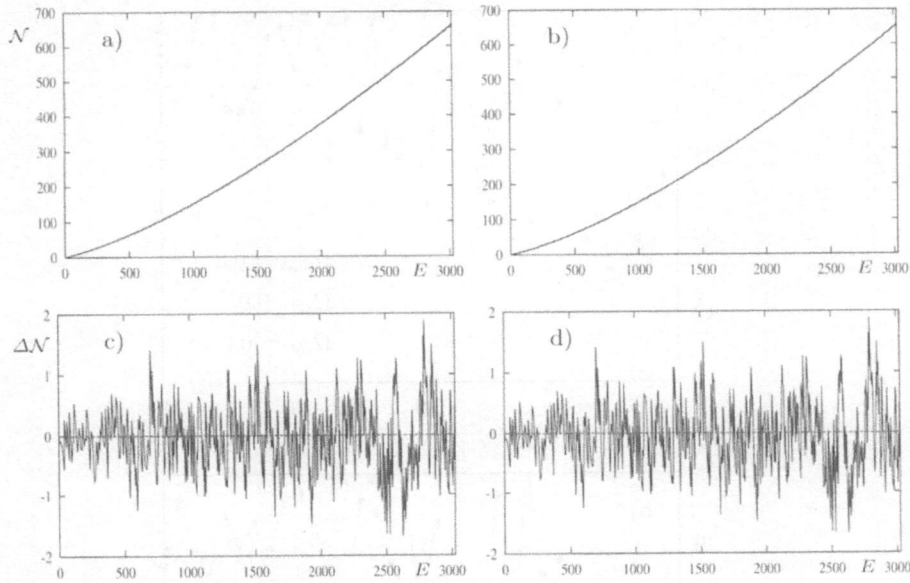

Fig. 2. The number of eigenvalues $\mathcal{N}(E)$ is compared with Weyl's law $\overline{\mathcal{N}}(E)$, eq.(10), for the Neumann parity class with positive parity in a) and negative parity in b). Both curves overlap in this resolution. In panels c) and d) the difference $\Delta\mathcal{N} := \mathcal{N}(E) - \overline{\mathcal{N}}(E) + \frac{1}{2}$ is shown for positive and negative parity, respectively. These difference plots reveal the completeness as well as the chaotic properties of the eigenvalue spectra.

CPU System time 1623 msec, CPU User time 10778712 msec and Vector User time 10280078 msec, i. e., again 95% is carried out in the vector modus.

The currently running programs compute the eigenfunctions belonging to the now known eigenvalues. Most eigenfunctions are already computed and it is expected that the computations will be finished in a few months.

5   Application to cosmology

Since the last "Tätigkeitsbericht 2000" already explained the applications to cosmology, i. e. the computations of the fluctuations of the cosmic microwave background radiation (CMB) for models of the universe with finite volume, I refer with respect to the computational details to that Tätigkeitsbericht. The computations in [6] deal with a hyperbolic model universe containing relativistic and non-relativistic energy, i. e. matter and radiation. The luminosity measurements of very distant Supernova Ia indicate [7,8] that the energy density of our Universe contains a large fraction of dark energy like the vacuum energy or the energy of a scalar field. The latter contribution is also known

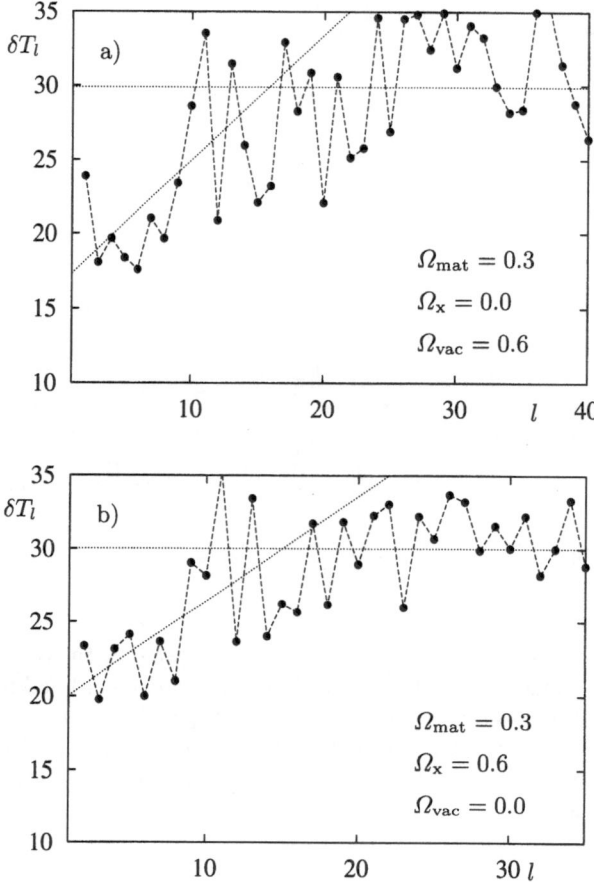

Fig. 3. The angular power spectrum $\delta T_l = \sqrt{l(l+1)C_l/2\pi}$ is shown for a hyperbolic universe with finite volume with vacuum energy and quintessence in panels a) and b), respectively.

as "quintessence". Thus the computations carried out in [6] where extended to incorporate also vacuum energy and a special version of quintessence with a fixed equation of state [9]. In that paper the fluctuations in the temperature δT of the CMB are studied. A quantitative measure of the scale of the fluctuations in δT is provided by the angular power spectrum C_l defined by

$$C_l = \frac{1}{2l+1} \sum_{m=-l}^{l} |a_{lm}|^2 \quad ,$$

where a_{lm} are the expansion coefficients of δT with respect to the spherical harmonics $Y_l^m(\theta, \phi)$. The models with finite volumes show a significant suppression at low multipoles l. As an example figure 3 shows two models with different contributions of dark energy. (For more details, see [9].) It is expected that the MAP satellite launched in June will measure the angular power spectrum C_l with sufficient accuracy such that a comparison with the C_l spectra of finite volume universes will be possible.

6   Application to quantum chaology

In quantum chaology one is interested in the properties of quantum mechanical systems whose classical counterparts display chaotic behaviour. The systems with the most chaotic behaviour, i. e., with so-called hard chaos, are Anosov systems [10]. The classical dynamic behaviour of a free point particle confined in a compact hyperbolic fundamental cell is exactly of this Anosov

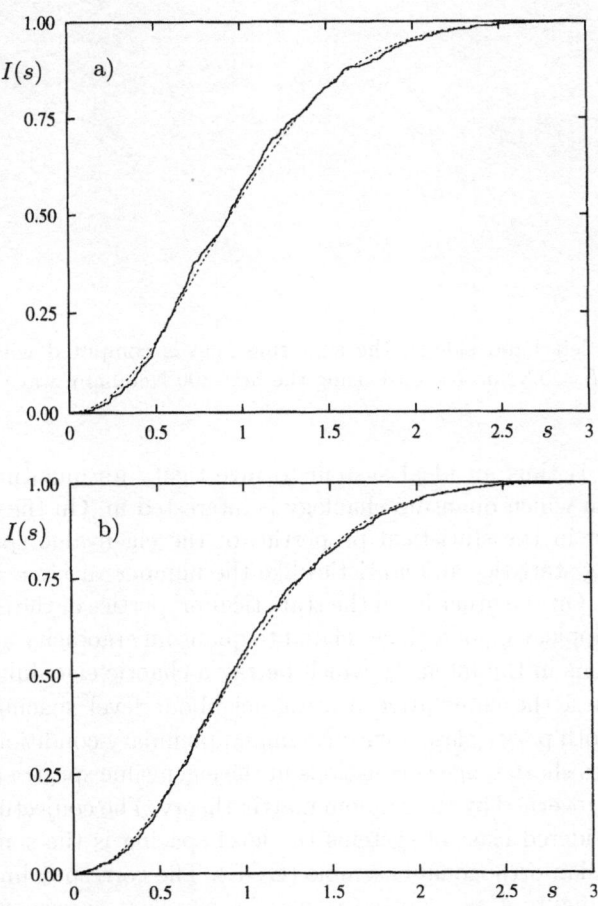

Fig. 4. The cumulative nearest neighbour level spacing $I(s)$ is shown for the eigenvalues with positive and negative parity in panels a) and b), respectively, with Neumann boundary conditions. The dotted curve is the corresponding GOE expectation.

Fig. 5. The left hand side of the sum rule (11) is computed with (13) for the parameters $L = 0.5$ and $t = 0.01$ using the first 300 Neumann wave functions.

type which is thus an ideal system to investigate its quantum mechanical properties in which quantum chaology is interested in. On the one hand one is interested in the statistical properties of the eigenvalue spectra such as level spacing statistics and statistics like the number variance and the spectral rigidity. On the other hand the statistical properties of the eigenfunctions are a main topic, i.e., questions related to quantum ergodicity and the nature of fluctuations in the intensity which betray a chaotic eigenfunction.

In figure 4 the cumulative nearest neighbour level spacing statistics is shown for both parity classes with Neumann boundary condition. This statistics measures short-range correlations in the eigenvalue spectra and is conjectured to be governed by the random matrix theory. The conjecture states that for the considered class of systems the level spacing is the same as the one of the Gaussian orthogonal ensemble (GOE). The corresponding GOE curve is shown in figure 4 as a dotted curve. An excellent agreement is observed confirming the conjecture also in this three-dimensional case.

In [11] (which can also be obtained at

http://www.physik.uni-ulm.de/theo/qc/ulm-tp/tp00-6.ps.gz   )

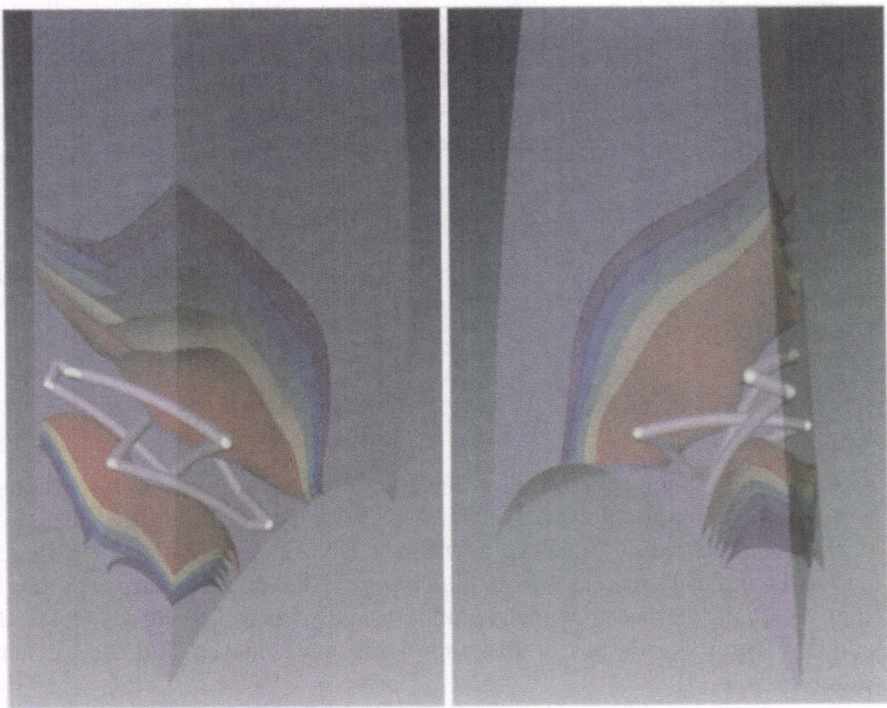

Fig. 6. The difference of the sum over Neumann and Dirichlet wave functions, respectively, is shown for $L = 1.6$ and $t = 0.01$, which gives large contributions in the neighborhood of the third shortest orbits shown as grey tubes.

the eigenfunctions, which are purely quantum mechanical quantities, are used to compute classical quantities like periodic orbits. Summing the eigenfunctions $\psi_n(z)$ weighted with specially chosen spectral functions $h(p)$ allows to extract the three-dimensional location of periodic orbits and the action of so-called elliptic points. Below an application of the general orbit sum rule is presented which reads in three dimensions [11]

$$\sum_{n=0}^{\infty} h(p_n)\,\psi_n(z)\,\psi_n^\star(z') = \frac{1}{2\pi^2} \sum_{g\in\Gamma} \frac{\chi(g)\,\tilde{h}(d(z,g(z')))}{\sinh d(z,g(z'))} \tag{11}$$

with

$$\tilde{h}(y) := \int_0^\infty p\,\sin(py)\,h(p)\,dp \quad . \tag{12}$$

Here the spectral function $h(p)$ has to fulfill some conditions to ensure the convergence of the sums in (11). The argument of $\tilde{h}$ is the hyperbolic distance $d(z, g(z'))$ between two points $z \in \mathfrak{H}_3$ and $g(z') \in \mathfrak{H}_3$. +he sum runs over the group elements $g \in \Gamma$ which define the structure of the fundamental cell and can be associated with classical orbits.

In order to extract the classical periodic orbits from the quantum mechanical wave functions in three dimensions (inverse quantum chaology in three dimensions), the spectral function is chosen to be

$$h(p) = \frac{\sin(pL)}{p} e^{-Et} \quad \text{with} \quad E = p^2 + 1 \quad \text{and} \quad t > 0, \, L \in \mathbb{R} \; .$$
(13)

This yields with (12)

$$\tilde{h}(y) = \frac{e^{-t}}{4} \sqrt{\frac{\pi}{t}} \left[e^{-\frac{(L-y)^2}{4t}} - e^{-\frac{(L+y)^2}{4t}} \right] \; .$$

In figure 5 a sum over the first 300 Neumann eigenfunctions reveals the contribution of elliptic points. The figure 6 shows highest intensities near the locations of the two periodic orbits with the third shortest length. (In these figures the intensity increases from blue to red.) The intensities are computed from the quantum mechanical eigenfunctions whereas the result is a classical quantity. For more details of the underlying mathematics and for further examples, see [11].

References

1. J. G. Ratcliffe, *Foundations of Hyperbolic Manifolds*, Graduate Texts in Mathematics 149, Springer (Berlin), 1994.
2. R. Aurich and J. Marklof, Trace Formulae for Three-Dimensional Hyperbolic Lattices and Application to a Strongly Chaotic Tetrahedral Billiard, Physica D **118**, 101–129 (1996).
3. F. Lannér, On complexes with transitive groups of automorphisms, Med. Lunds Univ. Math. Sem. **11**, 1 (1950).
4. C. Maclachlan and W. Reid, The Arithmetic Structure of Tetrahedral Groups of Hyperbolic Isometries, Mathematika **36**, 221–240 (1989).
5. C. Maclachlan, Triangle Subgroups of Hyperbolic Tetrahedral Groups, Pacific J. Math. **176**, 195–203 (1996).
6. R. Aurich, The Fluctuations of the Cosmic Microwave Background for a Compact Hyperbolic Universe, Astrophys. J. **524**, 497–503 (Oct. 1999).
7. A. G. Riess, A. V. Filippenko, P. Challis, A. Clocchiatti, A. Diercks, P. M. Garnavich, R. L. Gilliland, C. J. Hogan, S. Jha, R. P. Kirshner, B. Leibundgut, M. M. Phillips, D. Reiss, B. P. Schmidt, R. A. Schommer, R. C. Smith, J. Spyromilio, C. Stubbs, N. B. Suntzeff, and J. Tonry, Observational Evidence from Supernovae for an Accelerating Universe and a Cosmological Constant, Astronomical J. **116**, 1009–1038 (Sept. 1998).
8. S. Perlmutter, G. Aldering, G. Goldhaber, R. A. Knop, P. Nugent, P. G. Castro, S. Deustua, S. Fabbro, A. Goobar, D. E. Groom, I. M. Hook, A. G. Kim, M. Y. Kim, J. C. Lee, N. J. Nunes, R. Pain, C. R. Pennypacker, R. Quimby,

C. Lidman, R. S. Ellis, M. Irwin, R. G. McMahon, P. Ruiz-Lapuente, N. Walton, B. Schaefer, B. J. Boyle, A. V. Filippenko, T. Matheson, A. S. Fruchter, N. Panagia, H. J. M. Newberg, W. J. Couch, and The Supernova Cosmology Project, Measurements of Omega and Lambda from 42 High-Redshift Supernovae, Astrophys. J. **517**, 565–586 (June 1999).

9. R. Aurich and F. Steiner, The Cosmic Microwave Background for a Nearly Flat Compact Hyperbolic Universe, Monthly Notices of the Royal Astronomical Society **323**, 1016–1024 (2001).

10. D. V. Anosov, Geodesic flows on closed Riemannian manifolds with negative curvature, Proceedings of the Steklov Institute of Mathematics **90**, 1–235 (1967).

11. R. Aurich and F. Steiner, Orbit Sum Rules for the Quantum Wave Functions of the Strongly Chaotic Hadamard Billiard in Arbitrary Dimensions, Foundations of Physics **31**, 569–592 (2001).

Propagation of Herbig-Haro Jets Through Inhomogeneous Molecular Clouds

Markus Thiele and Max Camenzind

Landessternwarte
D-69117 Königstuhl, Heidelberg, Germany

Abstract. In this article we present the results of a simulation of an effectively 3–dimensional hydrodynamic jet. The parameters are chosen appropriate for a typical collimated outflow emanating from a young stellar object of low mass. It is the first step in a series of high resolution simulations that aim at separating the influence of the different processes that are thought to govern the physics of these objects. The simulations were performed on the SX-5 at HLRS.

1 Introduction

Many of the very youngest stars deeply embedded in molecular clouds eject highly collimated supersonic plasma flows, called Herbig-Haro jets (HH jets) which interact with the cloud gas that is to be transversed. Typically these jets have radii of the order 100 AU and lengths ranging from 0.01 pc up to 10 pc. The jet beam consists of a series of clearly separated regions of drastically enhanced line emission, called emission line knots. These knots effectively define the jet beam since they are its only visible part [3][10][22]. The overall geometry of by far the most of all HH jets strongly deviates from a straight line. This phenomenon is called wiggling. A prototypical example in this context is HH47 which changes its direction on comparatively small propagation lengths [14]. In general there are two possibilities to cause such deviations from a straight line: an inhomogeneous ambient medium and a precession of the jet at its base (near the young star and its surrounding accretion disk). Whereas there are only a few examples, where indications of a precessing behaviour could be found [18] it is definitely confirmed by observations that the clouds are usually very inhomogeneous on almost all scales and even turbulent [1][7][12][22]. As is to be expected, non-straight jets can be found in such non-uniform environments, while the few straight jets are located in an unusually homogeneous ambient medium [22]. Plasma flows as HH jets are subject to a number of instabilities. The most important among them are the hydrodynamic so-called Kelvin-Helmholtz-instabilities and magnetohydrodynamic instabilities. The spectrum of these instabilities comprises both axisymmetric and non-axisymmetric modes. It follows that in simulations of axisymmetric jets the non-axisymmetric modes are suppressed, although they may play an important if not decisive role concerning the stability and the overall structure of the jets. It is almost certain that they are

deeply involved in the processes which lead to the observed undulatory behaviour. Thus, simulations of fully 3–dimensional jets are an inevitable step
towards an understanding of the mechanisms, which are responsible for the
structure and the dynamics of these objects. Only little work has been done
that takes into account the non-uniform nature of the molecular clouds in
the context of jet simulations. So far they were restricted to the following
two cases:

1. a smoothly varying density distribution, depending only on one spatial
   coordinate (the radius r or the z-coordinate) [4][5][15]
2. a mass clump residing in an otherwise homogeneous medium [6][8][20]

Simulations of the first type have revealed a lot about the global reaction of
the jet on a stratified ambient medium whereas simulations of the latter type
can mainly be applied to the extragalactic case. But the non-uniformity on
scales smaller than the jet radius may play the decisive role in exciting the
instabilities that generate the characteristic structure of the jets. Especially
the question about the presence of dynamically important magnetic fields
inside of the jet beam is concerned by this problem, since pure hydodynamic
jets would eventually become totally disrupted by the instabilities (above all
by the small wavelength KH instabilities) already in an early stage [19].

In this first step we wanted to study the influence of a non-uniform density
distribution on the propagation and on the stability of a purely hydrodynamic
jet by performing simulations with a high resolution. A high resolution is
essential in this context since the small wavelength KH instabilities are the
more dangerous ones concerning the jet stability.

2   The jet model

2.1   Overall model

In this work the jet plasma is described within the (magneto–)hydrodynamic
framework. Thus the following set of equations is to be solved (Euler equations):

$$\frac{\partial \rho}{\partial t} + \nabla \cdot (\rho \boldsymbol{v}) = 0 \tag{1}$$

$$\rho \left(\frac{\partial \boldsymbol{v}}{\partial t} + \boldsymbol{v} \cdot \nabla \boldsymbol{v} \right) = -\nabla p \tag{2}$$

$$\frac{\partial e}{\partial t} + \nabla \cdot (e\boldsymbol{v}) = -p \, \nabla \cdot \boldsymbol{v} \tag{3}$$

$$p = (\gamma - 1)\, e \tag{4}$$

Here ρ denotes the density, $\boldsymbol{v}$ the velocity vector, e the internal energy and p
the thermal pressure. The ratio of the specific heats was chosen $\gamma = 5/3$. The

system of equations is solved numerically by using the 3D–MHD–code NIR-VANA [21]. This code is based on time explicit Finite-Differences/Finite-Volume-methods. The source and advection terms are operator splitted from each other and the calculation of the fluxes on the zone interfaces is of second order (Van-Leer-method). Furthermore, an artificial viscosity is included to damp numerical oscillations near discontinuities and to guarantee the correct jump conditions at shock fronts. The code was extended by time implicit methods for the calculation of the internal chemistry of plasmas in various astrophysical contexts [19]. The resulting 3D–RMHD code is NIRVANA_C, which will be used in the next step to study the influence of the chemistry (ionisation, recombination, cooling) and of the magnetic fields on the jet physics. From the time-explicit nature of the major part of the numerical scheme it follows that every integration time step has to be less than the so-called Courant time step which guarantees numerical stability and essentially fixes the number of time steps for the integration over a certain intervall of time.

For diagnostic purposes we have solved an additional continuity equation for the so-called tracer. The tracer is defined as the jet density normalized to the density at the jet inlet. This allows to separate the jet medium from the ambient medium. As a passively advected quantity the tracer has no influence on the dynamics.

This simulation corresponds physically to a situation where a homogeneous, overdense and approximately cylindrical jet propagates through an initially non-uniform and stationary ambient medium. We have used a 3–dimensional cartesian grid which spans a volume of $\Delta x \times \Delta y \times \Delta z = 2\,10^{16} \times 2\,10^{16} \times 1.2\,10^{17}$ cm^3. Since we have chosen the jet radius as $R_j = 1.5\,10^{15}$ cm, this corresponds to 80 jet radii in the z-direction and approximately 13 jet radii both in the x- and the y-direction. The whole domain was resolved by $N_x \times N_y \times N_z = 200 \times 200 \times 1200$ grid cells of equal size. Thus, the jet radius as the fundamental length scale of the problem is resolved by 15 cells. The simulation was followed until the jet had reached the upper z-boundary of the grid. This corresponds to a jet length which is in the range of the typical extensions of the observed protostellar jets.

2.2   Initial and boundary conditions

The system of equations to be solved is of first order in time and of hyperbolic type. Thus, the specification of appropriate initial and boundary conditions is required.

Boundary conditions On the whole surface of the cartesian grid standard Outflow boundary conditions were installed. The only exception from this is an approximately circular domain with a radius of $R \approx R_j = 1.5\,10^{15}$ cm, where we have instead installed Inflow boundary conditions, appropriate to represent the flow of jet matter into the ambient medium:

$$v_x = v_y = 0, v_z = v_j \qquad \sqrt{(x - x_0)^2 + (y - y_0)^2} \leq R_j, \; z = 0 \qquad (5)$$

Here the coordinates of the middle of the lower z-boundary are denoted by x_0, y_0 and define the position of the jet axis at the inlet. Like in the case of the velocity there is no radial variation of the density ρ_j and of the thermal pressure p_j (alternatively the temperature T_j) over the jet cross section at the inlet.

Initial conditions As was already mentioned, observations strongly indicate that the jet hosting molecular clouds are non–uniform on almost all scales and even turbulent. This has consequences for the theory and should result in appropriate initial conditions. Our model for the 3–dimensional jet is based on the idea, that the inhomogeneities of the ambient medium are the key factor in the excitation of different types of hydrodynamic and magnetohydrodynamic instabilities that lead

- to the undulatory behaviour of the observed jets,
- possibly to the destruction of the jet beam on long time scales ($t \approx$ 500 . . . 1000 years) in the pure hydrodynamic case
- and to the excitation of pinch-instabilities in the magnetohydrodynamic case that in the end result in a knot structure along the beam.

In this context it should be noted that the details of the density structure inside the star forming clouds are only poorly known and differ from one case to the next. Taking this into account a stochastic method was used to fix the initial density on each grid point individually. The only restriction of this method is that the stochastically determined values for the density at individual grid points have to be taken out of the following set of values (in units of the jet density at the inlet):

$$\rho_{i,j,k}^{(0)} \in \{0.3, 0.35, 0.4, 0.45, 0.5\} \qquad (6)$$

$$\forall i, j, k : 0 < i < N_x, 0 < j < N_y, 0 < k < N_z$$

In contrast to the spatially varying values of the density we have set the initial pressure constant on the whole grid. Its value was chosen equal to the value of the jet pressure at the jet inlet. This guarantees that there is a force equilibrium between the jet matter and the surrounding medium on the boundary. The initially non–uniform density distribution and the spatially constant pressure imply that the initial temperature has to vary spatially ($p \sim \rho T$). The main drawback of choosing initial conditions of this kind is that it lacks a true parametrization. But we think that it is inevitable to take into account the non-homogeneous nature of the molecular cloud to be transversed by the jet due to its strong influence both on the structure and the propagation of the jets as can be demonstrated by this simulation.

Having specified the boundary and initial conditions, the physics is then governed by the small set of the following parameters:

$$\eta := \langle \rho_{\rm j}/\rho_{\rm a} \rangle = 2.5 \qquad M_{\rm j} := v_{\rm j}/c_{\rm j} \approx 50 \qquad \zeta := (p_{\rm j}/p_{\rm a}) = 1 \ .$$

Here 'j' denotes the jet quantities and 'a' those of the ambient medium. $c_{\rm j}$ is the thermal sound speed in the jet. The jet number density and the jet velocity were chosen $n_{\rm j} \approx 500\,{\rm cm}^{-3}$ and $v_{\rm j} = 200\,{\rm km\,sec}^{-1}$, respectively.

3   Results

This simulation of a hydrodynamic jet already reveals that the inhomogeneities of the ambient medium are of major importance concerning the structure and the stability of HH jets. One obvious exception from this is that the hydrodynamic jet remains almost perfectly straight over its full propagation distance of almost $80\,R_{\rm j}$. This can clearly be seen in the density and the tracer plots (Figs. 1,2). Thus, in spite of the inhomogeneities as an effective means for generating fluid instabilities there is no definite sign of an undulatory behaviour. This is in accordance with jet simulations of lower resolution but larger propagation distance [19]. In the hydrodynamic case an undulatory behaviour can be caused by the non-axisymmetric modes of the Kelvin-Helmholtz instability (we exclude precession). For typical jet parameters the non-axisymmetric modes grow on relatively large time scales of the order of 100 years. So even under favourable circumstances as in this case it is doubtful that the wiggling is caused by these modes. Especially the changes of direction over small propagation distances as in the case of the prototypical HH47 jet can hardly be explained by the non-axisymmetric Kelvin-Helmholtz surface modes. In contrast to the non-axisymmetric surface modes, the internal structure of the jet beam is strongly influenced by the body modes which are excited at the boundary between the beam and the shocked jet gas ('cocoon'). In this context it is known that the small scale Kelvin-Helmholtz body modes are the more dangerous ones. Of course in numerical simulations the minimum unstable wavelength cannot be smaller than the size of the grid cell. Concerning our simulation this means $\mathrm{Min}(\lambda_{\rm KH}) \geq \Delta x = 10^{14}$ cm. This scale is quite small for the effectively 3-dimensional case and is of the same order of what is common in the case of simulations of axisymmetric jets although still a better resolution would be highly desirable. In our simulation a pinching of the beam (Figs. 1,2: $z \approx 4.8\,10^{16}$ cm, $t = 75.68$) is caused by the significant shear flow at the jet boundary in the ultimate neighbourhood of the jet inlet which is a characteristic feature of jet simulations of this type. The result is a X-like shock cell which is associated with a beam collimation by the shock front and with a rarefaction of the beam by corresponding waves behind the shock front. The net effect is an only moderate change of the beam radius. The overall stability of the jet is not affected.

Knot structures or knot-like structures as seen in some hydodynamical jet simulations with a stratified ambient medium [5] are absent in this simulation. This can be explained mainly by two effects: We use a higher Mach number ($M \approx 50$ instead of $M \approx 10 \ldots 20$) which is nevertheless in accordance with the observations. Thus, it takes a longer time for the Kelvin-Helmholtz modes to develop. Furthermore the initial density distribution of the ambient medium that we use has no symmetries in contrast to simulations with an only stratified ambient medium where there is still a cylindrical symmetry around the central axis. The internal oblique shocks driven by beam pinching with a typical separation of a few jet radii obtained in these simulations seem to require a cylindrical symmetry and probably disappear without this kind of symmetry. Thus, unless low Mach numbers and at most a stratified ambient medium is used, knots can hardly be explained by pure hydrodynamic effects. Since we were primarily interested in the influence of the inhomogeneous ambient medium on the stability properties and the propagation of the jet, this is nevertheless of secondary interest. The generation of knots will be studied in the next step, when dynamically important magnetic fields are included in our simulations. A second very important feature of this hydrodynamic jet is the structure of its surrounding cocoon, best seen in the tracer plots (Fig. 2). The outer radial boundary of the tracer effectively corresponds to the contact discontinuity that separates the shocked jet gas from the shocked ambient medium. The highly structured and even turbulent shocked gas and the strong deformation of the contact discontinuity already in the early stages (Fig. 1,2: $t = 25.22$) are very different from the corresponding structure of hydrodynamic jets propagating into an initially homogeneous medium [9][15][17][19] and are ultimately caused by the inhomogeneities of the ambient gas to be transversed by the jet. The high degree of structure in the cocoon represents a by far more effective means for the excitation of the Kelvin-Helmholtz instabilities. Apart from exciting fluid instabilities a second effect of the turbulent structure is a very strong mixing between the jet gas and the ambient gas. As can be seen in the tracer plots (Fig. 2), both media penetrate each other almost right from the beginning. But note that the mixing occurs mainly between the shocked portions of the jet plasma and the ambient gas whereas apart from the part in the ultimate neighbourhood of the working surface there is no sign of entrainment of shocked gas into the beam. Accordingly the time average of the propagation velocity, v_{bs}, comes out to deviate by less than 3% from the value obtained from eq. (7) for $\alpha = 1$ (see below). It follows that apart from the foremost part of the jet also in this case the transfer of momentum from the jet to the ambient gas mainly takes place in the working surface. This is also indicated by Fig. 3 where the radial profile of the tracer, the density and the axial velocity in the foremost quarter of the jet ($z = 9 \, 10^{16}$ cm) are shown. It can clearly be seen that there is a steep decline of the axial velocity at the beam boundary. Only the shocked jet gas in the ultimate neighbourhood of the beam has a significant axial velocity. But of course this may change when the jet becomes older since in its foremost part the jet beam is strongly disturbed and eventually becomes disrupted (Figs. 1,2,4: $t = 126.1, z \approx 1.17\,10^{17}$ cm). Observations can provide information about the mixing properties. So, the line broadening (FWHM ≤ 150 km/sec) seen along some jets as HH30 and HL Tau can be attributed to a turbulent mixing of jet gas with the gas of the ambient medium [11]. In this context it should be noted that a mixing of this kind is suppressed in jets that carry dynamically important magnetic fields or are subject to line cooling [19]. Especially cooling might act as a further mechanism that damps the large scale instabilities and helps to confine the beam significantly although on small scales fragmentation takes place [19]. Interestingly the development of the structure along the beam is closely related to what happens in the working surface of the jet ('jet head'). Also this part of the jet is strongly influenced by the inhomogeneities of the ambient medium. The shock jump conditions for high Mach number flows allow to derive a simple expression for the bow shock velocity,

$$v_{\mathrm{bs}} \approx v_{\mathrm{j}} \left(1 + \frac{1}{\sqrt{\alpha^2 \eta}} \right)^{-1} \quad , \tag{7}$$

where α is defined as the ratio of the width of the jet radius and the radius at the jet head, $\alpha := R_{\mathrm{j}}/R_{\mathrm{bs}}$. The other quantities have their usual meaning. From the inhomogeneity of the ambient medium it follows that there is a corresponding variation of the ram pressure of the ambient medium, $\rho_{\mathrm{a}}\,v_{\mathrm{bs}}^2$, both in the transverse and in the axial direction. This enters the expression for the bow shock velocity via a spatially varying η. The time dependence of the beam radius in the ultimate neighbourhood of the Mach disk is represented by a varying α. Together both effects cause the bow shock velocity to vary in space and time. From the data we obtain time variations of the bow shock velocity up to 30%. Thus, a significant acceleration and deceleration takes place. Oscillations of the propagation velocity of this strenght were known to occur only in jets which experience a significant loss of internal energy via cooling. Thus, we expect that the time dependence of the propagation velocity will become even stronger if we include cooling in our calculations that will be done in the next step. The spatial variation of the bow shock velocity results in a significant deformation of the working surface (Figs. 2,1: $t = 50.45, 75.68, 100.9$). The time evolution of the tracer reveals that even a penetration of ambient gas into the jet gas takes place. This leads to channels of ambient gas, streaming through the shocked jet gas. Since this streaming is associated with a strong shear Kelvin-Helmholtz modes are excited. Also in this case a turbulent pattern and a strong mixing develops. Correspondingly, there is a strong influence on the jet beam which is surrounded by this gas. Thus, the general outcome is a highly time dependent structure of both the beam and the surrounding shocked gas. This in general stands in clear contrast to hydrodynamic jets that propagate through a homogeneous medium and its effect on the structure and the propagation is much stronger than in the case of an only smoothly varying density of the ambient medium. In a next step we want to study the influence of the inhomogeneities on cooling and magnetized jets.

4 Computation

For our calculations we have used a cartesian grid that consists of $200 \times 200 \times 1200 = 4.8\,10^7$ cells. On this grid the values of the five fundamental variables (density, energy density, velocity) had to become sucessively updated during the simulation. In addition to the fundamental ones four more auxiliary variables were used due to the significant simplification of the coding. In the sequential version of our code the memory for these auxiliary variables was treated dynamically. To minimize the time for the allocation and deallocation of the variables we have changed the scheme so that both the memory for the auxiliary and the fundamentral variables is allocated once at the beginning of the run and deallocated at its end. Thus, effectively the number of variables is $5 + 4$. The resulting memory requirement on the SX-5 was approximately 3.8 GByte. We have performed our simulation on one processor of the SX-5 using a fully vectorized version of our code NIRVANA_C. The vectorization of the code which is written in the programming language C was achieved by the implementation of corresponding compiler directives of the cc-Compiler. The code consists mainly of a sequence of three nested loops over the three spatial coordinates. Since only the innermost loop can become vectorized we have always chosen the innermost loop as the largest. With this strategy we achieved a performance of about 672 MFlops. Test runs with $N_z = 1500$ loop counts for the innermost loop even yielded almost 1 GFlop. Nevertheless we have chosen $N_z = 1200$ loop counts since we want to use the same resolution also for the simulation of magnetized and cooling jets which comprise much more variables ($8 + N$; N=number of cooling species). In these cases, more than 1200 loop counts would result in prohibitive memory requirements. The whole simulation was run until the jet had reached a propagation time of 300 years. With an average Courant time step of $1.2\,10^6$ sec this took about 7900 time steps. The required CPU-time was approximately 330000 sec $\approx$ 3.85 CPU days with 87% spent in the vectorized area and a vector operation ratio of 99.32%. The total

Table 1. Listing of the essential parameters of the simulation

Number of variables	Number of grid cells	Memory requirement [GByte]
5	$4.8 \cdot 10^7$	3.8

Number of time steps	Megaflops	CPU time on SX–5 [days]
7900	672	3.85

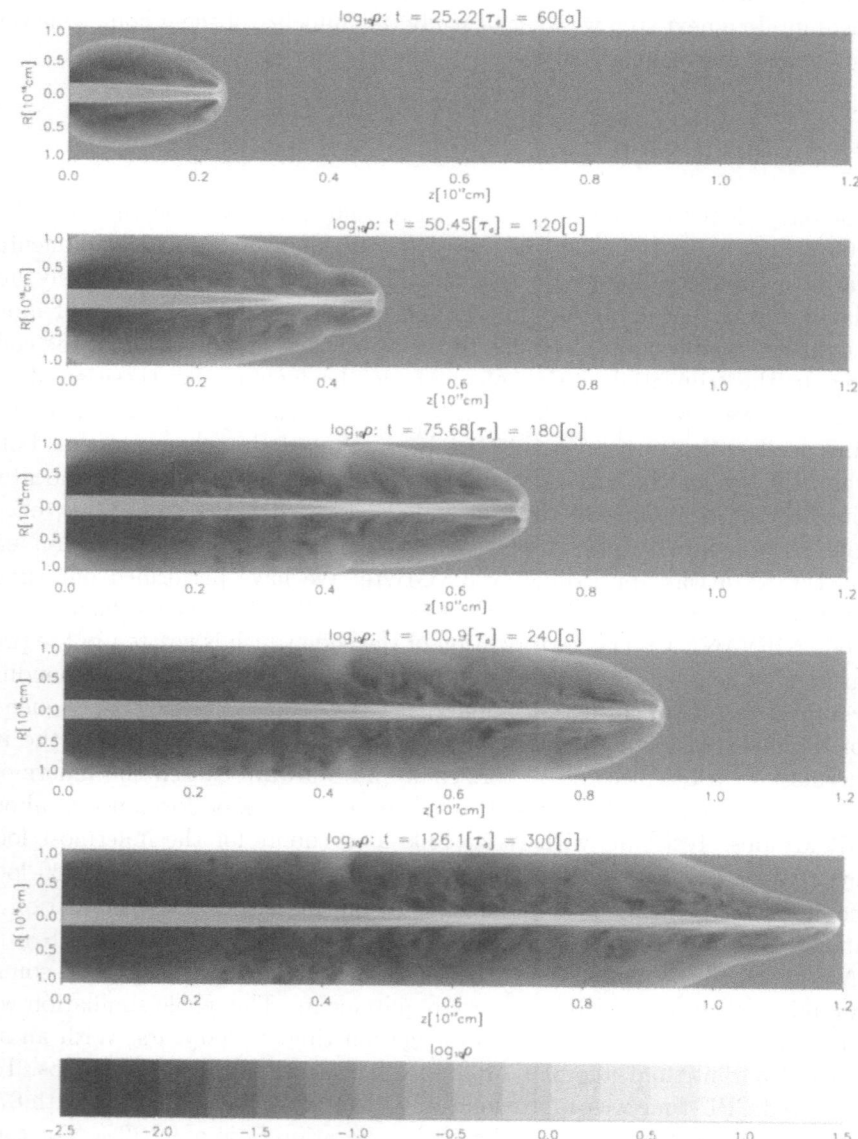

Fig. 1. Grey scale images of the logarithm of the density (in arbritrary units) at five times

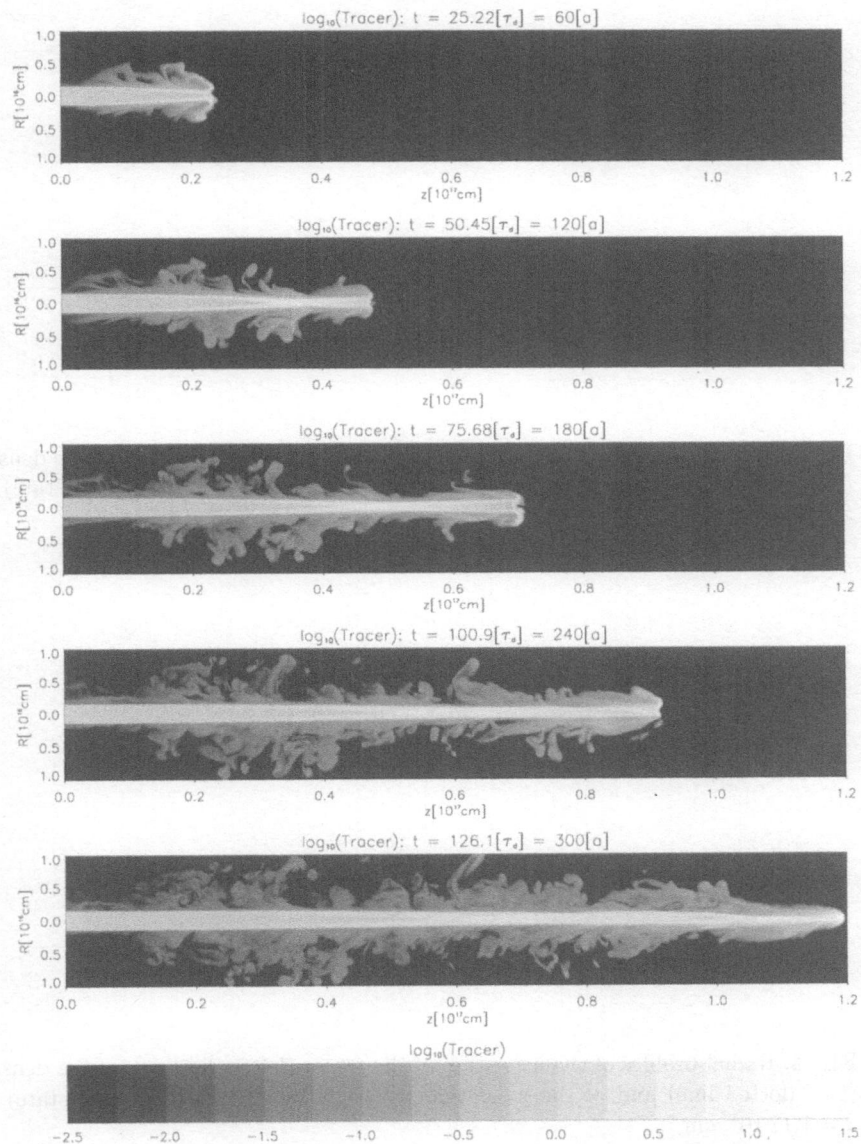

Fig. 2. Grey scale images of the logarithm of the tracer (in arbritrary units) at five times

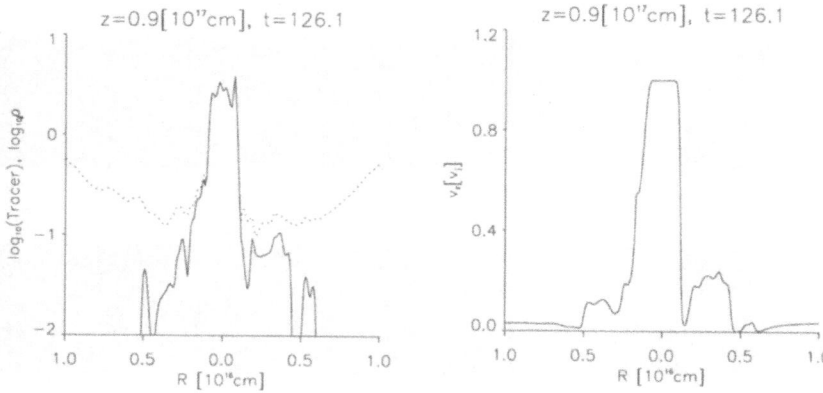

Fig. 3. Radial profiles of the logarithm of the tracer (left, solid line), of the density (left, dotted line) and of the axial velocity (right) for $t = 126.1$ (final state) at $z = 9\,10^{16}$ cm

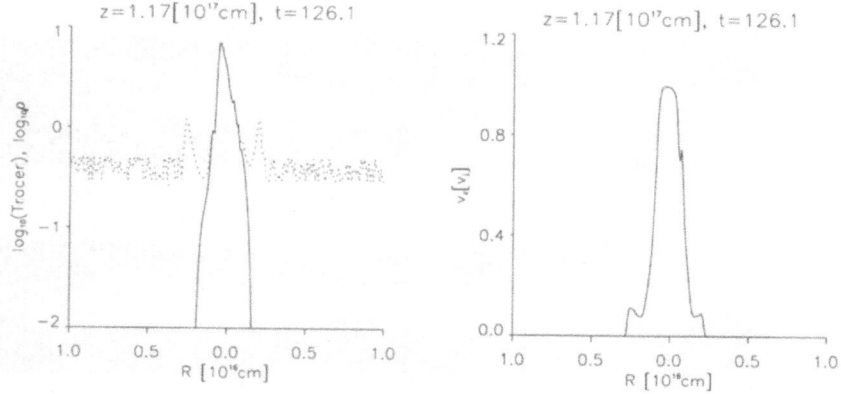

Fig. 4. Radial profiles of the logarithm of the tracer (left, solid line), of the density (left, dotted line) and of the axial velocity (right) for $t = 126.1$ (final state) at $z = 1.17\,10^{17}$ cm

computing time took about 6.85 days. For corresponding low resolution simulations of this kind with $3\,10^6$ grid cells on a Pentium III processor typically $1.5\ldots 2$ months of CPU time is required. Thus, vectorization alone already leads to a very effective reduction of the required computation time.

We did also apply microtasking techniques to achieve a fully parallelized version of our code where parallelisation was generally applied to the middle loop. In this case we have used only the vectorized version, since the results allow to check the different evolution of the numerical errors due to parallelization of a time-explicit scheme. The vectorized/parallized version will then be used for the simulations of magnetized and cooling jets, since in these cases the gross computation time could otherwise become prohibitive.

Acknowledgement

This work has been partially supported by the Deutsche Forschungsgemein-schaft (Schwerpunkt 'Physik der Sternentstehung').

References

1. Bally, J., Devine, D. 1994, ApJ, 428, L65
2. Blondin, J. M., Fryxell, B. A., Königl, A. 1990, ApJ, 360, 370
3. Bührke, T., Mundt, R., Ray, T. 1988, A&A, 200, 99
4. De Gouveia Dal Pino, E. M., Birkinshaw, M., Benz, W. 1996, ApJ, 460, L111
5. De Gouveia Dal Pino, E. M., Birkinshaw, M. 1996, ApJ, 471, 832
6. De Gouveia Dal Pino, E. M. 1999, ApJ, 526, 862
7. Falgarone, E., Puget, J.-L. & Pérault, M. 1992, A&A, 257, 715
8. Higgins, S. W., O'Brien, T. J., Dunlop, J. S. 1999, MNRAS, 309, 273
9. Kössl, D., Müller, E., Hillebrandt, W. 1990, A&A, 229, 37
10. Mundt, R., Brugel, E. W., Bührke, T. 1987, ApJ, 319, 275
11. Mundt, R., Bührke, T., Solf, J., Ray, T. P., Raga, A. C. 1990, A&A, 232, 37
12. Mundt, R., Ray, T. P., Raga, A. C. 1991, A&A, 252, 740
13. Payne, D. G., Cohn, H. 1985, ApJ, 291, 655
14. Reipurth, B. 1997, in 'Herbig-Haro Flows and the birth of Low Mass stars', IAU Symp. 182, eds. Reipurth, B. & bertout, C., p. 3
15. Stone, J. M., Norman, M. L. 1993, ApJ, 413, 198
16. Stone, J. M. & Norman, M. L. 1994, ApJ, 420, 237
17. Suttner, G., Smith, M. D., Yorke, H. W., Zinnecker, H. 1997, A&A, 318, 595
18. Terquem, C., Eislöffel, J., Papaloizou, J. C. B., Nelson, R. P. 1999, ApJL, 512, 131
19. Thiele, M. 2000, Dissertation, Universität Heidelberg
20. Todo, Y., Uchida, Y., Sato, T., Rosner, R. 1993 ApJ, 403,164
21. Ziegler, U. 1995, Dissertation, Universität Würzburg
22. Zinnecker, H., McCaughrean, M.J., Rayner, J.T. 1998, Nature, 394, 862

Phase Transitions and Quantum Effects in Systems with Reduced Geometry

J. Hoffmann and P. Nielaba

Physics Department (Theory), University of Konstanz, 78457 Konstanz, Germany

Abstract. We discuss phase transition and quantum effects in pore condensates as well as the determination of elastic constants and analyze the melting of hard disk systems. Particular emphasis is put on the efficiency of our parallel algorithm for our path integral Monte Carlo simulations.

1 Introduction

Nanostructures in reduced geometry have become an interesting research domain in the last years. At the University of Konstanz we have a corresponding "Sonderforschungsbereich" in this field, from which many of the studies presented here have been sponsored in long term projects.

Despite the fact that by experimental techniques many structural-, elastical-, electronic-, and phase-properties of systems in the size of a few nanometers have been obtained, the theoretical investigations and analyses are still in an initial stage. This is partly because of the fact that systems which are far away from the thermodynamic limit (with infinitely many particles) due to their finite size are difficult to handle by analytical methods which are suitable for systems with either few particles (2–5) or in the limit of infinitely many particles. In this field, in the last ten years, computer simulations have become more and more important, because nano-systems in reduced geometry contain about 10–10.000 particles, which is nearly ideal for the application of computer simulation methods. In particlar in the field of quantum simulations our group has been able to achieve many interesting contributions [6], which to a great deal have been obtained by computer time support by the HLRS.

Many of the effects take place at low temperatures so that the consideration of quantum mechanics is important. In the research domain of nanostructures in reduced geometry a relatively small amount of theoretical research has been done. We plan to contribute to bridge this gap by our project at the HLRS. By computer simulation methods we plan to study the structures and the elastic and electronic properties as well as phase "transitions" in systems of a the size of a few nanometers located at surfaces. In order to quantify quantum effects we perform path integral Monte Carlo (PIMC) simulations. As a result we have been able to publish many interesting aspects of our work in scientific journals, see reference list.

2   Performance of our path integral Monte Carlo simulations on the CRAY-T3E

Canonical averages $< A >$ of an observable A in a system defined by the Hamiltonian $\mathcal{H} = E_{kin} + V_{pot}$ of N particles in a volume V are given by:

$$\langle A \rangle = Z^{-1} \ \ \text{tr} \ \ [A \exp(-\beta \mathcal{H})] \ . \tag{1}$$

Here $Z = \text{tr} \ [\exp(-\beta \mathcal{H})]$ is the partition function and $\beta = 1/k_B T$ is the inverse temperature. Utilizing the Trotter–product formula,

$$\exp(\beta \mathcal{H}) = \lim_{P \to \infty} (\exp(-\beta E_{kin}/P) \exp(-\beta V_{pot}/P))^P \ \ , \tag{2}$$

we obtain the path integral expression for the partition function:

$$Z(N,V,T) = \lim_{P \to \infty} \left(\frac{mP}{2\pi \beta \hbar^2} \right)^{3NP/2} \prod_{s=1}^{P} \int d\{\mathbf{r}^{(s)}\} \cdot \tag{3}$$

$$\cdot \exp \left\{ -\frac{\beta}{P} \left[\sum_{k=1}^{N} \frac{mP^2}{2\hbar^2 \beta^2} (\mathbf{r}_k^{(s)} - \mathbf{r}_k^{(s+1)})^2 + V_{pot}(\{\mathbf{r}^{(s)}\}) \right] \right\}$$

Here, m is the particle mass, integer P is the Trotter number and $\mathbf{r}_k^{(s)}$ denotes the coordinate of particle k at Trotter-index s, and periodic boundary conditions apply, $P + 1 = 1$. This formulation of the partition function allows us to perform Monte Carlo simulations for increasing values of P approaching the true quantum limit for $P \to \infty$.

Thermal averages in the ensemble with constant pressure p are given via the corresponding partition function $\Delta(N, p, T) = \int_0^\infty dV \exp[-\beta pV] Z(N, V, T)$. In Eq. (1) we see that in the path integral formalism each quantum particle k can (for finite P- values) be represented by closed quantum chains of length P in position space where the classical coordinate of the point $\mathbf{r}_k^{(s)}$ on this chain at the Trotter index s has a harmonic interaction to its nearest neighbors at $\mathbf{r}_k^{(s+1)}$ and $\mathbf{r}_k^{(s-1)}$. An interaction between different quantum particles takes places only between particles $\{\mathbf{r}^{(s)}\}$ with the same Trotter index s. Due to this property the entire system with Trotter index s can be placed efficiently in one processor of a parallel computer with P processors, where the potential energy of all N particles can be computed for this Trotter index s (with an effort $\propto N(N-1)/2$). The different processors then have a physical coupling due to the kinetic energy term resulting in the harmonic interaction between nearest neighbor Trotter indices in the PIMC formalism. Thus only N interactions between the harmonically interacting particles have to be computed due to the harmonic interactions between nearest neighbor Trotter indices s and $s + 1$. It is even more efficient to place the system with two neighboring Trotter indices into one processor, since then only two neighboring processors communicate when a local Monte Carlo move (for one quantum particle k) is done.

In Fig. 1 the "speed up" with this algorithm is shown. We note that with a Trotter order of $P = 64$ (that means 32 processors) the running time of the simulation is increasing only by about 5 % when doubling the Trotter- order. In contrast to this very good scaling behaviour, in a scalar algorithm with a linear-P dependency the running time would increase by 100%. This shows that only computations on the parallel computer T3E make it possible to investigate systems

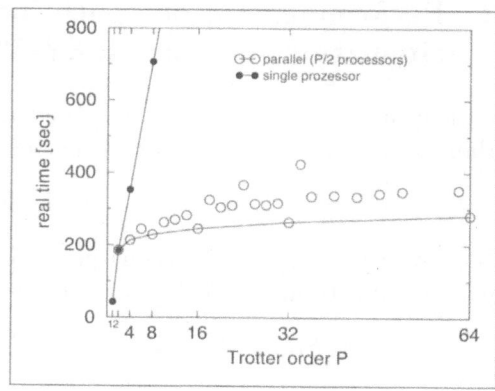

Fig. 1: Comparison of the running time in a typical PIMC simulation for various P- values. In each processor two Trotter-"particles" are located.

at temperatures at which the proper approximation of the quantum limit requires large P- values ($P/2$ processors). Many of the studies presented in this report thus in practice are only possible at the T3E.

Besides this inherent advantage of the parallel algorithm for the PIMC simulations we utilize the parallel machine as well most efficiently in running different replicas of the system in parallel in order to increase the statistics in the statistical averages.

3   Phase transitions in nano- pores

Phase transitions of pore condensates in nano-pores (i.e. Vycor) have been investigated by experimental methods recently [1,2]. Besides spinodal decomposition, phase transition temperature reductions have been studied for cylindrical nano-pores with small diameters.

With computer simulations (CRAY-T3E) we have analyzed [3,4] many interesting properties of "Ar"- and "Ne"- pore condensates recently (modeled as Lennard-Jones systems with particle diameter σ and interaction energy ϵ, in our computations we use particle masses $m^*=m\sigma^2\varepsilon/\hbar^2= 100$ and $m^*=1000$ for simplicity well approximating the particle masses of Ne and Ar ($m^*=112$ and $m^*=1160$)). These systems have - like the "bulk"- systems a gas- liquid phase transition at low temperatures, the precise shape of the phase diagram is strongly influenced by the system geometry (pore radius). In turns out that with increasing attractive wall interaction the critical density increases, the adsorbate density increases strongly, and the condensate density increases weakly. A meniscus is formed with increasing curvature, the configurations become less stable and the critical temperature decreases.

The critical temperature is reduced with decreasing pore diameter. Beginning from the wall a formation of layered shell structures is found which may favor or disfavor the occupancy of sites at the pore axis due to packing effects. In Figure 2 we present the density profiles (cylindrical average) as well as the radial density distribution in the condensate (center part of the system in the left picture). We note the layering structure and an oscillatory behavior of the density at the pore axis (r=0) as a function of pore radius with density maxima for pore radii of $n\sigma$ and minima for pore radii of $(n + 1/2)\sigma$.

For large pore diameters the density oscillations decay from the wall towards the pore axis and the system approaches the "bulk".

At lower temperatures we obtain [3,4] a phase transition into a solid phase with long ranged positional order. In this solid phase cylindrical shells are formed with triangular lattice structures in azimuthal direction. In Fig. 3 we show such unrolled layers for a pore with radius 5σ in a NVT- ensemble simulation indicating a meniscus shaped interface in the solid phase (see Fig.4).

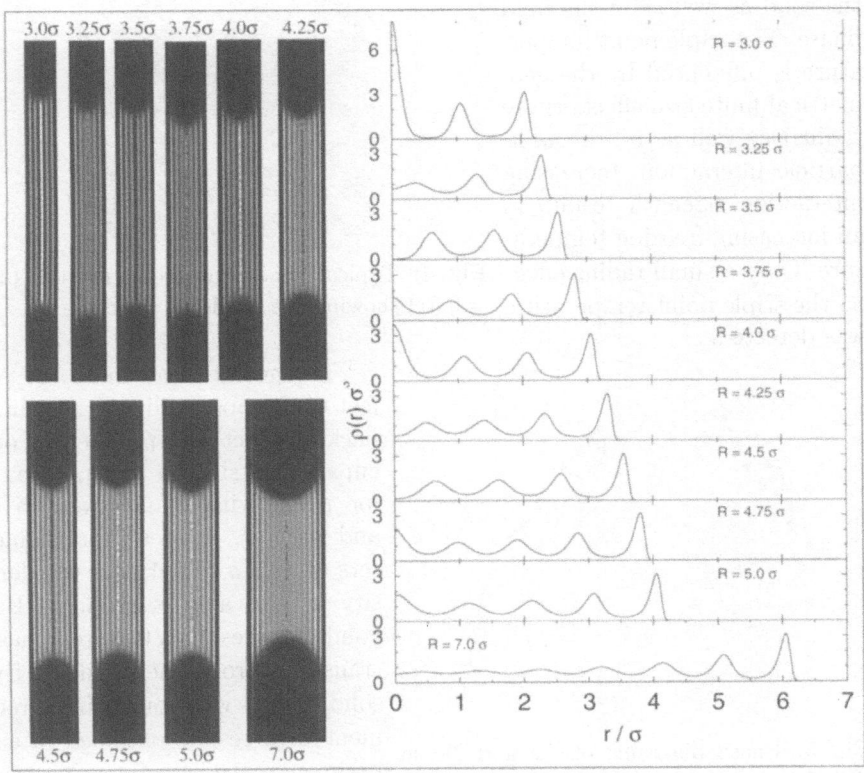

Fig. 2: Density distributions (cylindrical average) for Ar in a cylindrical pore at $T^* = 0.6$. Left: density profile for pores with various radii. Right: radial density distribution in the condensate (center part left).

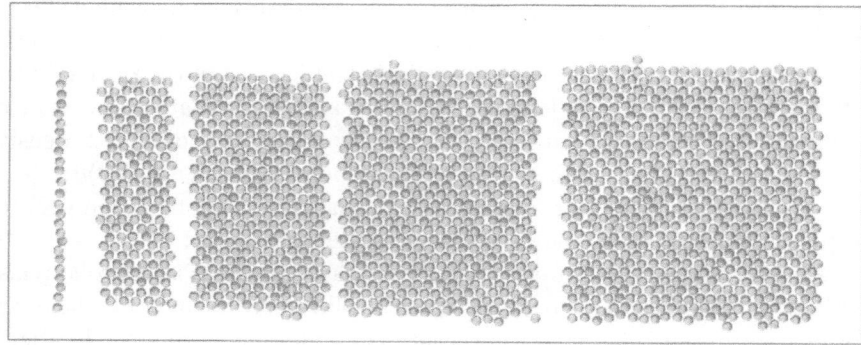

Fig. 3: Radial- layers- resolved (unrolled) pore Ar- condensate in cylindrical pore with diameter 5 σ (T*=0.34).

The meniscus curvature decreases with the temperature in the solid as well as in the fluid phase. The triple point temperature is influenced by the geometrical finite size effects (pore radius) as well as by the wall-particle interaction. Increasing interaction strength results in an increasing freezing temperature. Only a small radius effect on the triple point temperature was detected.

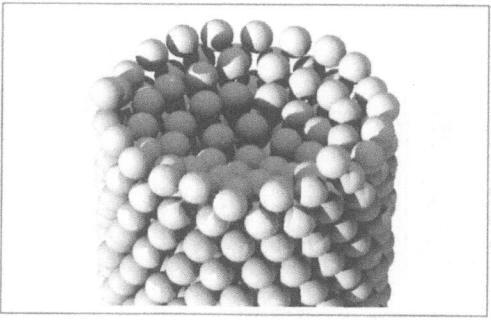

Fig. 4: Typical condensate configuration (T* = 0.31) showing the meniscus structure.

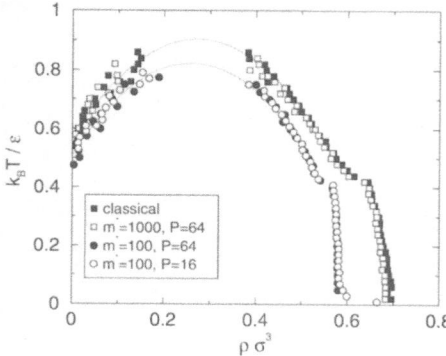

Fig. 5: Phase diagrams of Ar and Ne in cylindrical pores (radius 3.25 σ, length 37.7 σ). Comparison of PIMC and classical MC. Lines: quadratical fits through the data ($T^* > 0.6$) for m*=100, P=64 and the classical case.

The geometrical finite size effect of the pore radius results in a packing effect with preferential occupancy of sites at the pore axis for pores with diameters 3.75 σ and 4.75 σ, whereas for diameters of 3.25 σ and 4.25 σ the density at the axis is reduced. Beyond these results a two step phase transition from the fluid to the solid phase was found in agreement with results obtained in experimental studies of specific heat capacities at the freezing of Ar in Vycor [5].

By path integral Monte Carlo simulations [6–9] the effect of the quantum mechanics on the potential energy as a function of the temperature has been quantified [3,4]. In contrast to classical simulations we obtain by PIMC simulations for Ar- and Ne- condensates an horizontal temperature dependency of the energy resulting in a decrease of the specific heat to zero at small temperatures in agreement with the third law of thermodynamics. The resulting phase diagram for Ar- and Ne- condensates and a comparison with classical computations is shown in Figure 5. In the Ne- system (containing the lighter particles) a significant reduction (by about 5–10%) of the critical temperature is found due to quantum delocalizations as well as a strong reduction of the solid density and a crystal structure modification in comparison with the classical case.

4   Elastic constants from microscopic strain fluctuations

One is often interested in long length scale and long time scale phenomena in solids (eg. late stage kinetics of solid state phase transformations; motion of domain walls interfaces; fracture; friction etc.). Such phenomena are usually described by continuum theories. Microscopic simulations [10] of finite systems, on the other hand, like molecular dynamics, lattice Boltzmann or Monte Carlo, deal with microscopic variables like the positions and velocities of constituent particles and together with detailed knowledge of interatomic potentials, hope to build up a description of the macro system from a knowledge of these micro variables. How does one recover continuum physics from simulating the dynamics of N particles? This requires a "coarse-graining" procedure in space (for equilibrium) or both space and time for non-equilibrium continuum theories. Over what coarse graining length and time scale does one recover results consistent with continuum theories? We attempted to answer these questions [11] for the simplest nontrivial case, namely, a crystalline solid, (without any point, line or surface defects [12]) in equilibrium, at a non zero temperature far away from phase transitions. Fluctuations of the instantaneous local Lagrangian strain $\epsilon_{ij}(\mathbf{r}, \mathbf{t})$, measured with respect to a static "reference" lattice, are used to obtain accurate estimates of the elastic constants of model solids from atomistic computer simulations. The measured strains are systematically coarse - grained by averaging them within subsystems (of size L_b) of a system (of total size L) in the canonical ensemble. Using a simple finite size scaling theory we predict the behaviour of the fluctuations $< \epsilon_{ij}\epsilon_{kl} >$ as a function of L_b/L and extract elastic constants of the system *in the thermodynamic limit* at nonzero temperature. Our method is simple to implement, efficient and general enough to be able to handle a wide class of model systems including those with singular potentials without any essential modification.

Imagine a system in the constant NVT (canonical) ensemble at a fixed density $\rho = N/V$ evolving in time t. For any "snapshot" of this system taken from this ensemble, the local instantaneous displacement field $\mathbf{u_R}(t)$ defined over the set of lattice vectors $\{\mathbf{R}\}$ of a reference lattice (at the same density ρ) is: $\mathbf{u_R}(t) = \mathbf{R}(t) - \mathbf{R}$, where $\mathbf{R}(t)$ is the instantaneous position of the particle tagged by the reference lattice point $\mathbf{R}$. Let us concentrate only on perfect crystalline lattices; if topological defects such as dislocations are present the analysis below needs to be modified. The instantaneous Lagrangian strain tensor ϵ_{ij} defined at $\mathbf{R}$ is then given by [12],

$$\epsilon_{ij} = \frac{1}{2} \left(\frac{\partial u_i}{\partial R_j} + \frac{\partial u_j}{\partial R_i} + \frac{\partial u_i}{\partial R_k} \frac{\partial u_k}{\partial R_j} \right) \tag{4}$$

The strains considered here are always small and so we, hereafter, neglect the non-linear terms in the definition given above for simplicity. The derivatives are required at the reference lattice points $\mathbf{R}$ and can be calculated by any suitable finite difference scheme once $\mathbf{u_R}(t)$ is known. We are now in a position to define coarse grained variables $\epsilon_{ij}^{L_b}$ which are simply averages of the strain over a sub-block of size L_b. The fluctuation of this variable then defines the size dependent compliance matrices $S_{ijkl} = < \epsilon_{ij} \epsilon_{kl} >$. Before proceeding further, we introduce a compact Voigt notation (which replaces a pair of indices ij with one α) appropriate for two dimensional strains - the only case considered here. Using $1 \equiv x$ and $2 \equiv y$ we have,

$$ij = \quad 11 \quad 22 \quad 12 \tag{5}$$
$$\alpha = \quad 1 \quad \ 2 \quad \ 3$$

The nonzero components of the compliance matrix are

$$S_{11} = < \epsilon_{xx} \epsilon_{xx} > \ = S_{22} \tag{6}$$
$$S_{12} = < \epsilon_{xx} \epsilon_{yy} > \ = S_{21}$$
$$S_{33} = 4 < \epsilon_{xy} \epsilon_{xy} >$$

It is useful to define the following linear combinations

$$S_{++} = < \epsilon_+ \epsilon_+ > \ = 2(S_{11} + S_{12}) \tag{7}$$
$$S_{--} = < \epsilon_- \epsilon_- > \ = 2(S_{11} - S_{12})$$

where $\epsilon_+ = \epsilon_{xx} + \epsilon_{yy}$ and $\epsilon_- = \epsilon_{xx} - \epsilon_{yy}$. Once the block averaged strains $\epsilon_{ij}^{L_b}$ are obtained, it is straight-forward to calculate these fluctuations (for each value of L_b).

Once the finite size scaled compliances are obtained the elastic constants viz. the Bulk modulus $B = \rho \partial p / \partial \rho$ and the shear modulus μ are obtained simply using the formulae [13]

$$\beta B = \frac{1}{2S_{++}} \qquad \beta \mu = \frac{1}{2S_{--}} - \beta p \qquad \beta \mu = \frac{1}{2S_{33}} - \beta p \tag{8}$$

where we assume that the system is under an uniform hydrostatic pressure p.

As an example we present our results for elastic constants of the hard disk system in Fig. 6. The two expressions for the shear modulus in Eq. (8) give almost identical results and this gives us confidence about the internal consistency of our method. We have also compared our results to those of Wojciechowski et al. [14]. We find that while their values of the pressure and bulk modulus are in good agreement with ours (and with free volume theory) they grossly overestimate the shear modulus. This is probably due to the extreme small size of their systems and/ or insufficient averaging. Our results for the sub-block analysis shows that finite size effects are non -trivial for elastic strain

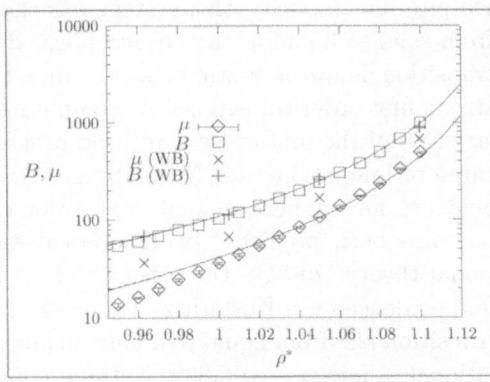

Fig. 6: The bulk (B) and shear (μ) moduli in units of $k_B T/\sigma^2$ for the hard disk solid. Our results for B (μ) are given by squares (diamonds). The values for the corresponding quantities from Ref. [14] are given by $+$ and $\times$. The line through the bulk modulus values is the analytical expression obtained from the free volume prediction for the pressure. The line through our shear modulus values is obtained from the free volume bulk modulus using the Cauchy relation $\mu = B/2 - p$.

fluctuations and they cannot be evaluated by varying the total size of the system from 24 to 90, an interval which is less than half of a decade. One immediate consequence of our results is that the Cauchy relation [14,22] $\mu = B/2 - p^*$ is seen to be valid upto $\pm 15\%$ over the entire density range we studied though there is a systematic deviation which changes sign going from negative for small densities to positive as the density is increased. This is in agreement with the usual situation in a variety of real systems [15] with central potentials and highly symmetric lattices and in disagreement with Ref. [14]. We have also compared our estimates for the elastic constants with the density functional theory (DFT) of Rhysov and Tareyeva [16]. We find that both the bulk and the shear moduli are grossly overestimated - sometimes by as much as 100%.

5   Melting of hard disks in two dimensions

One of the first continuous systems to be studied by computer simulations [17] is the system of hard disks of diameter σ $(=1)$ interacting with the two body potential,

$$V(r) = \begin{cases} \infty & r \le \sigma \\ 0 & r > \sigma \end{cases} \tag{9}$$

where, σ (taken to be 1 in the rest of the paper) the hard disk diameter, sets the length scale for the system and the energy scale is set by $k_B T = 1$.

Despite its simplicity, this system was shown to undergo a phase transition from solid to liquid as the density ρ was decreased. The nature of this phase transition, however, is still being debated. Early simulations [17] always found strong first order transitions. As computational power increased the observed strength of the first order transition progressively decreased! Using sophisticated techniques Lee and Strandburg [18] and Zollweg and Chester [19] found evidence for, at best, a weak first order transition. A first order transition has also been predicted by theoretical approaches based on density functional theory [20]. On the other hand, recent simulations of hard disks [21] find evidence for a Kosterlitz -Thouless -Halperin -Nelson -Young (KTHNY) transition [22] from liquid to a hexatic phase, with orientational but no translational order, at $\rho = 0.899$. Nothing could be ascertained, however, about the expected hexatic to the crystalline solid transition at higher densities because the computations became prohibitively expensive. The solid to hexatic melting transition was estimated to occur at a density $\rho_c \geq .91$. A priori, it is difficult to assess why various simulations give contradicting results concerning the order of the transition. In Ref. [23] we took an approach, complementary to Jaster's, and investigated the melting transition of the solid phase. We showed that the hard disk solid is unstable to perturbations which attempt to produce free dislocations leading to a solid $\rightarrow$ hexatic transition in accordance with KTHNY theory [22] and recent experiments in colloidal systems [24]. Though this has been attempted in the past [14,26], numerical difficulties, especially with regard to equilibration of defect degrees of freedom, makes this task highly challenging. The elastic Hamiltonian for hard disks is given by, $F = -P\epsilon_+ + B/2\epsilon_+^2 + (\mu + P)(\epsilon_-/2 + 2\epsilon_{xy})$, where B is the bulk modulus. The quantity $\mu_{eff} = \mu + p$ is the "effective" shear modulus (the slope of the shear stress vs shear strain curve) and p is the pressure.

The KTHNY- theory [22] is presented usually for a 2-d triangular solid under *zero external stress*. It is shown that the dimensionless Young's modulus of a two-dimensional solid, $K = (8/\sqrt{3}\rho)(\mu/\{1 + \mu/(\lambda + \mu)\})$, where μ and λ are the Lamé constants, depends on the fugacity of dislocation pairs, $y = \exp(-E_c)$, where E_c is the core energy of the dislocation, and the "coarse -graining" length scale l. This dependence is expressed in the form of the following coupled recursion relations for the renormalization of K and y:

$$\frac{\partial K^{-1}}{\partial l} = 3\pi y^2 e^{\frac{K}{8\pi}}[\frac{1}{2}I_0(\frac{K}{8\pi}) - \frac{1}{4}I_1(\frac{K}{8\pi})], \tag{10}$$

$$\frac{\partial y}{\partial l} = (2 - \frac{K}{8\pi})y + 2\pi y^2 e^{\frac{K}{16\pi}} I_0(\frac{K}{8\pi}).$$

where I_0 and I_1 are Bessel functions. The thermodynamic value is recovered by taking the limit $l \rightarrow \infty$.

We see in Fig. 7 that the trajectories in y-K plane can be classified in two classes, namely those for which $y \rightarrow 0$ as $l \rightarrow \infty$ (ordered phase) and those $y \rightarrow \infty$ as $l \rightarrow \infty$ (disordered phase). These two classes of flows are separated by lines called the separatrix. The transition temperature T_c (or ρ_c) is given by the intersection of the separatrix with the line of initial conditions $K(\rho, T)$ and $y = \exp(-E_c(K))$ where $E_c \sim cK/16\pi$. The disordered phase is a phase where free dislocations proliferate. Proliferation of dislocations however *does not* produce a liquid, rather a liquid crystalline phase called a "hexatic" with quasi- long ranged (QLR) orientational order but short ranged positional order. A *second* K-T transition destroys QLR orientational order and takes the hexatic to the liquid phase by the proliferation of "disclinations" (scalar charges). Apart from T_c there are several universal predictions from KTHNY-theory, for example, the order parameter correlation length and susceptibility has essential singularities ($\sim e^{bt^{-\nu}}, t \equiv T/T_c - 1$) near T_c. All these predictions can, in principle, be checked in simulations [21].

One way to circumvent the problem of large finite size effects and slow relaxation due to diverging correlation lengths is to simulate a system which is constrained to remain defect (dislocation) free and, as it turns out, without a phase transition. Surprisingly, using this data it is possible to predict the expected equilibrium behaviour of the unconstrained system. The simulation [23] is always started from a perfect triangular lattice which fits into our box – the size of the box determining the density. Once a regular MC move is about to be accepted, we perform a *local* Delaunay triangulation involving the moved disk and its nearest and next nearest neighbors. We compare the connectivity of this Delaunay triangulation with that of the reference lattice (a copy of the initial state) around the same particle. If any old bond is broken and a new bond formed (Fig. 7) we reject the move since one can show that this is equivalent to a dislocation - antidislocation pair separated by one lattice constant involving dislocations of the smallest Burger's vector.

Microscopic strains $\epsilon_{ij}(\mathbf{R})$ can be calculated now for every reference lattice point $\mathbf{R}$. Next, we coarse grain (average) the microscopic strains within a sub-box of size L_b, $\bar{\epsilon}_{ij} = L_b^{-d} \int^{L_b} d^d r \epsilon_{ij}(\mathbf{r})$ and calculate the (L_b dependent) quantities [11],

$$S_{++}^{L_b} = < \bar{\epsilon}_+ \bar{\epsilon}_+ >, \quad S_{--}^{L_b} = < \bar{\epsilon}_- \bar{\epsilon}_- >, \quad (11)$$
$$S_{33}^{L_b} = 4 < \bar{\epsilon}_{xy} \bar{\epsilon}_{xy} >$$

The elastic constants in the thermodynamic limit are obtained from, the set: $B = 1/2S_{++}^{\infty}$ and $\mu_{eff} = 1/2S_{--}^{\infty} = 1/2S_{33}^{\infty}$. We obtain highly accurate values of the unrenormalized coupling constant K and the defect fugacity y which can be used as inputs to the KTHNY recursion relations. Numerical solution of these recursion relations then yields the renormalized coupling K_R and hence the density and pressure of the solid to hexatic melting transition.

We can draw a few very precise conclusions from our results. Firstly, a solid without dislocations is stable against fluctuations of the amplitude of the solid order parameter and against long wavelength phonons. So any melting transition mediated by phonon or amplitude fluctuation is ruled out in our system. Secondly, the core energy $E_c > 2.7$ at the transition so KTHNY

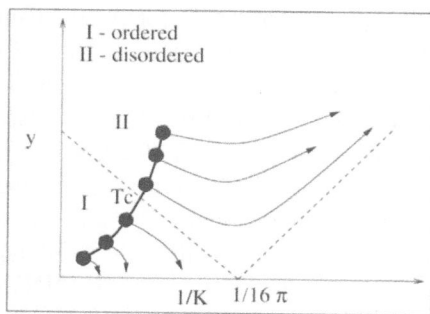
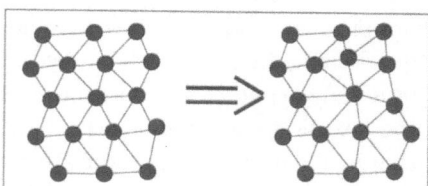

Fig. 7. (*Left*) Schematic flows of the coupling constant K and the defect fugacity y under the action of the KTHNY recursion relations. The dashed line is the separatrix whose intersection with the line of initial state (solid line connecting filled circles, $y(T, l = 0), K^{-1}(T, l = 0)$) determines the transition point T_c. (*Right*) Typical move which attempts to change the coordination number and therefore the local connectivity around the central particle. Such moves were rejected in our simulation.

perturbation theory is valid though numerical values of nonuniversal quantities may depend on the order of the perturbation analysis. Thirdly, solution of the recursion relations shows that a KTHNY transition at $p_c = 9.39$ *preempts* the first order transition at $p_1 = 9.2$. Since these transitions, as well as the hexatic -liquid KTHNY transition lies so close to each other, the effect of, as yet unknown, higher order corrections to the recursion relations may need to be examined in the future. Due to this caveat, our conclusion that a hexatic phase exists over some region of density exceeding $\rho = .899$ still must be taken as preliminary. Also, in actual simulations, cross over effects near the bicritical point, where two critical lines corresponding to the liquid - hexatic and hexatic -solid transitions meet a first order liquid -solid line may complicate the analysis of the data, which may, in part, explain the confusion which persists in the literature on this subject.

Acknowledgements

We thank S. Sengupta, K. Binder and M. Rao for cooperation and useful discussions and the HLRS for computer time. This work was supported by the SFB 513.

References

1. Z. Zhang, A. Chakrabarti, Phys. Rev. E **50**, R4290 (1994); J.C. Lee, Phys. Rev. Lett. **70**, 3599 (1993); A. Chakrabarti, Phys. Rev. Lett. **69**, 1548 (1992); A.J. Liu, G.S. Grest, Phys. Rev. **A 44**, R7894 (1991); A.J. Liu, D.J. Durian, E. Herbolzheimer, S.A. Safran, Phys. Rev. Lett. **65**, 1897 (1990); L. Monette,

A.J. Liu, G.S. Grest, Phys. Rev. **A 46**, 7664 (1992); Z. Zhang, A. Chakrabarti, Phys. Rev. **E 52**, 2736 (1995).

2. M.W. Maddox, K.E. Gubbins, M. Sliwinska- Bartkowiak, Soong- hyuck Suh, Mol. Simulat. **17**, 333 (1997); M.W. Maddox, K.E. Gubbins, J. Chem. Phys. **107**, 9659 (1997); R. Radhakrishnan, K.E. Gubbins, Phys. Rev. Lett. **79**, 2847 (1997); L.D. Gelb, K.E. Gubbins, Phys. Rev. **E 56**, 3185 (1997); P. Huber, K. Knorr, Phys. Rev.**B 60**, 12657 (1999).

3. J. Hoffmann, Ph.D.-thesis, University of Konstanz (in preparation).

4. J. Hoffmann, P. Nielaba, preprint.

5. D. Wallacher, K. Knorr, Phys. Rev. **B63**, 104202 (2001).

6. P. Nielaba, in: *Computational Methods in Surface and Colloid Science*, M. Borowko (Ed.), Marcel Dekker Inc., New York (2000), pp.77-134.

7. M. Presber, D. Löding, R. Martonak, P. Nielaba, Phys. Rev. **B 58**, 11937 (1998).

8. M. Reber, D. Löding, M. Presber, Chr. Rickward, P. Nielaba, Comp. Phys. Commun. **121-122**, 524 (1999).

9. C.Rickwardt, P.Nielaba, M.H.Müser, K.Binder, Phys.Rev.**B63**, 045204 (2001).

10. D.P. Landau, K. Binder, *A Guide to Monte Carlo Simulations in Statistical Physics*, Cambridge University Press (2000).

11. S. Sengupta, P. Nielaba, M. Rao, K. Binder, Phys.Rev.**E 61**, 1072 (2000).

12. P.M.Chaikin, T.C.Lubensky, *Principles of condensed matter physics*, (Cambridge University Press, Cambridge, 1995).

13. D. C. Wallace, in *Solid state physics*, eds. H. Ehrenreich, F. Seitz and D. Turnbull (Academic Press, New York, 1958); J. H. Weiner, *Statistical mechanics of elasticity* (Wiley, New York, 1983).

14. K.W. Wojciechowski, A.C. Brańka, Phys.Lett. **134A**, 314 (1988).

15. F. Seitz, *Modern Theory of Solids* (McGaraw-Hill, New York, 1940); M. Born and K. Huang, *Dynamical Theory of Crystal Lattices* (Oxford University Press, Oxford, 1954); For experimental temperature dependent elastic constants of solid argon see: H. Meixner, P. Leiderer, P. Berberich and E. Lüscher, Phys. Lett. **40A**, 257 (1972).

16. V.N. Ryzhov, E.E. Tareyeva, Phys.Rev.**B51**, 8789 (1995).

17. B.J. Alder, T.E. Wainwright, Phys.Rev.**127**, 359 (1962).

18. J. Lee and K. Strandburg, Phys. Rev. **B46**, 11190 (1992).

19. J.A.Zollweg, G.V.Chester, Phys.Rev.**B46**,11186 (1992).

20. T. V. Ramakrishnan, Phys. Rev. Lett. **42**, 795 (1979); X. C. Zeng and D. W. Oxtoby, J. Chem. Phys. **93**, 2692 (1990); Y. Rosenfeld, Phys. Rev. **A42**, 5978 (1990). V. N. Ryzhov and E. E. Tareyeva, Phys. Rev. **B51**, 8789 (1995).

21. A. Jaster, Phys. Rev. E **59**, 2594 (1999).

22. J. M. Kosterlitz, D. J. Thouless, J. Phys. **C 6**, 1181 (1973); B.I. Halperin and D.R. Nelson, Phys. Rev. Lett. **41**, 121 (1978); D. R. Nelson and B. I. Halperin, Phys. Rev. B **19**, 2457 (1979); A.P. Young, Phys. Rev. B **19**, 1855 (1979).

23. S. Sengupta, P. Nielaba, K. Binder, Phys. Rev. **E 61**, 6294 (2000).

24. K. Zahn, R. Lenke and G. Maret, Phys. Rev. Lett. **82**, 2721, (1999)

25. K. W. Wojciechowski and A. C. Brańka, Phys. Lett. **134A**, 314 (1988).

26. M. Bates and D. Frenkel, preprint.

Probing Hot Quantum Chromodynamics with a Complex Chemical Potential

A. Hart[1], M. Laine[2], and O. Philipsen[3,4]

[1] DAMTP, University of Cambridge, Cambridge CB3 0WA, UK
[2] CERN-TH, 1211 Geneva 23, Switzerland
[3] ITP Universität Heidelberg, 69120 Heidelberg, Germany
[4] Massachusetts Institute of Technology, Cambridge, MA 02139, USA

Abstract. Quantum Chromodynamics (QCD) at finite baryon density cannot be investigated using standard Monte Carlo simulation methods because its path integral measure is complex, leading to large cancellations in expectation values. This is often called the "sign problem". One suggestion for determining the properties of QCD at finite temperatures and densities is to carry out lattice simulations with an imaginary chemical potential whereby no sign problem arises, and to convert the results to real physical observables only afterwards. We test the feasibility of this for spatial correlation lengths in the quark-gluon plasma phase. Simulations with imaginary chemical potential followed by analytic continuation are compared with simulations with real chemical potential, using a dimensionally reduced effective action for hot QCD. We find that for imaginary chemical potential the system undergoes a phase transition at $|\mu/T| \approx \pi/3$, and thus observables are analytic only in a limited range. However, utilising this range, relevant information can be obtained for QCD with a real chemical potential. These results have been published in [1,2].

1   Introduction

Quantum Chromodynamics (QCD) describes the strong interactions of elementary particles. It is important to determine its properties at temperatures T and chemical potentials μ relevant for Early Universe cosmology and heavy ion collision experiments, with $T, \mu \sim 200$ MeV. For instance, one would like to know the locations of any phase transitions, and the properties of the quark-gluon plasma phase including its free energy density, or pressure, and the spatial and temporal correlation lengths felt by various types of excitations in the system.

QCD is a strongly interacting theory, and the only practical first principles method available for addressing these questions is lattice simulations. While there has been steady improvement in the accuracy of results at vanishing baryon density [3], the case of a non-vanishing density is still largely open, despite much work [3]. Indeed, introducing a non-zero density, or chemical potential, is difficult because it leads to a measure which is not positive definite (this is the so called sign problem), whereby standard Monte Carlo techniques fail.

In this project we focus on one of the suggestions for how a finite density system could eventually be addressed with practical lattice simulations.

One may first study an imaginary chemical potential, where there is no sign problem, and then relate this to the case of a real chemical potential. Let us denote by μ the chemical potential for quark number Q, and by μ_B the chemical potential for baryon number $B = Q/3$: then $\mu = \mu_B/3$. By $\mu_R, \mu_I \in \mathbb{R}$ we denote the real and imaginary parts of μ:

$$\mu = \mu_R + i\mu_I. \tag{1}$$

The idea ([4] and references therein) is that, away from possible phase transition lines, the partition function and expectation values for various observables should be analytic in their arguments, in particular in μ/T. Thus, we may attempt a general power series ansatz for the functional behaviour in μ/T. Assuming it be convergent, we can determine a finite number of coefficients with an imaginary chemical potential, and finally analytically continue the series to real values.

In this work we study this suggestion within the framework of a dimensionally reduced effective theory. As we shall review in the next section, at temperatures sufficiently above the phase transition, the thermodynamics of QCD can be represented, with good practical accuracy, by a simple three-dimensional (3d) purely bosonic theory. This can also be done for theories with a chemical potential, both real and imaginary [1]. We use this theory to measure the longest static correlation lengths in the system for both cases. We find that, for small $|\mu/T|$, the observables are well described by a truncated power series with coefficients determined by fits. We then inspect how well the analytically continued series describes the real data.

2 Effective theory

2.1 Action

The effective theory emerging from hot QCD by dimensional reduction [5]–[11],[1] is written in three, rather than the original four, dimensions. It is an SU(3) gauge field coupled to an adjoint Higgs scalar field, with action

$$S = \int d^3x \left\{ \frac{1}{2}\text{Tr}\, F_{ij}^2 + \text{Tr}\, [D_i, A_0]^2 + m_3^2 \text{Tr}\, A_0^2 + i\gamma_3 \text{Tr}\, A_0^3 + \lambda_3 (\text{Tr}\, A_0^2)^2 \right\}, \tag{2}$$

where $F_{ij} = \partial_i A_j - \partial_j A_i + ig_3[A_i, A_j]$, $D_i = \partial_i + ig_3 A_i$, F_{ij}, A_i, and A_0 are all traceless 3×3 Hermitian matrices ($A_0 = A_0^a T_a$, etc), and g_3^2 and λ_3 are the gauge and scalar coupling constants with mass dimension one, respectively. The physical properties of the effective theory are determined by the three dimensionless ratios

$$x = \frac{\lambda_3}{g_3^2}, \quad y = \frac{m_3^2(\bar{\mu}_3 = g_3^2)}{g_3^4}, \quad z = \frac{\gamma_3}{g_3^3}, \tag{3}$$

where $\bar{\mu}_3$ is the $\overline{\text{MS}}$ dimensional regularization scheme in 3d. These ratios are via dimensional reduction functions of the temperature $T/\Lambda_{\overline{\text{MS}}}$ and the chemical potential $\mu/\Lambda_{\overline{\text{MS}}}$, where $\Lambda_{\overline{\text{MS}}} \sim 200$ MeV is the characteristic QCD scale. They also depend on N_f, the number of massless quark flavours: for the case $\mu = 0$, we refer to [11]. The inclusion of quark masses is also possible in principle, but in the numerical part of this work we assume $N_f = 2$ massless dynamical flavours, the other flavours being approximated as infinitely heavy.

The mass parameter m_3^2, represented by y in Eq. (3), turns out to be positive [11]. This guarantees that the 3d theory tends to live in its symmetric phase, $A_0 \sim 0$, at least on the mean field level. We will return to this issue presently.

For vanishing chemical potential $\gamma_3 = 0$ and no term cubic in A_0 appears. When $\mu \neq 0$, the dominant changes in the action due to a small chemical potential are now [1]

$$z: 0 \to \frac{\mu}{T} \frac{N_f}{3\pi^2}; \quad y: y \to y\left(1 + \left(\frac{\mu}{\pi T}\right)^2 \frac{3N_f}{2N_c + N_f}\right), \tag{4}$$

where $N_c = 3$. Thus, one new operator is generated in the effective action, and one of the parameters which already existed, gets modified.

For real chemical potential, $\mu = \mu_R$, the effective action is thus complex, whereas for imaginary chemical potential, $\mu = i\mu_I$, it is real.

2.2   Ranges of validity

There are several requirements for the effective description in Eq. (2) to be reliable. They are all related to a sufficiently "weak coupling", or effective expansion parameter, for a given $T/\Lambda_{\overline{\text{MS}}}, \mu/\Lambda_{\overline{\text{MS}}}, N_f$. Let us briefly reiterate them here.

First, the perturbative expansions for the effective parameters in Eq. (3) have to be well convergent. Inspecting the actual series up to next-to-leading order, it appears that this requirement is surprisingly well met even at temperatures not much above the critical one [11].

Second, the higher dimensional operators arising in the reduction step which are not included in the effective action in Eq. (2), should only give small corrections. This condition is met if the dynamical mass scales described by the effective theory are smaller than the ones $\sim 2\pi T$ that have been integrated out. In pure Yang-Mills theory, there is evidence that this can be sufficiently satisfied at temperatures as low as $T \sim 2\Lambda_{\overline{\text{MS}}}$ [6],[11]–[15],[1]. However, when fermions are included and a *real* chemical potential is switched on, some of the mass scales increase (see below), and the effective description will become less accurate.

Third, the effective 3d theory represents the 4d theory reliably only when it lies in its symmetric phase [11,16], i.e. $A_0 \sim 0$. This is realised as long as the imaginary chemical potential satisfies $|\mu/T| < 1$ (see [2] for an explicit discussion).

In summary, the effective theory roughly loses its accuracy with a real chemical potential once even the longest correlation length is shorter than $\sim 1/(2\pi T)$, and with an imaginary chemical potential once $|\mu/T|$ exceeds unity. Fortunately, this range of validity contains the parameters that are phenomenologically most relevant. Indeed, heavy ion collision experiments at and above AGS and SPS energies can be estimated to correspond to $\mu_B/T \lesssim 4.0$ [17], or a quark chemical potential $\mu/T \lesssim 1.3$.

2.3 Observables and their parametric behaviour

As we have mentioned, the physical observables which we shall study are spatial correlation lengths: we consider operators living in the (x_1, x_2)-plane, and measure the correlation lengths in the x_3-direction.

In the presence of $\mu \neq 0$, there are only two different quantum number channels to be considered, distinguished by the two-dimensional parity P in the transverse plane. The lowest dimensional gauge invariant operators in the scalar $(J = 0)$ channels are:

$$J^P = 0^+ : \text{Tr } A_0^2, \text{Tr } F_{12}^2, \text{Tr } A_0^3, \text{Tr } A_0 F_{12}^2, ...$$
$$J^P = 0^- : \text{Tr } F_{12}^3, \text{Tr } A_0^2 F_{12}, \text{Tr } A_0 F_{12}, ... \tag{5}$$

The corresponding 4d operators can be found in [18]. We shall measure whole cross correlation matrices between all (smeared) operators in these channels, but mostly focus on their lowest eigenstates, corresponding to the longest correlation lengths in the 4d finite temperature system. We denote the "energies" of these eigenstates, viz. inverses of correlation lengths, by m. We also examine the overlap of operators of different field contents onto the eigenstates.

Since a change $\mu \to -\mu$ can be compensated for by a field redefinition $A_0 \to -A_0$ in Eq. (2), all physical observables must be even under this operation. In the original 4d theory the same statement follows from compensating $\mu \to -\mu$ by a C (or CP) operation. Moreover, since there are no massless modes at $\mu = 0$, we expect the masses to be analytic in μ away from phase transitions. For small values of μ/T, the inverse correlation lengths may thus be written as

$$\frac{m}{T} = c_0 + c_1 \left(\frac{\mu}{\pi T}\right)^2 + c_2 \left(\frac{\mu}{\pi T}\right)^4 + \mathcal{O}\left(\left(\frac{\mu}{\pi T}\right)^6\right). \tag{6}$$

We have chosen to include πT in the denominators, because the chemical potential appears with this structure in the effective parameters, cf. Eq. (4). Of course, the radii of convergence of such expansions are not known a priori.

Here we first check to what extent a truncated series of the type in Eq. (6) can accurately describe the data. In the range where this is possible, we determine the $\{c_i\}$ with $\mu = i\mu_I$, and check if the analytically continued result reproduces the independent measurements carried out with $\mu = \mu_R$.

3   Simulations

3.1   Simulation methods

The simulation algorithm is outlined in the appendix. We simulate the theory at several μ/T. The values chosen, together with the corresponding continuum parameters, are listed in Table 1. Discretization and lattice–continuum relations [19] are implemented as in [1]. As discussed there, finite volume and lattice spacing effects are expected to be smaller or at most of the same order as the statistical errors for the parameter values we employ.

For real $\mu = \mu_R$, the action in Eq. (2) with parameters as in Eq. (4) is complex, which precludes direct Monte Carlo simulations. We must thus carry out simulations using a reweighting technique, which has been explained in detail in [1]. There it was found that physically realistic lattice volumes may be simulated for chemical potentials up to $\mu_R/T \lesssim 4$. For imaginary $\mu = i\mu_I$, the action in Eq. (2) with parameters as in Eq. (4) is real, and correspondingly we simulate the full action using a Metropolis update.

3.2   Results

As a first result, let us note that, as has been the case in several related theories [20,13,1], we again observe a dynamical decoupling of operators, such that operators involving scalars ($\mathrm{Tr}\, A_0^2$, $\mathrm{Tr}\, A_0 F_{12}^2$ etc.) and purely gluonic operators ($\mathrm{Tr}\, F_{12}^2$ etc.) have a mutual overlap consistent with zero. The correlation matrix thus assumes an approximately block diagonal form. We find that the gluonic states remain extremely insensitive to μ/T, and agree well with the masses found in $d = 3$ pure gauge theory [21]. This situation is illustrated in Fig. 1.

Table 1. The parameters used for $\mu \neq 0$ (cf. Eq. (3)). All correspond to $T = 2\Lambda_{\overline{\mathrm{MS}}}$, $N_f = 2$. In addition, $x = 0.0919$, $g_3^2 = 2.92T$, $\beta = 21$, volume $= 30^3$, where β determines the lattice spacing (for the detailed relations employed here, see [1]).

		real μ		imaginary μ	
$\frac{|\mu|}{T}$	$\frac{|\mu|^2}{(\pi T)^2}$	y	z	y	z
0.50	0.0253	0.49218	0.0338	0.47382	0.0338i
0.75	0.0570	—	—	0.46235	0.0507i
1.00	0.1013	0.51970	0.0675	0.44630	0.0675i
1.25	0.1583	0.54035	0.0844	0.42565	0.0844i
1.50	0.2280	0.56558	0.1013	0.40042	0.1013i
1.75	0.3103	0.59540	0.1182	—	—
2.00	0.4053	0.62981	0.1351	—	—
3.00	0.9119	0.81333	0.2026	—	—
3.75	1.4248	0.99914	0.2533	—	—
4.00	1.6211	1.07026	0.2702	—	—

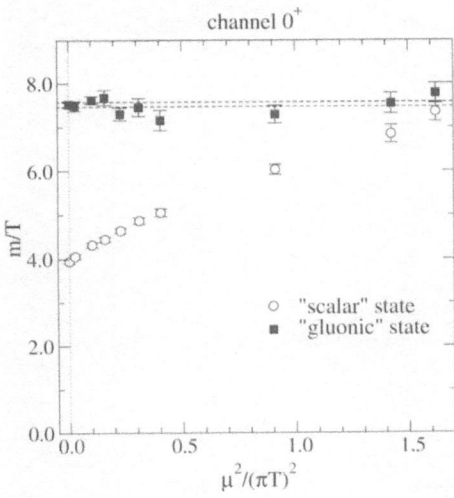

Fig. 1. Inverse correlation lengths in the channel 0^+, for real μ/T. "Scalar" states (Tr A_0^2 etc) do depend on μ/T, while "gluonic" states (Tr F_{12}^2 etc) are practically independent of it. For comparison, the horizontal band indicates the 3d pure glue result for Tr F_{12}^2 [21], converted to our units via $g_3^2 = 2.92T$.

The scalar states, on the other hand, show a marked dependence on μ/T, with their masses increasing for real μ and decreasing for imaginary values. For both small real and small imaginary μ/T, the ground state in each channel is scalar in nature, and we plot these states in Fig. 2.

Because of the different qualitative behaviours of 3d gluonic and scalar states, we may expect to observe a change in the nature of the ground state excitation at some μ_R. Indeed, Fig. 1 suggests a level crossing at $\mu/T \sim 4.0$. This would mean that the longest correlation length in the thermal system does not get arbitrarily short with increasing density, but rather stays at a constant level. Note that the value of m/T at this crossing is already so large that the effective theory may be inaccurate quantitatively, and in fact in the full 4d theory the flattening off could take place much earlier. However, the qualitative effect should be the same.

Next, let us discuss the applicability of the power series ansatz in Eq. (6). To this end we perform fits over a range $|\mu| = 0...\mu^{max}$ to the inverses of the longest correlation lengths, both for real and imaginary μ.

The results are shown for the 0^+ channel in Table 2, and for the 0^- channel in Table 3. Examining these fits we see that in all cases we have good fits, as demonstrated by the low χ^2/dof and good Q values. In the case of real μ we find stable and well constrained values for the coefficients as we increase the size of the fitting range. For imaginary μ, due to the breakdown of the effective theory at large values of μ_I/T, we have fewer significant data points, and consequently the coefficient of the quartic term is much less constrained.

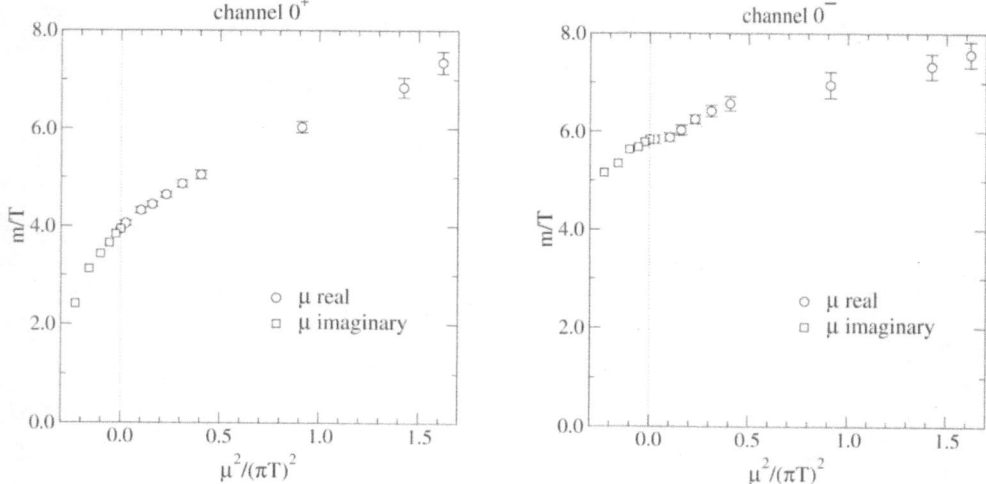

Fig. 2. (*Left*) Inverses of longest correlation lengths in the channel 0^+, for real and imaginary μ. (*Right*) the same for 0^-.

As our main result, we can now state that we observe good evidence for analytic continuation in the first non–trivial term, with c_1^R consistent with c_1^I in the 0^+ channel and similarly for the 0^- states. Unfortunately, the data is not accurate enough to make a similar statement for c_2^R, c_2^I. Extremely precise measurements would be needed, because the range in $|\mu/T|$ available to imaginary chemical potential simulations is very limited. On the other hand, from the phenomenological point of view the first non-trivial coefficient is sufficient, since the series expansion turns out to be in powers of $\mu^2/(\pi T)^2$, which is small in the most important practical applications.

4   Conclusions

In this project, we have studied the question as to what extent imaginary chemical potential simulations could be useful for determining the properties of the quark gluon plasma phase at high temperatures and finite densities. The physical observables we have measured are static bosonic correlation lengths.

The method we have used is based on a dimensionally reduced effective field theory. This way we can address both a system with a real and an imaginary chemical potential, as long as their absolute values are relatively small compared with the temperature. For larger absolute values of $\mu/T = i\mu_I/T$, there is a (first order) phase transition, and the effective description breaks down.

Despite the fact that we are only working in the quark-gluon plasma phase, we find an interesting structure in the longest correlation length, which

Table 2. Fitting the lowest masses in the channel 0^+ from $\mu = 0$ up to $\mu = \mu^{\max}$. The numbers in parentheses indicate the error of the last digit shown, the coefficients refer to Eq. (6), the sub and superscripts R, I denote real or imaginary μ, and Q is the quality of the fit.

$\mu_R^{\max}/T$	c_0^R	c_1^R	c_2^R	χ^2/dof	Q
1.50	3.952 (37)	3.89 (99)	−3.92 (449)	0.175	0.840
2.00	3.956 (35)	3.54 (52)	−2.06 (144)	0.145	0.965
3.00	3.965 (32)	3.22 (27)	−1.06 (33)	0.216	0.956
4.00	3.983 (30)	2.94 (20)	−0.61 (16)	0.607	0.751
$\mu_I^{\max}/T$	c_0^I	c_1^I	c_2^I	χ^2/dof	Q
1.25	3.952 (38)	4.73 (157)	−3.07 (933)	0.090	0.914
1.50	3.925 (35)	2.64 (96)	−16.89 (443)	1.004	0.390

Table 3. Fitting the lowest masses in the channel 0^-. The notation is as in Table 2.

$\mu_R^{\max}/T$	c_0^R	c_1^R	c_2^R	χ^2/dof	Q
1.50	5.839 (69)	−0.54 (167)	10.33 (722)	0.029	0.971
2.00	5.804 (63)	1.22 (91)	2.06 (246)	0.429	0.788
3.00	5.770 (57)	2.18 (47)	−0.90 (65)	0.655	0.658
4.00	5.782 (54)	2.01 (35)	−0.60 (23)	0.546	0.800
$\mu_I^{\max}/T$	c_0^I	c_1^I	c_2^I	χ^2/dof	Q
1.25	5.818 (71)	0.36 (195)	−16.36 (1087)	0.298	0.742
1.50	5.857 (65)	2.57 (116)	−2.53 (465)	0.858	0.462

decreases first but becomes constant beyond some real value of μ/T, which we estimate to be $\lesssim 4.0$.

Furthermore, in the region where the effective theory is applicable, we find that direct analytic continuation does seem to provide a working tool for determining correlation lengths. For phenomenological applications, only the first two coefficients in the power series are needed, since we find the expansion parameter to be $\lesssim \mu^2/(\pi T)^2$, which is small in heavy ion collision experiments.

We are thus encouraged to believe that in 4d simulations analytic continuation of imaginary chemical potential results would give physically relevant results if a good ansatz for the μ-dependence is available, and would allow to go closer to T_c determining, e.g., the free energy density and the spatial correlation lengths there.

A   Appendix

A.1   Monte Carlo Algorithm

Our Monte Carlo simulations were performed using the discretised version of eq. (2) (for details, see [1]),

$$S = \beta \sum_{\mathbf{x},i>j} \left(1 - \frac{1}{3} \operatorname{Re} \operatorname{Tr} U_{ij}(\mathbf{x})\right) + 2 \sum_{\mathbf{x}} \operatorname{Tr}\left(\varphi(\mathbf{x})\varphi(\mathbf{x})\right)$$
$$- 2\kappa \sum_{\mathbf{x},i} \operatorname{Tr}\left(\varphi(\mathbf{x})U_i(\mathbf{x})\varphi(\mathbf{x}+\hat{\imath})U_i^\dagger(\mathbf{x})\right) + \lambda \sum_{\mathbf{x}} (2\operatorname{Tr}\left(\varphi(\mathbf{x})\varphi(\mathbf{x})\right) - 1)^2.$$

For the update of the gauge variables we used a combination of heatbath and over-relaxation algorithms as described in detail in [21]. A Cabibbo-Marinari algorithm [22] is used to combine updates of SU(2) subgroups by standard Kennedy-Pendleton algorithms [23] into an update of the SU(3) field. The scalar fields were updated by a combination of heat bath and reflection steps. The algorithm was originally devised for SU(2)+Higgs models [24] and generalised to the SU(3) case in [1], using all eight real components of the scalar field $\varphi(x)$. Over-relaxation (reflection) steps in the update of the scalar field can be easily incorporated, provided the Higgs self-coupling λ is not too large, which would lead to a poor acceptance rate. In our simulations, we achieved acceptance rates of well over 90%.

In our simulations, one "compound" sweep consisted of heatbath (HB) and several reflection (REF) updates of the gauge and scalar fields,

$$1\text{HB}\{U\} + 1\text{HB}\{\phi\} + n_{\text{OR}} \left(\text{REF}\{U\} + \text{REF}\{\phi\}\right). \tag{A.1}$$

In accordance with ref. [24], we chose n_{OR} to be roughly equal to the inverse scalar mass in order to achieve maximum decorrelation. With this choice we found that the average integrated autocorrelation time estimated using the scalar mass was close to one, in agreement with [24]. Measurements were then taken after every compound sweep.

A.2   Platform and Performance

The code for our simulations was written in Fortran90 and run on the NEC-SX32/4 at the HLRS, Universität Stuttgart. From previous calculations [20], an efficient vector code for SU(2) Monte Carlo existed. Furthermore, our memory requirements were typically quite modest, since we simulated theories in three dimensions. For these reasons we felt that the expected gain from a fully parallelised version of the code would not outweigh the effort for its development. We therefore decided to simply convert the code to SU(3) while maintaining its general structure, and run it in single processor vector mode.

With a vector operation ratio (=number of operations processed by vector instructions / number of operations processed by all instructions) of over 99% the vectorisation of our code is nearly optimal, with an average vector length (= number of operations processed by vector instruction / number of executed vector instructions) of about 250. On a single processor, the code then runs at 0.9-1.2 GFLOPS. The memory required for our largest lattices was about 500 MB. Typical job run times were chosen to be about 12 hours, after which we monitored the simulations and checked the need for more statistics. The CPU time consumed for the complete set of calculations is estimated to total 500-700 hours.

References

1. A. Hart, M. Laine and O. Philipsen, Nucl. Phys. B 586 (2000) 443 [hep-ph/0004060].
2. A. Hart, M. Laine and O. Philipsen, Phys. Lett. B 505 (2001) 141 [hep-lat/0010008].
3. For a review, see F. Karsch, Nucl. Phys. B (Proc. Suppl.) 83 (2000) 14 [hep-lat/9909006].
4. M.-P. Lombardo, Nucl. Phys. B (Proc. Suppl.) 83 (2000) 375 [hep-lat/9908006].
5. P. Ginsparg, Nucl. Phys. B 170 (1980) 388; T. Appelquist and R.D. Pisarski, Phys. Rev. D 23 (1981) 2305.
6. S. Nadkarni, Phys. Rev. Lett. 60 (1988) 491; T. Reisz, Z. Phys. C 53 (1992) 169; L. Kärkkäinen, P. Lacock, D.E. Miller, B. Petersson and T. Reisz, Phys. Lett. B 282 (1992) 121; Nucl. Phys. B 418 (1994) 3 [hep-lat/9310014]; L. Kärkkäinen, P. Lacock, B. Petersson and T. Reisz, Nucl. Phys. B 395 (1993) 733.
7. K. Farakos, K. Kajantie, K. Rummukainen and M. Shaposhnikov, Nucl. Phys. B 425 (1994) 67 [hep-ph/9404201].
8. S. Huang and M. Lissia, Nucl. Phys. B 438 (1995) 54 [hep-ph/9411293]; Nucl. Phys. B 480 (1996) 623 [hep-ph/9511383].
9. K. Kajantie, M. Laine, K. Rummukainen and M. Shaposhnikov, Nucl. Phys. B 458 (1996) 90 [hep-ph/9508379]; Phys. Lett. B 423 (1998) 137 [hep-ph/9710538].
10. E. Braaten and A. Nieto, Phys. Rev. Lett. 76 (1996) 1417 [hep-ph/9508406]; Phys. Rev. D 53 (1996) 3421 [hep-ph/9510408].
11. K. Kajantie, M. Laine, K. Rummukainen and M. Shaposhnikov, Nucl. Phys. B 503 (1997) 357 [hep-ph/9704416]; K. Kajantie, M. Laine, J. Peisa, A. Rajantie, K. Rummukainen and M. Shaposhnikov, Phys. Rev. Lett. 79 (1997) 3130 [hep-ph/9708207].
12. M. Laine and O. Philipsen, Nucl. Phys. B 523 (1998) 267 [hep-lat/9711022]; Phys. Lett. B 459 (1999) 259 [hep-lat/9905004].
13. A. Hart and O. Philipsen, Nucl. Phys. B 572 (2000) 243 [hep-lat/9908041].
14. F. Karsch, M. Oevers and P. Petreczky, Phys. Lett. B 442 (1998) 291 [hep-lat/9807035].
15. P. Bialas, A. Morel, B. Petersson, K. Petrov and T. Reisz, Nucl. Phys. B 581 (2000) 477 [hep-lat/0003004].
16. K. Kajantie, M. Laine, A. Rajantie, K. Rummukainen and M. Tsypin, JHEP 9811 (1998) 011 [hep-lat/9811004].

17. J. Cleymans and K. Redlich, Phys. Rev. C 60 (1999) 054908 [nucl-th/9903063]; and references therein.
18. P. Arnold and L.G. Yaffe, Phys. Rev. D 52 (1995) 7208 [hep-ph/9508280].
19. M. Laine and A. Rajantie, Nucl. Phys. B 513 (1998) 471 [hep-lat/9705003].
20. O. Philipsen, M. Teper and H. Wittig, Nucl. Phys. B 469 (1996) 445 [hep-lat/9602006]; Nucl. Phys. B 528 (1998) 379 [hep-lat/9709145].
21. M.J. Teper, Phys. Rev. D 59 (1999) 014512 [hep-lat/9804008].
22. N. Cabibbo and E. Marinari, Phys. Lett. B 119, 387 (1982).
23. K. Fabricius and O. Haan, Phys. Lett. B 143 (1984) 459;
    A.D. Kennedy and B.J. Pendleton, Phys. Lett. B 156 (1985) 393.
24. B. Bunk, Nucl. Phys. B (Proc. Suppl.) 42 (1995) 566.

Solid State Physics

Prof. Dr. Werner Hanke

Institut für Theoretische Physik und Astrophysik, Universität Würzburg
Am Hubland, D-97074 Würzburg

In this book on high-performance computing solid state physics is presented with contributions from quite different areas of quantum many-body physics. The majority of computational physics projects were directed towards an interplay between electronic correlations, interactions between electron and lattice degrees of freedom and an interplay with disorder. As in the last few years the materials investigated were ranging from high temperature superconducting systems and other strongly correlated materials such as the manganites (interesting e.g. because of their magnetic recording possibilities) to semiconductors and their surfaces. The corresponding calculations can be taken as prime examples for the fact that the days of conventional workstations for the computational use in modern solid state physics are over and sophisticated large-scale supercomputors such as the architectures in Stuttgart and Karlsruhe are required.

The first contribution by K. Bernadet, G. Batrouni, M. Troyer and A. Dorneich, entitled "Destruction of Superfluid and Long Range Order by Impurities in Two Dimensional Systems", is a successful and stimulating example for high-performance computing in science.

It deals with interacting bosons in two dimensions in the presence of impurities. Why is this an interesting subject to model on high- performance supercomputers? Two dimensional bosonic models, in particular the so-called Hubbard model with an extremely short range boson-boson interaction have been the subject of intense research in recent years. This is due to the fact that many salient features of two dimensional superconductors and superfluids at very low temperatures can be described by such a theoretical model. In particular, the high temperature superconductors realized in the copper oxide systems are two dimensional superconductors, where the size of the Cooper pairs is small compared to a typical inter-pair distance. In this case the Cooper pairs, which carry the superconducting current, can be treated as bosons moving in a two dimensional environment. Another much studied example is the case of helium atoms absorbed on a surface which can be described by bosons moving on a two dimensional surface environment.

Since realistic physical systems always display disorder, of particular interest in these bosonic systems is then the interplay and competition between interaction and disorder. Of special interest is the case when both disorder and Bose-Bose interacions are strong. Here one much discussed question is whether disorder " always wins" and localizes the quantum mechanical eigenstates into an insulating state. Recently also fascinating experiments on the superconducting insulating transition suggesting the possibility of a material independent, so-called universal, conductance right at the transition have appeared. Ideas based on the disordered bosonic Hubbard model, which was extensively studied in the present project, have previously appeared to support these ideas qualitatively but have not yielded numerical values of the conductance in agreement with experiment. For these reasons numerical simulations of the disordered bosonic Hubbard model and the determination of its various phases are of high current interest. This model has for the first time been numerically exactly solved in the present project using a new Monte Carlo technique, the so-called SSE (Stochastic Series Expansion) technique.

Quantum Monte Carlo (QMC) methods have become very powerful and importants tools in many-body physics. Common basic idea of this method is to assess the thermodynamic properties of physical systems by starting from an arbitrary initial configuration and then using stochastic processes for sampling the statistically relevant regions of configuration space. In contrast to these earlier QMC procedures, the SSE technique is a numerically exact method without any discretization error. It uses so-called non-local updates, hence even large systems of several thousand sites and particles can be sampled efficiently. This is in particular relevant for extracting the above discussed phases in the thermodynamic limit. The SSE is much easier to implement and more general in applicability than other non-local QMC approaches like the so-called loop-algorithm.

On the basis of this new technique and with the help of the supercomputing facilities in Stuttgart many open questions already mentioned above have been answered. For example, the simulations have shown that the so-called checker-board solid (ordered solid) present at strong short-range bosonic interaction for half-filling is completely destroyed by any amount of disorder: "Disorder wins the competition with the interaction." This agrees e. g. with a famous simple earlier argument by Imry and Ma.

Two more many-body works are included in the present publication. One work deals with the density matrix renormalization group algorithm for the interaction of electrons with phonons in many-body electron-ion solid state systems. The main issue here was the phonon Hilbert space reduction in the numerical diagonalisation of quantum many-body systems. This is important, in particular when the Coloumb correlation between the electrons is very strong and not susceptible to any perturbation theory.

The last selected contribution by Ph. Brune and A. P. Kampf deals with a very old topic, namely the possible band to Mott insulator transition in the so-called ionic Hubbard model. The physical interest in this question stems again partly from the physics of the high temperature superconductors. There, parent compounds like $LaCuO_4$ are so-called Mott-Hubbard insulators. In these systems the insulating character is due to a very strong Coloumb correlation, the so-called Hubbard interaction on the copper sites, experienced by the corresponding Cu-(d) electron. This deviates from the band insulators experienced in semi-conductors where the insulating character is already observed in a non-interacting, free-electron like band picture. Here the insulating character is determined entirely by the electronic filling of a unit cell in the solid. The idea of the Brune/Kampf work was to provide a kind of mapping of the transition from this band to the interacting Mott insulating regime.

The "rapporteur" would like to stress that, quite generally, the solid-state projects of 2001 were of very high quality demonstrating the positive impact of supercomputing in this field.

Destruction of Superfluid and Long Range Order by Impurities in Two Dimensional Systems

Karim Bernardet[1], G. George Batrouni[1], Matthias Troyer[2], Ansgar Dorneich[3]

[1] Institut Non-Linéaire de Nice, Université de Nice–Sophia Antipolis, 1361 route des Lucioles, 06560 Valbonne, France
[2] Theoretische Physik, Eidgenössische Technische Hochschule Zürich, CH-8093 Zürich, Switzerland
[3] Inst. f. Theoretische Physik, Univ. Würzburg, 97074 Würzburg, Germany

Abstract. We use the stochastic series expansion (SSE) Quantum Monte Carlo algorithm to examine the effect of impurities, in the form of disordered chemical potential, on the phase diagram of the hardcore bosonic Hubbard model in two dimensions. This model is often used to study the properties of several physical systems such as Helium adsorbed on surfaces and granular superconductors. We show that in two dimensions, no matter how weak the disorder is, it will always destroy the long range density wave order (checkerboard solid) present at half filling and strong near neighbor (nn) repulsion. In addition part of the superfluid phase surrounding the checkerboard solid is also destroyed. We study properties of the glassy phase thus generated at strong nn coupling, and the possibility of other localized phases at weak nn repulsion, i.e. Anderson localization. The SSE algorithm is used to measure several physical quantities such as the superfluid density, energy gaps, equal and unequal time Green functions.

1 Introduction

The two dimensional bosonic Hubbard model has been the subject of intense study these past several years because it is thought to capture many of the important qualitative features of two dimensional superconductors and superfluids at very low temperature. For example, Helium atoms adsorbed on a surface[1] can clearly be described by bosons moving in a two dimensional environment. It is then natural to examine the role of disorder in localizing the wavefunction and producing exotic insulating phases. One such a phase, the bose glass, was predicted[2] and subsequently verified numerically[3,4].

Furthermore, in superconducting systems where the bosonic Cooper pairs are already formed, and where the correlation length is very short ("the size" of the pairs is small as in high T_c superconductors) the Cooper pairs can be treated as bosons without reference to their fermionic constituents.

Another reason for the increased interest in disordered bosonic systems is a set of fascinating experiments on the superconducting-insulating transition suggesting the possibility of a universal conductance right at the transition[5–11]. Several ideas, based on disordered bosonic Hubbard models, have been suggested[2] to explain these results. Extensive numerical simulations[1,12,14] appear to support these ideas qualitatively, although the numerical values of the conductance are not in agreement.

The question of existence of a normal conducting state at zero temperature has regained momentum with recent experimental discoveries[15]. Attempts to explain this phase proceed via models of disordered bosons, see for example [16,17] and references therein.

For these reasons, the numerical simulation of the disordered bosonic Hubbard model and the determination of its various phases are of high current interest.

The paper is organized as follows. In section 2 we present the bosonic Hubbard model and define some terms and in section 3 we introduce the stochastic series expansion (SSE) algorithm. We present some numerical results in section 4 while section 5 is reserved for the conclusions and remarks.

2   The Boson Hubbard model

The hardcore boson Hubbard Hamiltonian is given by

$$H = -t \sum_{\langle i,j \rangle} (a_i^\dagger a_j + a_j^\dagger a_i) - \mu \sum_i n_i$$
$$+ V_1 \sum_{\langle i,j \rangle} n_i n_j + V_2 \sum_{\langle\langle i,k \rangle\rangle} n_i n_k. \tag{1}$$

a_i ($a_i^\dagger$) are destruction (creation) operators of hard–core bosons on site i of a 2–d square lattice, and n_i is the density at site i while μ is the chemical potential. The hopping parameter is chosen to be $t = 1$ to fix the energy scale. V_1 (V_2) is the near (next near) neighbor interaction. In this paper we will concentrate on $V_2 = 0$ although the non-zero case is extremely interesting since it exhibits a supersolid phase[18–20]. This will be left for later.

At half filling, for weak couplings, the ground state of the Hamiltonian is a stable superfluid with a finite critical velocity. Increasing the near neighbor interaction strength, V_1, at half filling increases the cost of having near neighbors and drives a quantum phase transition to a checkerboard solid phase where the sites are alternately occupied and empty. This phase is characterized by a vanishing superfluid density, ρ_s, long range density–density correlations, and a gap in the energy spectrum exhibited, for example, as a vanishing compressibility, κ, and a corresponding plateau in a plot of density ρ versus chemical potential μ[18–20].

The behavior of the boson–Hubbard model away from half-filling is considerably more complex. While the compressibility surely becomes nonzero, so that the state is not a Mott insulator, it is possible that, despite doping, the charge correlations remain long ranged. If the doped holes are mobile and interspersed with the density ordered bosons, one has simultaneous superfluid order. On the other hand, the doped holes might phase separate leaving distinct regions of the lattice with superfluid and charge ordering.

To characterize the different phases, we need to measure several physical quantities. The superfluid phase is characterized by the absence of long range density order and a nonvanishing superfluid (SF) density. The SF density, ρ_s is given by

$$\rho_s = \langle W^2 \rangle / 2t\beta, \tag{2}$$

where W is the winding number of the phase of the boson wavefunction[3,19] and $\beta = 1/kT$. Long range density order (such as in the checkerboard solid) is charaterized by the density-density correlation function, $c(\mathbf{l})$, and the structure factor, $S(\mathbf{q})$, its Fourier transform. They are given by

$$c(\mathbf{l}) = \langle n_{\mathbf{j}+\mathbf{l}} n_{\mathbf{j}} \rangle$$
$$S(\mathbf{q}) = \sum_{\mathbf{l}} e^{i\mathbf{q}\cdot\mathbf{l}} c(\mathbf{l}), \tag{3}$$

where $n_{\mathbf{j}}$ is the occupancy at site $\mathbf{j}$. In the presence of long range order, $S(\mathbf{q})$ will diverge with the system size for a given ordering momentum, $\mathbf{q}^\star$, which characterizes the ordered phase. For example, for checkerboard order, $\mathbf{q}^\star = (\pi, \pi)$.

Two other very useful quantities are the equal time Green function

$$G(|\mathbf{j} - \mathbf{i}|) = \langle a_{\mathbf{j}} a_{\mathbf{i}}^\dagger \rangle, \tag{4}$$

and the Green function in imaginary time,

$$G(\tau) = \langle a_{\mathbf{i},\tau} a_{\mathbf{i},0}^\dagger \rangle. \tag{5}$$

In the superfluid phase, $G(|\mathbf{j} - \mathbf{i}|)$ saturates at a nonzero value for large separations, while $G(\tau)$ tends to zero exponentially thus yielding the excitation energy spectrum.

3   The Monte Carlo Algorithms

3.1   The SSE technique

Since their formulation in the early eighties[21,22] Quantum Monte Carlo (QMC) methods have become very powerful and important tools in many-body physics. The common basic idea of these methods is to access the thermodynamic properties of physical systems by starting from an arbitrary initial configuration and then using a stochastic process for sampling the statistically "relevant" regions of configuration space. The central quantity steering this sampling process is the partition function

$$Z = \mathrm{Tr}(e^{-\beta \hat{H}}), \tag{6}$$

where $\hat{H}$ is the Hamiltonian. Most QMC algorithms split β into discrete "imaginary time slices" $\Delta\tau$ and expand the exponential into a truncated Taylor series in $\Delta\tau$, thereby introducing a discretization error of order $\Delta\tau^n$. The elementary update steps during the sampling process are normally purely *local* in the sense that a new system configuration is generated by taking the current one and by modifying only the state of one single atom (or "site") of the system. This approach is easy to implement but quite inefficient for generating the *globally* uncorrelated configurations which are essential for accurate numerical measurements.

These drawbacks can be overcome using the "stochastic series expansion" (SSE) approach together with a loop-type updating scheme (see [23] and earlier work referenced therein):

- SSE is a numerically exact method without any discretization error.
- It uses non-local updates, hence even large systems of several thousand sites and particles can be sampled efficiently.
- It is much easier to implement and more general in applicability than other non-local QMC approaches like the loop-algorithm[24].

In SSE equation (6) is written as the infinite power series

$$Z = \sum_{\alpha} \sum_{n=0}^{\infty} \frac{(-\beta)^n}{n!} \langle \alpha | \hat{H}^n | \alpha \rangle. \tag{7}$$

where $\{|\alpha\rangle\}$ are arbitrarily chosen complete basis states of the linear space which describes the quantum mechanical properties of the system. It can be shown[23] that the statistically relevant terms of this series are sharply centered around

$$\langle n \rangle \propto V\beta. \tag{8}$$

Therefore, in practical computations, the sum in equation (7) can be truncated at a finite cut-off exponent L proportional to the system's volume V and β without introducing any systematic error. Now we assume that $\hat{H}$ can be decomposed into a sum of elementary "bond" interactions or "vertices" $\hat{H}_{s,t}^{(k)}$ affecting exactly two sites s and t. For example, $\hat{H}_{7,8}^{(1)}$ could symbolize the hopping of a particle from site 7 to site 8, and $\hat{H}_{7,8}^{(2)}$ could describe the electrostatic interaction of two charged particles at sites 7 and 8. Then equation (7) can be written as

$$Z = \sum_{\alpha} \sum_{\{S_L\}} \frac{(-\beta)^n (L-n)!}{L!} \langle \alpha | \prod_{i=0}^{L} \hat{H}_{s_i,t_i}^{(k_i)} | \alpha \rangle, \tag{9}$$

where $\{S_L\}$ denotes the set of all possible concatenations of $n \le L$ elementary bond interactions. The physics behind equation (9) can be visualized in a so-called "world-line" diagram, spanned by the system's spatial dimension(s) as the x-axis and the "propagation steps" (i.e. the positions 1 through L in the bond operator string) as y-axis (see figure 1). From figure 1 one can also see very nicely how the non-local configuration update of SSE works: a local change is inserted somewhere on a certain world-line (here it is the removal of a type-2 particle on propagation level 6 of world line 2). Then the change is moved through the web of world-lines and interaction vertices until the starting point is reached again. On every vertex it must be decided on which leg of this vertex to go on. This is implemented such that a path is more likely to be selected the lower the energy of the resulting vertex is ("detailed balance" criterion).

An efficient update mechanism is not the only "ingredient" for a successful QMC method; equally important are efficient estimators for the mesurable physical quantities to be extracted from the simulation. Efficient estimators for many *static* observables – such as energy, heat capacity, and certain correlations and susceptibilities – within the SSE mechanism have been derived by Sandvik[25]. In the new SSE codes used for the calculations presented here, much effort was spent to create fast estimators also for Green functions $G(k,\omega)$[26]. These $G(k,\omega)$ are the essential quantities for measuring *dynamical* properties of the system, i.e. the system's reaction when exposed

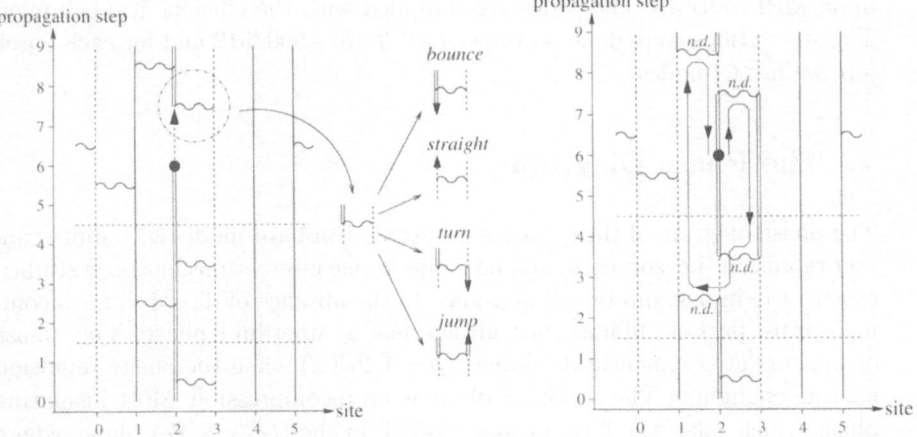

Fig. 1. World-line representation of a system with volume $V = 6$ and with cut-off exponent $L = 9$. The arbitrarily chosen start configuration carries a particle of type 1 on sites 1, 4 and 5 and a particle of type 2 on site 2; sites 0 and 3 are empty. The $L = 9$ propagation steps (i.e. operator string positions) are filled with $n = 7$ elementary bond interactions connecting adjacent sites and symbolized by wavy horizontal lines. In the loop-type update step a local change is inserted somewhere on a world line (left) and then moved through the world lines and interaction vertices. At each vertex a new direction is chosen such that the probability of the new path is proportional to the negative energy of the resulting new interaction vertex ("detailed balance" criterion). The loop closes when the starting position is reached again and the local change is healed up (right).

to external perturbations such as incident photons or neutrons transfering momentum k and energy ω.

To demonstrate the performance of SSE, we study the scaling of the computation time, C, needed to reach an accuracy of 4 digits in the energy as a functions of β and volume, V. The exponents $z_{(\beta)}$ in $C \propto \beta^{z(\beta)}$ and $z_{(N_s)}$ in $C \propto V^{z(V)}$ are found to be: $z_{(\beta)} = 0.34 \pm 0.05$ and $z_{(V)} = 0.48 \pm 0.05$. Both exponents are far below 1, indicating that the growth of computational effort is slower than linear with system size and inverse temperature. Consequently, systems of several thousand sites can be treated on a standard PC or workstation and even larger systems on high-performance computers.

3.2   Parallelization and implementation details

QMC algorithms are very suitable for massively parallel computer architectures since each node of the system can create its own starting configuration and stochastic process; the only inter-node communication is needed at the end of the simulation when the accumulated measurements are collected for the final statistical analysis and the calculation of mean values and variances. In our SSE codes we use the Alea library[27] by M. Troyer and B. Ammon as "parallelization management middleware"; Alea itself is a C++ library based upon MPI routines. The codes are compiled with the efficient Kaï compiler. The simulations were done on the $CRAY$ $T3E - 900/512$ and for each batch job, we use 64 nodes.

4   The Phase Diagram

The phase diagram of the disordered bosonic Hubbard model with finite contact repulsion (i.e. soft core) and no longer range interactions has been studied extensively in one and two dimensions. In the absence of disorder, for incommensurate particle fillings, one always has a superfluid phase. This phase disappears at commensurate fillings ($\rho = 1, 2, 3...$) when the onsite repulsion is large enough[2]. The resulting phase is an incompressible Mott insulating phase which takes the form of lobes[2,4,28] in the ($t/V_0, \mu/V_0$) plane, where V_0 is the onsite repulsion. This phase is gapped, there is a substantial energy cost (increase in the chemical potential) for adding a particle onto a commensurate phase. It was argued[2] that in the presence of any amount of disorder, a new compressible insulating (i.e. localized) phase, the bose glass, is produced at incommensurate fillings and that for strong enough disorder the gapped phase disappears entirely. This was subsequently confirmed numerically[3,4,13,29,30].

The picture changes for hardcore bosons with near neighbor repulsion, V_1. The bosons still form a superfluid for incommensurate particle filling and V_1 not too strong. Also, at full filling, the bosons are always frozen into a Mott insulator since hopping to a neighbor would produce double occupancy which is strictly forbidden. At half filling, increasing V_1 eventually freezes the bosons into an incompressible gapped checkerboard solid: Alternate sites are occupied since the presence of a neighbor costs too much energy and there is a big energy cost (gap) to add a particle. The phase diagram is shown in Fig. 2.

Now we introduce disorder in the form of a random site dependent chemical potential, μ_i. The disorder in μ_i is uniformly distributed between $\pm\Delta$ which is a tuneable parameter characterizing the strength of disorder. The average of μ_i is μ, the normal chemical potential. In other words, the disorder is a random site energy which could attract (repel) bosons for negative (positive) values.

The question is, how strong should disorder be to destroy the solid phase and what replaces this phase. One can try to answer this question with a simple argument based on energy balance (Imry-Ma). We start at half filling with a perfect checkerboard solid and introduce the site disorder. Suppose that at an empty site there is, due to μ_i, a deep potential well which pulls in a neighboring boson. This boson will now have near neighbors which it will try push away to rearrange its neighborhood in a local checkerboard solid which will, consequently, have a mismatch at its boundary with the original checkerboard. The likelihood of this happening depends on the disorder and dimensionality. The energy cost, in d dimensions, due to the mismatch at the boundary scales like L^{d-1} for a region of length L. On the other hand, the energy gained by the bosons by falling into favorable energy wells scales like $L^{d/2}$. For $d = 1$, disorder is always relevant and no matter how weak it is, it always destroys the solid order. For $d = 3$ or larger, the energy cost outweighs the gain and the system maintains checkerboard order. The $d = 2$

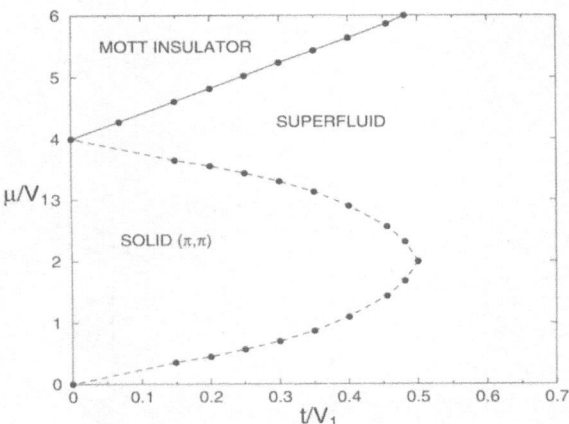

Fig. 2. The phase diagram of the hardcore bosonic Hubbard model in the absence of disorder (8×8, $\beta = 14$). Dashed lines indicate first order transitions, continuous lines second order transitions.

case is marginal since both, cost and gain, scale like L. Typically, in such marginal cases, the conclusion is that disorder will indeed destroy long range order but just barely. The correlation length, ξ, is very long and the system size should be larger than ξ to see the effect.

For very strong disorder ($\Delta/V_1 = 2$) we can show that the indeed the gapped checkerboard solid is destroyed on lattices of size $L = 12$ and is replaced by a compressible, glassy insulator with no energy gap. However our interest is in weak disorder and its ability to destroy order. We found that it is not possible with current algorithms and computers to study large enough systems to see this directly for very weak disorder. Instead we resort to finite size scaling.

In Figure 3 we show the density, ρ, as a function of the chemical potential, μ, for $L = 8, 12, 14$ and $\Delta/V_1 = 1$. For $L = 12$ and 14, we average 100 disorder realizations, for $L = 8$ we did 400. We see that the incompressible gapped region ($\kappa = \partial\rho/\partial\mu = 0$) gets smaller as L increases but does not quite reach zero. The inset shows the average gap versus L^{-1}. We see clearly that the gap tends to zero for a finite, but large, L. This suggests that, for these values of V_1 and Δ, the gap will disappear by $L = 20$. Since L should be greater than ξ to observe the destruction of the checkerboard order, we estimate from this that $\xi \sim 20$. We are in the process of calculating ξ directly to verify this conclusion.

To elaborate further this size effect, we show in Fig. 4 the distributions of the gap sizes for different disorder realizations for $L = 8, 14$. We see that for $L = 8$, the distribution is quite narrow and peaked at a nonzero value.

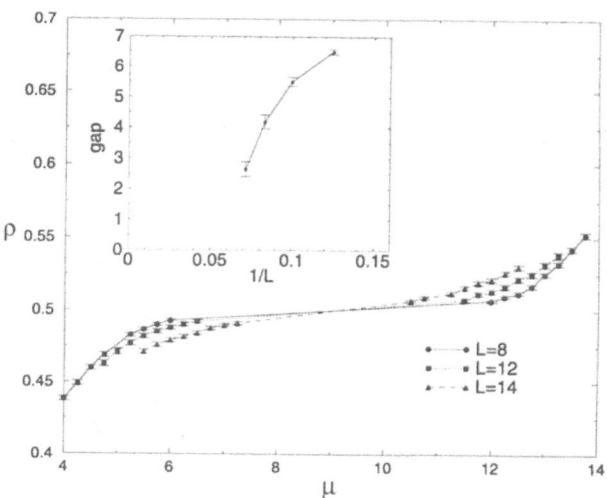

Fig. 3. ρ versus μ for different systems sizes showing the shrinking gap. Inset: The average gap versus L^{-1}.

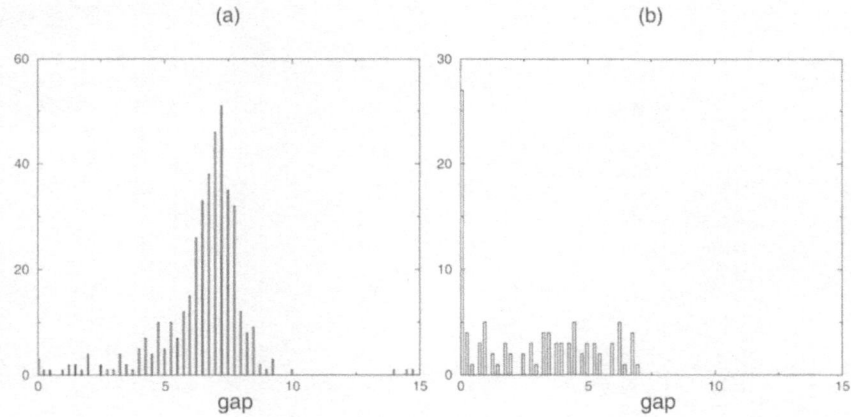

Fig. 4. Gap distribution for different realizations for $L = 8$ (400 realizations) (a) and $L = 14$ (100 realizations) (b). $V_1 = \Delta = 4.5$. As lattice size increases, the distribution gets very wide and peaked at 0.

However, for $L = 14$, the distribution is very wide and in fact peaked at zero indicating that the gap is vanishing for the larger sizes. In this case, it is meaningless to discuss the "average" of the gap, which is still non-zero. The figure shows clearly that in fact for this lattice, the gap is zero.

Clearly, it is very difficult to examine these issues with smaller couplings and disorder: The correlation length will be even longer and much larger sizes would be needed. However, the above results demonstrate that, whereas it might appear on a finite lattice that the gapped solid phase is still present, finite size scaling clearly shows the gap to disappear on large enough systems. We may conclude from this that the Imry-Ma argument is indeed correct and that disorder, no matter how weak, will produce a glassy, compressible, ungapped insulating phase at strong near neighbor couplings.

But what happens for very small (even vanishing) nn couplings ? For softcore bosons, a re-entrant behavior was observed for ρ_s as a function of the contact repulsion[3,4]. In other words, the bosons are localized at very large couplings (the bose glass phase), and as the coupling is reduced, the bosons delocalize and enter a superfluid phase. If the coupling is lowered further, the bosons localize again (Anderson localization). Figure 5 shows that this is not the case for hardcore bosons. While the bosons are localized by disorder for large nn coupling, V_1, as this coupling is reduced, the bosons become superfluid and stay that way even at $V_1 = 0$. In other words, there seems to be no analog of the Anderson localization for hardcore bosons. The simulations were done with a particle density $\rho \approx 0.5$. We have preliminary results that raise the possibility that at lower fillings ($\rho \approx 0.1$) the bosons may be localized also at very small V_1. However more finite size studies are needed to verify this. This is in progress.

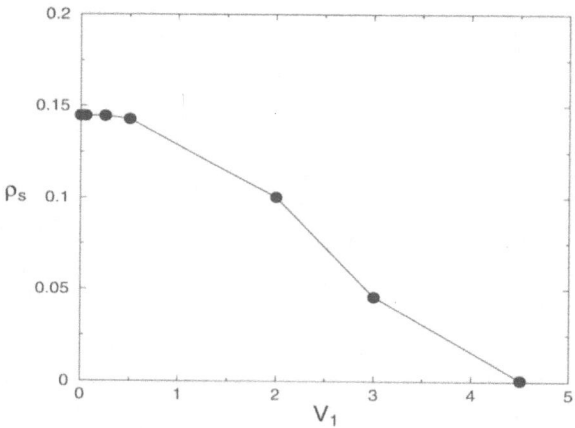

Fig. 5. ρ_s versus V_1 for a $10 X 10$ system. $\Delta = 4.5$, $\beta = 20$ and 250 realizations. ρ_s vanishes only for large V_1. The particle density is $\rho \approx 0.5$.

5   Conclusions and Discussion

We have used the SSE algorithm to simulate the disordered hardcore bosonic Hubbard model in two dimensions to try to understand the interplay and competition between interaction and disorder. We have found that the checker-board solid present at strong nn interactions for half filling is completely destroyed by any amount of disorder. This agrees with the simple Imry-Ma energy balance argument. However, it is still somewhat surprising that we were able to observe this numerically since the two dimensional case is marginal and the correlation length is very long. We had to use finite size scaling and relatively large systems. These simulations are very long since disorder slows down convergence (due to the glassy phase) and also due to the need for averaging over many disorder realizations.

One surprising result we found is that at very weak, even vanishing, nn coupling, disorder does not localize the hardcore bosons. For example, this would mean that Helium adsorbed on a very rough surface is always super-fluid (if one can approximate the $He - -He$ interaction by a pure hardcore). This is in contrast with the softcore case: Disorder appeared to localize the bosons up to very large contact repulsion[3,4] even in the absence of nn interactions. Naively, one would expect the hardcore case to be the extreme limit of the softcore case, and therefore be similarly localized. Why this is not so is not clear. As mentioned above, initial results hint that at lower densities, the bosons may be localized.

Currently we are studying the behavior of the Green functions in space and imaginary time. These will yield a direct estimate of the correlation length to compare with our estimate above. They will also yield the dispersion relations for excitations and superfluid susceptibility which is very interesting to have to compare with theoretical predictions.

Acknowledgments

We acknowledge useful conversations with R. T. Scalettar and H. Rieger.

References

1. G. T. Zimanyi, P. A. Crowell, R. T. Scalettar, G. G. Batrouni, Phys. Rev. **B50** 6515 (1994).
2. M. P. A. Fisher, P. B. Weichman, G. Grinstein, and D. S. Fisher, Phys. Rev. **B40**, 546 (1989).
3. R.T. Scalettar, G. Batrouni, and G.T. Zimanyi, Phys. Rev. Lett. **66**, 3144 (1991).
4. N. Trivedi, D. Ceperley, and W. Krauth, Phys. Rev. Lett. **67**, 2307 (1991), W. Krauth and N. Trivedi, Europhys. Lett. **14**, 627 (1991).
5. B.G. Orr, H.M. Jaeger, A.M. Goldman and C.G. Kuper, Phys. Rev. Lett. **56**, 378 (1986).
6. D.B. Haviland, Y. Liu, and A.M. Goldman, Phys. Rev. Lett. **62**, 2180 (1989).
7. H.M. Jaeger, D.N. Haviland, A.M. Goldman, and B.G. Orr, Phys. Rev. **B34**, 4920 (1986).
8. S.J. Lee and J.B. Ketterson, Phys. Rev. Lett. **64**, 3078 (1990).
9. A.F. Hebard and M.A. Paalanen, Phys. Rev. **B30**, 4063 (1984); Phys. Rev. Lett. **54**, 2155 (1985).
10. L.G. Geerligs, M. Peters, L.E.M. de Groot, A. Verbruggen, and J.E. Mooij, Phys. Rev. Lett. **63**, 326 (1989).
11. R.C. Dynes, J.P. Garno, G.B. Hertel, and T.P. Orlando, Phys. Rev. Lett. **53**, 2437 (1984); A.E. White, R.C. Dynes, and J.P. Garno, Phys. Rev. **B33**, 3549 (1986). R.C. Dynes, A.E. White, J.M. Graybeal, and J.P. Garno, Phys. Rev. Lett. **57**, 2195 (1986).
12. M.C. Cha, M.P.A. Fisher, S.M. Girvin, M. Wallin, and A.P. Young, Phys. Rev. **B44**, 6883 (1991).
13. G.G. Batrouni, B. Larson, R.T. Scalettar, J. Tobochnik and J. Wang, Phys. Rev. **B48** 9628 (1993).
14. M. Wallin, E. S. Søorensen, S. M. Girvin, and A. P. Young Phys. Rev. B 49, 12115 (1994).
15. S. V. Kravchenko et. al. Phys. Rev. **B 50**, 8039 (1994); S. V. Kravchenko et. al. Phys. Rev. **B 51**, 7038 (1995); D. PopevićA. B. Fowler, S. Washburn, Phys. Rev. Lett. **79**, 1543 (1997); S. J. Papadakis and S. J. Shayegan, Phys. Rev. **B 57**, R15068 (1998), E. Ribeiro et. el. Phys. Rev. Lett. **82**, 996 (1999).
16. D. das and S. Doniach, Phys. Rev. **B 60**, 1261 (1999).
17. I. F. Herbut, Phys. Rev. **B 60**, 14503 (1999).
18. G. G. Batrouni, R. T. Scalettar, G. T. Zimanyi and A. P. Kampf, Phys. Rev. Lett. **74** 2527 (1995); R. T. Scalettar, G. G. Batrouni, A. P. Kampf and G. T. Zimanyi, Phys. Rev. **B51**, 8467 (1995).
19. G. G. Batrouni, R. T. Scalettar, Phys. Rev. Lett. **84**, 1599 (2000).
20. G. G. Batrouni, R. T. Scalettar, M. Troyer and A. Dorneich, unpublished.

21. M. Suzuki, Prog. Theor. Phys. **65**, 1454 (1976).
22. J. E. Hirsch, R. L. Sugar, D. J. Scalapino and R. Blankenbecler, Phys. Rev. B **26**, 5033 (1982).
23. A. W. Sandvik, cond-mat/9902226.
24. H. G. Evertz, G. Lana and M. Marcu, Phys. Rev. Lett. **70**, 875 (1993).
25. A. W. Sandvik, R. R. P. Singh and D. K. Campbell, Phys. Rev. B **56**, 14510 (1997).
26. A. Dorneich, M. Troyer, cond-mat/0106471.
27. M. Troyer *et al.*, Lecture Notes in Computer Science 1505, 191 (1998).
28. G. G. Batrouni, R. T. Scalettar et G. T. Zimanyi, Phys. Rev. Lett. **65**, 1765 (1990).
29. K. G. Singh and D. S. Rokhsar, Phys. Rev. **B 46**, 3002 (1992); K. G. Singh and D. S. Rokhsar, Phys. Rev. **B 49**, 9013 (1994).
30. J. Kisker and H. Rieger, Phys. Rev. **B 55**, 11981R (1997), J. Kisker and H. Rieger, Physica **A 246**, 348 (1997).

Density-Matrix Algorithm for Phonon Hilbert Space Reduction in the Numerical Diagonalization of Quantum Many-Body Systems

Alexander Weiße[1], Gerhard Wellein[2], and Holger Fehske[1]

[1] Physikalisches Institut, Universität Bayreuth, D-95440 Bayreuth
[2] Regionales Rechenzentrum Erlangen, Universität Erlangen, D-91058 Erlangen

Abstract. Combining density-matrix and Lanczos algorithms we propose a new optimized phonon approach for finite-cluster diagonalizations of interacting electron-phonon systems. To illustrate the efficiency and reliability of our method, we investigate the problem of bipolaron band formation in the extended Holstein Hubbard model.

1 Introduction

Considerable work is currently focused on the experimental and theoretical study of strongly coupled electron-phonon (EP) systems, triggered by the recognition that the EP interaction plays an important role in understanding the physics of novel materials such as colossal magneto-resistance manganites [1] or the very recently discovered superconducting magnesium diboride [2]. From a theoretical point of view the challenge is to describe the partly exotic properties of these materials in terms of simplified microscopic models which take into account the complex interplay of charge, spin and lattice degrees of freedom.

As a generic model for systems with competing electron-electron and electron-phonon interactions the extended Holstein Hubbard model (EHHM),

$$\mathcal{H} = -t \sum_{\langle i,j \rangle; \sigma} c_{i\sigma}^{\dagger} c_{j\sigma} + U \sum_{i} n_{i\uparrow} n_{i\downarrow} - \sum_{i,l;\sigma} f_l(i) n_{i\sigma} x_0 (b_l^{\dagger} + b_l) + \omega_0 \sum_{i} (b_i^{\dagger} b_i + \tfrac{1}{2}),$$

(1)

is usually considered [3–5], where $c_{i\sigma}^{[\dagger]}$ and $b_i^{[\dagger]}$ annihilates [creates] a spin-σ electron and a phonon at Wannier site i, respectively, and $n_{i\sigma} = c_{i\sigma}^{\dagger} c_{i\sigma}$. The Hamiltonian (1) consists of a kinetic term describing the electronic motion on a discrete lattice (transfer amplitudes t), an extremely screened (on-site) Coulomb repulsion (Hubbard parameter U), and a "density-displacement" type non-screened EP coupling ($\propto \kappa x_0$)

$$f_l(i) = \frac{\kappa}{(|l - i|^2 + 1)^{3/2}} , \quad x_0 = \sqrt{1/2M\omega_0} , \quad \kappa x_0 = \sqrt{\varepsilon_p \omega_0} . \quad (2)$$

Here ω_0 is the bare phonon frequency of dispersionsless optical phonons, being polarized in the direction perpendicular to the chain (1D case). Defining the polaron binding energy as $\tilde{\varepsilon}_p = (x_0^2/\omega_0) \sum_l f_l^2(0) = 1.27\varepsilon_p$, the famous Holstein Hubbard model (HHM) results by setting

$$f_l(i) = \kappa\delta_{i,l}\,, \quad \tilde{\varepsilon}_p \to \varepsilon_p\,, \tag{3}$$

i.e., with respect to the EP coupling term the EHHM represents an extension of the Fröhlich model [6] to a discrete ionic lattice or of the Holstein model [7] including longer ranged EP interactions.

Adapting the EHHM to real physical situations one is frequently faced with the difficulty that the energy scales of electrons (t, U), phonons (ω_0) and their interaction ($\tilde{\varepsilon}_p$) are of the same order of magnitude, causing analytic methods, and especially adiabatic techniques, to fail in most of these cases. For this reason the most reliable results available so far come from powerful numerical calculations, such as finite-cluster exact diagonalizations (ED) [8–10] or (Quantum) Monte Carlo simulations [11,12], which are usually performed on supercomputers. However, even for these numerical approaches strong EP interactions are a challenging task, since they require some cutoff in the phonon Hilbert space. Starting with the work of White [13] in 1993, during the last years a class of algorithms became very popular, which is based on the use of a so-called density matrix for the reduction of large Hilbert spaces to manageable dimensions.

In the present paper, we will demonstrate that finite-cluster diagonalization methods also benefit substantially from these ideas. Along this line, in the next section we introduce an optimized phonon approach for the ED of electron-phonon problems. To exemplify this technique, we analyze the formation of bipolarons in Sec. III. Our main results are summarized in Sec. IV.

2   Optimized phonon approach

Let us first resume the connection between density matrices and optimized basis states. Starting with an arbitrary normalized quantum state

$$|\psi\rangle = \sum_{r=0}^{D_r-1} \sum_{\nu=0}^{D_\nu-1} \gamma_{\nu r}|\nu\rangle|r\rangle \tag{4}$$

expressed in terms of the basis $\{|\nu\rangle|r\rangle\}$ of the direct product space $H = H_\nu \otimes H_r$, we wish to reduce the dimension D_ν of the space H_ν by introducing a new basis,

$$|\tilde{\nu}\rangle = \sum_{\nu=0}^{D_\nu-1} \alpha_{\tilde{\nu}\nu}|\nu\rangle\,, \tag{5}$$

with $\tilde{\nu} = 0 \ldots (D_{\tilde{\nu}} - 1)$ and $D_{\tilde{\nu}} < D_\nu$. The projection of $|\psi\rangle$ onto the corresponding subspace $\tilde{H} = H_{\tilde{\nu}} \otimes H_r \subset H$ is given by

$$|\tilde{\psi}\rangle = \sum_{r=0}^{D_r-1} \sum_{\tilde{\nu}=0}^{D_{\tilde{\nu}}-1} \sum_{\nu'=0}^{D_\nu-1} \alpha_{\tilde{\nu}\nu'}^* \gamma_{\nu'r} |\tilde{\nu}\rangle |r\rangle$$

$$= \sum_{r=0}^{D_r-1} \sum_{\tilde{\nu}=0}^{D_{\tilde{\nu}}-1} \sum_{\nu,\nu'=0}^{D_\nu-1} \alpha_{\tilde{\nu}\nu} \alpha_{\tilde{\nu}\nu'}^* \gamma_{\nu'r} |\nu\rangle |r\rangle . \tag{6}$$

We call $\{|\tilde{\nu}\rangle\}$ an optimized basis, if $|\tilde{\psi}\rangle$ is as close as possible to the original state $|\psi\rangle$. Therefore we minimize $\||\psi\rangle - |\tilde{\psi}\rangle\|^2$ with respect to the elements $\alpha_{\tilde{\nu}\nu}$ of the transformation matrix α under the orthogonality condition $\langle \tilde{\nu}'|\tilde{\nu}\rangle = \sum_{\nu=0}^{D_\nu-1} \alpha_{\tilde{\nu}'\nu}^* \alpha_{\tilde{\nu}\nu} = \delta_{\tilde{\nu}'\tilde{\nu}}$. Applying the latter condition, we find

$$\||\psi\rangle - |\tilde{\psi}\rangle\|^2 = \langle\psi|\psi\rangle - \langle\tilde{\psi}|\psi\rangle - \langle\psi|\tilde{\psi}\rangle + \langle\tilde{\psi}|\tilde{\psi}\rangle$$

$$= 1 - \left[\sum_{r=0}^{D_r-1} \sum_{\tilde{\nu}=0}^{D_{\tilde{\nu}}-1} \sum_{\nu,\nu'=0}^{D_\nu-1} \gamma_{\nu r}^* \alpha_{\tilde{\nu}\nu} \alpha_{\tilde{\nu}\nu'}^* \gamma_{\nu'r} + \text{H.c.} \right]$$

$$+ \sum_{r=0}^{D_r-1} \sum_{\tilde{\nu},\tilde{\nu}'=0}^{D_{\tilde{\nu}}-1} \sum_{\nu,\nu',\nu''=0}^{D_\nu-1} \gamma_{\nu''r}^* \alpha_{\tilde{\nu}'\nu''} \alpha_{\tilde{\nu}'\nu}^* \alpha_{\tilde{\nu}\nu} \alpha_{\tilde{\nu}\nu'}^* \gamma_{\nu'r}$$

$$= 1 - \sum_{r=0}^{D_r-1} \sum_{\tilde{\nu}=0}^{D_{\tilde{\nu}}-1} \sum_{\nu,\nu'=0}^{D_\nu-1} \alpha_{\tilde{\nu}\nu} \gamma_{\nu r}^* \gamma_{\nu'r} \alpha_{\tilde{\nu}\nu'}^*$$

$$= 1 - \text{Tr}(\alpha\rho\alpha^\dagger), \tag{7}$$

where $\rho = \sum_{r=0}^{D_r-1} \gamma_{\nu r}^* \gamma_{\nu'r}$ is called the *density matrix* of the state $|\psi\rangle$ with respect to $\{|\nu\rangle\}$. We observe immediately that the states $\{|\tilde{\nu}\rangle\}$ are optimal if the rows of α are eigenvectors of ρ corresponding to its $D_{\tilde{\nu}}$ largest eigenvalues $w_{\tilde{\nu}}$.

Following Zhang et al. [14], we now apply these features to construct an optimized phonon basis for the eigenstates of an interacting electron/spin-phonon system. Consider a system composed of N sites, each contributing a phonon degree of freedom $|\nu_i\rangle$, $\nu_i = 0 \ldots \infty$, and some other (spin or electronic) states $|r_i\rangle$. Then, the Hilbert space of the model under consideration is spanned by the basis $\{\bigotimes_{i=0}^{N-1} |\nu_i\rangle |r_i\rangle\}$. Of course, to numerically diagonalize an Hamiltonian operating on this space, we need to restrict ourselves to a finite-dimensional subspace. To calculate, for instance, the lowest eigenstates of the HHM (1)-(3) we could limit the phonon space spanned by $|\nu_i\rangle = (\nu_i!)^{-1/2}(b_i^\dagger)^{\nu_i}|0\rangle$ by retaining only the states $\nu_i < D_i$. Most simply we can choose $D_i = M \, \forall \, i$ yielding $D_{\text{ph}} = M^N$ for the dimension of the total phonon space. However, if we think of the states $\{\bigotimes_{i=0}^{N-1} |\nu_i\rangle\}$ as eigenstates of the Hamiltonian $\mathcal{H}_{\text{ph}} = \omega \sum_{i=0}^{N-1} b_i^\dagger b_i$, it is more suitable for most problems to choose an energy cut-off instead. Thus for most of our previous numerical work (see e.g. Ref. [15]) we used the condition $\sum_{i=0}^{N-1} \nu_i < M$, leading to $D_{\mathrm{ph}} = \binom{N+M-1}{N}$. For weakly interacting systems already a small number M of phonon states is sufficient to reach very good convergence for ground states and low-lying excitations. However, with increasing coupling strength most systems require a large number of the above "bare" phonons, thus exceeding capacities of even large supercomputers. In some cases one can avoid these problems by choosing an appropriate unitary transformation of the Hamiltonian (e.g. by using center-of-mass coordinates), but, in general, it is desirable to find an optimized basis automatically.

Within the current density-matrix algorithm [14] for the construction of an optimal phonon basis the phonon subsystem is considered as a product of one "large" ($i = 0$) and a number of "small" sites ($i > 0$). The same optimized basis $\{|\mu_{i>0}\rangle\} = \{|\tilde{\nu}\rangle\}$ with $\tilde{\nu} = 0 \ldots (m-1)$ is used for each except the large site. The basis of the large site consists of the states $\{|\tilde{\nu}\rangle\}$ plus some bare states $\{|\nu\rangle\}$, $\{|\mu_0\rangle\} = \mathrm{ON}(\{|\tilde{\nu}\rangle\} \cup \{|\nu\rangle\})$, where $\mathrm{ON}(\ldots)$ denotes orthonormalization (see Figure 1). After a first initialization the optimized states are improved iteratively through the following steps

(1) calculating the requested eigenstate $|\psi\rangle$ of the Hamiltonian $\mathcal{H}$ in terms of the actual basis,
(2) replacing $\{|\tilde{\nu}\rangle\}$ with the most important (i.e., largest eigenvalues $w_{\tilde{\nu}}$) eigenstates of the density matrix ρ, calculated with respect to $|\psi\rangle$ and $\{|\mu_0\rangle\}$,
(3) changing the additional states $\{|\nu\rangle\}$ in the set $\{|\mu_0\rangle\}$,
(4) orthonormalizing the set $\{|\mu_0\rangle\}$, and returning to step (1).

A simple way to proceed in step (3) is to sweep the bare states $\{|\nu\rangle\}$ through a sufficiently large part of the infinite dimensional phonon Hilbert space. One can think of the algorithm as "feeding" the optimized states with bare phonons, thus allowing the optimized states to become increasingly perfect linear combinations of bare phonon states. Of course, the whole procedure converges only for eigenstates of $\mathcal{H}$ at the lower edge of the spectrum,

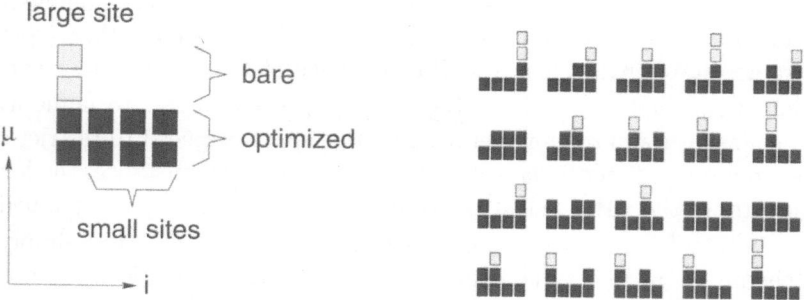

Fig. 1. Structure of the phonon basis in terms of the highest accessible μ_i. Left: as proposed by Zhang et al. [14]; Right: used within this work.

since usually the spectrum of a Hamiltonian involving phonons has no upper bound. The applicability of the algorithm was demonstrated in Ref. [14] with the Holstein model (i.e., $U = 0$ in Eq. (1)) as an example.

When we implemented the above algorithm together with a Lanczos ED method for our systems of interest, we found two objections against the above choice of an optimized basis: (i) the basis is not symmetric under the symmetry operations of the Hamiltonian (e.g. translations), and (ii) the phonon Hilbert space is still large ($D_{\mathrm{ph}} = M\, m^{N-1}$, where M is the dimension at the large site), since we usually need more than one optimized state per site.

The first problem is solved by including all those states into the phonon basis that can be created by symmetry operations (see Figure 1, right panel), and by calculating the density matrix in a symmetric way, i.e., by adding the density matrices generated with respect to every site, not just site $i = 0$. Concerning the second problem we note that the eigenvalues $w_{\tilde{\nu}}$ of the density matrix ρ decrease approximately exponentially, see Figure 3. If we interpret $w_{\tilde{\nu}} \sim \exp(-a\tilde{\nu})$ as the probability of the system to occupy the corresponding optimized state $|\tilde{\nu}\rangle$, we immediately find that the probability for the complete phonon basis state $\bigotimes_{i=0}^{N-1} |\tilde{\nu}_i\rangle$ is proportional to $\exp(-a \sum_{i=0}^{N-1} \tilde{\nu}_i)$. This is reminiscent of the energy cut-off discussed above, and we therefore propose the following choice of phonon basis states at each site,

$$\forall\, i:\ \{|\mu_i\rangle\} = \mathrm{ON}(\{|\mu\rangle\}) \tag{8}$$

$$|\mu\rangle = \begin{cases} \text{opt. state } |\tilde{\nu}\rangle, & 0 \le \mu < m \\ \text{bare state } |\nu\rangle, & m \le \mu < M \end{cases}, \tag{9}$$

and for the complete phonon basis $\left\{\bigotimes_{\Sigma_i \mu_i < M} |\mu_i\rangle\right\}$, yielding $D_{\mathrm{ph}} = \binom{N+M-1}{N}$.

To discuss the nature of the obtained optimized states, the convergence of the algorithm, and some variants in more detail, let us consider a special case of the HHM, Eq. (3), namely the Holstein model of spinless fermions in one dimension,

$$\mathcal{H} = -t \sum_i \left[c_i^\dagger c_{i+1} + \text{H.c.}\right] + g\omega_0 \sum_i (b_i^\dagger + b_i)(n_i - \tfrac{1}{2}) + \omega_0 \sum_i b_i^\dagger b_i. \tag{10}$$

The optimized phonon approach comes into play in the case of strong electron-phonon coupling g. Then the systems develops lattice distortions which accompany the itinerant fermions. These finite elongations need to be expressed in terms of Harmonic oscillator states $|\nu_i\rangle = (\nu_i!)^{-1/2}(b_i^\dagger)^{\nu_i}|0\rangle$, that are centered around the equilibrium position. Hence, for the numerical diagonalization either a large number of these bare states or some other states embodying a finite distortion are required. By sweeping through the large space of bare phonons, the above optimization procedure automatically creates a small number of basis states $|\tilde{\nu}\rangle$, which are sufficient for a good approximation of eigenstates of $\mathcal{H}(t, g, \omega_0)$.

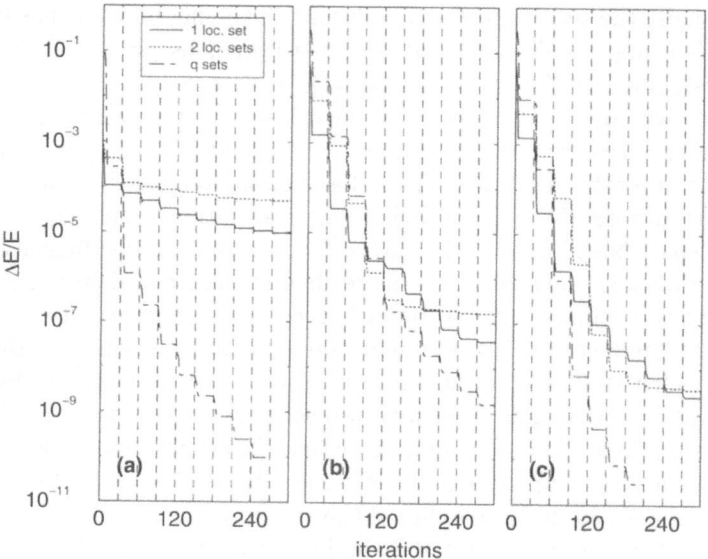

Fig. 2. Convergence of the ground-state energy for the Holstein model, Eq. (10), with electron-phonon coupling $g = 5$ and different phonon frequencies: (a) $\omega_0 = 0.1\,t$, (b) $\omega_0 = t$, and (c) $\omega_0 = 10\,t$. Solid lines: one local set, dotted lines: two local sets for each fermion state, dot-dashed lines: momentum dependent sets.

In Figure 2 the convergence of the ground state energy of a two-site system at half-filling is shown for an increasing number of iterations (solid lines). We compare the results of an ordinary diagonalization using up to $M = 80$ bare phonons per site (i.e., $D_{\mathrm{ph}} = \binom{N+M-1}{N} = 3240$) and of the optimized approach. In the latter case the phonon basis consists of 6 optimized and 4 bare states, i.e., $M = 10$ and $D_{\mathrm{ph}} = 55$. Each optimized state is chosen to be a linear combination of the first 120 bare states. Initially we set $|\tilde{\nu}\rangle = |\nu\rangle$ and then sweep the 4 bare states through the states $|\nu\rangle$ with $\nu = 0 \ldots 119$. Vertical dashed lines denote the end of each sweep. The plateau structure is due to the fact that states of high ν are less important for the optimized basis $|\tilde{\nu}\rangle$. Note that every iteration involves the calculation and diagonalization of the density matrix, an update of the operators $b_i^{(\dagger)}$, which need to be transformed to the current basis, and, most expensive, a Lanczos iteration to obtain a new approximation for the requested eigenstate of $\mathcal{H}$. It is therefore recommended to use the optimized states obtained for a small cluster as the initial basis for a larger cluster, and to restart the Lanczos procedure with the previous eigenvector (although it is expressed in a slightly different basis).

The figure also includes data for two variants of the algorithm, namely the construction of *two* sets of optimized states, one for each local fermion state, or a transfer of the calculation into *momentum space* and the use of

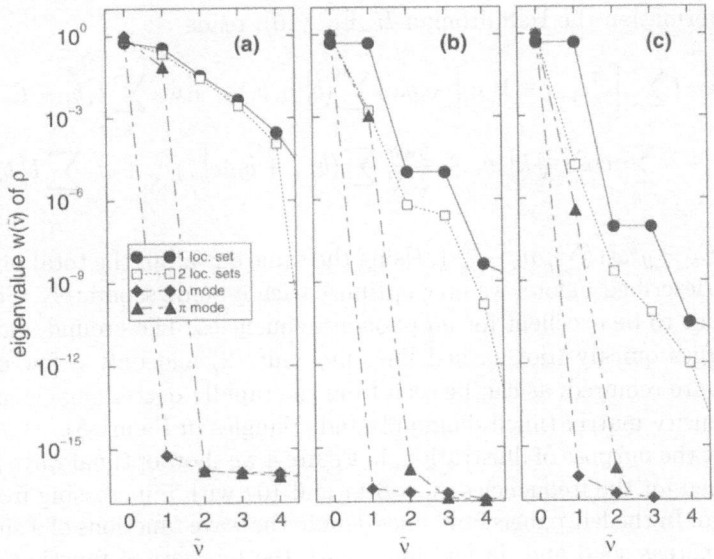

Fig. 3. Eigenvalues $w_{\tilde{\nu}}$ of ρ calculated with the ground state of the Holstein model for $g = 5$ and frequencies: (a) $\omega_0 = 0.1\,t$, (b) $\omega_0 = t$, and (c) $\omega_0 = 10\,t$. Filled circles: one local set; open squares: two local sets for each fermion state; filled diamonds and triangles: momentum dependent sets.

different optimized states for each momentum q. The advantages of these ideas become clear by looking at Figure 3. Here the eigenvalues $w_{\tilde{\nu}}$ of the density matrix ρ are given for different phonon frequencies ω_0 and fixed coupling $g = 5$. In the anti-adiabatic regime of high frequencies we observe pairs of equally important eigenstates of ρ [filled circles in panels (b) and (c)]. This indicates that the lattice immediately follows the fermion (small polaron). It is locally distorted to one side or the other, if the site is occupied by an electron or not. Hence, only half of the optimized states couple to one of the local fermion states, and it appears reasonable to use different basis sets for the two situations. This results in step free exponential decrease of the eigenvalues of the density matrices for both basis sets (open squares in Figure 3), which allows for a smaller cut-off M in the total phonon space of the cluster.

However, at small frequencies the lattice is slow compared to the fermions, and distortions are long ranged. Obviously, there is no gain using different local basis sets, the convergence as well as the decay of the eigenvalues $w_{\tilde{\nu}}$ of the density matrix is slow (see panel (a) in Figures 2 and 3). The performance of each *local* optimization seems to be poor and switching to momentum space is recommended. After a shift of operators, $b_i^{(\dagger)} \rightarrow b_i^{(\dagger)} + \frac{g}{2}$, and Fourier transformation the Hamiltonian $\mathcal{H}$, Eq. (10), reads

$$\mathcal{H} = -t \sum_i \left[c_i^\dagger c_{i+1} + \text{H.c.} \right] + g\omega_0 \sum_i (b_i^\dagger + b_i) n_i + \omega_0 \sum_i b_i^\dagger b_i + E_s \qquad (11)$$

$$= -2t \sum_k \cos\left[\tfrac{2\pi}{N} k \right] n_k + \frac{g\omega_0}{\sqrt{N}} \sum_{k,q} (b_{-k}^\dagger + b_k) c_{q+k}^\dagger c_q + \omega_0 \sum_k b_k^\dagger b_k + E_s ,$$

with $E_s = g^2 \omega_0 \left(\sum_k n_k - \tfrac{N}{4} \right)$. Using the same cut-off in the total phonon basis as described before, we now optimize each q-mode separately. The results turn out to be excellent for *all* phonon frequencies. The ground-state energy converges quickly (dot-dashed lines in Figure 2) and only a few optimized states are required, as can be seen from the rapidly decreasing eigenvalues of the density matrix (filled diamonds and triangles in Figure 3).

For the purpose of illustration, in Figure 4 we show optimal wave functions obtained for the frequencies $\omega_0 = 0.1\,t$ and $10\,t$ with $\tilde{\nu}$ increasing from top to bottom. In the left panels bold lines denote the wave functions of a single local set, whereas solid and dashed lines mark the two sets of functions that depend on the fermion occupation number. The right panels show optimal wave functions in momentum space. In all cases x denotes the normalized elongation $\frac{1}{\sqrt{2}}(b^\dagger + b)$, i.e., the expansion of the optimized states $|\tilde{\nu}\rangle = \sum_\nu \alpha_{\tilde{\nu}\nu} |\nu\rangle$ is plotted using Harmonic oscillator eigenfunctions in elongation space,

$$\langle x | \nu \rangle = \frac{e^{-x^2/2}}{\sqrt{2^\nu \nu! \sqrt{\pi}}} H_\nu(x) , \qquad (12)$$

with Hermite polynomials $H_\nu(x)$. In momentum space only one or two states have a non-negligible eigenvalue $w_{\tilde{\nu}}$. Therefore higher optimized states, which do not contribute to the ground state of $\mathcal{H}$, are not expected to be converged. In all cases the most important states ($\tilde{\nu} = 0$) resemble the eigenstates of shifted Harmonic oscillators. If we use a single local set, the states correspond to symmetric and anti-symmetric combinations of left- and right-shifted oscillator functions (recall the step structure in Figure 3). For $\omega_0 = 0.1\,t$ the shift of the real space functions decreases with higher $\tilde{\nu}$. This reflects the fact, that the lattice is slow and fermions move within a distortion. In comparison, the case of high phonon frequency $\omega_0 = 10\,t$ is much simpler, as we deal only with states of almost fixed shift.

To sum up, the example of the two-site Holstein model provides good insight into the properties of the optimized phonon approach. This approach is very efficient for determining an optimal basis within a given phonon Hilbert space. Nevertheless, we are not relieved from choosing the most appropriate decomposition of our model under consideration. Here physical intuition and some knowledge of the model is necessary to decide between real and momentum space, or other special choices for the phonon Hilbert space. As is demonstrated above, the structure of the optimized wave functions and the behavior of the eigenvalues of the density matrix may give some hints.

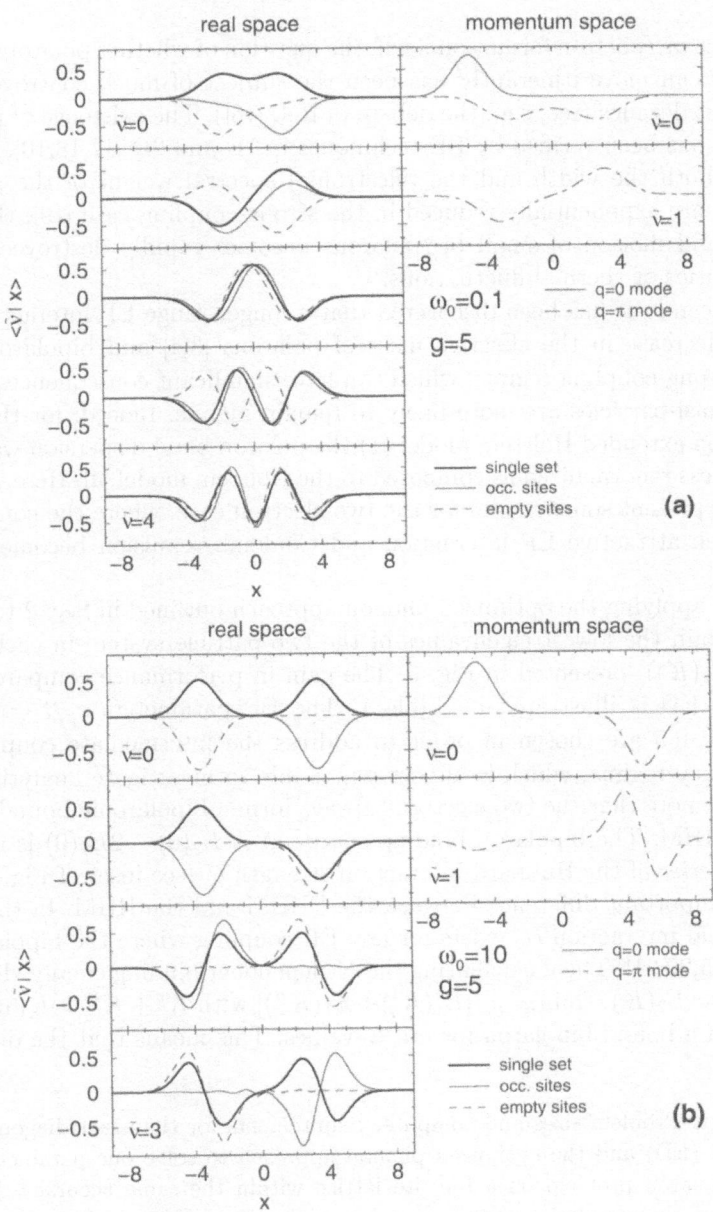

Fig. 4. Optimized phonon states in elongation space for (a) $\omega_0 = 0.1\,t$, and (b) $\omega_0 = 10\,t$, and different choices of the optimized basis (left panels: real space; right panels: momentum space).

3   Bipolaron band formation in the EHHM

Besides bi-/polaron formation itself, the question of whether polarons or bipolarons can move itinerantly has been the subject of much controversy over the last decades (see, e.g., the debate in Ref. [16]). The existence of polaronic bands has been verified by ED techniques in 1D and 2D [17,18,10], however, since both the width and the (electronic) spectral weight of the polaronic bands are exponentially reduced in the strong-coupling case [19], the coherent band motion of small bi-/polarons becomes rapidly destroyed, e.g. by impurities or thermal fluctuations.

Recently it has been discovered that a longer-range EP interaction leads to a decrease in the effective mass of polarons [3,4] and bipolarons [5] in the strong-coupling regime, which can have significant consequences because the quasi-particles are more likely to remain mobile. Indeed, for the single-electron extended Holstein model (1) the polaron band dispersion was shown to be less renormalized as compared to the Holstein model [4]. Here we would like to present some results for the two-electron case, where the competition between attractive EP interaction and Coulomb repulsion becomes important.

By applying the optimized phonon approach outlined in Sec. 2 to EHHM we obtain the lowest eigenvalues of the two-particle system in each K sector $(E_2(K))$, presented in Fig. 5. The gain in performance compared to ordinary ED is illustrated in Table 1. The EP parameters $\varepsilon_p/t = 3.0$ and $\omega_0/t = 0.5$ are chosen in order to address the intermediate coupling and frequency regime, which is almost impossible to investigate analytically. At first we note that the two electrons always form a bipolaronic bound state in the EHHM: The bipolaron binding energy $\Delta = E_2(0) - 2E_1(0)$ is negative, *irrespective* of the Hubbard interaction strength U (see inset of Fig. 5). This is an important difference between the EHHM and the HHM. In the HHM, a critical interaction U_c exists for any EP coupling where the bipolaron unbinds [5,20,21]. Then, calculating the K-dependent binding energy, defined as $\Delta(K) = E_2(K) - \min_{K',K''}[E_1(K') + E_1(K'')]$ with $K' + K'' = K(\bmod 2\pi)$, we find a bound bipolaron for *all* K-values. This means that the dispersion

Table 1. Problem sizes and computer requirements for the exact diagonalization method (ED) and the optimized phonon approach to solve one parameter set of the bipolaron problem on a ten-site lattice within the same accuracy. For both strategies the calculation of the ground-state in a fixed K-sector is assumed.

	phonon cut-off	matrix dimension	Lanczos diagonalizations	memory requirements	CPU time per run
ED	$M = 81$	$\sim 5 \times 10^{13}$	1	$\gg 1000$ TBytes	???
optimized phonons	$M = 13$ $m = 7$	1.8×10^6	50-300	~ 500 MBytes	$\sim 20 - 200$ CRAY T3E hrs

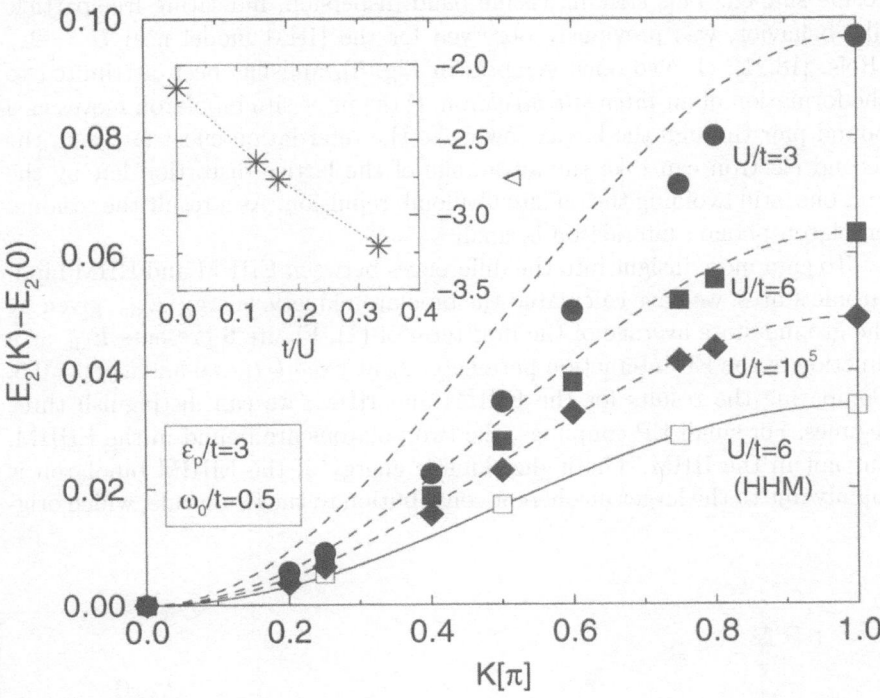

Fig. 5. Bipolaron band structure in the 1D extended Holstein Hubbard model. Data points (filled symbols) are obtained from exact diagonalizations of the EHHM on eight- and ten-site chains (employing periodic boundary conditions) with different Hubbard interaction strengths; dotted lines give the corresponding rescaled cosine bands having the same bandwidths. At $U/t = 6$, the bipolaron band dispersion of the Holstein Hubbard model is included for comparison (open symbols). The inset displays the bipolaron binding energy Δ as a function of the inverse Hubbard interaction strength.

curves depicted in Fig. 5 can be deemed to be well-defined bipolaron quasiparticle bands. Of course, due to the "dressing" with phonons, the effective mass of the bipolaron is substantially enhanced. Accordingly the coherent bandwidth of the EHHM bipolaron, $\Delta E = E_2(\pi) - E_2(0)$, is reduced as compared to the free electron case, but, on the other hand, it is notably larger than that of the HHM bipolaron, indicating that the EHHM bipolaron is a rather mobile quasiparticle. As U increases ΔE monotonously decreases, which clearly is a correlation effect (have in mind that U hinders double occupancy). At last we notice that the band structure of the bipolaron significantly deviates from a rescaled cosine band only for U much smaller than $2\tilde{\varepsilon}_p$. In this case the on-site and nearest-neighbor density-density correlation functions are of about the same size. For $U \gtrsim 2\tilde{\varepsilon}_p$ the band becomes almost cosine shaped. This striking cosine band dispersion, indicating free-particle like behavior, was previously observed for the HHM model near $U = 2\varepsilon_p$ (Refs. [18,21], cf. also open symbols in Fig. 5), and has been attributed to the formation of an *inter-site bipolaron*. If the inter-site bipolaron moves as a bound pair through the lattice, owing to the retardation effect ($\omega_0 < t$), the second electron can take the advantage of the lattice distortion left by the first one, still avoiding the on-site Coulomb repulsion. As a result the residual bipolaron-phonon interaction is small.

To gain more insight into the differences between EHHM and HHM bipolaronic states, we have calculated the bipolaron kinetic energy, E_{kin}, given by the ground-state average of the first term of (1). Figure 6 presents E_{kin} as a function of the EP interaction parameter ε_p at fixed $U/t = 6$ and $\omega_0/t = 0.5$. Comparing the results for the EHHM and HHM, we can distinguish three regimes. For small EP couplings, the two polarons are bound in the EHHM, but not in the HHM. The higher kinetic energy of the EHHM bipolaron is mainly due to the larger incoherent contribution to the f-sum rule, which orig-

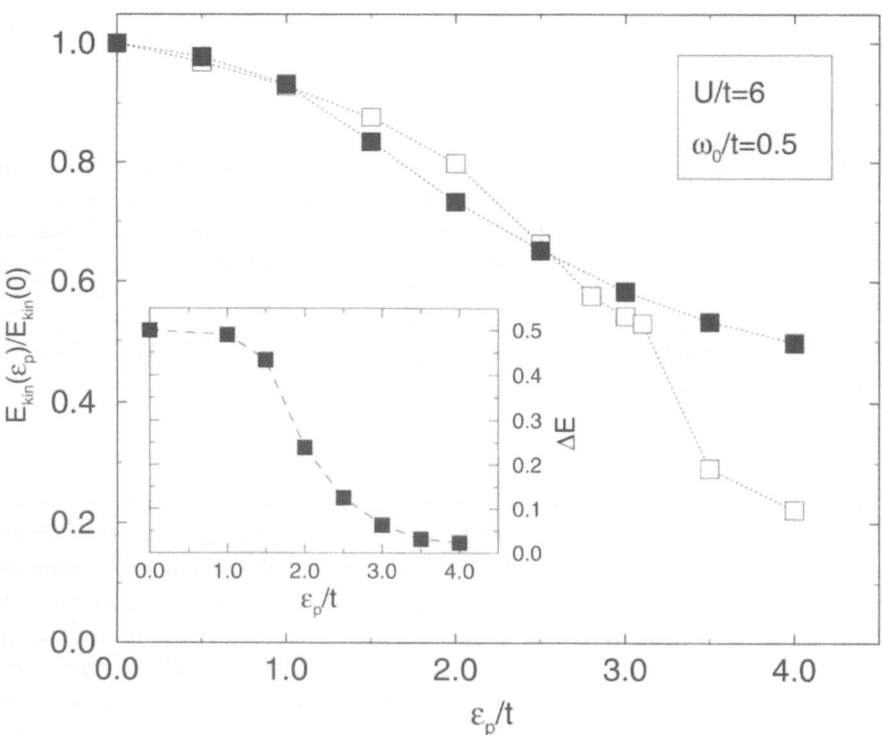

Fig. 6. Renormalized bipolaron kinetic energy in the eight-site EHHM (filled symbols) and HHM (open symbols). The inset gives the coherent bandwidth of the bipolaron in dependence on the EP coupling strength.

inates from incoherent hopping processes of the two charge carriers within the joint lattice distortion spread over the whole lattice. In the intermediate coupling regime predominantly inter-site bipolarons are formed in both models. Now the coherent part to the f-sum rule being proportional to the (inverse) effective mass plays a decisive role. The EHHM bipolaron has to drag a larger phonon cloud than the HHM bipolaron coherently through the lattice and therefore acquires a larger effective mass. At strong EP couplings, in the HHM, a second transition to an on-site bipolaronic state takes place at about $\varepsilon_p/t = 3(\simeq U/2)$, whereas the EHHM bipolaron stays in a spatially much more extended state. Consequently we observe a very gradual decrease of the bipolaron kinetic energy for the EHHM. Finally we comment on the renormalization of the coherent bandwidth of the EHHM bipolaron. As in the single polaron case, at small EP couplings the band structure is flattened near the zone boundary by the intersection with the dispersionless phonon branch. Hence we find $\Delta E \simeq 0.5$ for $\varepsilon_p/t \ll 1$. The strong reduction of the bandwidth starting to come in at about $\varepsilon_p/t \gtrsim 1$ can be traced back to the formation of a bipolaron with pronounced nearest-neighbor correlations.

4 Summary

The objective of this work was the presentation of an advanced phonon optimization algorithm for application in Lanczos diagonalizations. At the heart of our procedure an optimized phonon basis is created automatically, improving the most important eigenstates of the density matrix by means of gradual sweeps through a sufficiently large part of the infinite dimensional phonon Hilbert space. Within this scheme the density matrix is calculated in compliance with the symmetries of the underlying model. Depending on the physical problem under consideration, the efficiency of the proposed method might be improved considerably using more than one set of optimized phonon states or by performing the optimization procedure in momentum space.

The reliability of our approach was demonstrated by calculating the band dispersion of bipolarons in the framework of the extended Holstein Hubbard model. In comparison with the Holstein Hubbard model where with increasing strength of the EP interaction a sequence of transitions from two unbound large polarons to an inter-site bipolaron and finally to a self-trapped on-site bipolaron takes place, in the EHHM the two electrons form a bipolaronic bound state for all EP couplings, irrespective of the magnitude of the Hubbard interaction. As an effect of the longer ranged non-screened EP coupling included in the EHHM, the EHHM bipolaron is a rather mobile, spatially extended quasiparticle even in the extreme strong-coupling regime.

Acknowledgment

The work was supported by the *Competence Network for Technical and Scientific High Performance Computing in Bavaria* (KONWIHR). Special thanks go to LRZ München, NIC Jülich and HLR Stuttgart for the generous granting of their parallel computer facilities. The authors acknowledge the hospitality at the Theoretical Division of Los Alamos National Laboratory where part of this work was performed.

References

1. S. Jin, T. H. Tiefel, M. McCormack, R. A. Fastnach, R. Ramesh, and L. H. Chen, Science **264**, 413 (1994).
2. J. Nagamatsu, N. Nakagawa, T. Muranaka, Y. Zenitani, and J. Akimitsu, Nature **410**, 63 (2001).
3. A. S. Alexandrov and P. E. Kornilovitch, Phys. Rev. Lett. **82**, 807 (1999).
4. H. Fehske, J. Loos, and G. Wellein, Phys. Rev. B **61**, 8016 (2000).
5. J. Bonča and S. A. Trugman, arXiv:cond-mat/0103457 (2001).
6. H. Fröhlich, Adv. Phys. **3**, 325 (1954).
7. T. Holstein, Ann. Phys. **8**, 325 (1959).
8. J. Ranninger and U. Thibblin, Phys. Rev. B **45**, 7730 (1992).
9. A. S. Alexandrov, V. V. Kabanov, and D. K. Ray, Phys. Rev. B **49**, 9915 (1994).
10. G. Wellein and H. Fehske, Phys. Rev. B **56**, 4513 (1997).
11. H. D. Raedt and A. Lagendijk, Phys. Rev. Lett. **49**, 1522 (1982).
12. E. Berger, P. Valášek, and W. v. d. Linden, Phys. Rev. B **52**, 4806 (1995).
13. S. R. White, Phys. Rev. B **48**, 1993 (1993).
14. C. Zhang, E. Jeckelmann, and S. R. White, Phys. Rev. Lett. **80**, 2661 (1998).
15. B. Bäuml, G. Wellein, and H. Fehske, Phys. Rev. B **58**, 3663 (1998).
16. Y. Bar-Yam, T. Egami, J. M. de Leon, and A. R. Bishop, *Lattice Effects in High-T_c Superconductors*, World Scientific, (Singapore 1992).
17. W. Stephan, Phys. Rev. B **54**, 8981 (1996).
18. G. Wellein, H. Röder, and H. Fehske, Phys. Rev. B **53**, 9666 (1996).
19. H. Fehske, J. Loos, and G. Wellein, Z. Phys. B **104**, 619 (1997).
20. J. Bonča, T. Katrašnik, and S. A. Trugman, Phys. Rev. Lett. **84**, 3153 (2000).
21. A. Weiße, H. Fehske, G. Wellein, and A. R. Bishop, Phys. Rev. B **62**, R747 (2000).

Single Hole Dynamics in Correlated Insulators

Michael Brunner[1,2], Catia Lavalle[1], Sylvain Capponi[1,3], Martin
Feldbacher[1], Fakher F. Assaad[1], and Alejandro Muramatsu[1]

[1] Institut für Theoretische Physik III, Universität Stuttgart, Pfaffenwaldring 57,
D-70550 Stuttgart, Germany
[2] School of Physics, The University of New South Wales, Sydney, NSW 2052,
Australia
[3] Université Paul Sabatier, Laboratoire de Physique Quantique, 118 route de
Narbonne, 31062 Toulouse, France

Abstract. We present recent quantum Monte Carlo results for two canonical
model Hamiltonians of strongly correlated electrons, the t-J and the Kondo lattice models. Ground state properties on systems sizes up to 24×24 are computed
numerically. Both models at a particular band filling are correlated insulators. Here,
we concentrate on the dynamics of single hole doped into this state. For the t-J
model we show that two-leg ladder systems can be well understood from a strong
coupling limit along the rungs. In the three-leg ladder system, the low-energy spectrum can be described as an effective chain and an effective two-leg ladder for
the symmetric and antisymmetric band, respectively. In the Kondo lattice model
competing interactions – the Ruderman-Kittel-Kasuya-Yossida (RKKY) interaction and Kondo effect – lead to a quantum phase transition between ordered and
disordered magnetic states. Analysis of the single-hole dispersion relation shows
that the RKKY interaction and Kondo effects coexist: impurity spins are partially
screened and the remnant magnetic moment orders.

1 Introduction

The quantum mechanical many body problem remains one of the outstanding
issues in solid state physics. In many cases, the interaction leads to elementary excitations which cannot be understood in terms of the excitations of
the corresponding non-interacting system. For example in the limiting one-
dimensional case spin and charge excitations separate. The elementary excitations are not quasiparticle carrying spin 1/2 and charge 1 but spinons
carrying spin 1/2 and no charge and holons with charge 1 and no spin. In
this article we will consider two canonical models of correlated electron systems. In both models, the complication arises due to the interplay of spin
and charge degrees of freedom. It has been suggested that this interplay can
lead to anomalous metallic states (i.e. non Fermi liquid) as well as to superconductivity as realized in high temperature superconductors.

The models we consider here are the t-J model and Kondo lattice model.
The t-J model reads:

$$H_{t-J} = -t \sum_{<i,j>,\sigma} \tilde{c}_{i,\sigma}^{\dagger} \tilde{c}_{j,\sigma} + J \sum_{<i,j>} \left(\boldsymbol{S}_i \cdot \boldsymbol{S}_j - \frac{1}{4} \tilde{n}_i \tilde{n}_j \right). \tag{1}$$

Here, $\tilde{c}_{i,\sigma}^\dagger$ are projected fermion operators $\tilde{c}_{i,\sigma}^\dagger = (1 - c_{i,-\sigma}^\dagger c_{i,-\sigma})c_{i,\sigma}^\dagger$, $\tilde{n}_i = \sum_\alpha \tilde{c}_{i,\alpha}^\dagger \tilde{c}_{i,\alpha}$, $\boldsymbol{S}_i = (1/2)\sum_{\alpha,\beta} c_{i,\alpha}^\dagger \boldsymbol{\sigma}_{\alpha,\beta}c_{i,\beta}$, $\boldsymbol{\sigma}_{\alpha,\beta}$ are the Pauli matrices, and the sum runs over nearest neighbors only. In two dimensions, on a square lattice, this model is believed to capture the physics of high temperature super-conductors. The t-J model corresponds to the limiting case of the Hubbard model when the onsite Coulomb repulsion is assumed to be large in comparison to the band-width. At half-filling (one-electron per lattice site) the model describes an insulator since charge degrees of freedom are localized due to the Coulomb repulsion. The remaining spin degrees of are freedom are described by the Heisenberg model. Depending upon the lattice topology the spin degrees of freedom will be gapped (even-leg ladders) or ungappedd (odd-leg ladders and two dimensional case). The nature of the metallic state evolving by doping this correlated or Mott insulator is of great interest. As a first step we will consider here the single-hole problem.

The Kondo lattice model (KLM) is defined as:

$$H_{KLM} = \sum_{\boldsymbol{k},\sigma} \varepsilon(\boldsymbol{k})c_{\boldsymbol{k},\sigma}^\dagger c_{\boldsymbol{k},\sigma} + J\sum_i \boldsymbol{S}_i^c \cdot \boldsymbol{S}_i^f. \tag{2}$$

The physics under consideration is that of a lattice of magnetic spin-1/2 impurities ($\boldsymbol{S}_i^f$) embedded in a metallic host described by the conduction band ($\varepsilon(\boldsymbol{k})$). In the above equation, $\boldsymbol{S}_i^c = (1/2)\sum_{\alpha,\beta} c_{i,\alpha}^\dagger \boldsymbol{\sigma}_{\alpha,\beta}c_{i,\beta}$ and $c_{\boldsymbol{k},\sigma}^\dagger$ creates an electron with z-component of spin σ and crystal momentum $\boldsymbol{k}$. Experimentally the magnetic impurity stems from atoms with partially filled f or d shells such as Ce^{3+} which has a single electron in the f shell. In much the same way as the mapping of the Hubbard model to the t-J model, the KLM may be derived from the periodic Anderson model when the Coulomb interaction is strong enough to inhibit charge fluctuations on, for example, singly occupied f shells. This charge localization generates the magnetic impurity. The KLM is expected to describe the physics of heavy-electron materials such as $CeCu_6$ and so-called Kondo insulators. Those materials are characterized by an effective mass which is up to three orders of magnitude larger than in generic Fermi liquids as realized in Au. Due to this large effective mass those materials are equally characterized by a low coherence temperature. Heavy fermions show an extremely rich variety of ground states ranging from insulators, anomalous metallic states, to superconducting states with large transition temperatures in comparison to the coherence temperature. It is at present not clear if the above simplified model can account for all the above mentioned phases. Here, we will concentrate on a specific issue which is the interplay and competition between the RKKY interaction which leads to magnetic ordering and Kondo screening which leads to a non-magnetic ground state.

2   Algorithm and implementation

Before discussing the physical properties of both models, let us discuss some technical aspects of the numerical simulations.

For the KLM, we use the auxiliary field QMC algorithms [1]. In this approach, the many body interaction is decoupled with a Hubbard-Stratonovitch field and the summation over the field is carried out with Monte-Carlo methods. Auxiliary field QMC simulations of the KLM as well as the two-impurity Kondo model have already been carried out by Fye and Scalapino [2,3]. However, their formulation leads to a sign problem even in the half-filled case where the model is invariant under a particle-hole transformation. We have been able to find an alternative formulation which does not lead to the infamous sign problem.

In order to achieve our goal, we take a detour and consider the Hamiltonian:

$$H = \sum_{k,\sigma} \varepsilon(k) c_{k,\sigma}^\dagger c_{k,\sigma} - \frac{J}{4} \sum_i \left[\sum_\sigma c_{i,\sigma}^\dagger f_{i,\sigma} + f_{i,\sigma}^\dagger c_{i,\sigma} \right]^2 . \tag{3}$$

At vanishing chemical potential this Hamiltonian has all the properties required to formulate a sign-free auxiliary field QMC algorithm. Here, we are interested in ground-state properties of H which we obtain by filtering out the ground state $|\Psi_0\rangle$ by propagating a trial wave function $|\Psi_T\rangle$ along the imaginary time axis:

$$\frac{\langle \Psi_0 | O | \Psi_0 \rangle}{\langle \Psi_0 | \Psi_0 \rangle} = \lim_{\Theta \to \infty} \frac{\langle \Psi_T | e^{-\Theta H} O e^{-\Theta H} | \Psi_T \rangle}{\langle \Psi_T | e^{-2\Theta H} | \Psi_T \rangle} \tag{4}$$

The above equation is valid provided that $\langle \Psi_T | \Psi_0 \rangle \neq 0$ and O denotes an arbitrary observable.

To see how H relates to H_{KLM} we compute the square in Eq. (3) to obtain:

$$H = \sum_{k,\sigma} \varepsilon(k) c_{k,\sigma}^\dagger c_{k,\sigma} + J \sum_i S_i^c \cdot S_i^f - \frac{J}{4} \sum_{i,\sigma} \left(c_{i,\sigma}^\dagger c_{i,-\sigma}^\dagger f_{i,-\sigma} f_{i,\sigma} + \text{H.c.} \right)$$

$$+ \frac{J}{4} \sum_i \left(n_i^c n_i^f - n_i^c - n_i^f \right) . \tag{5}$$

As apparent, there are only pair-hopping processes between the f- and c-sites. Thus the total number of doubly occupied and empty f-sites is a conserved quantity:

$$[H, \sum_i (1 - n_{i,\uparrow}^f)(1 - n_{i,\downarrow}^f) + n_{i,\uparrow}^f n_{i,\downarrow}^f] = 0. \tag{6}$$

If we denote by Q_n the projection onto the Hilbert space with $\sum_i (1 - n_{i,\uparrow}^f)(1 - n_{i,\downarrow}^f) + n_{i,\uparrow}^f n_{i,\downarrow}^f = n$ then:

$$H Q_0 = H_{KLM} + \frac{JN}{4} \tag{7}$$

since in the Q_0 subspace the f-sites are singly occupied and hence the pair-hopping term vanishes. Here, N corresponds to the number of unit cells. Thus, it suffices to choose

$$Q_0|\Psi_T\rangle = |\Psi_T\rangle \tag{8}$$

to ensure that

$$\frac{\langle\Psi_T|e^{-\Theta H}Oe^{-\Theta H}|\Psi_T\rangle}{\langle\Psi_T|e^{-2\Theta H}|\Psi_T\rangle} = \frac{\langle\Psi_T|e^{-\Theta H_{KLM}}Oe^{-\Theta H_{KLM}}|\Psi_T\rangle}{\langle\Psi_T|e^{-2\Theta H_{KLM}}|\Psi_T\rangle}. \tag{9}$$

Since it is possible to compute the left hand side of Eq. 9 without generating a sign problem it follows that we are able to simulate the KLM. The reader is referred to Refs. [4–6] for the details of the algorithm.

The Monte Carlo algorithm used for the simulation of the t-J model is based on a world-line-loop algorithm for the spins, and an exact propagation of the holes in each given spin configuration. The quality of the results in imaginary is sufficient to obtain reliable dynamical information via the *Maximum Entropy* method [7]. More details about the algorithm can be found in Ref. [8], the implementation on massive parallel computers is illustrated in the previous issue [9] of this series.

3   Single hole dynamics in t-J ladder systems

Ladder systems can be considered as a bridge between one-dimensional chains and two-dimensional planes. Well known examples for such systems are cuprate compounds like $Sr_xCu_yO_z$. Depending on their actual composition, these compounds can build chains ($SrCuO_2$ [10]), two-leg ladders ($SrCu_2O_3$ [11]) and three-leg ladders ($Sr_2Cu_3O_5$ [12]). According to the work of F. Zhang and T.M. Rice, the t-J model is a simple, but well suited effective model for the low-energy physics of these cuprates [13].

It is important to note, that in undoped spin systems, the crossover from the chain to the plane is not smooth at all [14–16], but a strong odd-even effect can be observed. At half filling the major difference is the existence of a spin gap for even-leg ladders (i.e. the lowest spin excitation has a finite energy in the thermodynamic limit) and its absence for odd-leg ladders, including the single chain, as has been confirmed by quantum Monte Carlo simulations [17]. It is further known, that there is no spin gap in the two-dimensional system.

The repercussion of this effect on the dynamics of a single hole moving in a spin background will be examined in the following. Here we will concentrate on the spectral function

$$A(\boldsymbol{k},\omega) = \sum_{f,\sigma}|\langle f, N-1|c_{\boldsymbol{k},\sigma}|0,N\rangle|^2 \delta\left(\omega - E_0^N + E_f^{N-1}\right), \tag{10}$$

and the quasiparticle weight

$$Z(\boldsymbol{k}) = |\langle 0, N-1|c_{\boldsymbol{k},\sigma}|0,N\rangle|^2. \tag{11}$$

Here $|0,N\rangle$ is the ground-state at half filling with energy E_0^N and $|f, N-1\rangle$ are states in the $N-1$ particle Hilbert space with energy E_f^{N-1}.

3.1   The spectral function

In Fig. 1, we show our results for the two-leg ladder with $J_\perp/t_\parallel = 1.6$, $J_\parallel/t\parallel = 0.4$, and $t_\perp/t_\parallel = 2$ ($t_\perp$, $J_\perp$, and $J_\parallel$, $t_\parallel$ refers to coupling along the rung and chains, respectively). The low energy physics of the limit where the coupling along the rung is much stronger than along the chain can be well understood by perturbative calculations, starting with isolated singlets on each rung [20–25]. As expected from this limit, the results for the spectral function show a cosine-like dispersion, both in the bonding and in the anti-bonding band (see Fig.1). We find additionally clear evidence of a shadow band in the antibonding channel, which can be derived from the bonding band by a momentum shift of π and an energy shift approximately given by the spin gap ($\sim 0.64t$ [26,27]). The weight of this shadow band increases at the expense of the other band in the antibonding channel, when the isotropic case ($J_\perp = J_\parallel, t_\perp = t_\parallel$) is approached. Additionally, the minimum of the lower edge of the spectral function shifts from $k = \pi$ and $k = 0$ respectively towards $k = \pi/2$ [20,25,28,29].

In Fig. 2 we address the spectral function of the three-leg ladder, with the same values of J and t as used for the two-leg ladder above. In the strong coupling limit, perturbative treatment suggests, that the low-energy behavior corresponds to an effective t-J model, and that these lowest excitations are in the antisymmetric [30] channel. This can, to some extend, be confirmed by our results [see Fig. 2(a)], although the agreement with the expected dispersion of a single chain is not very good, and worsens when the isotropic case is considered [29]. More agreement can be seen with the spectrum of the two-dimensional t-J model, both at low and high energies [8,28,9].

Even more interesting are the results for the two symmetric bands of the three-leg ladder [Fig. 2(c,d)]. They are well separated from the antisymmetric one in energy, and can be well fitted by the bonding band dispersion for the

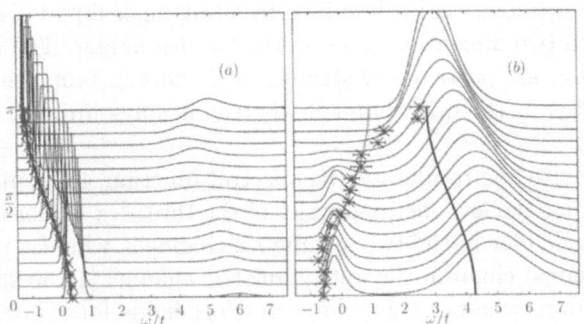

Fig. 1. Quasiparticle dispersion and full spectral function for the two-leg ladder with $J_\perp/t_\parallel = 1.6$, $J_\parallel/t\parallel = 0.4$, and $t_\perp/t_\parallel = 2$. The solid lines correspond to $(0.75\cos k_\parallel - 0.47)t_\parallel$ for the bonding band (a), and $(0.75\cos k_\parallel - 0.47 + 4)t_\parallel$ for the antibonding band (b).

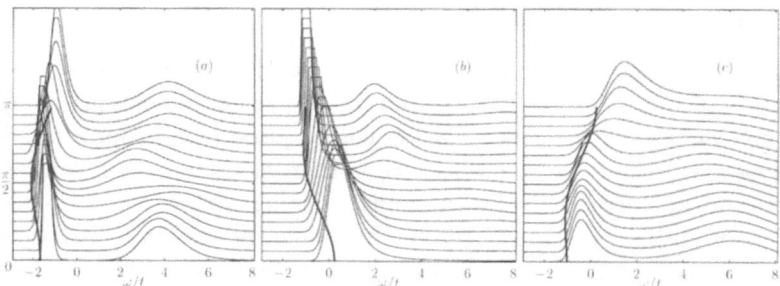

Fig. 2. Spectral function for the three -leg ladder with $J_\perp/t_\| = 1.6$, $J_\|/t_\| = 0.4$, and $t_\perp/t_\| = 2$. The lines correspond to the charge-spin separation *Ansatz* in one dimension [antibonding,(a)][18,19], and to the quasiparticle dispersion obtained for the two-leg ladder [20].

two-leg ladder from perturbation theory [20]. As the two symmetric bands are connected to each other by gapless spin excitations [31,15], there is no shift in energy between them, as it was the case for the two-leg ladder.

3.2   The quasiparticle weight

In this section, we summarize recent results [19,8,29] for the quasiparticle weight in different geometries. The quasiparticle weight is probably the clearest evidence, whether a fermion system shows *normal* Fermi-liquid behavior (finite quasiparticle weight), or non-Fermi-liquid behavior (vanishing quasiparticle weight).

As can be seen from Fig. 3, the quasiparticle weight in the t-J model can exhibit both behaviors, depending on the dimension or topology of the system. It can be either vanishing as in the one-dimensional chain, where its low energy excitations can be described by a Luttinger-liquid picture, or it can be finite, as in two dimensions and on the two-leg ladder. The results on the three-leg ladder are particularly striking, as in this system, the quasiparticle weight vanishes in the odd channel, whereas it stays finite in the two even channels.

Together with the results for the spectral function, this circumstantiates the picture, that the low-energy physics of the three-leg ladder is very similar to the two-leg ladder in the two symmetric channels, whereas it is similar to a one-dimensional chain in the antisymmetric channel [15,30,28,29].

The question, whether this odd-even effect holds for arbitrary wide ladders, and how the crossover to the finite two-dimensional quasiparticle weight occurs in this case can not be answered conclusively. A possible scenario is, that only the lowest band in the odd-leg ladders shows a vanishing quasiparticle weight, and thus this effect disappears with the inverse width of the ladders considered.

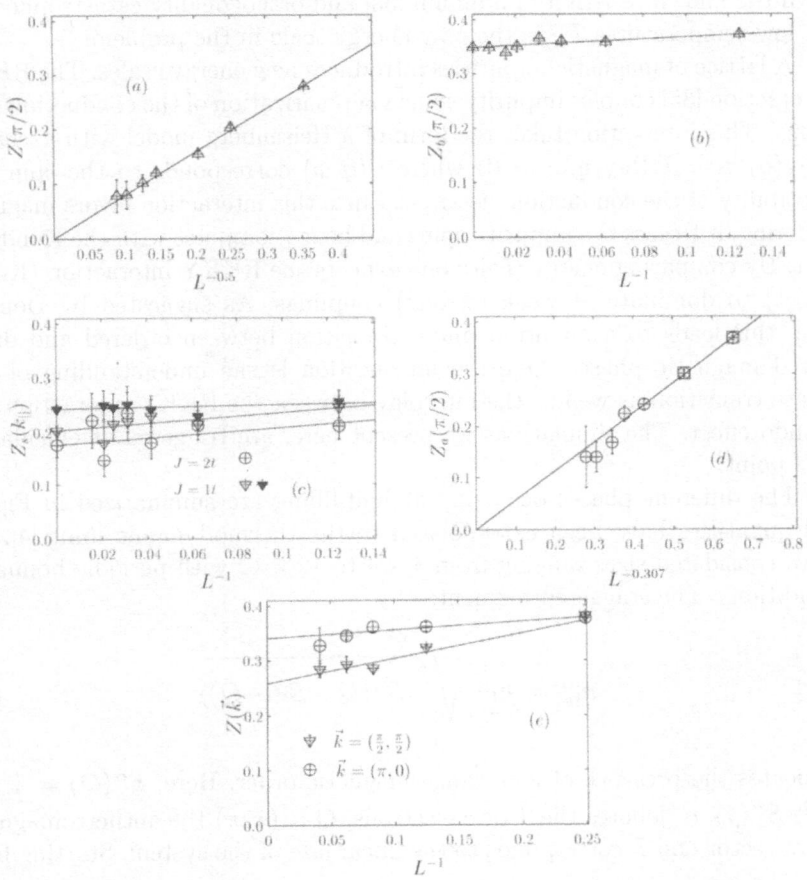

Fig. 3. Quasiparticle weight for a chain (a) at $k = \pi/2$, a two-leg ladder in the bonding band [$k = \pi/2$,(b)], a three-leg ladder both in the symmetric channels [$k = 3\pi/4$ for the bonding band (solid circles) and $k = \pi/4$ in the antibonding band (open circles),(c)] and in the antisymmetric band [$k = \pi/2$,(d)], and in the two-dimensional t-J model at $\boldsymbol{k} = (\pi/2, \pi/2)$ and $\boldsymbol{k} = (\pi, 0)$ (e). $J/t = 2$ in all systems, if not given explicitly. All values of k are at or near the minimum of the dispersion.

4   The half-filled Kondo lattice model

The physics of the single magnetic impurity embedded in a metallic host is well understood [32]. At high temperatures the impurity spin is essentially free thus yielding a Curie law for the impurity spin susceptibility. Below the Kondo temperature $T_K \propto \varepsilon_f e^{-1/JN(\varepsilon_f)}$ the impurity spin is screened by the conduction electrons. Here, ε_f is the Fermi energy and $N(\varepsilon_f)$ the density of states taken at the Fermi energy. The transition from high to low temperatures is non-perturbative and corresponds to the Kondo problem with the known resistivity minimum [33] and orthogonality catastrophe [34]. At low temperatures T_K is the only energy scale in the problem.

A lattice of magnetic impurities introduces new energy scales. The RKKY interaction [35] couples impurity spins via polarization of the conduction electrons. This interaction takes the form of a Heisenberg model with exchange $J_{eff}(q) \propto -J^2 \mathrm{Re}\chi(q, \omega = 0)$ where $\chi(q, \omega)$ corresponds to the spin susceptibility of the conduction electrons. Since this interaction favors magnetic ordering, it freezes the impurity spins and hence competes with the Kondo effect. By comparing energy scales one expects the RKKY interaction (Kondo effect) to dominate at weak (strong) couplings. As suggested by Doniach [36], this leads to a quantum phase transition between ordered and disordered magnetic phases. One crucial question is the understanding of this phase transition as well as the interplay between the RKKY interaction and Kondo effect. The simulations we present here, are triggered at elucidating this point.

The different phases occurring at half-filling are summarized in Fig. 4. All quantities have been extrapolated to the thermodynamic limit [4]. We have considered sizes ranging from 4×4 to 12×12 with periodic boundary conditions. The staggered moment:

$$m_s^\alpha = \lim_{L \to \infty} \sqrt{\frac{4}{3} \langle S^\alpha(Q) \cdot S^\alpha(-Q) \rangle} \tag{12}$$

indicates the presence of long-range magnetic order. Here, $S^\alpha(Q) = \frac{1}{L} \sum_j e^{iQ \cdot j} S^\alpha(j)$, α denotes the f or c-electrons, $Q = (\pi, \pi)$ the antiferromagnetic wave vector and L corresponds to the linear size of the system. Starting from

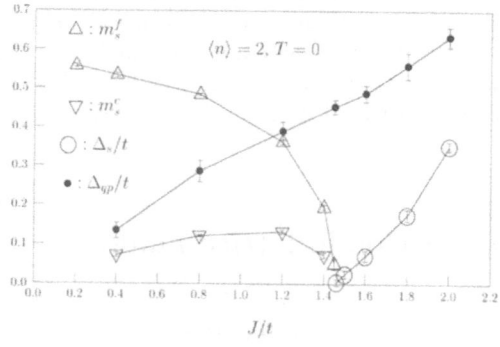

Fig. 4. Staggered moment, m_s, spin gap Δ_s and quasiparticle gap for the ferromagnetic and antiferromagnetic KLM. All quantities have been extrapolated to the thermodynamic limit based on results on lattice sizes up to 12×12. For the staggered moment both conduction and impurity spins are considered separately.

the weak coupling, this quantity vanishes at $J_c/t \sim 1.45$ thus signaling the occurrence of a quantum phase transition.

The onset of a spin gap,

$$\Delta_s = \lim_{L \to \infty} E_0^L(S = 1, N_p = 2N) - E_0^L(S = 0, N_p = 2N), \qquad (13)$$

is observed when magnetic order disappears. Here, $E_0^L(S, N_p)$ is the ground state energy on a square lattice with $N = L^2$ unit cells, N_p electrons and spin S. Finally, the system remains an insulator for all considered coupling constants. This is supported by a non-vanishing quasiparticle gap,

$$\Delta_{qp} = \lim_{L \to \infty} E_0^L(S = 1/2, N_p = 2N + 1) - E_0^L(S = 0, N_p = 2N). \qquad (14)$$

Hence, Fig. 4 shows the competition of the RKKY interaction and Kondo screening indeed leads to a quantum phase transition between ordered and disordered magnetic phases. The origin of the charge and spin gaps is easy to understand in the limit of large J/t where each conduction electron is trapped by an impurity spin to form a Kondo singlet. In contrast in the weak coupling limit the charge gap orginates from magnetic ordering [5]. It is interesting to note that magnetism occurs on a energy scale J^2 in accordance with the RKKY interaction but generates an effect (the charge gap) on a energy scale J.

We now consider the single particle dynamics across the phase transition. Let us start with the strong coupling case, $J/t >> 1$. In this limit the ground state is a direct product of Kondo singlets. This state is very similar to that of the $t - J$ ladder in the strong rung coupling limit. In this limit mean-field calculations give a dispersion relation for the quasiparticle:

$$E_\pm(\mathbf{k}) = \frac{1}{2}\left(\varepsilon(\mathbf{k}) \pm E(\mathbf{k})\right), \text{ with } E(\mathbf{k}) = \sqrt{\varepsilon(\mathbf{k})^2 + \Delta^2}. \qquad (15)$$

The origin of this functional form is Kondo screening. The QMC data in $J/t = 2$ (Fig. 5a) is well reproduced by this form. As we enter the ordered phase, $J/t = 1.2$ (Fig. 5c) one sees that this functional form survives. Clearly this dispersion relation is supplemented by shadows features which originate from the magnetic ordering. The only way to understand the insensitivity of the single particle spectral function across the phase transition is to assume that Kondo screening is still present in the magnetically ordered phase. This means that the magnetic impurities are partially screened and that the remnant magnetic moment orders with the RKKY interaction. It is this mechanism which generates arbitrarily small magnetic moments in the vicinity of the phase transition. We can cross-check this idea by considering the ferromagnetic coupling, $J/t < 0$. In this case Kondo screening is absent. As apparent from Fig. 5d one obtains a completely different functional form for the quasiparticle dispersion relation. The QMC result may be well accounted for by assuming that the RKKY interaction freezes the impurity spin antiferromagnetically. Thus the conduction electrons are effectively subject to an

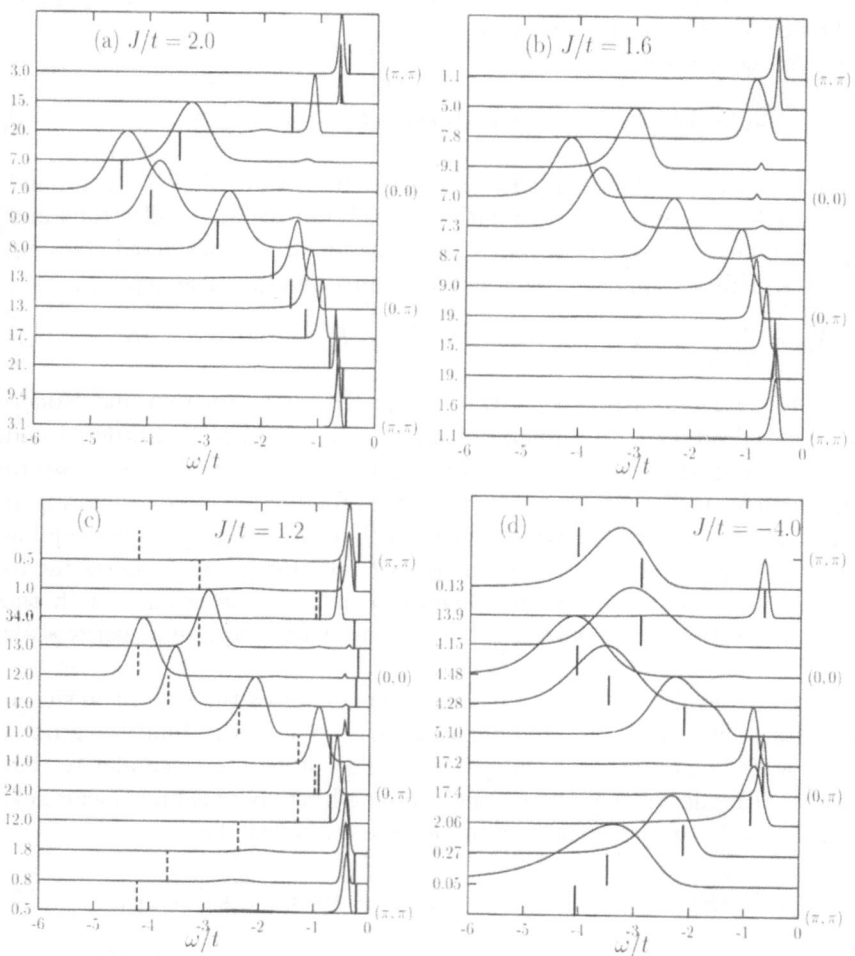

Fig. 5. Single particle spectral function at $T = 0$ for the ferromagnetic and anti-ferromagnetic KLM. We have normalized the peak heights to unity. The numbers on the left hand side of the figures correspond to the normalization factor. The vertical bars are mean-field fits to the data (see text). Figures (a)-(c) correspond to the antiferromagnetic case and figure (d) to ferromagnetic couplings.

external staggered magnetic field which yields the dispersion relation

$$E_\pm(\mathbf{k}) = \sqrt{\varepsilon(\mathbf{k})^2 + (J/4)^2}. \tag{16}$$

The vertical bars in Fig. 5d follow this functional form. It is worth noting that both the KLM at $0 < J < J_c$ and $J < 0$ have long range antiferromagnetic ordering and that magnetic excitations are nothing but spin waves. Hence on the magnetic side, both ferro and antiferromagnetic couplings are indistinguishable. Nevertheless, due to the coupling between the added hole and spin excitations radically different hole dynamics is obtained.

5 Conclusions

In conclusion, we have used two QMC algorithms to compute spectral and static properties of the t-J model and KLM. We have demonstrated that the numerical approach provides invaluable insight into the problem. From the technical point of view the algorithm used are optimal for parallel machines in the sense that a maximal speedup is obtained as a function of nodes. In accordance to the central limit theorem, the errorbars scale as $1/\sqrt{T_{CPU}N_{Nodes}}$ where T_{CPU} is the CPU time per node and N_{Nodes} the number of nodes. (We typically use 128 nodes.) This enables us to carry out very precise calculations on extremely large cluster sizes (up to 576 sites) given the complexity of the problem.

For the t-J model, we have summarized our results for the two- and three- leg ladder topologies. Single hole dynamics in the two-leg ladder may be well accounted for by the strong rung coupling limit. In particular, we have shown that the quasiparticle weight does not vanish. In contrast the low energy properties of the three-leg ladder have aspects of the single chain as well as of two-leg ladder systems. The two symmetric channels have a non-vanishing quasiparticle weight in accordance to the two-leg ladder system. In contrast, the antisymmetric channel has a vanishing quasiparticle weight as in the single chain case. Hence the conjecture that the low energy physics of the three-leg ladder system splits into those of chains and ladders has been clarified numerically. The key to this success, is the ability of carrying out simulations on large lattice sizes so as to control finite size effects.

Our results on the KLM are based on a technical innovation which allows us to avoid the sign problem at least in the half-filled case. We have shown that the competition between the RKKY interaction and Kondo effects lead to a quantum phase transition between ordered and disordered magnetic states. Based on the calculation of the single particle spectral function, we have demonstrated that the Kondo effect and RKKY interaction coexist. Impurity spins are partially screened and the remnant magnetic moment orders magnetically. This mechanism enables one to generate magnetic states with arbitrarily small magnetic moments.

Acknowledgment

This work was supported by the Deutsche Forschungsgemeinschaft and the Australian Research Council. The numerical calculations were performed at HLRS Stuttgart. We thank the above institutions for their support.

References

1. R. Blankenbecler, D. J. Scalapino, and R. L. Sugar, Phys. Rev. D **24**, 2278 (1981).
2. R. M. Fye and D. J. Scalapino, Phys. Rev. Lett. **65**, 3177 (1990).
3. R. M. Fye and D. J. Scalapino, Phys. Rev. B **44**, 7486 (1991).
4. F. F. Assaad, Phys. Rev. Lett. **83**, 796 (1999).
5. S. Capponi and F. F. Assaad, Phs. Rev. B **63**, (2001).
6. M. Feldbacher and F. F. Assaad, Phys. Rev. B **63**, 73105 (2001).
7. M. Jarrell and J. Gubernatis, Phys. Rep. **269**, 133 (1996).
8. M. Brunner, F. F. Assaad, and A. Muramatsu, Phys. Rev. B **62**, 15480 (2000).
9. C. Lavalle, M. Brunner, F. F. Assaad, and A. Muramatsu, *High Performance Computing in Science and Engineering'00* (Springer-Verlag, Berlin, 2001), pp. 143–154.
10. C. Kim, A. Y. Matsuura, Z.-X. Shen, N. Motoyama, H. Eisaki, S. Uchida, T. Tohyama, , and S. Maekawa, Phys. Rev. Lett **77**, 4054 (1996).
11. Z. Hiroi, M. Azuma, and Y.Bando, J. Solid State Chem. **95**, 230 (1991).
12. M. Azuma, Z. Hiroi, M. Takano, K. Ishida, and Y. Kitaoka, Phys. Rev. Lett. **73**, 3463 (1994).
13. F. Zhang and T. M. Rice, Phys. Rev. B **37**, 3759 (1988).
14. E. Dagotto and T. M. Rice, Science **271**, 618 (1996).
15. T. M. Rice, Z. Phys. B **103**, 165 (1997).
16. E. Dagotto, Rep. Prog. Phys. **62**, 1525 (1999).
17. B. Frischmuth, S. Haas, G. Sierra, and T. Rice, Phys. Rev. B **55**, R3340 (1997).
18. H. Suzuura and N. Nagaosa, Phys. Rev. B **56**, 3548 (1997).
19. M. Brunner, F. F. Assaad, and A. Muramatsu, Eur. Phys. J. B **16**, 209 (2000).
20. O. P. Sushkov, Phys. Rev. B **60**, 3289 (1999).
21. A. V. Chubukov, Pis'ma Zh. Éksp. Teor. Fiz. **49**, 108 (1989), JETP Lett. 49, 129 (1989).
22. S. Sachdev and R. Bhatt, Phys. Rev. B **41**, 9323 (1990).
23. S. Gopalan, T. M. Rice, and M. Sigrist, Phys. Rev. B **49**, 8901 (1994).
24. H. Endres, R. M. Noack, W. Hanke, D. Poilblanc, and D. Scalapino, Phys. Rev. B **53**, 5530 (1996).
25. J. Oitmaa, C. J. Hamer, and Z. Weihong, Phys. Rev. B **60**, 16364 (1999).
26. T. Barnes, E. Dagotto, and E. S. Swanson, Phys. Rev. B **47**, 3196 (1993).
27. S. R. White, R. M. Noack, and D. J. Scalapino, Phys. Rev. Lett. **73**, 886 (1994).
28. M. Brunner, Ph.D. thesis, Universität Stuttgart, 2000, http://elib.uni-stuttgart.de/opus/volltexte/2000/597.
29. M. Brunner, S. Capponi, F. F. Assaad, and A. Muramatsu, to appear in Phys. Rev. B (Rapid communication) .
30. M. Y. Kagan, S. Haas, and T. Rice, Physica C **318**, 185 (1999).
31. B. Frischmuth, B. Ammon, and M. Troyer, Phys. Rev. B **54**, R3714 (1996).
32. A. C. Hewson, *The Kondo Problem to Heavy Fermions* (Cambridge University Press, Cambridge, 1993).
33. J. Kondo, Prog. Theor. Phys. **32**, 37 (1964).
34. P. W. Anderson, Phys. Rev. **164**, 352 (1967).
35. C. Kittel, *Quantum Theory of Solids* (Wiley, New York, 1963).
36. S. Doniach, Physica B **91**, 231 (1977).

Impurities in a Hubbard-chain

Cosima Schuster, Philipp Brune, and Ulrich Eckern

Institut für Physik, Universität Augsburg, Universitätsstr. 1, D-86135 Augsburg, Germany

Abstract. Using the density matrix renormalization group method we study the quantum coherence of one-dimensional interacting Fermi systems. We investigate the the effects of several kinds of impurities on the ground-state of a Hubbard chain in detail. Thereby we look at the transition from a metallic to an insulating ground-state caused by a local potential, a locally modified interaction or hopping. Unfortunately the preliminary results show that the successful treatment of a system of interacting spinless fermions, using the phase sensitivity as the observable of the phase transition, is unsuitable in the disordered Hubbard-chain. Nevertheless the data lead to new insight in the level structure. The investigation of the optical conductivity is still in progress. In addition we determine the exponent of the algebraic decay of Friedel oscillations at the boundary and around an impurity in the middle of the chain. These results are very useful for the characterization of the above mentioned impurities.

1 Introduction to the physical model

Disorder versus interaction induced metal-insulator-transition is a general question in solid state physics. The influence of electron-electron interaction on Anderson localization has attracted a lot of interest for several years. Hereby attention is also directed to randomly distorted one-dimensional systems, because the investigation of low dimensional systems can provide important insights. In contrast to higher dimensions, a detailed theoretical (analytical and numerical) description is possible in one dimension. In several cases analytical solutions are available; either exact, with the help of the Bethe-Ansatz, or approximative, using the bosonization technique. The latter is also suitable for distorted – hence non-integrable – systems. The density matrix renormalization group method [1] is a quasi-exact numerical method used to determine ground-state properties of long one-dimensional but non-integrable systems with excellent accuracy. Recent studies of a spinless fermion model have been a first step in a detailed understanding [2]. Whereas non-interacting fermions localize immediately in the presence of infinite small disorder in one dimension, a strong attractive interaction leads to a metal-insulator transition at a finite value of the disorder in this particular model. Special realizations of disorder, i. e. quasiperiodic potentials, likewise lead to a metal-insulator transition at finite value – even in the non-interacting case [3].

In this project we investigate the Hubbard model, the simplest model of interacting electrons with spin $\sigma =\uparrow, \downarrow$ on a chain. In the clean case, a Bethe-ansatz solution is known [4]. We consider $N_f = N_\uparrow + N_\downarrow$ electrons on a chain with length $L = Na$, where a is the lattice spacing,

$$H_{\text{Hubb}} = -\sum_{i,\sigma}^{N} t_i \left(c_{i,\sigma}^+ c_{i+1,\sigma} + \text{h. c.}\right) + \sum_{i}^{N} U_i n_{i,\uparrow} n_{i,\downarrow} + \sum_{i}^{N} \epsilon_i n_i, \quad (1)$$

where the operators $c_{i,\sigma}^+$, $c_{i,\sigma}$, $n_{i,\sigma}$, and $n_i = [n_{i,\sigma} + n_{i,\sigma}]/2$ denote Fermi creation, annihilation and number operators. The magnetization is given $M = N_\uparrow - N_\downarrow$. Disorder can be introduced by local potentials of random strength, $\epsilon_i \in [-W/2; W/2]$, randomly distributed $U_i \in [-W/2i+U; W/2+U]$, or $t_i \in [-t - W/2; -tW/2]$. Initially we consider single impurities, given by a local potential at site n, $\epsilon_i = \delta_{in}\epsilon_n$, a locally modified interaction, $U_i = U(1 - \delta_{in}u)$, and tow symmetric modified bonds, $t_i = t(1 - \delta_{in}b - \delta_{in+1}b)$.

The clean model ($W = 0$) shows three phases. Phase one appears for $U < 0$, where the spin-excitation spectrum has a gap and the the low-lying charge-excitations can be described by those of a Luttinger liquid [5]. Phase two for $U \geq 0$ and away from half filling, is characterized by gapless spin- and charge-excitations. The last phase occurs for $U > 0$ and half filling, where the charge excitations have a gap and the spin-excitations are those of a Luttinger liquid.

In the previous studies of an equivalent model for interacting spinless fermions in periodic, quasiperiodic and random potentials [2,3,6], the metal-insulator transition was determined with the help of the phase sensitivity, $N\Delta E = [E(\pi) - E(0)]$, [7], i.e. the energy difference between periodic, $E(\Phi = 0)$, and anti-periodic boundary conditions, $E(\Phi = \pi)$. The different boundary condition are modeled via a magnetic flux enclosed in the ring, $c_{i+L} = \exp(i\Phi)c_i$. In a metal, exact solutions based on the Bethe-ansatz or conformal field theory show that this energy difference decreases with $1/N$. Thus the phase sensitivity is constant in this case. In an insulator, on the other hand, the system cannot react to the twist in the boundary condition, i.e. the phase sensitivity is expected to decrease with system size.

The situation is more complicated in the Hubbard-model. The phase sensitivities for up- and down-spins can be related to the spin- and charge-stiffness, see [7]. By considering again periodic and anti-periodic boundary conditions for the spin-up and spin-down electrons, we can write

$$E(\Phi_\uparrow, \Phi_\downarrow) = \frac{D_c}{2}[2\pi(J_\uparrow + J\downarrow) + (\Phi_\uparrow + \Phi_\downarrow)]^2 + \frac{D_s}{2}[2\pi(J_\uparrow - J\downarrow) + (\Phi_\uparrow - \Phi_\downarrow)]^2,$$
$$(2)$$

where the $J_{\uparrow,\downarrow}$ are the quantum numbers of the topological excitations. Evaluating the above formula, we find

$$E(\pi,0) - E(0,0) = \frac{\pi^2}{2}(D_c + D_s^E) \tag{3}$$

$$E(\pi,\pi) - E(0,0) = 2\pi^2 \min(D_c, D_s^E). \tag{4}$$

The spin-stiffness D_s can also be obtained – in the clean system away from half filling and $U \geq 0$, i. e. in the Luttinger liquid phase – from the energy difference between the ground-state energy for different magnetizations,

$$E(N, N_f, M) - E(N, N_f, M-1) = 2\pi^2 D_s^\Delta / N. \tag{5}$$

In the case of attractive interaction, this energy difference gives, together with a finite size scaling, the gap in the spin excitations. The identity $D_s^E = D_s^\Delta$ was checked numerically with reasonable accuracy. In addition the numerical values of D_c and D_s are correctly produced for the clean system. As far as these tests are concerned, the phase sensitivities seems again to be a suitable observable.

For the study of the disordered Hubbard-chain, we need to compute the ground-state energy very accurately for different boundary conditions, long system sizes and, in addition, for different fillings. Since numerical methods like exact diagonalization (ED) and quantum Monte Carlo (QMC) simulations are restricted either to small systems or to finite temperatures, we use the density matrix renormalization group technique (DMRG) [1].

To complete the investigations of the disordered Hubbard-chain we calculate the Friedel oscillations at the boundary and around an impurity in the middle of the chain. In the case of $N_\uparrow \neq N_\downarrow$ they were already determined, [8]. We therefore consider the case $N_\uparrow = N_\downarrow$. In this case, logarithmic corrections are expected. Using conformal field theory, the local behavior of the density at a boundary is given by the density-density-correlation-function [9]. The Friedel oscillations are then given by [10]

$$n(x) - n_0 \propto \frac{\cos(2k_F x)}{x^{(1+g_c)/2} \ln^{3/4} x}, \tag{6}$$

where $g_c(U) \approx 1 - U/(2t\pi)$ is the Luttinger-parameter.

2   The density matrix renormalization group method

2.1   Numerical and Computational Aspects

The DMRG is an improvement of the blocking scheme of the numerical real space renormalization treatment (NRG) [11]. The method is used to determine the ground-state properties of long systems (about a hundred lattice sites, compared to about thirty using ED). The complete lattice is ithereby built up out of smaller subsystems A, B – by considering in each step only the m states of the subsystems which contribute mostly to the ground-state, $|\Psi_C\rangle = \sum_{ij} c_{ij} |\psi_i^A\rangle |\psi_j^B\rangle$, of the whole system C. This most important states are gained in the DMRG with the help of the density matrix. The dimension of the new Hamilton matrix in the NRG is in contrast reduced by considering only the m states with the lowest energy. Thus each step of the algorithm works as follows: Starting with a system of length M, one site is added – this system is called A. C is built from A and $B = A$. Now the ground-state of C and the coefficients c_{ij} are calculated by diagonalizing the Hamilton matrix using the Davidson algorithm for sparse matrix diagonalization [12]. The density matrix, $\rho = \sum_j c_{ij} c_{mj}^*$, a fully occupied matrix, is diagonalized using standard library (ESSL) routines. Finally all states and operators have to be transformed to the new basis which consists of the m states corresponding to the m lowest eigenvalues of the density matrix. For the implementation, see [13]. The number of states which have to be kept for reasonable accuracy (about 10^{-4}) depends mostly on the boundary conditions and the criticality of the system. We typically kept $m = 200 \ldots 400$ states in the calculations of the disordered system where the system sizes varied from $N = 30 \ldots 60$. The demands on the memory lay in the region of about 500-1000 MB. In the finite lattice algorithm all states and operators for all sizes of the intermediate subsystems must be stored. Thus, we need about 100 MB temporary disk space.

2.2   Vectorization, Parallelization and Performance

According to several symmetries of the Hamiltonian and conserved quantities – such as the fermion number or the magnetization – the block structure of the Hamilton matrix is used to save space and time. For this reason the blocks are relatively small (about 500×500) and the program was optimized for a serial processor and run on the IBM/RS6000 SP. The code was written in C++. The program spend most of the time in the subroutine where the Hamiltonian is diagonalized. Hence, the Davidson algorithm has to be optimized. It was possible to reach a performance of about 100-200 Mflops on a single processor. Parallelization (using MPI) was mostly carried out to obtain the results for various parameters in a reasonable time.

3   Results for the metal-insulator transition in a Hubbard-chain

As mentioned above, we start our studies of disordered systems by considering only one impurity. In this case the averaging over many disorder configurations can be omitted. Furthermore, the renormalization group (RG) treatment of [14] can be used to check the numerical data. The most important result is that even a weak impurity destroys the structure of the energy-levels with the external flux.

3.1 A local potential

The RG treatment for a local potential shows that the local potential leads to insulating behavior (decrease of the phase sensitivity or stiffness with system size) for all fillings, impurity strengths and interactions.

In Fig. 1 we show the numerical results for a local potential at half and third filling ($\rho = N_f/N$). We begin by discussing the left hand side of Fig 1. At half filling the Hubbard-model is symmetric in spin and charge. $N[E(\pi,\pi) - E(0,0)]$ shows for weak impurity strength the expected symmetric behavior for $U > 0$ and $U < 0$ (spin-gap versus charge-gap behavior) and a decrease of the phase sensitivity with system length. But this decrease is clearly related to the localization due to the interaction, U. The overall decrease of the phase sensitivity is due to the impurity. A local potential couples spin and charge. Thus this symmetry vanishes by adding the impurity. In our case we find a small increase near $U = 0.5t$ which was already found by [15]. This maximum is shifted to $U = 1t$ for stronger impurity strength. The same behavior is found for $N[E(\pi,0) - E(0,0)]$, where the unexpected (wrong?)

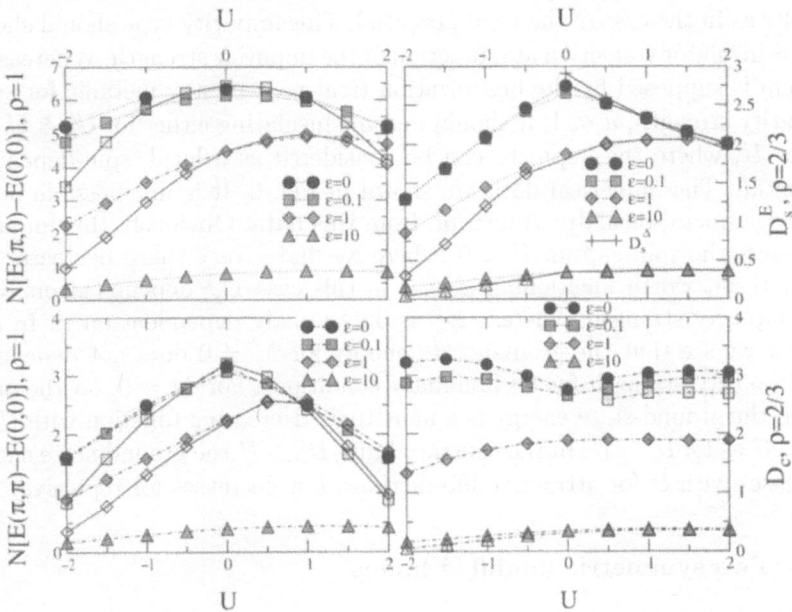

Fig. 1. Phase sensitivity or extracted stiffness versus interaction for a Hubbard-chain with a local potential. On the left side we show the results for half filling, on the right for third filling, respectively. The circles denote the clean systems at $N = 10$ or $N = 9$, respectively. The squares denote an impurity of strength $\epsilon_1 = 0.1t$, the diamonds an impurity of strength $\epsilon_1 = 1t$, and the triangles an impurity of strength $\epsilon_1 = 10t$. The dark shaded symbols correspond to $N = 10$ or $N = 9$, respectively, and the light shaded symbols to $N = 30$ or $N = 27$.

metallic behavior is between $-0.5t < U < 1t$ for small impurity strength and between $0t < U < 1t$ for stronger. The assignment of the phase sensitivities to the spin- and charge-stiffness for the whole interaction regime is possiblei for third filling. The spin-stiffness, to be more precise D_S^E, shows partly the expected behavior: A clear decrease with system size for $U < 0$ in the spin-gap-phase is seen. From the renormalization group treatmenti, however, we also expect a decrease of the spin-stiffness with system length due to the impurity for repulsive interaction. This is not found in the numerical data. In addition the spin-stiffness deduced from the spin-gap (marked with a $+$ in the upper right corner of Fig. 1) is completely independent of the impurity strength. Thus it is not possible to make a statement of a possible transition in the spin-sector with help of the phase sensitivity. As far as the charge stiffness is concerned, we find at least the decrease for small impurity strength and repulsive interactions.

3.2   Locally modified interaction

The numerical studies of a system with a modified interaction $U_i = U' = U(1 - u)$ at the impurity site and $U_i = U$ otherwise lead to no significant results as in the case of the local potential. This impurity type should show a metal-insulator transition as a function of the impurity strength. Whereas the system is supposed by the bosonization treatment to stay metallic for small impurity strength, $u \ll 1$, it should become insulating either for $U' \ll U$ and $U' \gg U$, where the impurity can be considered as a local, spin-dependent potential. The numerical data are shown in Fig. 2. It is not possible to extract the metal-insulator transition from this data. Obviously this impurity enhances the spin-gap for $U < 0$, where we find a very sharp decrease. Another result worth mentioning is that in this case D_s^Δ depends strongly on the impurity strength, whereas D_s^E is only weakly dependent on it. In particular we see that the ground-state energy for $M \neq 0$ does not depend on the impurity strength for all boundary conditions. For $M = 0$, on the other hand, the ground-state energy is a monotonic decreasing function with U for $U' \ll U$ as for $U' = U$. In the contrary limit, $U' \gg U$ the ground-state energy increases with U for attractive interactions, but decreases for repulsive.

3.3   Two symmetric modified bonds

In this case we find for the non-interacting system that the backscattering contributions from this impurity cancel out for half filling in first order, and are reduced for third filling.

The numerical data shown in Fig. 3, show clearly the slight reduction of the phase sensitivity for weak impurities for half filling at fixed chain length. Stronger impurities lead to a larger decrease which is, in addition, strengthened by increasing repulsive interaction. In the case of third filling, D_S^E is – due to the incomplete cancellation of the backscattering – distinctly reduced

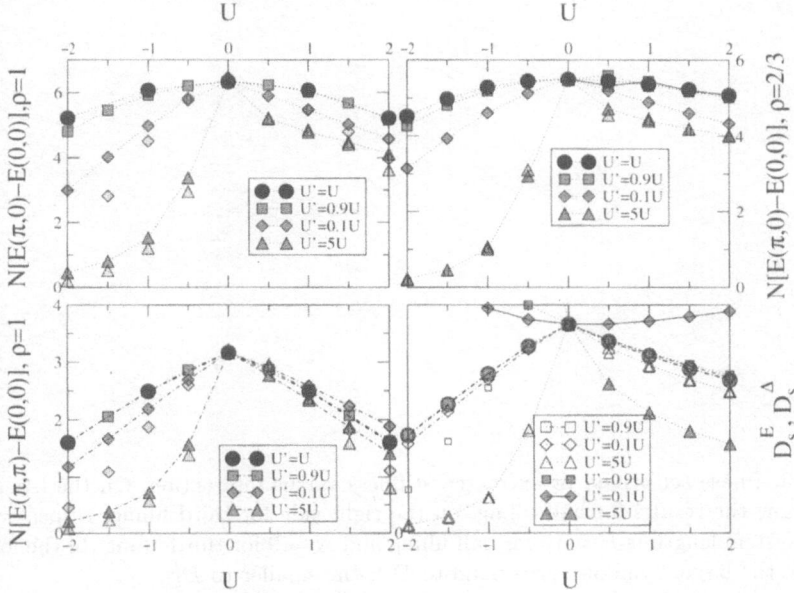

Fig. 2. Phase sensitivity or extracted stiffness versus interaction for a Hubbard-chain with a modified interaction. On the left side we show the results for half filling, on the right for third filling, respectively. The circles denote the clean systems at $N = 10$ or $N = 9$, respectively. The squares denote an impurity of strength $U' = 0.9U_0$, the diamonds an impurity of strength $U' = 0.1U_0$, and the triangles one of strength $U' = 5U_0$. The dark shaded symbol correspond to $N = 10$ or $N = 9$, respectively, and the light shaded symbols to $N = 30$ or $N = 27$. In the lower right plot D_s^E is marked with the open symbols and D_s^Δ with shaded symbols.

in the non-interacting system. Increasing the interaction the localizing effects of the impurity are weakened. The behavior of D_s^Δ is complementary. The identity $D_s^E = D_s^\Delta$ holds above a critical interaction, $U_c = 5b$.

4   Friedel-oscillations

In Fig. 4 we show results for the decay of the Friedel oscillations at the boundary. In this case, for half filling, an algebraic decay rather than the pronounced algebraic decay with logarithmic corrections is found. Nevertheless, the exponents obtained from the numerical data are not in agreement with the predictions from bosonization. The data for third filling show no clear behavior.

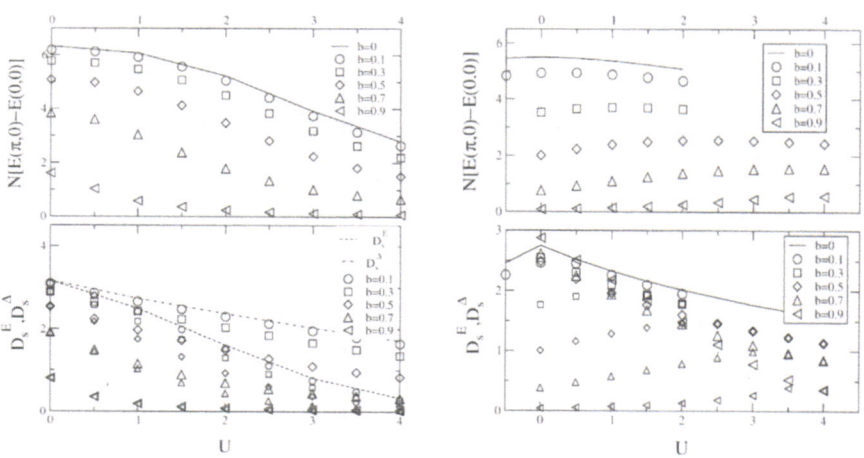

Fig. 3. Phase sensitivity or extracted stiffness versus interaction. On the left side we show the results for half filling, on the right side for third filling, respectively. The system length is $N = 10$ for half filling and $N = 9$ for third filling. In the lower panel, the larger symbols correspond to D_s^{Δ}, the smaller to D_s^{E}.

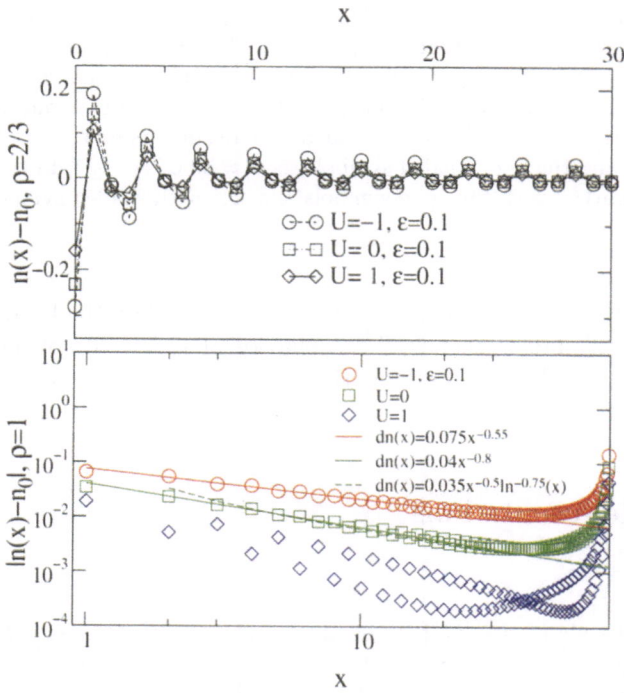

Fig. 4. Local density-distribution versus distance for half and third filling and $U = -1, 0, 1$. An additional potential scatterer is located at the ends of the chain, $\epsilon_1 = \epsilon$, $\epsilon_N = -\epsilon$. In the upper plot, the lines are connecting the data points.

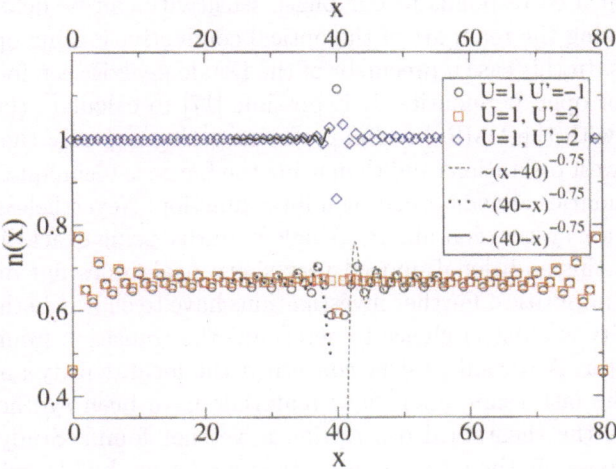

Fig. 5. Local density-distribution versus distance for half and third filling and $U = 1$. The interaction in the middle of the chain is modified, $U' = -U$ or $U' = 2U$.

They can be described either with or without logarithmic corrections. For example we find,

$$n(x) - \rho = -0.35 \cos(\rho\pi x + 0.85)x^{-0.5} \log^{-0.75}(x) \quad U = -1$$
$$= -0.6 \cos(\rho\pi x + 0.85)x^{-1} \log^{-0.75}(x) \quad U = 0$$
$$= -0.25 \cos(\rho\pi x + 0.85)x^{-0.75} \log^{-0.75}(x) \quad U = 1.$$

Considering the logarithmic corrections, the exponents agree well with the predictions from bosonization. The prefactor depends only weakly on the additional local potential, as was found in the case of $N_\uparrow \neq N_\downarrow$ using the Bethe-ansatz, [8]. Fig. 5 shows the results for the decay around an impurity which is given by a modified interaction in the middle of the chain. Again, we find oscillations at the boundary, now without an additional potential. The decay around the impurity in the middle of the chain shows the logarithmic corrections more clearly.

5   Summary and outlook

Contrary to the model of spinless fermions on a ring, where it was possible to characterize the metallic and insulating phases as well for different kinds of impurities as well as for random, quasi-periodic and periodic potentials, the phase sensitivity is not a suitable observable in the Hubbard model. The second degree of freedom in the Hubbard model changes most features essentially. Only the case of two modified bonds is comparable in both systems. To overcome these problems concerning the level structure, we try to gain reasonable results using the optical conductivity. As reported in [16], the Drude weight which corresponds to our phase sensitivity can be determined easily by calculating the real part of the optical conductivity using open boundary conditions. In this case a precursor of the Drude peak is seen for small energy transfer for open boundaries. It is possible [17] to calculate the optical conductivity with the DMRG technique by targeting not only the ground-state but the lowest excitations and then using the Lanczos technique. Effects of the open boundaries are smoothed by a filter function. Nevertheless, preliminary studies of the system containing a single impurity using exact diagonalization show that this additional, indeed very sharp peak, does not disappear even for strong impurities. Further investigations have to show whether the optical conductivity is a better choice to determine the transition from the metal to the insulator. A second project concerned the local density and magnetization. In the last years, many new materials have been synthesized, but in some cases the theoretical description is yet not found. Studying the local behavior, possibly the relevant theoretical model can be identified. We found that the logarithmic corrections occur in the case of a bulk impurity but are maybe absent in the case of the decay near a boundary at the open end of the chain. Thus, further investigations are necessary.

References

1. S. R. White, Phys. Rev. Lett. **69** (1992), 2863
   Ingo Peschel, Xiaqun Wang, Matthias Kaulke, and Karen Hallberg, *Density-Matrix Renormalization: A New Numerical Method in Physics* , Lecture Notes in Physics, Springer Verlag (1999)
2. P. Schmitteckert, T. Schulze, C. Schuster, P. Schwab, and U. Eckern, Phys. Rev. Lett. **80** (1998), 560-563
3. A. Eilmes, R. A. Römer, C. Schuster, and M. Schreiber, Technische Universität Chemnitz, SFB393-Preprint 2001-05, 2001
4. E. H. Lieb and F. Y. Wu, Phys. Rev. Lett. **20** (1968)
5. F. D. M. Haldane, Phys. Rev. Lett. **47** (1981), 1840
6. C. Schuster, *Random and periodic lattice distortions in one-dimensional Fermi and spin systems*, Shaker Verlag, Aachen (1999)
7. B. S. Shastry and B. Sutherland, Phys. Rev. Lett. **65** (1990), 243
8. G. Bedürftig, B. Brendel, H. Frahm. R. M. Noack, Phys. Rev. B **58** (1998), 10225
9. Y. Wang, J. Voit, and Fu-Cho Pu, Phys. Rev. B **54** (1996), 8491
10. H. J. Schulz, Phys. Rev. Lett. **64** (1990), 2831
11. K. G. Wilson, Rev. Mod. Phys. 47 (1975), 773
12. E. R. Davidson, J. Comp. Phys. **17** (1975), 87
13. P. Brune, PhD thesis, Universität Augsburg (2001)
14. C. L. Kane and M. P. A. Fisher, Phys. Rev. B **46** (1992), 15 233
15. R. A. Römer and A. Punnoose, Phys. Rev. B **52** (1995), 14 809
16. R. M. Fye, M. J. Martins, D. J. Scalapino, J. Wagner, and W. Hanke, Phys, Rev B **44** (1991), 6909
17. T. D. Kühner and S. R. White, Phys. Rev. B **60** (1999), 335

Band to Mott Insulator Transition in the Ionic Hubbard Model

Ph. Brune and A.P. Kampf

Theoretische Physik III, Elektronische Korrelationen und Magnetismus,
Institut für Physik, Universität Augsburg, 86135 Augsburg, Germany

Abstract. We investigate the ground state phase diagram of the one-dimensional "ionic" Hubbard model with an alternating periodic potential at half filling by numerical diagonalization of finite systems with the Lanczos and DMRG methods. In addition, we present results for the optical conductivity obtained by the recently developed dynamical DMRG (DDMRG) method. The band insulator to Mott insulator phase transition and its characteristics are discussed.

1 Introduction

The so-called "ionic" Hubbard model was originally proposed about 20 years ago in the context of organic charge transfer crystals consisting of a sequence of alternating donor (D) and acceptor (A) molecules $(\ldots D^{+\rho} A^{-\rho} D^{+\rho} A^{-\rho})$ [1,2]. These stacks form quasi one-dimensional (1D) insulating or semiconducting chains, and are classified into two categories depending on the amount of charge transfer ρ: quasi-neutral for $\rho < 0.5$, and quasi-ionic for $\rho > 0.5$. Torrance *et al.* [1] found that several of these systems undergo a reversible neutral-to-ionic phase transition (NIT), i.e. a discontinuous jump in the ionicity ρ upon changing temperature or pressure. In a different context, the ionic Hubbard model has also been used recently to describe the ferroelectric transition in perovskite materials such as $BaTiO_3$ [4].

Explicitely, the 1D ionic Hubbard model is defined by the Hamiltonian

$$H = -t \sum_{i,\sigma}(1 + (-1)^i \delta) \left(c_{i\sigma}^\dagger c_{i+1\sigma} + h.c. \right)$$

$$+ U \sum_i n_{i\uparrow} n_{i\downarrow} + \frac{\Delta}{2} \sum_i (-1)^i n_i \quad , \tag{1}$$

where $c_{i\sigma}^\dagger$ creates an electron on site i with spin σ, $n_{i\sigma} = c_{i\sigma}^\dagger c_{i\sigma}$, U the on-site Hubbard interaction, Δ an on-site potential; in (1) we have included an additional Peierls modulation δ of the hopping matrix element t. The limit $\Delta = 0$ and $\delta > 0$ is called the *Peierls-Hubbard model*.

The NIT at finite temperatures has been intensively studied theoretically in a series of articles by Nagaosa *et al.* [2] using quantum Monte Carlo (QMC) simulations. NIT has been investigated. The 1D ionic Hubbard model (1) with $\Delta > 0$ and $\delta = 0$ at half filling has served as an appropriate model for a D-A chain. The model parameters U and Δ in (1) are used for an effective description of the microscopic parameters, like e.g. the electron affinity of the acceptor molecules, the ionization potential of the donors, and the Madelung energy of ionized D^+A^- pairs. Within this picture, Δ could be interpreted as the energy necessary to move an electron from the donor to the acceptor.

For an understanding of the existence of a phase transition in the 1D ionic Hubbard model with $\delta = 0$ the best starting point is the atomic limit [8]. For $t = 0$, it is immediately seen that for half filling and $U < \Delta$ the ground state of (1) has two electrons on the B (i odd), and no electrons on the A sites (i even) ("neutral" phase). This corresponds to a charge density wave (CDW) ordering with maximum amplitude. On the other hand, for $U > \Delta$ each site is occupied by one electron ("ionic" phase). Obviously, for $t = 0$ a transition occurs at a critical value $U_C = \Delta$. This transition is expected to persist for $t > 0$, where the alternating potential still defines two sublattices A and B, opening up a band gap Δ for $U = 0$ at $k = \pm\pi/2$. For $t > 0$ the critical coupling shifts to $U_C(t) > \Delta$, with U_C monotonically increasing with increasing Δ. For $U, \Delta \gg t$ the system is close to atomic limit, and U_C approaches Δ from above.

For $U = 0$ the ground state at half-filling is a CDW band insulator (BI); its elementary excitation spectrum consists of particle-hole excitations over the band gap. We consider a system to be in a BI phase when the criterion $\Delta_S = \Delta_C$ holds where the spin (Δ_S) and the charge gap (Δ_C) are given by

$$\Delta_S = E_0(N = L, S_z = 1) - E_0(N = L, S_z = 0) \tag{2}$$

$$\Delta_C = E_0(N = L + 1, S_z = 0) + E_0(N = L - 1, S_z = 0) \tag{3}$$
$$- 2E_0(N = L, S_z = 0) \quad .$$

$E_0(N, S_z)$ is the ground state energy, L the system length in units of the lattice constant, N the number of electrons in the system, and S_z the z-component of the total spin.

On the other hand, in the ionic phase for $U > U_C$ the charge gap is dominated by the Coulomb interaction U; thus the system is a correlated insulator with $\Delta_C > \Delta_S$. However, in contrast to the cases with $\Delta = 0$ or $t = 0$, the CDW order is present in this phase for all finite values of U, except for $U \to +\infty$. If also $\Delta_S = 0$, i.e. if the system is a true Mott insulator (MI) or $\Delta_S > 0$, is an important question closely related to the physics of the intermediate region around $U \approx \Delta$, as well as the nature of the phase transition from the band to the correlated or Mott insulator.

The ionic Hubbard model is different from the Peierls-Hubbard model which is also a BI at $U = 0$, but in contrast to the ionic Hubbard model ($\delta = 0$) $\Delta_C > \Delta_S > 0$ is realized for any value $U > 0$, i.e. the phase transition from the Peierls BI to the correlated insulator occurs at $U_C = 0$.

2   Symmetry analysis

The Hamiltonian (1) for $\delta = 0$ is invariant with respect to *inversion* at a site i and *translation* by two lattice sites. Thus, any nondegenerate eigenstate of H is also an eigenstate of the operators that generate the corresponding symmetry transformation. We denote the site inversion operator by P, as defined by Gidopoulos *et al.* [8] through

$$P c_{i\sigma}^\dagger P^\dagger = c_{L-i\sigma}^\dagger \quad \text{for } i = 0, 1, \cdots, L-1 \ . \quad , \tag{4}$$

Then, any nondegenerate eigenstate $|\psi_n\rangle$ of H must obey $P|\psi_n\rangle = \pm|\psi_n\rangle$. The alternating charge order (CDW) in the ground state at $U = 0$ will therefore persist up to $U \to +\infty$.

For $U = 0$, the ground state is a direct product of spin up and spin down Slater determinants, both formed from the same occupied spatial wavefunctions having the same parity $P_\sigma = \pm 1$, so the parity eigenvalue of the total wavefunction is given by their product $P = P_\downarrow \times P_\uparrow = 1$ for the two spin projections. Hence, the groundstate at $U = 0$ is even under site inversion. On the other hand, in the large U limit of the Hubbard model the mapping to the Heisenberg Hamiltonian can be used to show that for $L = 4n$ with periodic boundary conditions (PBC) or $L = 4n + 2$ with antiperiodic boundary conditions (APBC) the ground state obeys $P|\psi_0\rangle = -|\psi_0\rangle$ (for details see [8]). As a consequence, upon increasing the number of sites L, the ground state for $U \gg t$ will be odd with respect to P as long as $k = \pi/2$ is an allowed k value. This shows that for the ordinary Hubbard model the ground state has $P = +1$ only for $U = 0$, and $P = -1$ for any $U > 0$.

However, in the ionic Hubbard model ($\Delta > 0, \delta = 0$) the phase transition from a BI to a correlated insulator occurs at some finite $U_C > 0$. The site parity of the ground state remains even for all $U < U_C$ in the BI phase. At U_C, a ground state level crossing occurs, verified by exact diagonalization studies (see below), connected with a site parity change. Similar to the situation with $\Delta = 0$, for $U \gg t$ the ground state has odd site parity for $U > U_c$.

3   Lanczos exact diagonalization results

Using the Lanczos diagonalization method [12] the energies of the few lowest eigenstates were obtained for finite chains with $L = 4n$ and PBC or $L = 4n + 2$ with APBC. The purpose of these calculations was to extend earlier exact diagonalizations [5,8] for a better understanding of the symmetries of the eigenstates involved in the level crossing. We first analyze a short chain without a finite size scaling analysis. The thermodynamic limit will be adressed below by analyzing DMRG results.

In Fig. 1, the lowest eigenvalues of the ionic Hubbard model for $\Delta = 0.5t$, $L = 8$ and PBC are shown as a function of U. At $U = 1.3t$, a level crossing of the two lowest eigenstates occurs signaling a discontinuous phase transition.

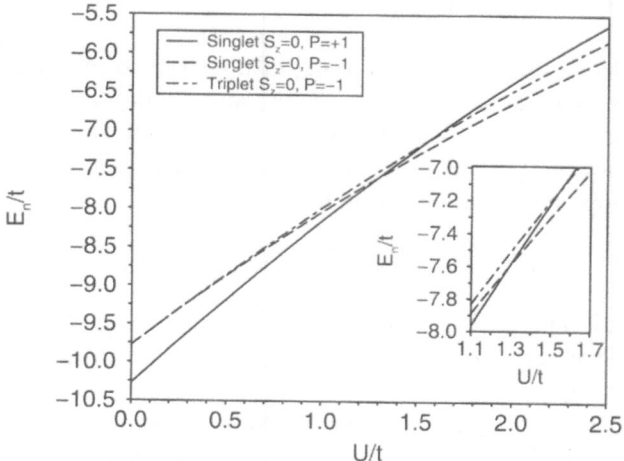

Fig. 1. Lowest energy eigenvalues of the ionic Hubbard model at half filling for $L = 8$ sites, periodic boundary conditions, $\Delta = 0.5t, \delta = 0$.

As pointed out before, a non-degenerate eigenstate of the ionic Hubbard model can only have site parity eigenvalues ± 1, so the groundstate level crossing transition corresponds to a parity change.

For $U = \delta = 0$, the Hamiltonian (1) is easily diagonalized obtaining two energy bands $E_{1/2}(k) = \pm\sqrt{4\cos^2(k) + (\Delta/4)^2}$. This is done by introducing new fermionic creation and annihilation operators $\{\gamma^{\dagger}_{k\sigma b}, \gamma_{k\sigma b}\}$ with an additional band index $b = 1, 2$ denoting the lower and upper bands, respectively. The first two degenerate excited states have negative site parity, because the ground state has positive parity and

$$P\gamma^{\dagger}_{q\sigma 2}\gamma_{q\sigma 1} = -\gamma^{\dagger}_{q\sigma 2}\gamma_{q\sigma 1}P \tag{5}$$

with $q = \pi/2$. The first two excited states shown in Fig. 1 are the spin singlet (total spin $S = 0$, $S_z = 0$) and triplet excitations ($S = 1$, $S_z = 0$) created from the ground state by applying the operators

$$\frac{1}{\sqrt{2}}\left(\gamma^{\dagger}_{q\uparrow 2}\gamma_{q\uparrow 1} - \gamma^{\dagger}_{q\downarrow 2}\gamma_{q\downarrow 1}\right) \quad , \quad \frac{1}{\sqrt{2}}\left(\gamma^{\dagger}_{q\uparrow 2}\gamma_{q\uparrow 1} + \gamma^{\dagger}_{q\downarrow 2}\gamma_{q\downarrow 1}\right) \quad . \tag{6}$$

Thus, both excited states have total momentum $k_{tot} = 0$ and negative site parity. For $U > 0$, these degenerate states split and the singlet state turns into the groundstate at U_c. Obviously, from symmetry considerations and exact diagonalizations of finite rings already a quite clear picture of the phase diagram of the model emerges. At a finite $U_C > 0$ a level crossing phase transition occurs connected with a change of parity of the ground state.

4   DMRG results

Δ_C and Δ_S for longer chains have been calculated using the DMRG method [13,14]. In Fig. 2 extrapolated results are shown as a function of U. Calculations were performed with open boundary conditions (OBC) for $L = \{30, 40, 50, 60\}$ for all values of U (main figure), and additionally for $L = 200$ and $L = 300$ in the transition region around the estimated U_C (inset). In contrast to the definition (4), the charge gap in the DMRG calculations was obtained using

$$\Delta_C = \frac{1}{2}\left[E_0(N = L + 2, S_z = 0) + E_0(N = L - 2, S_z = 0) \right. \tag{7}$$

$$\left. - 2E_0(N = L, S_z = 0) \right]$$

in order to calculate the energies E_0 in the subspace of total spin $S_z = 0$. We assume the finite size scaling behaviour to be of the form $(i \in \{S, C\})$ [15]

$$\Delta_i(L) = \Delta_i^\infty + \frac{A_i}{L} + \frac{B_i}{L^2} + \cdots \quad , \tag{8}$$

which is the usual choice for extrapolations of data obtained with OBC.

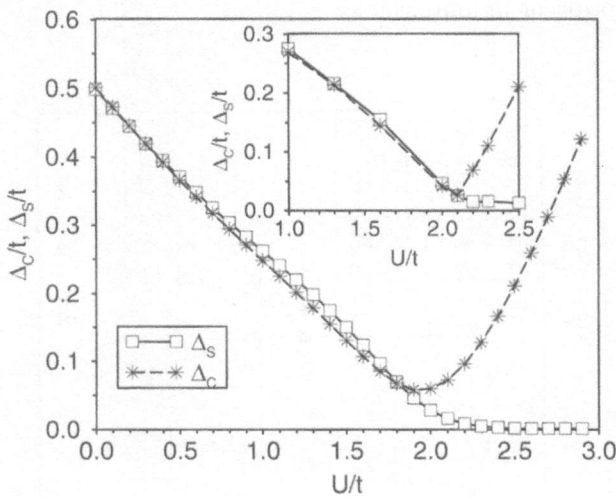

Fig. 2. Results for the spin (Δ_S) and charge (Δ_C) gap of the 1D ionic Hubbard model at half-filling with an on-site energy modulation $\Delta = 0.5t$ vs. U. Energies were obtained by DMRG calculations with $L = \{30, 40, 50, 60\}$ (main plot) and $L = \{30, 40, 50, 60, 200, 300\}$ (inset), and extrapolated to the infinite chain length limit.

The fact that the transition at U_C is connected with inversion symmetry requires some caution when open boundary conditions are used. In this case for $L = 2n$ the Hamiltonian (1) with $\delta = 0$ is not reflection symmetric at any site. Thus, the corresponding groundstate does not have a well defined site parity, and the level crossing transition does not occur. Instead, only a smooth crossover is observed. To overcome this problem, one might try to use chains with OBC and an *odd* number of sites $L = 2n + 1$, since the Hamiltonian in this case is reflection symmetric with respect to the central site, and a site parity operator is well defined by

$$Pc_{i\sigma}^\dagger P = c_{L-1-i\sigma}^\dagger . \tag{9}$$

The site parity eigenvalue of the $U = 0$ ground state for $L = 2n + 1$ is given by

$$P|\psi_0\rangle = (-1)^n |\psi_0\rangle . \tag{10}$$

On the other hand, in the large U limit the mapping to the effective spin Hamiltonian could be used again. By extending the idea of Gidopoulos *et al.* [8] to chains with $L = 2n + 1$, for $U \gg t$ we obtain

$$P|\psi_0\rangle = (-1)^{\sum_{m=1}^{L-1} m} |\psi_0\rangle = (-1)^n |\psi_0\rangle . \tag{11}$$

Thus, the parity eigenvalue of the ground state is the same at $U = 0$ and $U \gg t$ for a given chain length, and no level crossing occurs. Thus we conclude that also for odd numbers of sites and OBC the level crossing in the ground state does not occur in finite chains.

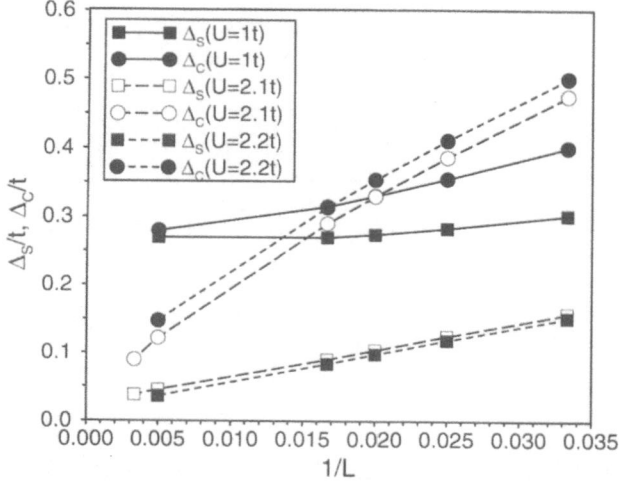

Fig. 3. Chain length dependence of the DMRG results for the spin (Δ_S) and charge (Δ_C) gap for $\Delta = 0.5t$ for selected values of U.

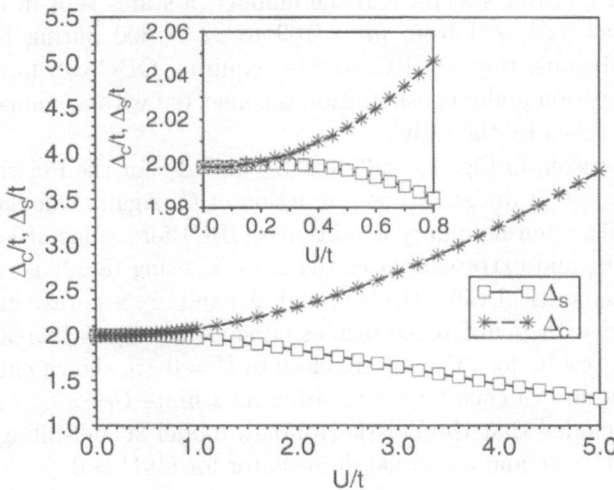

Fig. 4. Results for the spin (Δ_S) and charge (Δ_C) gap of the Peierls Hubbard model at half-filling with a modulation of the hopping amplitude $\delta = 0.1t$ as a function of the on-site Coulomb repulsion U. Energies were obtained by DMRG calculations with $L = \{30, 40, 50, 60\}$.

Obviously, results for any choice of boundary conditions should recover the level crossing scenario when extrapolated to the thermodynamic limit. Due to the fact that the sharp transition at a well defined U_C does not exist in the finite chain results with OBC, the extrapolation is quite a subtle problem. This requires the use of very long chains in the critical region (up to $L = 300$ sites or more). As can be seen from the main plot in Fig. 2, extrapolating the results for $L = \{30, 40, 50, 60\}$ does not give a sharp transition behaviour of the gaps. Up to approximately $U \sim 0.4t$, the band insulator with $\Delta_C = \Delta_S$ persists, and for $U > 2.3t$ the system is a correlated insulator with $\Delta_C > \Delta_S > 0$. Within the numerical accuracy $\Delta_S = 0$, which suggests that this phase represents a true Mott insulator. The intermediate region does not show a clear sharp transition, the charge gap never closes, and one observes $\Delta_C \neq \Delta_S$ already below the estimated U_C.

As illustrated in the inset, adding results for $L = 200$ and even $L = 300$ in the critical region changes the picture considerably, approaching the behaviour expected from the symmetry arguments and exact diagonalization results. This allows the conclusion that the above level crossing scenario will be recovered in the thermodynamic limit for the DMRG results on chains with an even number of sites. To illustrate the dependence of the gaps on the chain lenths L, a scaling plot for Δ_C and Δ_S at selected values of U as a function of $1/L$ is shown in Fig. 3. The changes in the behaviour occuring for longer chains are clearly visible. The DMRG calculations were performed with four finite lattice sweeps, and the number of states kept in the DMRG projection was increased from $m = 100$ to $m = 300$ during the sweeps. It becomes obvious, that DMRG studies requiring OBC can face problems when the transition under consideration is connected with a change in spatial symmetries broken by the OBC.

For comparison, in Fig. 4 results for Δ_C and Δ_S for the Peierls-Hubbard model with $\delta = 0.5t$ are shown as a function of U. Again, calculations were performed with open boundary conditions (OBC) for chains of lengths $L = \{30, 40, 50, 60\}$, and extrapolated to the $L \to \infty$ using (8). It is clearly seen that there is no critical value $U_C > 0$, and Δ_C and Δ_S separate immediately for $U > 0$, with their difference increasing smoothly upon increasing U. In the inset, the results for Δ_C and Δ_S close to $U = 0$ are shown enlarged, but again there is no evidence for a transition at a finite $U_C > 0$. Thus, it can be safely concluded that the Peierls-Hubbard model at half-filling is a band insulator for $U = 0$, and a correlated insulator for all $U > 0$.

5 Optical Conductivity

To further analyze the band and correlated insulating phases we have calculated the frequency-dependent optical conductivity within the dynamical DMRG (DDMRG) approach making use of the Lanczos vector technique described in [16,17]. Calculations were performed for $L = 100$ sites with OBC using a Parzen filter to suppress the influence of the chain boundaries [17].

The real part of the longitudinal optical conductivity $\sigma(\omega)$ is determined by the linear response of the system to an external electromagnetic field. In the Kubo formalism, $\sigma(\omega)$ is connected to the imaginary part of the retarded current-current correlation function

$$\chi_{jj}(q, \omega) = \frac{i}{La} \int_0^\infty d\tau \, e^{i\omega\tau} \langle \psi_0 | [j_{-q}(\tau), j_q(0)] | \psi_0 \rangle \quad , \tag{12}$$

with a the lattice constant and $|\psi_0\rangle$ denoting the exact groundstate of (1). For $\omega > 0$ the imaginary part of χ_{jj} can be obtained numerically from [19]

$$\mathrm{Im}\,\chi_{jj}(q, \omega) = -\frac{\hbar}{La} \mathrm{Im} \langle \psi_0 | j_{-q} \frac{1}{\hbar\omega + E_0 - H + i0^+} j_q | \psi_0 \rangle \quad . \tag{13}$$

Here, the current operator is defined by

$$j_q = -it\frac{ea}{\hbar} \sum_{i,\sigma} e^{iqx_i} \left(c_{i+1\sigma}^\dagger c_{i\sigma} - c_{i\sigma}^\dagger c_{i+1\sigma} \right) \quad , \tag{14}$$

where e is the electron charge. The real part of the optical conductivity for $q = 0$ is then given by

$$\sigma_1(\omega) = D\delta(\omega) + \sigma_1^{reg}(\omega > 0) = D\delta(\omega) + \frac{1}{\hbar\omega} \mathrm{Im}\,\chi_{jj}(q = 0, \omega) \quad . \tag{15}$$

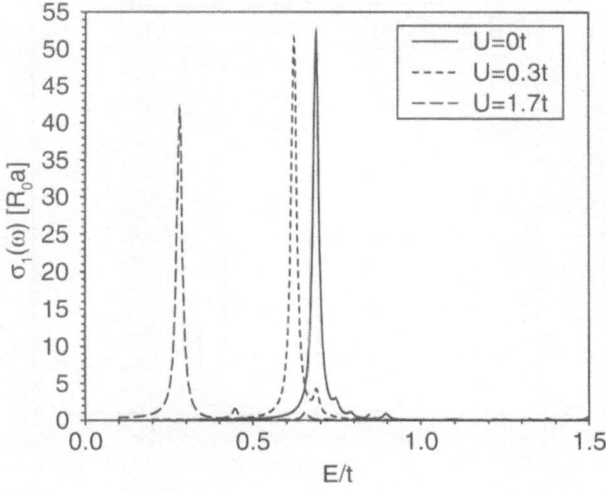

Fig. 5. Real part of the optical conductivity $\sigma_1(\omega)$ in units of $R_0 a = e^2 a/\hbar$ vs. energy $E = \hbar\omega$ for the half-filled 1D ionic Hubbard model with on-site energy modulation $\Delta = 0.5t$. A finite broadening $\eta = 0.01t$ of the peaks has been used. The selected values of the on-site Coulomb repulsion U shown all lie within the band insulating phase ($U < U_C$). The results where obtained by DDMRG calculations using the Lanczos vector method for a chain with $L = 100$ sites and open boundary conditions.

For the band and correlated insulating phases we always have a vanishing Drude weight $D = 0$.

In Fig. 5 DDMRG results for $\sigma_1(\omega)$ are plotted for $\Delta = 0.5t$ and different values of $U < U_C$ in the band insulator phase. In the infinite chain for $U = 0$ $\sigma_1(\omega)$ vanishes below the band gap Δ, and diverges as $\sigma_1(\omega) \sim \frac{1}{\sqrt{\omega - \Delta}}$ for $\omega \to \Delta$ [18]. This divergence is anticipated by the very strong dominant lowest excitation peak in the finite chain results. The picture does not change qualitatively upon increasing U over the whole band insulator phase. It suggests that the physics of the excitations in the band insulating phase remains similar to $U = 0$ for any finite $U < U_C$.

For $U > U_C$, the situation is different, as demonstrated in Fig. 6. There we show the DDMRG result for $\sigma_1(\omega)$ for $U = 2.9t$, with the other parameters as in Fig. 5. Here, the properties of the charge excitations are expected to be similar to the Hubbard model where $\sigma_1(\omega) \sim \sqrt{\omega - \Delta_C}$ near the charge gap Δ_C [19].

Due to the strong finite size effects, the square root behaviour itself is not visible in Fig. 6, but the lowest energy peak of $\sigma_1(\omega)$ is clearly not the one with the highest spectral weight anymore. Instead, the spectral weight increases with ω above the gap, and decreases again after reaching a maxi-

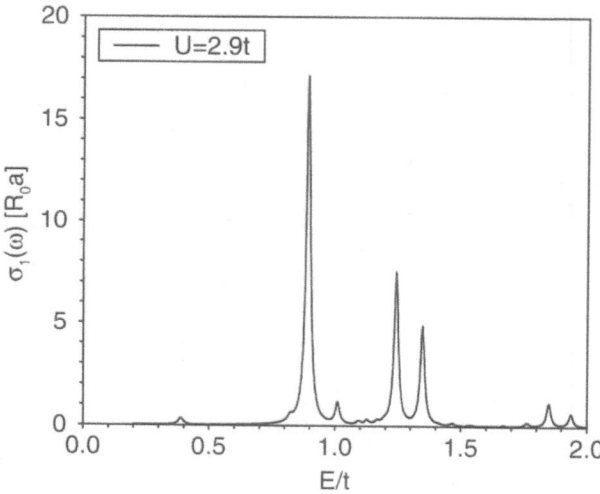

Fig. 6. Same as Fig. 5, but for a value of U corresponding to the correlated insulating phase $(U > U_C)$.

mum value. The qualitative behaviour is consistent with the results obtained for the Hubbard chain at $U = 3t$ [19].

Whenever the site parity of an eigenstate is well defined quantity, the current operator $j_{q=0}$ will change parity when acting on that state. Thus, the matrix element $\langle \psi_n | j_{q=0} | \psi_0 \rangle$ in (12) vanishes whenever $|\psi_n\rangle$ and $|\psi_0\rangle$ have the same parity. In the ionic Hubbard model, the first two excited states have parity opposite to the ground state, and the singlet state contributes to the optical conductivity. Precisely at U_C, the ground state and the first excited state cross, so the ground state is degenerate. This critical point has been commonly referred to in the literature as "metallic", due to the vanishing charge gap $\Delta_C = 0$. This implies the existence of optical excitations at arbitrary small energies $\omega > 0$. However, per definition a metal is characterized not only by gapless excitations, but also by a finite Drude weight $D > 0$, i.e. a finite dc conductivity. Thus, it follows that the ionic Hubbard model for $U = U_C$ has a gapless optical spectrum, but it cannot be considered "metallic" in the usual sense.

6   Calculations on the IBM RS/6000 SP and SP-SMP at the SSC

All calculations were performed on the IBM RS/6000 SP and the new SP-SMP supercomputers at the SSC Karlsruhe using parallelized DMRG and dynamical DMRG (DDMRG) programs. The programs are written in the C++ programming language and make use of additional library routines written in FORTRAN 77 (BLAS, LAPACK, ESSL, and the Davidson exact diagonalization routine). The programs are parallelized using the MPI communication protocol to simultaneously diagonalize systems with different parameter sets.

Due to the nature of the required operations as well as to the lower efficiency of C++ compilers in vectorizing the code, our programs do not benefit very strongly from vector CPUs; the general purpose POWER RISC architecture of the IBM RS/6000 computers is much better suited to our needs. Especially, the state-of-the-art POWER3 CPUs of the new SP-SMP at the SSC have enabled us to perform the resource-intensive DDMRG calculations of the optical conductivity.

On the RS/6000 SP at the SSC, we typically ran our programs in the production class using 10-30 processors for each job, making use of the maximum available 500 MB per node. Depending on the length of the chains the necessary CPU time were close to the maximum available 240 min. in this class. On the new RS/6000 SP-SMP, we also used the production class with the maximum available 2 GB per node to perform the DDMRG calculations. These runs also took the maximum 240 min. CPU time available for a job submitted to this class.

Acknowledgements

We thank G. Japaridze for helpful contributions. This work was partially supported by the DFG through SP 1073.

References

1. J. B. Torrance *et al.*, Phys. Rev. Lett. **46**, 253 (1981); ibid. **47**, 1747 (1981).
2. N. Nagaosa and J. Takimoto, J. Phys. Soc. Jpn. **55**, 2735, 2747 (1986); N. Nagaosa, ibid. **55**, 2754 (1986).
3. A. Girlando and A. Painelli, Phys. Rev. B **34**, 2131 (1986).
4. T. Egami, S. Ishihara, M. Tachiki, Science **261**, 1307 (1993).
5. R. Resta and S. Sorella, Phys. Rev. Lett. **74**, 4738 (1995).
6. G. Ortiz *et al.*, Phys. Rev. B **54**, 13515 (1996).
7. M. Fabrizio *et al.*, Phys. Rev. Lett. **83**, 2014 (1999).
8. N. Gidopoulos, S. Sorella, and E. Tosatti, Eur. Phys. J. B **14**, 217 (2000).
9. Y. Takada and M. Kido, J. Phys. Soc. Jpn. **70**, 21 (2001).
10. T. Wilkens and R. M. Martin, preprint cond-mat/0007472 (2000).
11. Y. Anusooya-Pati *et al.*, preprint cond-mat/0009153 (2000).
12. C. Lanczos, J. Res. Natl. Bur. Stand. **45**, 255 (1950).
13. S. R. White, Phys. Rev. Lett. **69**, 2863 (1992); Phys. Rev. B **48**, 10345 (1993).
14. I. Peschel *et al.*, (Eds.): *Density-Matrix Renormalization*, Springer (1999).
15. R. Noack, private communication.
16. K. A. Hallberg, Phys. Rev. B **52**, R9827 (1995).
17. T. D. Kühner and S. R. White, Phys. Rev. B **60**, 335 (1999).
18. F. Gebhard *et al.*, Phil. Mag. B **75**, 1, 13, 47 (1997).
19. E. Jeckelmann *et al.*, Phys. Rev. Lett. **85**, 3910 (2000).

GaAs and InAs (001) Surface Structures from Large-scale Real-space Multigrid Calculations

W.G. Schmidt, P.H. Hahn, and F. Bechstedt

Computational Materials Science Group
Institut für Festkörpertheorie und Theoretische Optik
Friedrich-Schiller-Universität, Max-Wien-Platz 1, 07743 Jena
(Email: W.G.Schmidt@ifto.physik.uni-jena.de)

Abstract. There has been a renewed interest in the structure of III-V compound semiconductor (001) surfaces caused by recent experimental and theoretical findings, which indicate that geometries different from the seemingly well-established dimer models describe the surface ground state for specific preparation conditions. We investigate large GaAs and InAs (001) surface reconstructions by means of accurate *first-principles* total-energy calculations based on a real-space multigrid method. The formation of a $\alpha2(2\times4)$ surface model containing single anion dimers in the first and third atomic atomic layers is predicted for a balanced surface stoichiometry for both GaAs and InAs. This structure is stabilized by its favorable electrostatics. Very complex (4×2) reconstructions consisting of three-fold coordinated surface anions and cations and subsurface dimers describe the surface ground state for cation-rich GaAs and InAs surfaces. The electronic properties of this so-called $\zeta(4\times2)$ structure are discussed in some detail. Several structural models for the Ga-rich GaAs(001)(4×6) surface are investigated, but dismissed on energetic grounds.

1 Introduction

Electronic structure calculations based on density functional theory (DFT) have been enormously successful in predicting and explaining the properties of a wide range of materials. The computational treatment of large and complex structures consisting of hundreds of atoms is still very demanding, though. "Real-space" methods are a very promising technique to reduce the computational cost and thus to enable even larger calculations.

Real-space methods avoid the use of plane waves, which extend throughout the entire system, so that the vast majority of operations are "local". That means on the one hand that parallelization via domain decomposition becomes very effective, since each processor can be assigned a given region of space and only little communication between the processors is needed. On the other hand, advanced mathematical techniques such as the multigrid algorithm can be used to deal efficiently with the various length scales present in the problem. Real-space techniques allow also for a natural implementation of mesh refinement or cluster boundary conditions and seem to be a particularly suitable starting point for the development of $O(N)$ techniques

(N is the number of atoms), which overcome the $O(N^3)$ scaling of traditional electronic structure calculations [1].

We work on an efficient real-space code to calculate surface optical properties highly accurately from *first principles*, i.e., without any empirical or adjustable parameters. The essential building blocks for the code, allowing the calculation of optical properties in independent quasiparticle approximation [2,3] with the inclusion electron-hole interaction effects have been completed [4]. To carry out successfully our project "Simulation of optical spectra for real-time monitoring of semiconductor growth", however, the precise knowlegde of the surface atomic and electronic structure is the necessary prerequisite. This report describes our recent progress in understanding the geometry of the prototypical GaAs and InAs(001) surfaces. The (001) oriented substrates of III-V semiconductors are commonly used in growth technologies like molecular beam epitaxy (MBE) or metal-organic chemical vapour deposition (MOVPE). The growth of GaAs and InAs is routinely monitored by means of optical spectroscopy. Nevertheless, the observed spectral features are only poorly understood. This makes them ideal candidates for our study.

2   Method

Density-functional theory in the local-density approximation (DFT-LDA) together with nonlocal norm-conserving pseudopotentials [5] is used to determine the structurally relaxed ground states of the surface structures. The Ga $3d$ and In $4d$ electrons, respectively, are partially taken into account by means of a nonlocal core correction to the exchange and correlation energy.

A massively parallel, real-space finite-difference method [6] is used to deal efficiently with the large unit cells needed to describe the surface. Thereby a real-space mesh is used to represent the wavefunctions, the charge density, and the ionic pseudopotentials. The density functional equations are discretized using a generalized eigenvalue form:

$$H_{mehr}[\psi_n] = \frac{1}{2}A_{mehr}[\psi_n] + B_{mehr}[V_{eff}\psi_n] = \epsilon_n B_{mehr}[\psi_n] \ ,$$

where A_{mehr} and B_{mehr} are the components of the *Mehrstellen* discretization, which is based on Hermite's generalization of Taylor's theorem. It uses a weighted sum of the wavefunction and potential values to improve the accuracy of the discretization of the entire differential equation. Only nearest and next-nearest neighbour points are used in the discretization. This short-ranged representation of the equation leads to an efficient domain-decomposition-based implementation on massively parallel computers (for details see Ref. [6]). The efficiency of the massively parallel implementation is illustrated in Fig. 1, which shows the speedup in execution time for a given problem as the number of PE's is increased. Typical jobs utilize 32 to 256 PE's and run from 2 to 12 hours. To efficiently solve the discretized equations, we use multigrid iteration techniques that accelerate convergence by

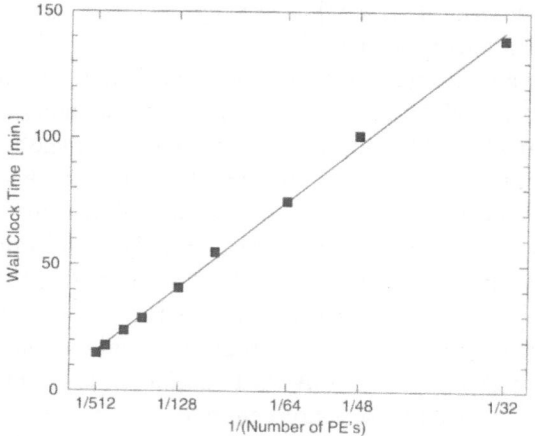

Fig. 1. Execution time needed to determine the structurally relaxed ground state of the GaAs(001)c(4×4) surface on the Cray-T3E of the Höchstleistungsrechenzentrum in Stuttgart is plotted vs number of processors. The surface is modeled by a super cell containing 70 atoms. The solid line is a guide to the eye.

employing a sequence of grids of varying resolutions. The solution is obtained on a grid fine enough to accurately represent the pseudopotentials and the electronic wavefunctions. We find that the structural and electronic properties of GaAs and InAs are converged for a spacing corresponding to 4% of the respective bulk lattice constants. However, the iterations are accelerated by solving both the Poisson and the density functional equations on coarser grids and transferring the resulting corrections to the fine grid. This procedure provides excellent preconditioning for all length scales present in our systems and leads to very fast convergence rates. The operation count to converge one wave function with a fixed potential is $O(N_{grid})$.

The surfaces are modeled by using periodic supercells. They contain material slabs consisting of 60 to 270 atoms which are about 12 Å thick, separated by 12 Å of vacuum. The surface dangling bonds at the bottom layer are saturated with fractionally charged pseudohydrogens. We relax the investigated geometries until all calculated forces are below 20 meV/Å. The atoms in the lowest bilayer remain in the their ideal bulk configuration. Integrations in the surface Brillouin zone are performed over four special **k** points in its irreducible part. In order to compare energetically surface structures representing different stoichiometries one has to take into account the chemical potentials μ of the surface constituents. Since the surface is in equilibrium with the bulk material, they are related to each other: their sum equals the chemical potential of the bulk semiconductor. Consequently, the surface formation energy may be written as a function of a single variable, which we take to be the relative chemical potential of the cation with respect to its bulk phase, $\Delta\mu$(Ga,In). The computational accuracy in determining the chemical potentials is of the order of 0.1 eV. The uncertainty of the calculated surface energies is less than 0.01 eV per surface atom.

3   Results and Discussion

3.1   GaAs

The GaAs(001) surface is known to exhibit a very rich variety of ordered phases whose occurrence depends on the preparation conditions (see, e.g., Ref. [7]). Among them, the As-rich (2×4) reconstructions were extensively investigated in the past, due to their importance for the MBE growth of GaAs. Three different (2×4) surfaces phases, called α, β and γ, are known.

The α phase occurs at the highest substrate temperature and is commonly assumed to correspond to a geometry combining two As dimers in the uppermost atomic layer with Ga-Ga bonds in the layer underneath. The same stoichiometry, however, can be realized with a slightly modified geometry, called $\alpha2(2\times4)$, as shown in Fig. 2 [8]. We find this structure to be 0.034 eV per (1×1) unit cell lower in energy than the α model. As both structures have the same stoichiometry there is no dependence on the chemical potentials of the surface constituents. The α structure will be instable with respect to $\alpha2$ irrespective of the surface preparation conditions. That result is somewhat surprising, as the α model is seemingly well established [7]. The reflectance anisotropy spectrum of the α phase of GaAs(001) [9] indicates a co-existence of anion dimers oriented along the [$\bar{1}$10] direction with cation-cation bonds parallel to [110]. These features are present, however, also for the $\alpha2(2\times4)$ geometry. It still remains to be seen whether the $\alpha2$ structure can account for the RHEED spot intensities assigned to the α phase of GaAs(001).

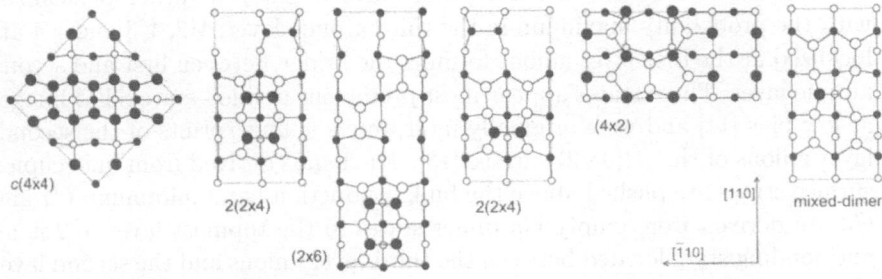

Fig. 2. Top view of relaxed GaAs(001) surface structures, ordered according to the Ga coverage. Empty (filled) circles represent Ga (As) atoms. Positions in the uppermost two atomic layers are indicated by larger symbols.

The bonding configuration of the $\alpha 2$ structure is very similar to the one of the α model. In order to understand its higher stability we analyzed the electrostatic interactions at the surface. Point charges were assigned to the surface atoms according to the *electron counting rule*, i.e., all acceptor-like states such as anion dangling bonds are filled, wheras cation dangling bonds are empty. Based on such a charge distribution one can then perform a Madelung summation for a periodic lattice of point charges,

$$S = \frac{1}{2} \sum_{i,j} \frac{q_i q_j}{|\mathbf{r}_i - \mathbf{r}_j|},$$

where the vectors $\mathbf{r}_i$ are the positions of the atoms which have been assigned charge q_i. The screening can be approximated by dividing S by the static dielectric constant of GaAs ($\epsilon \sim 13$). That rough estimate yields a difference in electrostatic energies between α and $\alpha 2$ of 0.038 eV per (1×1) surface unit cell. That is remarkably close to the energy difference of 0.034 eV obtained from *first principles*. The good agreement indicates that the $\alpha 2$ structure is indeed stabilized with respect to α by its more favourable electrostatics. The actual occurrence of the predicted structure, however, remains to be proven experimentally. Our results for the geometry of the more As-rich GaAs(001)$\beta 2(2 \times 4)$ and $c(4 \times 4)$ reconstructions (see. Fig. 2) agree with earlier findings.

The atomic structure of the Ga-rich GaAs(001)(4×2) reconstruction has been puzzling surface scientists for several years (see, e.g., Ref. [10]). Very recently a rather complex surface geometry, a so-called ζ structure (cf. Fig. 2) has been proposed on the basis of total-energy calculations [11]. The stability of the ζ structure with respect to all previously suggested (4×2) geometries was confirmed by two of the present authors in Refs. [8,12]. The calculated band structure of the $\zeta(4 \times 2)$ surface together with the orbital character of the relevant surface states is shown in Fig. 3. The occupied surface states are energetically below the bulk valence band maximum.

The bound surface valence bands are derived from surface anion related orbitals. Interestingly, $V1$, the highest surface state, is rather delocalized with the probability maximum in the third atomic layer. $V2$, $V3$ and $V4$ are localized at the first-layer anions forming the bridge between first and second atomic layer. These states appear most prominent in filled-state STM images at low bias [11] and were originally interpreted as fingerprints of the second-layer anions of the $\beta 2(4 \times 2)$ surface [13]. The bands derived from unoccupied surface states are pushed above the bulk conduction band minimum. $C1$ and $C3$ are derived from empty Ga dimer states in the topmost layer. $C2$ is an antibonding state located between the third-layer anions and the second layer Ga-dimer atoms. $C4$ comprises both anion and cation empty dangling bonds in the topmost atomic layer.

When discussing the band structure in Fig. 3, a word of caution is in order. The presented band structure suffers from the well-known DFT-LDA gap

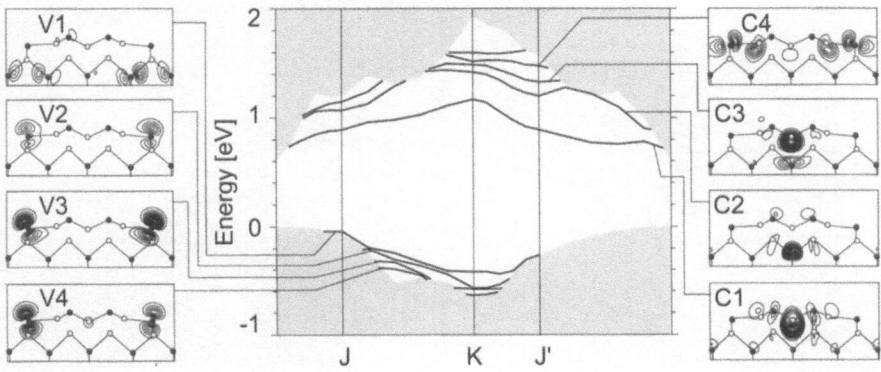

Fig. 3. Surface band structure (bound states) for the GaAs(001)ζ(4×2) surface. Gray regions indicate the projected bulk band structure. In the left and right panel, the corresponding squared wave functions at K are shown. The contour spacing is $\frac{1}{3}10^{-3}$ Bohr $^{-3}$.

problem, i.e., the underestimation of the excitation energies. The inclusion of electronic self-energy effects may lead to different energy shifts for bulk and surface state energies, as we have demonstrated for GaP(001) and InP(001) surfaces [14,15]. Experience shows, however, that the relative ordering and orbital character of the surface states are usually reliably described within DFT-LDA.

Structural models were also proposed for GaAs(001) surface reconstructions larger than (2×4) or (4×2). A (1×6) symmetry was observed by electron diffraction [7] as a non-equilibrium or transient phase between As and Ga-rich surfaces. STM images by Biegelsen and co-workers [16] and Kuball *et al.* [17] revealed that (2×6) and (6×6) reconstructed domains are responsible for the observed LEED patterns. The energy of Biegelsen's model for the (2×6) surface has been approximated by a linear combination of structural motifs (LCSM) method by Zhang and Zunger [18]. They found it to be higher in energy than the α(2×4) structure by 0.034 eV per (1×1) surface unit cell. More precise *ab initio* calculations [12] indicate that the energy differences is actually much smaller, 0.006 eV. By a simple rearrangement of the As dimers in blocks instead of zig-zag chains, resulting in the (2×6) structure shown in Fig. 2, a further lowering of the surface energy below the value of the α(2×4) structure is possible [12]. Still, the energy gain is not sufficient to make this (2×6) geometry more favourable than the newly proposed α2(2×4) structure (see calculated phase diagram in Fig. 4). The surface energy of the (6×6) model suggested in Ref. [17] will very likely be in the same range as the one calculated for the (2×6) reconstruction, as both models are rather similar. The stability of the stoichiometric α2(2×4) surface is limited to a very small window of preparation conditions as can be seen from Fig. 4. Given the higher

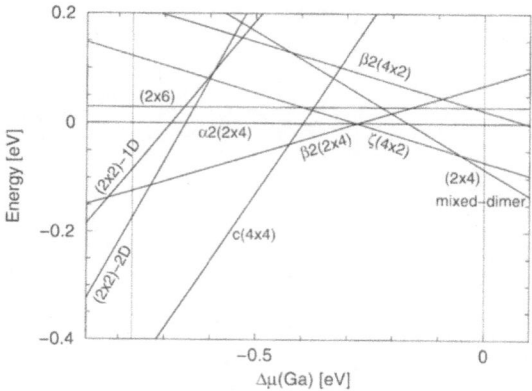

Fig. 4. Relative formation energy per (1×1) unit cell for GaAs surface reconstructions vs the cation chemical potential. Dashed lines mark the approximate anion- and cation-rich limits of the thermodynamically allowed range of $\Delta \mu(\text{Ga})$.

energy of the (2×6) and probably also the (6×6) surface, these structures are only metastable. This may explain their low longe-range order.

The calculated phase diagram for GaAs(001) in Fig. 4 indicates that for extreme Ga-rich conditions the $\zeta(4 \times 2)$ surface should transform into a (2×4) reconstructed mixed-dimer structure. Such a phase transition is not observed experimentally. Rather a (4×6) surface forms. The atomic structure of this GaAs(001)(4×6) surface is extensively discussed in the literature (see Ref. [7] for a comprehensive review). Xue *et al.* point out that one has to discriminate between a *genuine* (4×6) reconstruction which is more Ga-rich than the Ga-rich (4×2) reconstruction, and a *pseudo* (4×6) phase, which actually consists of a mixture of the (1×6), the (4×2), and the *genuine* (4×6) phase. To explain the surface structure of the genuine GaAs(001)(4×6) surface Xue *et al.* [7,10] propose a regular array of Ga clusters containing 6 to 8 atoms on top of a (4×2) reconstructed GaAs surface (see Fig. 5).

The concept of 'magic' numbers of electrons in free metal clusters allows an initial guess on the size of stable Ga clusters. Both experiment and model calculations indicate that clusters of simple, monovalent metals are particularly stable if they contain a so-called magic number of electrons such as 8, 20, 40,.. [19]. In the present case, of course, the number of electrons in the cluster is modified by the charge transfer from the substrate. If one assumes the cluster to be bonded to four surface anions as indicated in Fig. 5, each anion donating the electrons from its doubly occupied dangling bond into the cluster, a minimum stable metal cluster containing 20 electrons would consist of four Gallium atoms. In the present work a number of adsorbed clusters ranging in size from four to eight atoms (C4...C8) with a variety of shapes and orientations were studied by means of *first-principles* total-energy calculations. The adsorption energies of the most favourable configuration for

Table 1. Relative surface energies for adsorbed Ga cluster configurations (C4...C8) as well as for substitutional models (S1 and S4). The energies refer to a (1×1) surface unit cell and are given with respect the Ga-rich GaAs(001)$\zeta(4\times2)$ surface under extreme Ga-rich conditions $(\mu(\text{Ga}) = \mu(\text{Ga})_{bulk})$.

structure	C4	C5	C6	C8	S1	S4
energy in meV	119	161	153	225	29	121

each cluster size are compiled in Tab. 1. All investigated cluster geometries are higher in energy than the Ga-rich GaAs(001)$\zeta(4\times2)$ surface even at the most Ga-rich conditions, where the Ga chemical potential assumes its bulk value. Unless a specific cluster configuration has been missed, the structural model proposed by Xue *et al.* is thus not borne out by the *ab initio* calculations. Moreover, considering the trend in the surface energies, no indication for the energetic preference of a certain magic cluster size can be recognized. Instead, the larger the clusters get, the more unfavourable their adsorption seems to be.

As an alternative geometry the substitution of one (S1) or four (S4) of the surface anions marked gray in Fig. 5 by Ga atoms was investigated. The substitutional sites are three-fold coordinated and should therefore be favoured by sp^2-like bonded group-III atoms. Nevertheless, the two surface structures considered are higher in energy than the GaAs(001)$\zeta(4\times2)$ surface.

Recently, a very different interpretation of what appears to be Ga clusters in the STM images of (4×6) surfaces has been given. Kruse and co-workers [20] showed that the bright spots disappear upon chemical titration with just $7.5 \cdot 10^{-4}$ monolayers of O_2. The spots do also not show up when empty states are imaged. This led Kruse *et al.* to conclude that the bright spots

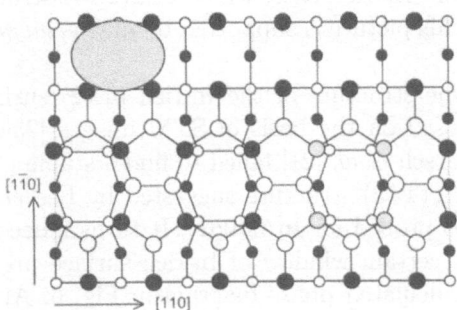

Fig. 5. Top view of investigated GaAs(001)(4×6) surface. Empty (filled) circles represent Ga (As) atoms. Positions in the uppermost two atomic layers are indicated by larger symbols. The large oval shape indicates trial Ga clusters. Positions of surface anions which were substituted by Ga in the substitution models are filled gray.

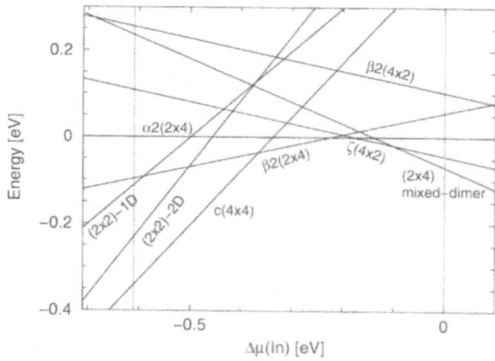

Fig. 6. Relative formation energy per (1×1) unit cell for InAs surface reconstructions vs the cation chemical potential. Dashed lines mark the approximate anion- and cation-rich limits of the thermodynamically allowed range of $\Delta\mu(\text{In})$.

arise from surface excess charge localized in the gallium dangling bonds. Repulsion between the trapped negative charge leads to what appears a surface reconstruction. That hypothesis is very interesting but rather speculative. The geometry of the (4×6) reconstructed GaAs(001) surface thus remains an open question.

3.2   InAs

The surface phase diagram of InAs(001), while less investigated, appears to be similar to the one of GaAs. (4×2), (2×4) and $c(4×4)$ reconstructions are observed with increasing surface As content. STM images of (2×4) reconstructed InAs surfaces are interpreted in terms of the $\beta2(2×4)$ geometry (cf. Fig. 2) for the more As-rich case, while a $\alpha2(2×4)$ structure seems to form after annealing. This picture is supported by *first-principles* total-energy calculations [21,22].

Less clear is the structure of the In-rich (4×2) surface. A $\beta3(4×2)$ geometry was proposed on the basis of STM images [23], while total-energy calculations by Ratsch *et al.* [21] failed to find a stable (4×2) surface. In the present study the $\zeta(4×2)$ structure suggested by Lee *et al.* [11] for Ga-rich GaAs surfaces was probed for InAs(001). It turns indeed out to be energetically stable for a certain window of In-rich surface preparation conditions as shown in the calculated phase diagram in Fig. 6. Also a very recent X-ray diffraction analysis of $c(8×2)$ reconstructed InAs(001) indicates a ζ-like geometry [24]. Therefore it seems that the ζ structure solves the structural puzzle of the In-rich InAs(001)(4×2) surface.

For extreme In-rich conditions the calculations predict a change of reconstruction again. A (2×4) mixed-dimer structure should then be favoured. Such a second phase transition is not observed experimentally. The theoretical finding may well be an artefact of the limited accuracy in the calculation of the surface phase diagram, in particular with respect to the boundaries of the thermodynamically allowed region. Also the occurrence of a modified, even more In-rich $\zeta(4\times2)$ structure, for example the adsorption of additional cations as suggested in Ref. [24], would explain the apparent contradiction.

For As-rich surfaces, finally, the three-dimer $c(4\times4)$ structure known from GaAs has the lowest energy.

4 Summary

We presented *first-principles* total-energy calculations on the structure and stability of (001) surfaces of GaAs and InAs. A $\alpha2(2\times4)$ surface model containing single anion dimers in the first and third atomic atomic layers is predicted for a balanced surface stoichiometry. Our results for cation-rich $(4\times2)/c(8\times2)$ reconstructed surfaces support the newly proposed ζ geometry. The surface electronic structure of the GaAs(001)$\zeta(4\times2)$ surface is dominated by anion dangling bonds energetically close to the bulk valence band maximum and empty Ga dimer states above the bulk conduction band minimum. A massively parallel implementation of DFT-LDA has enabled us to study for the first time GaAs surface reconstructions as large as (4×6). Our results seem to exclude the formation of Ga clusters and As-Ga substitution models as possible explanation for the observed (4×6) symmetries on Ga-rich GaAs surfaces.

Grants of computer time from the Höchstleistungsrechenzentrum Stuttgart, the Leibniz-Rechenzentrum München and the John von Neumann-Institut Jülich are gratefully acknowledged.

References

1. J. L. Fattebert and J. Bernholc, Phys. Rev. B **62**, 1713 (2000).
2. J. Bernholc, E. L. Briggs, C. Bungaro, M. B. Nardelli, J. L. Fattebert, K. Rapcewicz, C. Roland, W. G. Schmidt, and Q. Zhao, phys. stat. sol. (b) **217**, 685 (2000).
3. W. G. Schmidt, F. Bechstedt, and J. Bernholc, J. Vac. Sci. Technol. **18**, 2215 (2000).
4. P. H. Hahn, Master's thesis, Friedrich-Schiller-Universität Jena, 2001.
5. M. Fuchs and M. Scheffler, Comput. Phys. Commun. **119**, 67 (1999).
6. E. L. Briggs, D. J. Sullivan, and J. Bernholc, Phys. Rev. B **54**, 14362 (1996).
7. Q.-K. Xue, T. Hashizume, and T. Sakurai, Prog. Surf. Sci. **56**, 1 (1997).
8. W. G. Schmidt, S. Mirbt, and F. Bechstedt, Phys. Rev. B **62**, 8087 (2000).
9. A. I. Shkrebtii, N. Esser, W. Richter, W. G. Schmidt, F. Bechstedt, B. O. Fimland, A. Kley, and R. Del Sole, Phys. Rev. Lett. **81**, 721 (1998).
10. Q.-K. Xue, T. Hashizume, J. M. Zhou, T. Sakata, T. Ohno, and T. Sakurai, Phys. Rev. Lett. **74**, 3177 (1995).
11. S.-H. Lee, W. Moritz, and M. Scheffler, Phys. Rev. Lett. **85**, 3890 (2000).

12. W. G. Schmidt, S. Mirbt, and F. Bechstedt, in *25th Int. Conf. on the Physics of Semiconductors 2000* (Springer-Verlag, Berlin, p. 279).
13. T. Ohno, Surf. Sci. **357/358**, 265 (1996).
14. W. G. Schmidt, J. L. Fattebert, J. Bernholc, and F. Bechstedt, Surf. Rev. Lett. **6**, 1159 (1999).
15. W. G. Schmidt, N. Esser, A. M. Frisch, P. Vogt, J. Bernholc, F. Bechstedt, M. Zorn, T. Hannappel, S. Visbeck, F. Willig, and W. Richter, Phys. Rev. B. **61**, R16335 (2000).
16. D. K. Biegelsen, R. D. Bringans, J. E. Northrup, and L. E. Swartz, Phys. Rev. B **41**, 5701 (1990).
17. M. Kuball, D. T. Wang, N. Esser, M. Cardona, and J. Zegenhagen, Phys. Rev. B **51**, 13880 (1995).
18. S. B. Zhang and A. Zunger, Phys. Rev. B **53**, 1343 (1996).
19. W. A. de Heer, Rev. of Mod. Phys. **65**, 611 (1993).
20. P. Kruse, J. G. McLean, and A. C. Kummel, J. Chem. Phys. **113**, 2060 (2000).
21. C. Ratsch, W. Barvoza-Carter, F. Grosse, J. H. G. Owen, and J. J. Zinck, Phys. Rev. B **62**, R7719 (2000).
22. R. H. Miwa and G. P. Srivastava, Phys. Rev. B **62**, 15778 (2000).
23. S. Ohkouchi and N. Ikoma, Jpn. J. Appl. Phys. **33**, 3710 (1994).
24. C. Kumpf, L. D. Marks, D. Ellis, D. Smilgies, E. Landemark, M. Nielsen, R. Feidenhans'l, J. Zegenhagen, O. Bunk, J. H. Zeysing, Y. Su, and R. L. Johnson, Phys. Rev. Lett. **86**, 3586 (2001).

The Role of the Geometric Structure for Electronic Excitations of Molecules and Surfaces

Michael Rohlfing

Institut für Festkörpertheorie, Universität Münster, Wilhelm-Klemm-Str. 10, 48149 Münster, Germany

Abstract. We investigate the spectrum of excited electronic states of various systems by a combination of three ab-initio techniques: density-functional theory, *GW* quasiparticle calculations, and the Bethe-Salpeter equation for coupled electron-hole excited states. Results for three different materials are discussed: the optical spectrum of quartz, the optical response of the Ge(111)-(2×1) surface, and excited states of methane. The particular focus is on the interrelation of the excitations with the atomic geometry. In the case of Ge(111)-(2×1), the calculated spectra allow to distinguish between two nearly isoenergetic isomers of the surface. In the case of methane, it is shown that the geometry changes drastically during the transition from the ground state to the excited state.

1 Introduction

Excited electronic states of materials and the corresponding spectral properties play a central role in physics and chemistry. They are used to characterize and manipulate materials and they form the basis for a vast variety of spectroscopic methods and technological applications, like photovoltaics, photochemistry, dye chemicals, light-emitting devices, etc. On this background, the detailed understanding, calculation, and prediction of excited electronic states is of great importance [1,2].

One particular issue concerns the interrelation between the electronic excitations and the underlying geometric structure. On the one hand, different structural arrangements (polytypes or isomers) of a system yield different excitation spectra. This is extremely important since it allows to distinguish between different structures by spectroscopic techniques. We will discuss one examples, i.e. the isomeric behavior of the Ge(111)-(2×1) surface [3]. On the other hand, dynamic coupling of the electronic and geometric degrees of freedom results in characteristic features of the spectra, in particular vibrational side bands and inhomogeneous broadening. As an example, we will present results for the methane molecule (CH_4) [4].

The accurate determination and analysis of such excitations by highly reliable *ab-initio* techniques, dealing with the quantum-mechanical nature on the atomic scale, is the subject of the present work. Two issues have to be kept in mind concerning *ab-initio* methods: On the one hand, an exact treatment of the quantum-mechanical many-electron problem is impossible. Instead, accurate approximations must be employed that allow for a reliable description of the electronic structure. Depending on the physical quantity in question and on the required accuracy, various techniques have been developed. On the other hand, even within such approximations an enormous numerical effort may result, which requires to use efficient algorithms and a large amount of computational power. This holds in particular for large molecules and for (quasi-)periodic systems with large supercells, since the numerical demand of most ab-initio methods increases at least as the third power of the number of atoms involved.

In the present work we use a hierarchy of three successive techniques to address the problem of optical excitations and the related spectra:

A. The first quantity to be calculated is the electronic ground state $|N, 0\rangle$ of the N-electron system, which forms the basis for all excitations. We calculate $|N, 0\rangle$ within density-functional theory (DFT) [5]. Norm-conserving ab-initio pseudopotentials are employed to eliminate the core electrons from the calculations. The exchange-correlation functional is calculated within the local-density approximation (LDA) or local spin-density approximation (LSDA), depending on the nature of the ground state. If necessary, the geometric equilibrium structure of the system is calculated from the DFT total energy.

B. The excitations to the $(N \pm 1)$-electron states are described by the single-particle Green function G_1. To evaluate the corresponding excitation spectrum one has to solve the equation of motion for G_1, including the corresponding interaction kernel (self-energy operator) which describes the many-particle effects in the system. In the present work we employ the GW approximation (GWA) for the electron self-energy operator [6].

C. The charge-neutral excitations to higher states within the N-electron system, that are relevant for the optical spectrum, are described by the (electron-hole) two-particle Green function G_2. Again, its equation of motion (Bethe-Salpeter equation, BSE) must be solved to evaluate the spectrum of G_2, and an interaction kernel (electron-hole interaction) must be included to account for many-particle effects [7]. The electron-hole interaction is related to the self-energy operator and is again calculated within the GW approximation. Note that the procedure in this step requires information about G_1. Therefore step B cannot be passed, but has to be carried out before addressing G_2. In the following, the approach will often be referred to as "GW+BSE".

All three steps have been turned into numerical algorithms by us [8–10,1] and by other groups (see, e.g., Refs. [6,11–14]), making use of powerful multiprocessor computers. While the first two steps (in particular, the DFT) are nowadays well-established approaches, the third step (BSE) is still relatively new and has only be evaluated by a few groups within the ab-initio framework.

We briefly introduce the theoretical background of the three steps (in particular, of the second and third step (GW+BSE)), and discuss some technical details concerning the computational requirements. Thereafter, recent results for a number of interesting physical questions[1–4] related to electronic excitations are discussed.

2 Theoretical and computational framework

In this section the theoretical framework of our approach is sketched out. It consists of the three computational ab-initio techniques A–C as discussed in the introduction.

(**A**): The starting point of our calculations is a density-functional theory (DFT) calculation to obtain the electronic ground state. A set of basis functions $\phi_\alpha(\mathbf{k}, \mathbf{r})$ must be chosen to represent the real-space behavior of the electronic wave functions $\psi_{n\mathbf{k}}^{\mathrm{DFT}}(\mathbf{r})$:

$$\psi_{n\mathbf{k}}^{\mathrm{DFT}}(\mathbf{r}) = \sum_\alpha C_{n\mathbf{k}}^\alpha \phi_\alpha(\mathbf{k}, \mathbf{r}) \qquad . \tag{1}$$

In most cases we employ Gaussian-orbital (GO) basis sets, that are especially useful for surfaces and molecules. In many studies, however, we also use a plane-wave (PW) code to determine the atomic structure, followed by a GO expansion as a starting point for the investigation of the spectral properties.

(**B**): Based on the electronic ground state, as described by DFT, the quasiparticle (QP) excitations of the electronic system are investigated by many-body perturbation theory [11]. The quasiparticles, that are long-lived excitations (=poles) of the single-particle Green function G_1, result from the Dyson equation

$$(H_0 + \Sigma(E_{n\mathbf{k}}^{\mathrm{QP}}))\psi_{n\mathbf{k}}^{\mathrm{QP}} = E_{n\mathbf{k}}^{\mathrm{QP}}\psi_{n\mathbf{k}}^{\mathrm{QP}} \qquad . \tag{2}$$

In here, H_0 is the Hamilton operator in the Hartree approximation of uncorrelated electrons. The self energy operator Σ describes the exchange and correlation effects among the electrons. An exact determination of Σ is basically impossible for most systems. Instead, reliable approximations must be employed. A very successful approach is given by the GW approximation (GWA) [6]:

$$\Sigma(1, 2) = iG_1(1, 2)W(1, 2) \tag{3}$$

with W being the screened Coulomb interaction, $W = \epsilon^{-1}v$, as resulting from the random-phase approximation [6,11].

The quantities occurring in the GW method must again be expanded in a suitable set of basis functions which can be chosen independent of the DFT basis. To this end, we employ again Gaussian orbitals, that will be labelled $\chi_\beta(\mathbf{q}, \mathbf{r})$. Since the functions P, ϵ, etc. are *two-point functions* in real space, the basis-set representation is a double expansion, leading to a matrix

Table 1. Data of the numerical requirements for the GW approach on the Cray T3E for the systems studied in the present work, i.e. the SiO_2 bulk crystal, the Ge(111)-(2×1) surface (containing 12 layers of Ge), and the CH_4 molecule. Given are the numbers of valence bands, the number of conduction bands to be considered, the number of GO's ($\{\beta\}$) in Eq. (4), the required total memory, and typical CPU times for the calculation of the polarizability P (4) and of the self energy Σ (3) (single-processor times).

	bulk SiO_2	Ge(111)-(2×1)	CH_4
valence bands	24	48	4
cond. bands	240	480	60
GO basis functions	270	720	220
memory	800 KB	900 MB	2 MB
CPU time: P	30 h	120 h	2 h
Σ	20 h	50 h	4 h

representation of each function. The polarizability P, e.g., is then given by matrix elements

$$P_{\beta\beta'}(\mathbf{q}, \omega) = 2\frac{1}{V} \sum_{\mathbf{k}} \sum_{v \in \mathrm{Val}} \sum_{c \in \mathrm{Con}} M_\beta^{vc}(\mathbf{k}, \mathbf{q}) \left[M_{\beta'}^{vc}(\mathbf{k}, \mathbf{q}) \right]^*$$

$$\times \left[\frac{1}{E_{v\mathbf{k}} - E_{c,\mathbf{k}+\mathbf{q}} - \omega} + \frac{1}{E_{v\mathbf{k}} - E_{c,\mathbf{k}+\mathbf{q}} + \omega} \right] \tag{4}$$

with integrals

$$M_\beta^{vc}(\mathbf{k}, \mathbf{q}) = \int \psi_{v\mathbf{k}}^*(\mathbf{r}) \chi_\beta^*(\mathbf{q}, \mathbf{r}) \psi_{c,\mathbf{k}+\mathbf{q}}(\mathbf{r}) d^3 r \quad . \tag{5}$$

We discuss these equations here to illustrate a number of important issues concerning the numerical realization of the approach: (*i*): Unlike DFT, where most quantities can be expressed by vector expansions [like Eq. (1)], the GW method is characterized by two-point functions (like P) and corresponding *matrix* expansions. This is one of the main reasons why GW calculations are much more computationally demanding than DFT calculations. (*ii*): The calculation of P requires a sum over all valence bands (v) and over a wide range of conduction bands (c) of the material. The number of both the valence bands and of the relevant conduction bands scales with the size of the system (i.e., the number of atoms in the unit cell). This means that the computational effort (in terms of memory and especially CPU time) increases drastically when complex materials are investigated, or when surfaces or other low-dimensional systems are studied by supercell calculations. Similar scaling properties with increasing number of atoms are found for molecules. (*iii*): The computational effort depends sensitively on the number of basis functions $\{\beta\}$. Therefore, an appropriate choice of the basis can help enormously to keep the computational effort moderate and make the calculations feasible. In many cases, the GO basis set used by us can be kept much smaller than a corresponding PW basis set of the same accuracy would have to be. This is especially advantegeous for molecules.

The most time-demanding part of the GW approach (more than 95 % of the total computation time) is consumed by calculating the integrals given by Eq. (5) and carry out the summations in Eq. (4). The integrals in Eq. (5) are composed of three-center Gaussian-orbital integrals

$$\int \phi_\alpha(\mathbf{r} - \mathbf{R}_\alpha)\chi_\beta(\mathbf{r} - \mathbf{R}_\beta)\phi_{\alpha'}(\mathbf{r} - \mathbf{R}_{\alpha'})d^3r \tag{6}$$

that we calculate iteratively by a method proposed by Obara and Saika [15]. $\mathbf{R}_\alpha$, $\mathbf{R}_\beta$ and $\mathbf{R}_{\alpha'}$ are the positions of the atoms at which the orbitals are centered. Since the integrals from different combinations of $\mathbf{R}_\alpha$, $\mathbf{R}_\beta$ and $\mathbf{R}_{\alpha'}$ are independent of each other, the calculation of the integrals can easily be distributed over different nodes of a parallel computer. The GW method is thus an ideal case for efficient parallelization. In most calculations, we use 16-32 parallel nodes on the Cray T3E machine at HLRS.

(C): After calculating the QP wave functions and energies of electron and hole states, the third step can be addressed, i.e. the calculation of the electron-hole interaction and the solution of the Bethe-Salpeter equation (BSE) for coupled electron-hole excitations. These excitations $|S\rangle$ are given by

$$|S\rangle = \sum_{\mathbf{k}}\sum_{v}\sum_{c}^{\text{hole elec}} A_{vc}^S|vc\mathbf{k}\rangle \tag{7}$$

where $|vc\mathbf{k}\rangle := \hat{a}_{v\mathbf{k}}^\dagger \hat{b}_{c\mathbf{k}}^\dagger|0\rangle$ denotes the simultaneous creation of a hole $(\hat{a}_{v\mathbf{k}}^\dagger)$ and an electron $(\hat{b}_{c\mathbf{k}}^\dagger)$ as independent particles. The sum, which involves coupling coefficients $A_{vc\mathbf{k}}^S$, indicates that the coupled excitations $|S\rangle$ are coherently superposed from the independent-particle transitions $|vc\mathbf{k}\rangle$.

The BSE for the excitations (7) is given by [7]

$$(E_{c\mathbf{k}}^{\text{QP}} - E_{v\mathbf{k}}^{\text{QP}})A_{vc\mathbf{k}}^S + \sum_{v'c'\mathbf{k}'}\langle vc\mathbf{k}|K^{eh}|v'c'\mathbf{k}'\rangle A_{v'c'\mathbf{k}'}^S = \Omega_S A_{vc\mathbf{k}}^S \quad . \tag{8}$$

Solving this equation yields the excitation energy Ω_S and the coupling coefficients $A_{vc\mathbf{k}}^S$ of each excited state S. The first term in Eq. (8) consists of the QP energy differences between the occupied and empty levels (corresponding to independent-particle transitions). The second term contains the electron-hole interaction $\langle vc\mathbf{k}|K^{eh}|v'c'\mathbf{k}'\rangle$, which is responsible for the coupling between the particles. The algorithm to evaluate these matrix elements has a structure which is extremely similar to the matrix elements of the GW self-energy operator in step (B). The numerical codes are partly identical, and the numerical demand is similar. In many cases, however, the number of interaction matrix elements is larger than the number of self-energy matrix elements, leading to a much higher total numerical demand that the GW calculation itself. Again, the largest part of the BSE problem can be distributed over parallel nodes in a highly efficient way. Also in these calculations, 16-32 parallel nodes are typically used.

The size of the two-particle problem (8) depends very much on the physical quantity to be investigated. If only the lowest-energy excited states are of interest (as in the case of the Ge(111)-(2×1) surface in Sec. 3.2), it may be sufficient to restrict the BSE to only one occupied and one empty band. If, on the other hand, the entire spectrum is relevant (as in the case of SiO_2), more bands must be included, thus covering a much larger spectral energy range. This drastically increases the size of the problem (see Table 2).

Table 2. Data of the numerical requirements for the BSE approach on the Cray T3E for a number of typical systems, i.e. the SiO_2 bulk crystal, the Ge(111)-(2×1) surface (containing 12 layers of Ge), and the CH_4 molecule. Given are the numbers of valence bands, conduction bands, and **k** points to be considered for the two-particle Hamiltonian, the size of the two-particle Hamilton matrix, and the typical total memory and total CPU time requirement.

	bulk SiO_2	Ge(111)-(2×1)	CH_4
valence bands	15	1	4
cond. bands	6	1	40
k points	100	200	1
size of H_2	9000	200	200
memory	1500 MB	900 MB	3 MB
CPU time	160 h	250 h	5 h

For details, we refer the reader to Ref. [1].

3   Results and discussion

In this section a number of recent results for various systems are discussed. One system is given by the SiO_2 bulk crystal; the understanding of the spectrum of this material had been limited before, and our calculations allow to reveal in detail the optical properties [2]. In addition, we investigate the interrelation between the electronic excitations and the geometric structure of two low-dimensional semiconductor systems, the Ge(111)-(2×1) surface [3] and the CH_4 molecule [4].

3.1   Optical spectrum of an insulators: α-Quartz

Insulators are characterized by a large band gap and a concomitantly weak dielectric screening. α-Quartz (SiO_2), e.g., has a dielectric constants of only

2.4. The electron-hole interaction and resulting excitonic effects are thus very strong in these materials. As an example, Fig. 1 shows the macroscopic dielectric functions of α-SiO$_2$ (see also Ref. [2]).

One observes a dramatic modification of the interband spectrum due to the interaction. The spectrum is completely changed at all energies. Most importantly, a sharp peaks occurs in $\epsilon_2(\omega)$ at 10.1 eV. In addition to the low-energy exciton, further peaks are found at higher energies. In $\epsilon_2(\omega)$, three additional peaks are identified at 11.3 eV, around 14 eV, and at 17.5 eV. All peaks are in good agreement with experimental data. This illustrates once more that excitonic effects are crucial for a detailed understanding of the spectrum at all energies. The free-particle interband spectrum, on the other hand, does not yield these structures: they do not correspond to critical points in the band structure.

The nature of the first peak in SiO$_2$ is slightly different from the bound exciton state found in direct semiconductors and insulators. The peak corresponds to a resonant exciton with a strong dipole due to constructive inter-

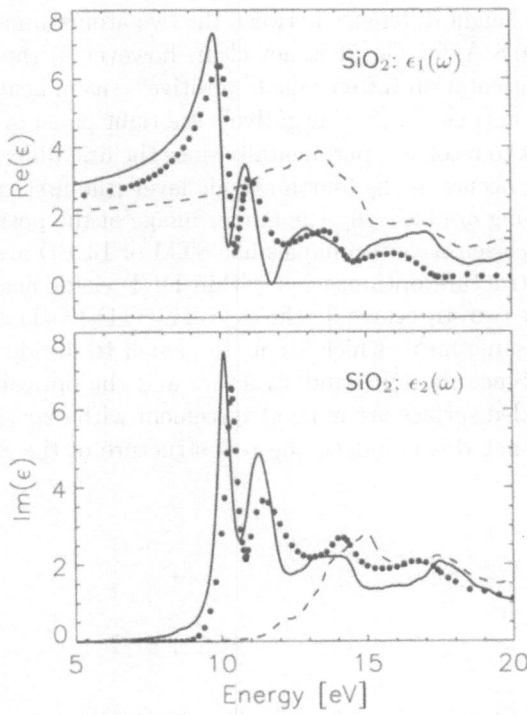

Fig. 1. Macroscopic dielectric function of α-quartz (SiO$_2$), calculated with (—) and without (- - -) electron-hole interaction. The experimental data ($\bullet$) are taken from Ref. [16].

ference. Its excitation energy of 10.1 eV is *above* the fundamental band gap of 9.5 eV. The reason for this behavior lies in the band structure of SiO_2, which is indirect, i.e. the valence-band maximum is at the boundary of the Brillouin zone while the conduction-band minimum is at the Γ point. The highest valence band has positive instead of negative curvature at the Γ point, which significantly influences the formation of the excitations. Furthermore, many bands contribute to the excited states in α-SiO_2 (due to the complex unit cell), which makes a detailed analysis of the individual excited states difficult.

For details, we refer the reader to Ref. [2].

3.2   Spectrum of the Ge(111)-(2×1) surface

Similar to the well-known Si(111)-(2×1) surface, the Ge(111)-(2×1) surface is also terminated by π-bonded zigzag chains, as proposed by Pandey [17]. The buckling direction of the chain, however, is unclear. In Ref. [17], Pandey considered a flat chain without buckling. Both for Si(111)-(2×1) and for Ge(111)-(2×1), the surface gains energy by *buckling*, i.e. by outward relaxation of one surface atom per (2×1) cell (up atom) and by inward relaxation of the other (down atom). The height difference between the two atoms amounts to about 0.5 Å for Si and 0.8 Å for Ge. It is not clear, however, if the buckling occurs in clockwise orientation (often called "positive"), as indicated in the left panel of Fig. 2, or anti-clockwise ("negative"; see right panel of Fig. 2). This question is difficult to resolve experimentally since the first difference between the two structures occurs in the fourth atomic layer [the first three layers of the negative buckling are basically the mirror image of the positive buckling (cf. Fig. 2)]. Surface-sensitive techniques like STM or LEED are not capable to identify this. Structure optimization within DFT yields nearly the same total energy for the two structures. In the case of Si(111)-(2×1), the difference is in the order of some meV, which seems too small to decide which structure is favorable. Since the QP band structure and the optical spectrum of the positively buckled surface are in good agreement with experimental data, it was concluded that this is indeed the real structure of the Si(111)-(2×1)

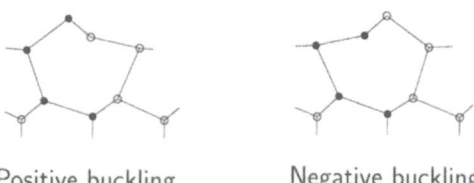

Positive buckling   Negative buckling

Fig. 2. Side view of the two almost isoenergetic structures of the (111)-(2×1) surface of Si or Ge, displaying 'positive' (clockwise) and 'negative' (anti-clockwise) buckling of the Pandey chains.

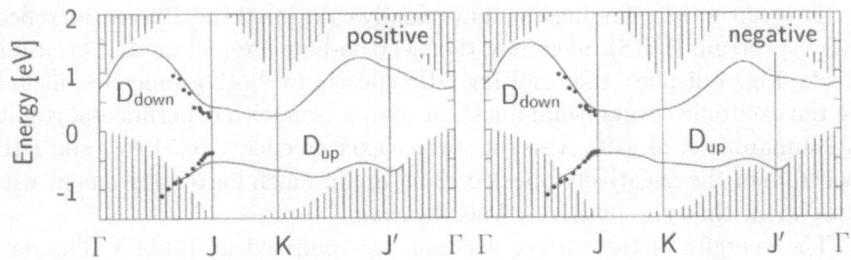

Fig. 3. *GW* QP band structure of the Ge(111)-(2×1) surface. The left (right) panel is for the positively (negatively) buckled chain (see text). The experimental data are from Ref. [18].

surface (cf. Ref. [10]). The spectra of the negatively buckled Si(111)-(2×1) surface, on the other hand, differ from experiment (see Table 3 below).

The situation is different for the Ge(111)-(2×1) surface. Also for this system, a Pandey chain model seems to be the real surface structure. Different from the silicon case, however, the negatively buckled chain is now lower in energy by 26 meV per surface unit cell. Still, this small energy difference may not allow for a definite conclusion concerning the direction of the buckling, and additional information on the spectra is required. Fig. 3 depicts the QP surface-band structure of the positively and negatively buckled surface, together with experimental data by Nicholls et al. [18]. Both band structures are very similar, but the negatively buckled chain produces a smaller surface gap of 0.66 eV than the positively buckled chain (0.88 eV). The former is in better agreement with the experimental findings of 0.61 eV [18].

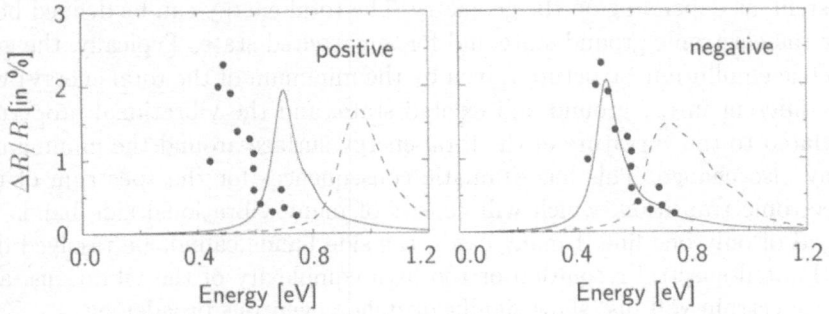

Fig. 4. Differential reflectivity spectrum of the Ge(111)-(2×1) surface, calculated for normal incidence. The left (right) panel is for the positively (negatively) buckled chain (see text). The dots denote experimental data by Nannarone *et al.* [19].

To confirm these findings, we have finally calculated the differential reflectivity spectrum (DRS), including the electron-hole interaction and excitonic effects. Fig. 4 displays the resulting DRS spectra for both geometries, including the excitonic effects (solid lines), in comparison with experimental results by Nannarone et al. [19]. Also the surface exciton energy (0.51 eV) and DRS spectrum of the negatively buckled chain are in much better agreement with experiment than the positively buckled chain.

The energies of the surface excitons are compiled in Table 3. The data indicate that the Si(111)-(2×1) surface is terminated by *positively* buckled Pandey chains while the Ge(111)-(2×1) surface is terminated by *negatively* buckled chains.

Table 3. Calculated and measured surface exciton transition energies at the Ge(111)-(2×1) and Si(111)-(2×1) surface. The calculated data are for the positively and the negatively buckled chain, respectively (see text).

[eV]	"+" buckling	"−" buckling	Exp.
Ge(111)-(2×1)	0.70	0.51	0.49[a]
Si(111)-(2×1)	0.43	∼0.25	0.45[b]

[a]Ref. [19].   [b]Refs. [20,21].

For details, we refer the reader to Ref. [3].

3.3   Excited states in methane

Another issue concerning the interrelation between the excitations and the geometric structure is the *dynamic* coupling between the geometric and electronic degrees of freedom. This is closely related to the total energy of the system, as depending on the geometry. The total energy can be defined both for the electronic ground state and for any excited state. Typically, the geometric equilibrium structure (given by the minimum of the total energy) will be different in the ground and excited state, and the vibrational properties (related to the curvature of the total-energy surface around the minimunm) may also change. This has dramatic consequences for the spectrum of the electronic transition, which will consist of many vibrational side bands instead of only one line. I many cases, the side bands cannot be resolved due to limited spectral resolution or too high complexity of the vibrations, and the spectrum will just show significant inhomogeneous broadening.

Here we present results for the geometry-dependent total energy of the methane molecule, CH_4. The equilibrium structures of the ground and excited state are very different. In particular, the bond angles change by 15°; this is accompanied by a gain in total energy of 1.5 eV in the excited state. The main reason for this behavior is the high symmetry of the molecule in the electronic ground state. The molecule has tetrahedral symmetry. Its electronic ground state (see left panel of Fig. 5) is characterized by a three-fold degenerate HOMO level (fully occupied by 6 electrons) and a non-degenerate LUMO level. Because of the full occupation of the HOMO levels the many-body state is non-degenerate. An excitation requires to either remove one HOMO electron from the system (ionization) or (in a simplified picture) to transfer it to an empty level, e.g. to the LUMO level (electron-hole excitation). In any case, there are three possibilities to excite one of the three HOMO levels (not counting for the spin degree of freedom), i.e. the excited state is three-fold degenerate.

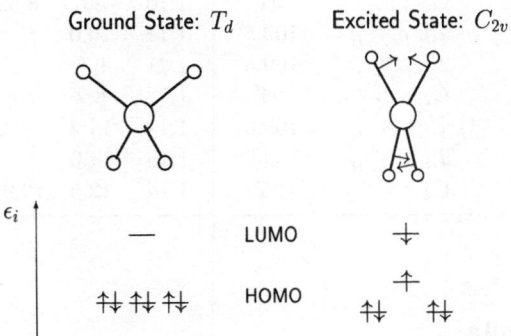

Fig. 5. Illustration of the geometric and electronic structure of the CH_4 molecule (see text).

Since this degeneracy (resulting from the degeneracy of the HOMO levels) is a direct consequence of the tetrahedral symmetry of the molecule, it is subject to a symmetry-breaking Jahn-Teller distortion of the structure, which lifts the degeneracy and leads to an additional gain in total energy. In the present case, one of the HOMO levels is shifted to higher energy while the others are lowered. Since this upwards-shifted new HOMO level is the one which is involved in the excitation, the transition energy is reduced and the total energy of the excited state is lowered. The distortion leads to a significant change of the H-C-H bond angles by 15° and to a strong reduction of the total energy by 1.5 eV. Details are given in Table 4.

Note that such an investigation represents a significant numerical effort although the calculation for one geometric structure only requires a few hours of computation time (see Tab. 1 and 2). The effort arises from the requirement to carry out this calculation many times, i.e. for a large number of geometric configurations. This becomes extremely complicated for polyatomic molecules that have a configuration space of $3(N-1)$ with N being the number of atoms.

For more details, we refer the reader to Ref. [4].

Table 4. Interrelation between the atomic geometry and electronic excitations of the CH_4 molecule (see text). d_{C-H}^{GS} denotes the ground-state equilibrium C-H distance of 1.11 Å. d_{C-H}^{opt} denotes the C-H distance optimized in the excited state. The bond angle denotes the smaller of the two different bond angles in the C_{2v} symmetry. The fourth column contains the total energy of the excited states relative to the ground state, which is set to 0 eV. The measured excitation energies are from Ref. [22].

State	Geometry	Bond angle	d_{C-H} [Å]	Energy [eV]	Exp. [eV]
$|0\rangle$	$T_d;\ d_{C-H}^{GS}$	109.5°	1.11	0	
$|S\rangle$	$T_d;\ d_{C-H}^{GS}$	109.5°	1.11	10.4	
	$T_d;\ d_{C-H}^{opt}$	109.5°	1.18	10.2	
	$C_{2v};\ d_{C-H}^{opt}$	94°	1.16	8.6	8.52
$|T\rangle$	$T_d;\ d_{C-H}^{GS}$	109.5°	1.11	10.0	
	$T_d;\ d_{C-H}^{opt}$	109.5°	1.21	9.6	
	$C_{2v};\ d_{C-H}^{opt}$	94°	1.18	8.2	
$|N-1\rangle$	$T_d;\ d_{C-H}^{GS}$	109.5°	1.11	14.2	
	$T_d;\ d_{C-H}^{opt}$	109.5°	1.16	14.0	
	$C_{2v};\ d_{C-H}^{opt}$	95°	1.14	12.5	12.99

Acknowledgments

The author thanks E. Chang, J.C. Grossman, S.G. Louie and J. Pollmann for fruitful discussions. This work was supported by the Deutsche Forschungsgemeinschaft (Bonn, Germany) under Grant No. Ro-1318/2-1. Computational resources have been provided by the Bundes-Höchstleistungsrechenzentrum Stuttgart (HLRS).

References

1. M. Rohlfing and S. G. Louie, Phys. Rev. B **62**, 4927 (2000).
2. E. Chang, M. Rohlfing, and S. G. Louie, Phys. Rev. Lett. **85**, 2613 (2000).
3. M. Rohlfing, M. Palummo, G. Onida, and R. Del Sole, Phys. Rev. Lett. **85**, 5440 (2000).
4. J. C. Grossman, M. Rohlfing, L. Mitas, S. G. Louie, and M. L. Cohen, Phys. Rev. Lett. **86**, 472 (2001).
5. P. Hohenberg and W. Kohn, Phys. Rev. **136**, B864 (1964); W. Kohn and L. J. Sham, Phys. Rev. **140**, A1133 (1965); G. P. Srivastava and D. Weaire, Advances in Physics **36**, 463 (1987); G. B. Bachelet, D. R. Hamann, and M. Schlüter, Phys. Rev. B **26**, 4199 (1982).
6. L. Hedin, Phys. Rev. **139**, A796 (1965); L. Hedin and S. Lundqvist, Solid State Physics **23**, 1 (1969).
7. L. J. Sham and T. M. Rice, Phys. Rev. **144**, 708 (1966); W. Hanke and L. J. Sham, Phys. Rev. Lett. **43**, 387 (1979); Phys. Rev. B **21**, 4656 (1980); G.

Strinati, Phys. Rev. Lett. **49**, 1519 (1982); Phys. Rev. B **29**, 5718 (1984); Rivista del Nouvo Cimento **11**, 1 (1988).

8. M. Rohlfing, P. Krüger, and J. Pollmann, Phys. Rev. Lett. **75**, 3489 (1995); Phys. Rev. B **52**, 1905 (1995).

9. M. Rohlfing and S. G. Louie, Phys. Rev. Lett. **80**, 3320 (1998); *ibid.* **81**, 2312 (1998); *ibid.* **82**, 1959 (1999); M. Rohlfing and S. G. Louie, Phys. Rev. B **62**, 4927 (2000).

10. M. Rohlfing and S. G. Louie, Phys. Rev. Lett. **83**, 856 (1999); Phys. Stat. Sol. (a) **175**, 17 (1999).

11. M. S. Hybertsen and S. G. Louie, Phys. Rev. Lett. **55**, 1418 (1985); Phys. Rev. B **34**, 5390 (1986); F. Aryasetiawan and O. Gunnarsson, Reports on Progress in Physics **61**, 237 (1998).

12. G. Onida, L. Reining, R. W. Godby, R. Del Sole, and W. Andreoni, Phys. Rev. Lett. **75**, 818 (1995).

13. S. Albrecht, G. Onida, and L. Reining, Phys. Rev. B **55**, 10 278 (1997); S. Albrecht, L. Reining, R. Del Sole, and G. Onida, Phys. Rev. Lett. **80**, 4510 (1998); Phys. Stat. Sol. A **170**, 189 (1998).

14. L. X. Benedict, E. L. Shirley, and R. B. Bohn, Phys. Rev. Lett. **80**, 4514 (1998); Phys. Rev. B **57**, R9385 (1998); Phys. Rev. B **59**, 5441 (1999).

15. S. Obara and A. Saika, J. Chem. Phys. **84**, 3963 (1986).

16. H.R. Philipp, Solid State Comm. **4**, 73 (1966).

17. K.C. Pandey, Phys. Rev. Lett. **49**, 223 (1982).

18. J.M. Nicholls, G.V. Hansson, U.O. Karlsson, R.I.G. Uhrberg, R. Engelhardt, K. Seki, S.A. Flodstrom, and E.E. Koch, Phys. Rev. Lett. **52**, 1555 (1984); J.M. Nicholls and B. Reihl, Surf. Sci. **218**, 237 (1989).

19. S. Nannarone, P. Chiaradia, F. Ciccacci, R. Memeo, P. Sassaroli, S. Selci, and G. Chiarotti, Solid State Comm. **33**, 593 (1980).

20. P. Chiaradia, A. Cricenti, S. Selci, and G. Chiarotti, Phys. Rev. Lett. **52**, 1145 (1984); F. Ciccacci, S. Selci, G. Chiarotti, and P. Chiaradia, Phys. Rev. Lett. **56**, 2411 (1986).

21. F. Ciccacci, S. Selci, G. Chiarotti, and P. Chiaradia, Phys. Rev. Lett. **56**, 2411 (1986).

22. G. Herzberg, *Molecular Spectra and Molecular Structure*, Vol. III: *Electronic Spectra and Electronic Structure of Polyatomic Molecules* (Van Nostrand, New York 1966).

Structural and Vibronic Properties of the Dihydride-terminated Si(001) Surface

Ulrich Freking, Albert Mazur, and Johannes Pollmann

Westfälische Wilhelms-Universität Münster
Institut für Festkörpertheorie
Wilhelm-Klemm-Str. 10
48149 Münster
Germany

Abstract. We have used density functional theory to study the electronic and structural properties and density functional *perturbation* theory to study the vibronic properties of the dihydride-terminated Si(001) surface. A theoretical treatment of the *electronic* and *structural* properties of this particular system by first principles methods is possible on conventional workstations. Corresponding results have been published in the last years. The *ab-initio* calculation of *surface phonons* within a reasonable time is not possible on conventional workstations but calls for the use of more sophisticated parallel architectures like the IBM RS/6000 SP/256 at the Scientific Supercomputing Center Karlsruhe or the CRAY T3E-900/512 at the High-Performance Computing Center Stuttgart.

1 Introduction

The clean Si(001) surface does not exist in nature in its ideal configuration, because it would be highly reactive. The easiest possibility to passivate such a reactive surface retaining the ideal (1×1) structure is the adsorption of atomic hydrogen. This surface, i. e. 2H:Si(001)-(1×1), has been prepared for the first time in 1976 by SAKURAI and HAGSTRUM [Sak76] exposing the clean Si(001) to hydrogen atoms. In 1993, a scanning-tunneling-microscopy study of BOLAND confirmed such a structure [Bol93].

Since the underlying theory, the physical-technical parts of the implementation like the plane-wave expansion and the computational aspects like the efficiency of the parallelization have been discussed in detail in [Fre00,Fre01], this article focusses primarily on physical results of 2H:Si(001)-(1×1). The contribution is organized as follows. In Section II we briefly describe the theoretical method, in particular the main idea of the Density Functional Perturbation Theory. In Section III we present results concerning the structural properties and selected characteristic vibronic features of the dihydride-terminated Si(001) surface. Section IV concludes this article.

2 Theory

In our calculations, the 2H:Si(001)-(1×1) surface has been modeled by super-cells containing eight atomic layers, which were separated by vacuum regions corresponding to four layers. The total energy was calculated by means of Density Functional Theory in Local Density Approximation [Hoh64,Koh65]. For the exchange-correlation potential we have chosen the CEPERLEY-ALDER form [Cep80]. The single-particle wave functions occuring in the KOHN-SHAM equations are expanded into plane waves, where we considered a cut-off energy of at least 15 Ryd. The crystal potential has been described by norm-conserving first-principle pseudopotentials. Both semilocal pseudopotentials of BACHELET, HAMANN and SCHLÜTER [BHS82] and nonlocal potentials of KLEINMAN and BYLANDER [KB82] have been used. The equilibrum positions were determined with the help of the HELLMANN-FEYNMAN-theorem [Fey37].

In order to obtain vibronic properties, it is necessary to calculate the changes of the self-consistent potential V^{SCF} due to a perturbation of a given periodicity $\mathbf{q}$:

$$V^{SCF}(\mathbf{r}) \Rightarrow V^{SCF}(\mathbf{r}) + \Delta V_{\mathbf{q}}^{SCF}(\mathbf{r}). \tag{1}$$

The response of the electronic charge density $\Delta n_{\mathbf{q}}(\mathbf{r})$ to this change of the potential is obtained by first-order perturbation theory. It is possible to avoid the very time consuming sums over all unoccupied conduction bands by the projection technique introduced by BARONI et al. [Bar87,Gia91]. The self-consistent calculation of these density changes is the most time consuming part of the calculation, so that it can be done only on high-performance computers. For further details with regard to the physical-technical aspects of the implementation like the plane-wave expansion see [Fre01].

Concerning the implementation of the Density Functional Perturbation Theory, we have been able to realize a parallelization of 98 % of our computer code. For an exploitation of sixteen nodes, this corresponds to a speed up of 12.3. On the IBM RS/6000 SP at the Supercomputing Center in Karlsruhe, the calculation of surface phonons with a given wavelength for a supercell containing eight atomic layers (with ten atoms because of the dihydride structure) takes about 12 hours CPU time on each node. This time refers to the use of thin P2SC nodes with 120 MHz P2SC processors and 512 MB main memory, each. To calculate the complete phonon dispersion curves for the supercell mentioned above, about 1500 CPU hours are required. For further details concerning the parallelization see [Fre00].

3 Results

3.1 Structural Properties

We will discuss two different structures of the (1 × 1) surface [Cir84]. They are shown in Fig. 1. The symmetric structure, shown in Fig. 1 a), results

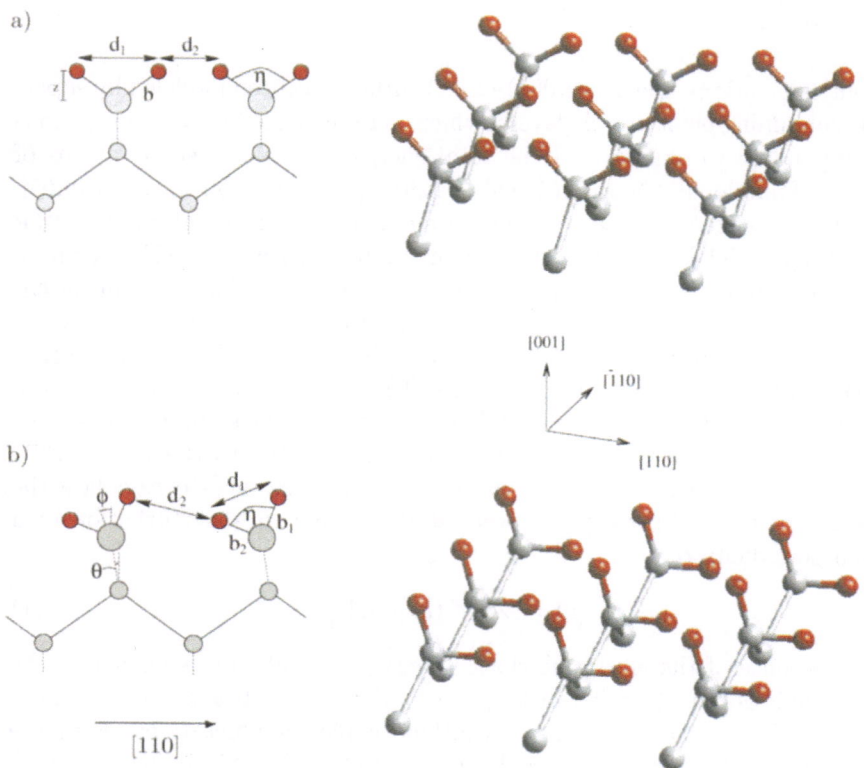

Fig. 1. a) Symmetric and b) canted-row configuration of 2H:Si(001)-(1 × 1). The red atoms symbolize the chemisorbed hydrogen atoms, grey atoms the underlying substrate.

through merely saturating the dangling bonds of the ideal surface by one H-atom each. The canted-row structure (see Fig. 1 b)) can be obtained from the symmetric structure by a rotation of the 2HSi groups around an axis which passes through the silicon atoms of the second substrate layer.

The most important structural parameters of the symmetric structure are listed in Table 1. They are in good agreement for the different pseudopotentials (after KLEINMAN and BYLANDER (KB) respectively after BACHELET, HAMANN and SCHLÜTER (BHS)) on the one hand and confirm theoretical data obtained by a DFT-LDA calculation of NORTRHUP [Nor91] on the other hand.

The canted-row structure ($\phi > 0$) and the equivalent mirror-inverted structure ($\phi < 0$) can be regarded as the two minima of a double-well potential, for which the symmetric configuration corresponds to the unstable maximum at $\phi = 0$. The structure with the rotated 2HSi group is found to be energetically more favorable than the symmetric structure. It allows

Table 1. Different structural parameters of 2H:Si(001)-(1 × 1) in the symmetric configuration including a comparison with other theoretical data. The definition of the parameters corresponds to Fig. 1 a).

	d_1	d_2	b	z	η
KB	2.33 Å	1.44 Å	1.50 Å	0.96 Å	101.0°
BHS	2.32 Å	1.46 Å	1.51 Å	0.96 Å	101.5°
[Nor91]	2.33 Å	1.51 Å	1.50 Å	0.95 Å	102°

a reduction of the surface energy by 0.2 eV per unit cell. The reason is the possibility of enlarging the distance d_2 between neighboring hydrogen atoms leading to a reduction of the repulsive interaction of their electrons. As can be seen from the structural parameters given in Table 2, again our results confirm the calculations of NORTHRUP. Also the energy gain mentioned above is equal to the value given in [Nor91].

Table 2. Different structural parameters of 2H:Si(001)-(1 × 1) in the canted-row configuration including a comparison with other theoretical data. The definition of the parameters is corresponding to Fig. 1 b).

	d_1	d_2	b_1	b_2	θ	ϕ	η
KB	2.43 Å	2.11 Å	1.53 Å	1.54 Å	15.6°	31.5°	104.7°
BHS	2.42 Å	2.12 Å	1.52 Å	1.54 Å	15.5°	31.4°	104.4°
[Nor91]	2.46 Å	2.21 Å	1.54 Å	1.54 Å	15°	30°	

As the canted-row structure turned out to be energetically more favorable, we have restricted our studies to this configuration.

3.2   Vibronic Properties

The most prominent surface modes can easily be understood by regarding the bonding geometry of the adsorbed H-atoms. They are named – corresponding to their physical origin – as *scissor*, *rocking*, *wagging*, *twisting* and *stretching* modes. The displacement patterns of these modes at the Γ-point, i. e. $\mathbf{q} = \mathbf{0}$, are shown in Fig. 2. The *scissor* mode S and the *rocking* mode R describe movements of the hydrogen atoms in the plane of the 2HSi group, in fact like scissors against each other out of phase on the one hand and like a rocking chair in phase on the other hand. The hydrogen atoms can move also perpendicular to this plane both in phase and out of phase; these modes are called the *wagging* mode W and the *twisting* mode T. Last but not least, there are two *stretching* modes desribing the oscillation of each of the hydrogen atoms along the direction of the respective H-Si-bond.

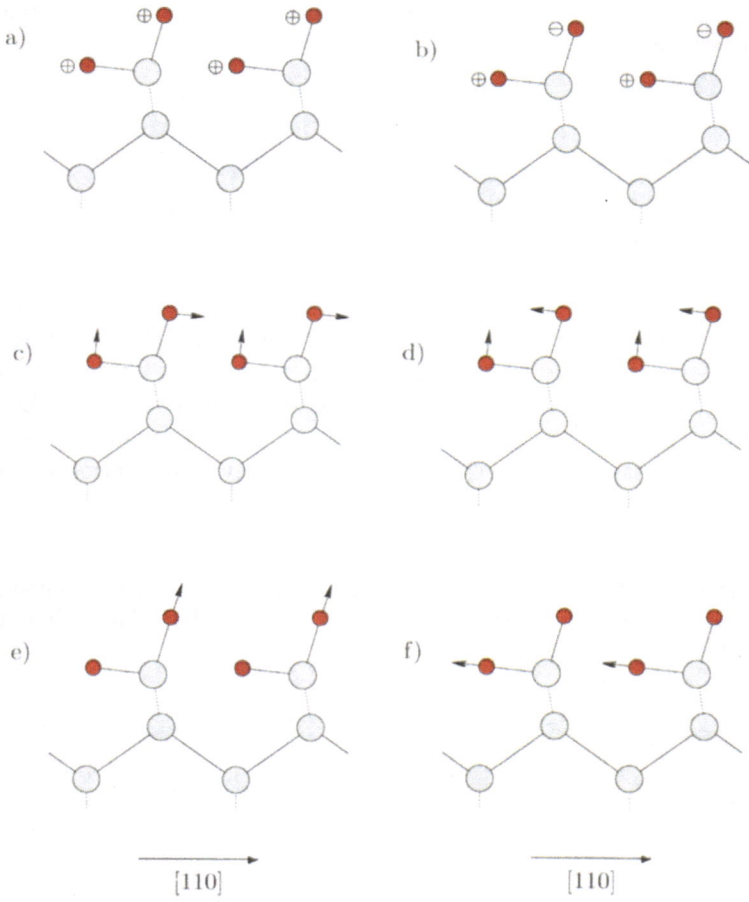

Fig. 2. Displacement patterns of the six characteristical surface modes at Γ. a) *wagging* mode W, b) *twisting* mode T, c) *rocking* mode R, d) *scissor* mode S as well as *stretching* modes e) S_{up} of the *up* and f) S_{down} of the *down* hydrogen atom.

Table 3 lists the frequencies of these six surface modes at the Γ-point as resulting for two different cut-off energies used in the calculations in comparison with experimental results. The results clearly demonstrate the necessity of a large cut-off energy in the plane-wave expansion.

To conclude this short discussion, one further aspect of the surface phonons shall be presented. As mentioned above, all surface modes give rise to characteristic changes in the electronic charge density. Let $n(\mathbf{r})$ be the charge density of the surface system and $n_{\mathbf{q}}^{(j)}(\mathbf{r})$ be the one in the presence of the mode $(\mathbf{q}, j)$:

$$\Delta n_{\mathbf{q}}^{(j)}(\mathbf{r}) = n_{\mathbf{q}}^{(j)}(\mathbf{r}) - n(\mathbf{r}). \tag{2}$$

Table 3. Frequencies of the six surface modes at the Γ-point for a cut-off energy of 15 and 18 Ryd in comparison with experimental data.

	S_{up}	S_{down}	W	T	R	S
15 Ryd	60.6	56.7	18.0	16.0	12.0	24.3
18 Ryd	62.6	59.7	18.4	15.8	12.7	24.7
[Exp.][a]	62.3–63.3		18.6–19.6		14.5–14.7	26.8–27.4

[a] The intervals given here correspond to different experimental results given in [Ang96,But84,Fro85,Stu83].

Furthermore, let $u_{\alpha,l,\nu}$ be the displacement of the atom that is located at $\mathbf{R}_l + \boldsymbol{\tau}_\nu$ ($\mathbf{R}_l$ indicate the unit cells, $\boldsymbol{\tau}_\nu$ the atoms within these cells) in the space direction α. Then, within first order perturbation theory, the charge density change induced by the phonon j with the wave vector $\mathbf{q}$ turns out to be

$$\Delta n_{\mathbf{q}}^{(j)}(\mathbf{r}) = \sum_{\alpha,l,\nu} u_{\alpha,l,\nu} \frac{\partial n(\mathbf{r})}{\partial u_{\alpha,l,\nu}}. \tag{3}$$

This change reflects both the polarization and the localization of the mode $(\mathbf{q}, j)$. As an example, we show in Fig. 3 the charge density change for the mode S_{up}. Since a wave vector $\mathbf{q} = \mathbf{0}$ corresponds to an infinite wave length of the phonon, the change in the electronic charge density in neighboring cells is identical at the Γ-point (see Fig. 3 a)). In contrast, at the J'-point the change has an opposite sign from cell to cell, because the wavelength of the mode roughly corresponds to the length of two primitive cells at this high symmetry point (see Fig. 3 b)). At 0.5 J', the wave length of the phonon is equal to the length of four primitive cells. As a consequence, in the snapshot shown in Fig. 3 c) the hydrogen atoms show a maximum in the electronic density response in every second primitive cell with opposite signs, while the atoms in between do not contribute.

4 Summary

High performance computing on powerful parallel computer architectures allows *ab-initio* calculations of surface phonons in a reasonable time. We were able to implement Density Functional Perturbation Theory with a very high parallelization efficiency being independent on the particular surface considered.

In this contribution, we have presented the structural and vibronic properties of 2H:Si(001)-(1 × 1). Especially for a convergent description of the vibronic properties it is necessary to take a large cut-off energy in the plane-wave expansion into account. We also presented changes in the valence-charge

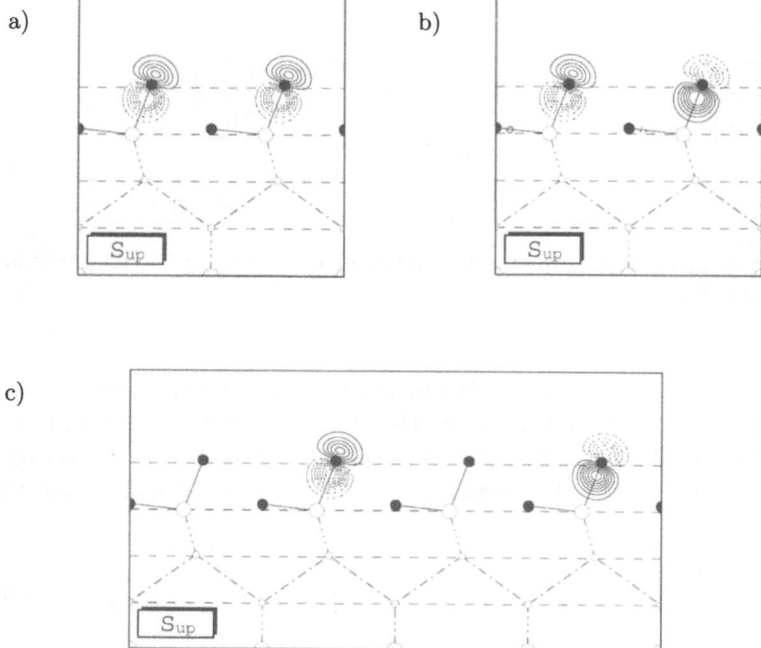

Fig. 3. Density change of the *stretching* mode S_{up} a) at the Γ-point, b) at the J'-point and c) at 0.5 J' in the optimized geometry. Solid lines show an increase of electronic charge density while the dashed lines show a decrease of charge density.

density induced by particular surface phonons, because these changes give a very intuitive picture of the physical nature of the different modes.

Acknowledgements

This work was supported in part by the BMBF, Bonn, Germany, under project No. 05 SE8 PMA. In addition, it is our great pleasure to acknowledge the Supercomputing Center in Karlsruhe for a grant of computer time on IBM RS/6000 SP/256. Especially we wish to thank Clemens Howar for his continuous competent and friendly help.

References

[Ang96] T. Angot, D. Bolmont, and J. J. Koulmann, High resolution electron energy loss spectroscopy study of the Si(001) 3 × 1 hydrogenated surface, Surf. Sci. **352-354**, 401 (1996).

[Bar87] S. Baroni, P. Giannozzi, and A. Testa, Green's-Function Approach to Linear Response in Solids, Phys. Rev. Lett. **58**, 1861 (1987).

[BHS82] G. B. Bachelet, D. R. Hamann, and M. Schlüter, Pseudopotentials that work: From H to Pu, Phys. Rev. B **26**, 4199 (1982).

[Bol93] J. J. Boland, Scanning tunneling microscopy of the interaction of hydrogen with silicon surfaces, Advances in Physics **42**, 129 (1993).

[But84] R. Butz, E. M. Oellig, H. Ibach, and H. Wagner, Mono- and dihydride phases on silicon surfaces – a comparative study bi EELS and UPS, Surf. Sci. **147**, 343 (1984).

[Cep80] D. M. Ceperley and B. J. Alder, Ground State of the Electron Gas by a Stochastical Method, Phys. Rev. Lett. **45**, 566 (1980).

[Cir84] S. Ciraci, R. Butz, E. M. Oellig, and H. Wagner, Chemisorption of hydrogen on the Si(001) surface: Monohydride and dihydride phases, Phys. Rev. B **30**, 711 (1984).

[Fey37] R. P. Feynman, Forces in Molecules, Phys. Rev. **56**, 340 (1939).

[Fre00] U. Freking, A. Mazur, and J. Pollmann, Vibronic Studies of adsorbate-covered semiconductor surfaces with the help of HPC, Eds.: E. Krause and W. Jäger, High Performance Computing in Science and Engineering '99, Springer-Verlag, Berlin, 2000.

[Fre01] U. Freking, A. Mazur, and J. Pollmann, Electronic, structural and vibrational properties of chalcogenides on Si(001) and Ge(001) surfaces, Eds.: E. Krause and W. Jäger, High Performance Computing in Science and Engineering 2000, Springer-Verlag, Berlin, 2001.

[Fro85] H. Froitzheim, U. Köhler, and H. Lammering, Surf. Sci. **149**, 537 (1985).

[Gia91] P. Giannozzi, S. de Gironcoli, P. Pavone, and S. Baroni, *Ab initio* calculation of phonon dispersions in semiconductors, Phys. Rev. B **43**, 7231 (1991).

[Hoh64] P. Hohenberg and W. Kohn, Inhomogeneous Electron Gas, Phys. Rev. B **136**, 864 (1964).

[KB82] L. Kleinmann and D. M. Bylander, Efficacious Form for Model Pseudopotentials, Phys. Rev. Lett. **48**, 1425 (1982).

[Koh65] W. Kohn and L. J. Sham, Self-consistent equations including exchange and correlation effects, Phys. Rev. A **140**, 1133 (1965).

[Nor91] J. E. Northrup, Structure of Si(100)H: Dependence on the H chemical potential, Phys. Rev. B **44**, 1419 (1991).

[Sak76] T. Sakurai and H. D. Hagstrum, Interplay of the monohydride phase and a newly discovered dihydride phase in chemisorption of H on Si(100)2 × 1, Phys. Rev. B **14**, 1593 (1976).

[Stu83] F. Stucki, J. A. Schaefer, J. R. Anderson, G. J. Lapeyre, and W. Göpel, Monohydride and dihydride formation at Si(001) 2 × 1: a high resolution electron energy loss spectroscopy study, Sol. State Commun. **47**, 795 (1983).

Interplay of Phase Fluctuations and Electronic Excitations in High-Temperature Superconductors – A Monte Carlo Simulation

Thomas Eckl[1], Enrico Arrigoni[1], Werner Hanke[1], and Douglas J. Scalapino[2]

[1] Institut für Theoretische Physik und Astrophysik, Universität Würzburg, am Hubland, D-97074 Würzburg, Germany
[2] Department of Physics, University of California, Santa Barbara, California 93106

Abstract. Although the binding of electrons into Cooper pairs is necessary in forming the superconducting state, its remarkable properties like zero resistance and perfect diamagnetism require long-range phase coherence among the pairs as well. When coherence is lost due to thermal fluctuations of the phase at the transition temperature T_c, pairing remains, together with short-range phase correlations. In conventional metals, Cooper pairs with short-range phase coherence are destroyed less than $1K$ above T_c. In underdoped high-T_c copper oxides, however there is spectroscopic evidence for some form of pairing up to a temperature T^*, which is more than $100K$ above T_c [1–3]. How this pairing above T_c and Cooper-pair formation are related is one of the most challenging problems in high-T_c superconductivity. Here we report first numerical results of the dynamical properties of a model Hamiltonian which explicitly takes these phase-fluctuations into account by using a new Monte-Carlo (MC) technique. The phases are thereby controlled by a classical XY-action.

1 Introduction

The superconducting order parameter of a metal is characterized by an amplitude and a phase. In the BCS-Eliashberg mean field theory [4], which is a very good approximation for conventional metals, the phase of the order parameter is unimportant for determining the value of the transition temperature T_c and the behavior of many physical quantities around T_c. Recently Emery and Kivelson [5] argued that superconductors with low superconducting carrier density (such as the organic and high-T_c oxide superconductors) are characterized by a relatively small phase 'stiffness' and poor screening, both of which imply a significantly larger role for phase fluctuations. As a consequence, in these materials the transition to the superconducting state may not display typical mean-field behavior, and phase fluctuations, both classical and quantum, may have a significant influence on low temperature properties. For some quasi-two-dimensional materials, in particular underdoped high-temperature superconductors, the onset of long-range phase order controls T_c as well as its systematic material dependency.

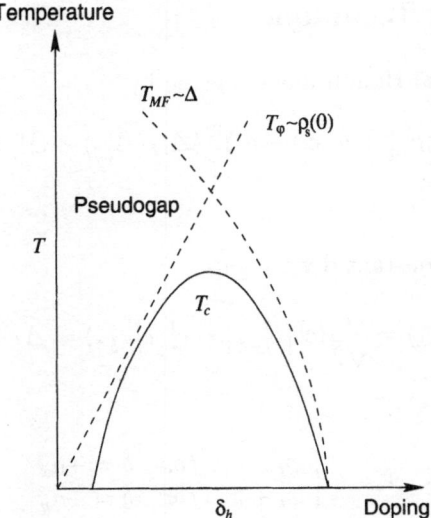

Fig. 1. A sketch of the phase diagram of high-temperature superconductors as a function of temperature T and doping δ_h. T^{MF} is the mean-field transition temperature, T_φ the upper bound on the phase ordering temperature, and T_c the actual transition temperature.

The fundamental reason why phase fluctuations are so important in the oxide superconductors is that they are doped insulators with very low superfluid density (ϱ_s). Empirically, when the concentration δ_h of doped holes in the CuO_2 planes is not too large, $\varrho_s(0)$ is proportional to δ_h. The temperature at which phase order would disappear is given by $T_\varphi \sim \varrho_s(0)$, thus $T_\varphi \to 0$ as $\delta_h \to 0$, even though T^{MF}, which is the mean-field transition temperature predicted by BCS-Eliashberg theory could remain finite. Figure 1 shows a schematic phase diagram for high-temperature superconductors. The superconducting phase is bounded by T_φ and T^{MF}. If $T_c \ll T_\varphi$, phase fluctuations are relatively unimportant, and T_c is close to T^{MF}. On the other hand, if $T_\varphi \approx T_c$, then the value of T_c is determined primarily by phase ordering, and T^{MF} is simply the characteristic temperature below which pairing becomes significant locally ('Pseudogap', T^*).

The above described theory gained further support when Corson et al. [6] were able to track the phase correlation time τ in the normal state, above T_c, with measurements of the high-frequency conductivity in the underdoped copper oxide superconductor $Bi_2Sr_2CaCu_2O_{8+\delta}$.

In the following section, we introduce a model Hamiltonian which has a local d-wave gap whose phases fluctuate freely. We then compare in detail the numerical results with experiments and finally state our conclusions and give a short outlook.

2   Model and Technique

Our grand-canonical Hamiltonian is given by

$$\widehat{K}_{el} = \sum_{\langle i\,j\rangle,\sigma} -t(c_{i\,\sigma}c_{j\,\sigma}^\dagger + h.\,c.) - g\sum_{i\,\delta}(\Delta_{i\,\delta}\langle\Delta_{i\,\delta}^\dagger\rangle + \Delta_{i\,\delta}^\dagger\langle\Delta_{i\,\delta}\rangle) - \mu\sum_{i,\sigma} n_{i\,\sigma},$$

(1)

which has a local constant d-wave gap

$$\langle\Delta_{i\,\delta}^\dagger\rangle = \frac{1}{\sqrt{2}}\langle c_{i\,\uparrow}^\dagger c_{i+\delta\,\downarrow}^\dagger - c_{i\,\downarrow}^\dagger c_{i+\delta\,\uparrow}^\dagger\rangle = \Delta\,e^{i\,\Phi_{i\delta}},$$

(2)

with phase

$$\Phi_{i\delta} = \begin{cases} \varphi_i & for \quad \delta = +a_x \\ \varphi_i + \pi & for \quad \delta = +a_y \end{cases}$$

(3)

and where the phases φ_i fluctuate around the d-wave mean-field saddle-point. Here $\langle i\,j\rangle$ stands for next neighbor sites on a $N \times N$ two-dimensional lattice, t is the hopping-rate of fermions with creation- (destruction-) operator $c_{i\,\sigma}^\dagger$ ($c_{i\,\sigma}$), and chemical potential μ. The coupling constant g controls the strength of the d-wave pairing-interaction.

In contrast to *standard* Quantum Monte Carlo simulations, where a determinant expression is used to sample the phases, we use an effective ('minus-sign free') action, which, in principle, can be calculated self-consistently from the Hamiltonian (1):

$$e^{-S_{eff}(\varphi_i(\tau_l))}.$$

(4)

As a first step, we carried out a Monte Carlo (MC) simulation using a classical XY-action for the phases:

$$S_{eff} = \frac{-1}{\beta}J\sum_{\langle i\,j\rangle} \cos(\varphi_i - \varphi_j).$$

(5)

In a following step quantum fluctuations should also be included by taking a time-dependent phase action. The XY-model displays a Kosterlitz-Thouless (KT) transition at $T_{KT} \approx J$. The parameters were chosen such that we get the correct critical behavior, appropriate to the underdoped cuprate superconductors:

1. interaction: $g = 0.6t \Rightarrow T_{MF} \approx 0.2t$
2. phase-stiffness: $J = 0.04t \Rightarrow T_\varphi \approx 0.04T$
3. particle number: $\langle n\rangle = 0.9$.

Most part of the calculations was done on the Cray T3E-900 at the HLRS-Stuttgart, using at an average 128 processors and a job-runtime of 12h for each parameter set.

3   Results

In this section we present preliminary numerical results and compare them
with the available experiments. Figure 2 displays the density of states. There
one can see, that the gap starts filling in with increasing T, but its magnitude
remains unchanged for $T < T_{KT}$ ($T_{KT} \approx 0.04t$). The results were obtained
from a 20×20 lattice with twisted boundary conditions to get an effective
80×80 lattice in order to smooth our curves. Despite the finite-size effects
that are still visible in Fig. 2, one can clearly see that the gap starts filling
with increasing temperature whereas its magnitude remains constant which
agrees with scanning tunneling microscopy experiments. Only for very high
temperatures ($T > T_{KT}$), the gap seems to get smaller.

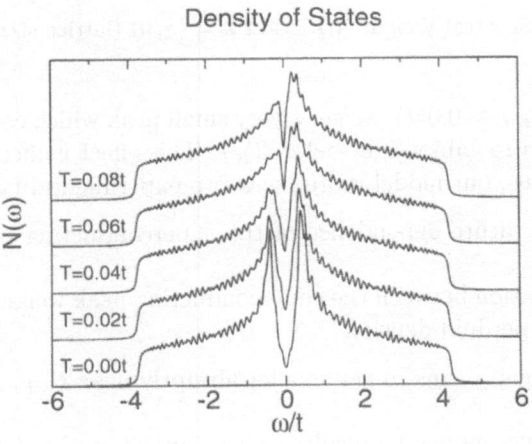

Fig. 2. Local density of States $N(\omega)$ (lattice size: 80×80; see text).

Figure 3 shows numerical results for the spectral weight $A(\boldsymbol{k}, \omega)$ at $\boldsymbol{k} =$
$(\pi, 0)$ on a 18×18 lattice. Experimental results for the weight of the superconducting peak [1] at $\boldsymbol{k} = (\pi, 0)$ clearly show [7], that there is a correlation
between the superconducting peak weight (SPW) at $\boldsymbol{k} = (\pi, 0)$ and the superfluid density as a function of doping and temperature from the underdoped
to the slightly overdoped region. Moreover, they show that the peak intensity
increases abruptly near T_c rather than near the temperature where the gap
opens (T^*).

Our numerical results for the spectral weight $A(\boldsymbol{k}, \omega)$ at $\boldsymbol{k} = (\pi, 0)$ indicate
that the peak not only gets broader as a function of T but also looses weight.

---

[1] This is the peak which forms below ϵ_F, which is present only in the superconducting case (in an ideal BCS-superconductor), since $\boldsymbol{k} = (\pi, 0)$ is outside of the
   Fermi surface.

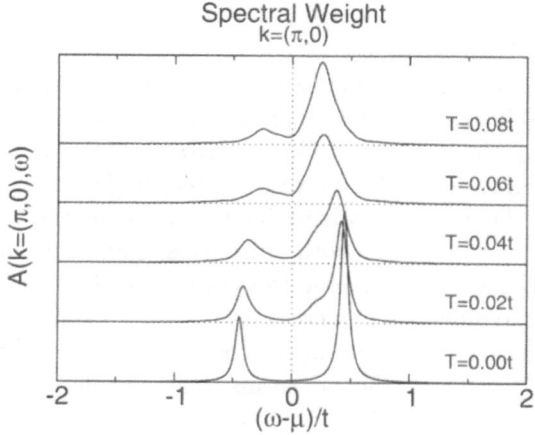

Fig. 3. Spectral Weight $A(\boldsymbol{k}, \omega)$ at $\boldsymbol{k} = (\pi, 0)$ (lattice size: 18×18).

Above T_{KT} ($T_{KT} \approx 0.04t$), we get a very small peak which considerably starts to sharpen and to gain weight below T_{KT}. This effect is most dominant near $\boldsymbol{k} = (\pi, 0)$. Thus, our model reproduces two experimental facts:

1. The temperature dependence of the superconducting peak weight near $\boldsymbol{k} = (\pi, 0)$.
2. The correlation between the superconducting peak weight near $\boldsymbol{k} = (\pi, 0)$ and the superfluid density.

Note that the gap seems to get smaller abruptly near T_{KT}, as in the density of states.

Figure 4 gives numerical results for the optical conductivity as a function of temperature for fixed frequency ($\omega \approx 0.2t$) on a 8×8 lattice. In high-frequency experiments [6], the optical conductivity at finite frequency shows a characteristic peak near T_c as a function of T on top of a background, which decreases with T. The peak was interpreted by the authors of [6] as signature of the partial coherent cooper pairs at finite frequency and the background as contribution from the normal conducting electrons. The imaginary part, which was also measured, decreases with T and is observable even $25K$ above T_c and has no obvious onset temperature.

In our simulations (Fig. 4), the real part shows a distinct maximum near $T \approx 0.06t$. The imaginary part decreases with temperature until $T \approx 0.06t$, where it has a minimum and then increases again. The reason for this behavior lies in the fact that our model undergoes a transition between a d-wave superconductor and an *ideal* metal, due to the absence of interaction when the superconducting gap vanishes. Thus, apart for the fact that our charge carriers have no finite lifetime τ, which would produce the background in $\sigma_1(\omega)$ and cause $\sigma_2(\omega)$ to go to zero above T_c, our model reproduces the salient features of the experiment.

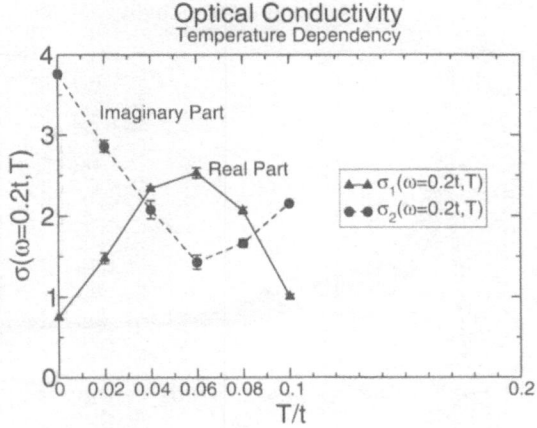

Fig. 4. Temperature dependence of the optical conductivity (lattice size: 8×8).

Figures 5 and 6 show details of the temperature dependence of $\sigma(\omega)$. In Fig. 5 (top), the real part $\sigma_1(\omega)$ is plotted for different temperatures and $0 < \omega < t$. Here, one can see that $\sigma_1(\omega)$ looses weight at $\omega \approx 0$, whereas $\sigma_1(\omega)$ is increasing with temperature for finite frequencies up to $\Delta \approx 0.4t$. To see what in detail happens in this frequency range, Fig. 5 (bottom) shows only the range $0.1t < \omega < 0.6t$. For $T < 0.06t$, $\sigma_1(\omega)$ is increasing monotonously over the whole frequency range. Above $T \approx 0.06t$ the situation changes considerably. At lower frequencies, $\sigma_1(\omega)$ starts to decrease with temperature, whereas at higher frequencies $\sigma_1(\omega)$ remains constant or is decreasing much more slowly with T. The reason for this behavior is that, above T_{KT} and for lower frequencies, we get a crossover to an ideal metal (which only shows a Drude-peak at $\omega = 0$), whereas at higher frequencies we can still see the partial phase coherent electrons.

The difference between the maximum in $\sigma_1(T)$ at $T \approx 0.06t$ and $T_{KT} \approx 0.04t$ is due to finite size effects: phase ordering sets in at a higher temperature on a smaller lattice. Please note that we use a finite imaginary part for the delta-function in the calculation ($\eta = \frac{\Delta}{10} = 0.04t$), thus we see no delta peak at $\omega = 0$ for $T = 0$ but the rather broad wings of the Lorentz-function.

Figure 6 shows the imaginary part $\sigma_2(\omega)$ in the range $0 < \omega < t$ (top) and $0.1 t < \omega < 0.6 t$ (bottom). For $T < 0.06t$, $\sigma_2(\omega)$ decreases monotonously, while for $T > 0.06t$ we see again the frequency-dependent behavior. At lower frequencies, we observe the crossover to the ideal metal, i. e. $\sigma_2(\omega)$ increases and approaches again the $\frac{1}{\omega}$ form, whereas for higher frequencies $\sigma_2(\omega)$ further decreases.

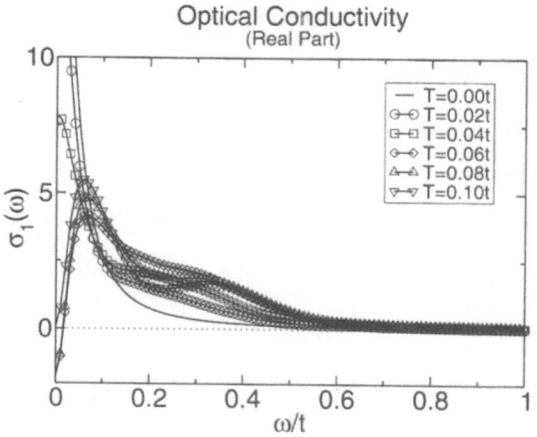

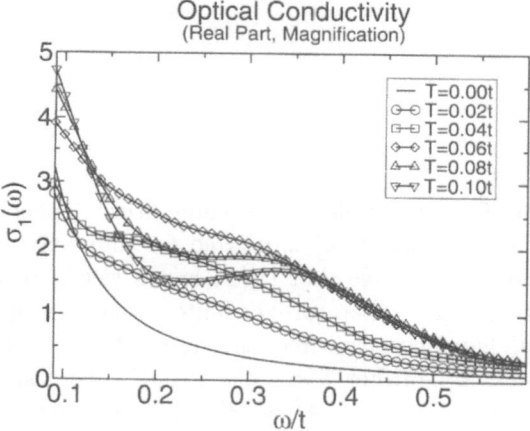

Fig. 5. Optical conductivity (Real Part; lattice size: 8×8).

Finally, Fig. 7 shows the way to extract the limit giving the superfluid weight

$$D_s = \langle -k_x \rangle + \lim_{q_y \to 0} D_{xx}(q_x = 0, q_y, \omega = 0), \qquad (6)$$

where $\langle -k_x \rangle$ is the local kinetic energy in x-direction and D_{xx} is the current-current correlation function. The dashed lines are only "guides to the eye" and give an idea of what the limit could be. Although the lattice size is only 8×8, one can clearly see the difference between the curves with $T < 0.05t$ and those with $T > 0.05t$ and thus conjecture that $T_{KT} \approx 0.05t \pm 0.01t$, which is consistent with the theoretical value of $T_{KT} \approx 0.04t$.

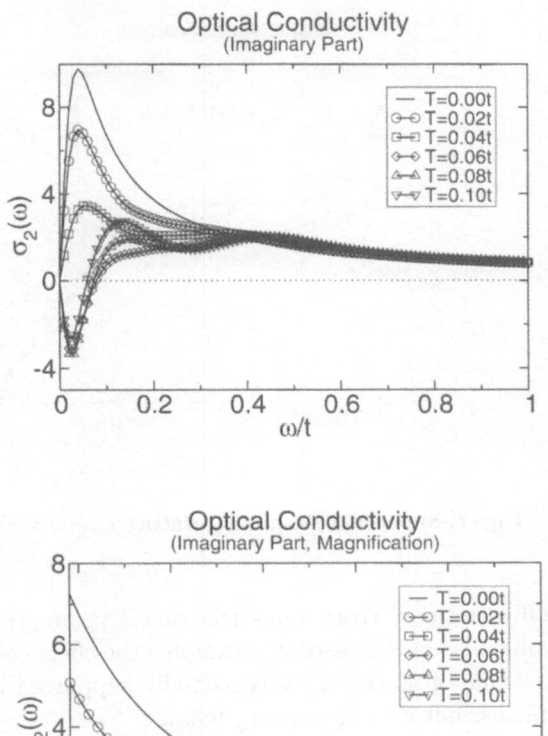

Fig. 6. Optical conductivity (Imaginary Part; lattice size: 8 × 8).

4   Conclusion

Although our simulations were carried out on relatively small lattices, we can already see some quite interesting physics. In particular, our model Hamiltonian (1) is able to reproduce salient features of the underdoped cuprates: it describes correctly the qualitative behavior of the density of states $N(\omega)$, the spectral weight $A(\mathbf{k}, \omega)$ at $\mathbf{k} = (\pi, 0)$ and the optical conductivity $\sigma(\omega)$ as a function of temperature T. Thus our results are quite promising with respect to the idea, that phase fluctuations play an important role in the underdoped high-T_c superconductors. Nevertheless, finite size effects certainly play an important role close to the KT transition. Therefore simulations on larger lattices are needed in order to incorporate accurately the effects of

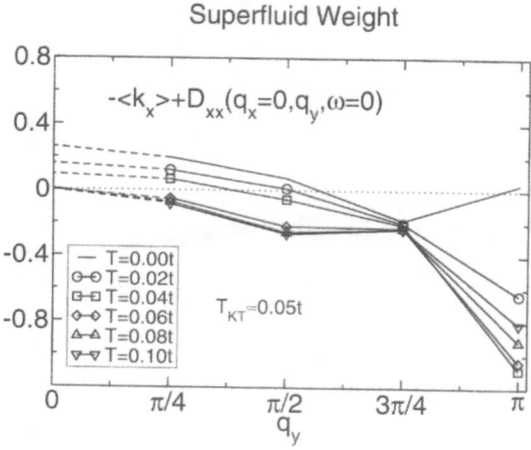

Fig. 7. Superfluid Weight D_s (lattice size: 8 × 8).

critical phase fluctuations. Work along this line is in progress. For low temperatures it would also be interesting to examine the effects of *quantum* phase fluctuations. — One of us (E. A.) was partially supported by the Deutsche Forschungsgemeinschaft as a Heisenberg fellow.

References

1. Loeser, A. G., Shen, Z.-X., Dessau, D. S., Marshall, D. S., Park, C. H., Fournier, P., and Kapitulnik, A.: Excitation Gap in the Normal State of Underdoped $Bi_2Sr_2CaCu_2O_{8+\delta}$. Science **273** (1996) 325–329
2. Ding, H., Yokoya, T., Campuzano, J. C., Takahashi, T., Randeria, M., Norman, M. R., Mochika, T., Kadowaki, K., and Giapintzakis, J.: Spectroscopic Evidence for a Pseudogap in the Normal State of Underdoped High-T_c Superconductors. Nature **382** (1996) 51–54
3. Renner, Ch., Revaz, B., Genoud, J.-Y., Kadowaki, K., and Fischer, Ø.: Pseudogap Precursor of the Superconducting Gap in Under- and Overdoped $Bi_2Sr_2CaCu_2O_{8+\delta}$. Phys. Rev. Lett. **80** (1998) 149–152
4. Schrieffer, J. R.: Theory of Superconductivity. Benjamin, New York (1964)
5. Emery, V. J., and Kivelson, S. A.: Importance of Phase Fluctuations in Superconductors with Small Superfluid Weight. Nature **374** (1995) 434–437
6. Corson, J., Mallozzi, R., Orenstein, J., Eckstein, J. N., Bozovic, I.: Vanishing of Phase Coherence in Underdoped $Bi_2Sr_2CaCu_2O_{8+\delta}$. Nature **398** (1999) 221–223
7. Feng, D. L. , Lu, D. H., Shen, K. M., Kim, C., Eisaki, H., Damascelli, A., Yoshizaki, R., Shimoyama, J.-i., Kishio, K., Gu, G. D., Oh, S., Andrus, A., O'Donnell, J., Eckstein, J. N., and Shen, Z.-X.: Signature of Superfluid Density in the Single-Particle Excitation Spectrum of $Bi_2Sr_2CaCu_2O_{8+\delta}$. Science **289** (2000) 277–281

Chemistry

Bernd A. Hess

Chair of Theoretical Chemistry, University of Erlangen–Nuremberg
Egerlandstr. 3, D–91058 Erlangen, Germany

The contributions of computational chemistry to the report on High Performance computing at the HLRS focus this year on the calculation of transition metal complexes and on the determination of the properties of hydrogen bonds.

Among the various intermolecular interactions, i. e., the interactions between two separate molecules or molecular fragments, the hydrogen bond is special in several aspects: Although in general weaker than covalent bonds (which are instrumental in defining the chemical bond and thus, by and large, are the bonds which form a molecule), hydrogen bonds with their bond energies of 30–100 kJ/mol are considerably stronger than other intermolecular interactions, which are usually called van-der-Waals interactions. Moreover, they generally lead to specific geometric arrangements of their constituent atoms, in the generic case an almost linear arrangement of the A–H...B moiety, which defines the hydrogen bridge between the two heavy atoms A and B. These two properties of the hydrogen bond are mostly responsible for the importance of hydrogen bonds in biological systems, where they lead to well-defined organization of the polymeric structures of proteins and DNA.

The contribution of Brutschy and Hobza deals with the structure of hydrogen bonded complexes which are formed by fluorobenzene derivatives and haloform molecules. Actually, the authors have attacked the problem by means of experimental and computational methods, an approach which often yields insight that could not be obtained by either method alone. Whereas usually the vibration frequency of CH bond is shifted to lower frquencies upon hydrogen bond formation, the present study has found an example of an "improper blue-shifting hydrogen bond".

The papers by Kirchner and Marx and Kreitmeier *et al.* make use of the Car–Parrinello quantum-mechanical molecular dynamics method to study hydrogen-bonded molecules in liquids. The first contribution reports on the solvation of a hydrogen radical in liquid water. This system is studied as a model system for hydrophobic solvation effects in aquaeous solution. This simulation for the first time indicates that a hydrophobic molecule shows *faster* diffusion in water that a water molecule itself. The second contribution studies the influence of hydrogen bonds on liquid hydrogen fluoride at several temperatures of the thermodynamic system. Since the intermolecular interactions in a hydrogen bond are to a large extent of electrostatic origin, a molecular dynamics treatment using classical forces is often considered sufficient; the present paper, however, makes the point that in the present case inductive forces are decisive, and thus a quantum-mechanical model is required in order to take the polarizabilities of the constituents properly into account.

A second important issue in the computational study of biological systems is the realistic quantum-mechanical description of metal centers and their surroundings, which are, for example, instrumental in the functionality of enzyme cofactors. In general, transition metal complexes are the decisive structural motif in all kind of catalysts. The paper by Frunzke and Frenking reports careful studies on the structure and reactivity of various transition metal complexes and gives insight in the nature of the bond between the metal and its ligands.

Improper, Blue-shifting Hydrogen Bond Between Fluorobenzene and CHX$_3$ (X=F,Cl)

Bernhard Brutschy[1] and Pavel Hobza[2]

[1] Institut für Physikalische und Theoretische Chemie der J.W. Goethe-Universität Frankfurt, Marie-Curie-Strasse 11, 60439 Frankfurt am Main,Germany
[2] J. Heyrovský Institute of Physical Chemistry, Academy of Sciences of the Czech Republic and Center for Complex Molecular Systems and Biomolecules, 182 23 Prague 8, Czech Republic

Abstract. Weakly bonded 1:1 complexes between fluorobenzene (Fb) and fluoroform (Ff) and chloroform (Chl) were investigated spectroscopically by infrared ion-depletion spectroscopy (IR/R2PI) and theoretically by correlated *ab initio* methods. CH stretches of the Fb.Ff complex are blue-shifted by 12 cm^{-1} and 21 cm^{-1}, respectively relative to the CH stretch of isolated fluoroform. Each IR band is assigned to a different hydrogen bonded fluorobenzene . fluoroform isomer. The isomer with the most blue-shifted CH stretching vibration (21 cm^{-1}) is assigned to a sandwich type structure, exhibiting a CH hydrogen bond. In the case of Fb.Chl only one structure (sandwich) exist for which blue-shift of 14 cm^{-1} was found. The cluster structures have been calculated by Counterpoise (CP)-corrected gradient optimization combined with anharmonic vibrational analysis using the CP- corrected Hessians. The predicted blue-shifts are 21 and 20.5 cm^{-1} for the CH stretching frequencies of fluoroform upon formation of a sandwich and a planar structure respectively, and 12 cm^{-1} for the CH stretching frequency of chloroform. The theoretical and experimental shifts for both complexes are thus well comparable. It is shown that the nature of the improper, blue- shifting H-bond in these complexes differs from that in a standard H-bond.

1 Introduction

Hydrogen bonded (H-bonded) complexes belong to the most frequent intermolecular complexes and play a very important role in nature[1-3]. Many H-bonds are of the XH-Y type, where X is an electronegative atom and Y is either an electronegative atom having one or more lone electron pairs or a region of excess electron density (e.g. π-electrons of aromatic systems). H-bonds having X = F, O and N are the most common ones. The concept of H-bonds was extended to CH-Y bonding and both, CH-electronegative atom as well as CH- π types of H-bonds have been observed. The formation of a H-bond of the XH -Y type is accompanied by a weakening of the XH bond, which is manifested by the elongation of the XH bond with a concomitant decrease of the XH stretching frequency (the so called red-shift). The red-shift of the XH stretching frequency is a characteristic feature of the formation of H-bonds. Recently we found theoretically a new type of intermolecular bonding in carbon hydrogen donor complexes, which is characterized by the contraction of the CH bond and a blue-shift of the respective CH stretching frequency[4,5]. Because the spectroscopic manifestation of this new type of intermolecular bonding is opposite to the classical hydrogen bonds, we called it the "improper, blue- shifting H-bond". The proper theoretical description of this phenomenon requires the use of highly accurate theoretical methods including electron correlation, corrections of the basis set superposition error (BSSE) and anharmonic corrections. In this paper we report theoretical analysis of the improper, blue-shifting H-bond between fluoroform (Ff) and fluorobenzene (Fb) and chloroform (Chl) and Fb; theoretical data will be compared with experimental results obatined by resonant two-photon ionization (R2PI) and IR/R2PI ion depletion spectroscopy.

2   Calculations

All *ab initio* calculations were performed at the second-order Møller-Plesset (MP2) level at three different basis sets, 6-31G* , 6-31G** and 6-31++G**. The structures of monomers and the 1:1 cluster were determined using a gradient optimization on either BSSE uncorrected or BSSE corrected potential energy surfaces. In the former case the BSSE was included a posteriori by applying the Boys-Bernardi counterpoise method[6] and also the deformation energy was considered. In the later case the BSSE corrections, again using the CP method, were applied not only to the energy calculations, but also to the gradients of energy, and the BSSE was eliminated at each optimization cycle[7,8]. The CP-uncorrected procedure is not correct since it does not consider the CP-correction effects on the geometry and vibrational frequencies. The standard gradient optimization was considered converged if the gradient norm was lower then 0.0003 au, for CP-corrected gradient optimization the limit was 0.0006 au. Also the harmonic vibrational frequencies have been evaluated for the optimal structures found by the standard as well as CP-corrected optimizations, in the latter case using the CP-corrected Hessians. To take the anharmonic effects into account, we performed one-dimensional (1-D) anharmonic calculations (Fb.Ff) and (1-D) and (2-D) anharmonic calculations (Fb.Chl). The C-H bonds of haloforms are treated as independent oscillators and all other (intermolecular and intramolecular) coordinates are optimized using a CP-corrected gradient optimization. The coupling with other coordinates is effectively covered via *ab initio* optimization of the 1:1 structures for a selected set of the C-H distances. The natural bond orbital analysis (NBO)[9] was performed using the MP2 electron density. To eliminate the basis set extension effects the NBO analysis of monomers was performed using the basis set of the 1:1 complex (with the proper amount of the ghost atoms). All the calculations were done using the GAUSSIAN 98 code[10] and our own program[8] for CP-corrected optimization and Hessian evaluation (using Gaussian 98 as a generator of the first and second energy derivatives).

The CP-corrected optimization is time-consuming what is due to i) necessity to evaluate 5 gradients (A.B, A.B(G), A(G).B, A, B) instead of 1 (A.B) in the standard optimization; ii) very poor convergency what requires large number of gradient optimization steps. In order to make the calculation feasible even for large clusters the use of fast CPU, large memory (at least 1Gb) and possibility for parallelization is inevitable.

3   Results and Discussion

3.1   Structures, energies and vibrational frequencies of Fb·Ff and FB·Chl complexes

For the Fb.Ff complex two different structures were considered, a sandwich (stacked) structure having CH-π contact and a planar structure with one or two CH-F contacts. For the Fb.Chl complex only the sandwich structure was taken into account. The structures found by standard as well as CP-corrected gradient optimizations are for both complexes similar. First, the Fb.Ff complex will be discussed. While the 6-31G* calculations (standard optimization) prefer the sandwich structure over the planar one, the opposite result is obtained when extending the basis set to 6-31++G**. The more reliable CP-corrected gradient optimizations gave unambiguous answers concerning the relative stability of both structures. At both levels the sandwich structure is more stable than the planar one. The energy difference calculated at the 6-31G* level is 0.56 kcal.mol^{-1} in favor of the sandwich complex, while at the 6-31++G** level the energies are comparable (0.08 kcal.mol^{-1}). In the case of CP-corrected gradient optimization of th e sandwich structure at the MP2/6-31++G** level we were not able to reduce the gradient norm below the value of 0.016 au. The obtained stabilization energy (2.38 kcal/mol) represents therefore a lower bound and the actual stabilization energy should be larger by an unknown amount. The sandwich structure of the Fb.Ff complex thus corresponds to the global minimum and the planar structure to a local one. However, the energy difference between both structures is small and both structures should coexist in a supersonic beam. We are aware of the fact that performing calculations at higher theoretical level (larger basis set and larger fraction of correlation energy) might change the ordering of complex isomers. However, the CP-corrected optimization exhibits faster convergency to complete basis set limit than standard optimization. An interesting effect was found for the planar structure of the complex possessing two CH-F contacts. Both CH bond lengths were reduced upon complex formation. This reduction is larger for the CH bond of Ff. A contraction of both bond lengths is accompanied by a blue-shift of both CH stretch frequencies which is again larger for the CH stretch of Ff. This finding is fully supported by experimental data. In the sandwich complex the CH bond length of Ff is contracted upon complex formation. The contraction of the CH bond is accompanied by the increase of the CH stretch frequency, i.e. by the blue-shift of this frequency. The blue- shift obtained by the standard calculation is larger than with the CP- corrected one. The differences in the blue shifts of the CH stretches in the sandwich and planar structure are larger in calculations with standard optimization; more reliable CP-corrected (harmonic) calculations yield a small difference of 3 cm^{-1} in favor of the sandwich structure. The same tendency was found for the corresponding complex with chloroform. The standard, CP-uncorrected, calculation predicted a blue-shift of 67 cm^{-1} for the sandwich structure of the Fb.Chl complex, which was reduced to 21 cm^{-1} when CP-corrected calculations were performed. The CP-corrected calculations predict blue-shifts of 31 and 34 cm^{-1} for the CH stretch frequencies of the Ff upon formation of the sandwich and planar structures, respectively of the Fb.Ff complex. Because the C-H vibration (as well as other vibrational modes) is anharmonic we should go beyond the limits of the harmonic approximation. Performing one-dimensional anharmonic calculations on the basis of the CP-corrected potential energy surface the calculated blue-shifts for the sandwich and planar structures were reduced to 20.5 and 21 cm^{-1}. These values nicely agree with the experimental results of 21 and 12.5 cm^{-1}. The present level of theory is, however, unable to explain the small diference of 8 cm^{-1} for the second shift. The Fb.Ff and Fb.Chl complexes were investigated at the same theoretical level (CP-corrected optimization and harmonic vibrational analysis on the basis of CP-corrected Hessians), which gives the chance to compare the theoretical characteristics of both clusters and also compare theoretical with experimental data. The Fb.Chl complex is more stable than Fb.Ff and the same difference in the stabilization energy is found between the standard and CP-corrected levels. However, the relative values of the blue-shifts evaluated at the standard and CP-corrected level differ. Standard level calculations predicted a larger blue-shift for the Fb.Chl complex while CP- corrected calculations give a larger blue-shift for the Fb.Ff complex. The experimental results unambiguously support the CP-corrected values. This is a strong argument supporting the use of CP-corrected gradient optimization and CP-corrected Hessians.

3.2   The Nature of improper H-bonding in the Fb.Ff and Fb.Chl complexes

Both complexes exhibit an EDT from Fb to haloforms. In the case of the sandwich structure of the Fb.Ff complex 4 me are transfered from Fb to Ff; this transfer is slightly smaller for the planar structure (2.4 me). Further, the same orbitals are involved in EDT of both isomers. The value of EDT is sensitive to the optimization procedure – in the case of standard gradient optimization (instead of CP-corrected optimization) the EDT is reduced from 4.0 me to 2.8 me. The largest part of this EDT goes to the lone electron pairs of all fluorine atoms and there is no EDT to the CH region. The increase of electron density in fluorine lone pairs results in a structural reorganization of the CHF$_3$ subsystem. If all geometrical changes in fluoroform are considered, the harmonic vibrational analysis for this deformed structure results in the blue-shift of 28 cm^{-1} of the CH stretch frequency with respect to the CH frequency in the fully optimized fluoroform. This means that the largest part of the overall blue-shift of 31 cm^{-1} calculated at the harmonic level is due to the structural reorganization of fluoroform and the rest is due to the polarization of fluoroform by fluorobenzene.

We can conclude by stating that the blue-shift of the CH stretch frequency of Ff is due to the structural reorganization of the Ff, which results from an electron density transfer to the fluorine atoms. The primary effect is thus the EDT to the fluorine atoms, which subsequently leads to the structural reorganization of Ff, including the contraction of the CH bond. Very similar situation was found also in the case of Fb.Chl complex. This conclusion is important since in the classical H-bonds of XH-Y type the primary effect is due to EDT to the antibonding orbital of the XH bond. The increase of electron density in this orbital results in the elongation of the XH bond. Also the smaller blue-shift in the Fb.Chl complex, possessing a larger stabilization energy and a larger amount of EDT than in the case of the Fb.Ff complex, can be explained on the basis of the NBO analysis. Contrary to the Fb.Ff complex where a dominant part of EDT is directed to the fluorine atoms, in the Fb.Chl complex the enhanced electron density (6.1 me; CP-corrected gradient optimization) is distributed almost equally among all atoms of the chloroform. The increase of electron density at the chlorine atoms and in the CCl antibonding orbitals induces a structural reorganization in the Chl including the CH contraction. On the other hand the increase of electron density in the CH antibonding orbital and in the hydrogen Rydberg orbital leads to the elongation of the CH bond. The final net effect is the contraction of the CH bond which is, however, smaller than in the case of Fb.Ff. Let us finally remind that neither the stabilization energy nor the amount of EDT correlate with the magnitude of the blue-shift in both complexes, which provides evidence about the complex character of the binding. The improper, blue-shifting H-bond has some features similar to a standard H-bond and some completely different. The similar features, which justify the use of the term "H-bond", are the following: i) the hydrogen is placed between X and Y (in the sense of XH-Y); ii) the electron density is transferred from the proton acceptor (Y) to the proton donor (XH); iii) the stability and directionality. The different features, which support the use of the term "improper, blue-shifting" are the following: i) the formation of the improper H-bond leads to a strengthening of the XH bond and not to its weakening as in the case of the classical H-bond; ii) upon complex formation the CH bond is contracted and CH stretching frequency exhibits a blue-shift; iii) the electron density is not transferred to the XH region but to the remote part of the proton donor, which results in its structural reorganization.

4 Conclusions

From the experimental results it was found that the fluoroform CH frequency was shifted to a higher frequency by 21 and 12 cm^{-1}, respectively (blue- shift) upon complexation; the larger blue-shift was assigned to the sandwich- type structure and the smaller one to the planar structure of the complex. In the case of Fb.Chl complex the CH frequency was blue-shifted by 14 cm^{-1}. On the basis of a CP-corrected gradient optimization we demonstrated that fluorobenzene forms a sandwich-type and a planar complex with fluoroform. The CH bond of fluoroform is contracted upon complex formation (both in the sandwich and planar structure) and its stretching frequency is blue-shifted by 31 and 34 cm^{-1}. Taking anharmonic corrections into account these shifts are reduced to 21 and 20.5 cm^{-1}. These values agree well with the experimental results. The same type of calculations performed for the Fb.Chl sandwich structure shown contraction of the CH bond of chloroform accompanied by a blue shift of 12 cm^{-1}. Also this value nicely agree with experiment. The relative values of the blue-shifts in fluorobenzene.fluoroform and fluorobenzene.chloroform determined experimentally are reproduced only if the CP-corrected gradient optimization combined with anharmonic vibrational analysis using CP-corrected Hessians is applied. On the basis of the NBO analysis using the MP2 electron density it was shown that the dominant part of the electron density transfer from fluorobenzene is directed to the fluorine atoms of fluoroform and none to the CH region. The structural reorganization of fluoroform, which results from this EDT to the fluorine lone pairs, is responsible for the blue-shift of the CH stretching frequency. The improper, blue-shifting H-bond in the complex studied is thus the secondary effect, while in the classical (red-shifted) H-bonds it is the primary effect. The same mechanism was found also in the case of Fb.Chl complex.

Acknowledgment

The financial support provided by Center for Complex Molecular Systems and Biomolecules, Czech Republic (project LN00A032 of MSMT CR), from the J.W. Goethe Universität Frankfurt and from the "Fonds der Chemischen Industrie" is acknowledged. We also gratefully acknowledge CPU time on the NEC-SX4 at the Rechenzentrum der Universität Stuttgart (RUS) (Grant No 11738); without this support the present study could not be realized.

References

1. Hobza, P., Zahradník, R. Intermolecular Complexes; Elsevier: Amsterdam, 1988.
2. Scheiner, S. Hydrogen Bonding; Oxford University Press: New York, 1997.
3. Desiraju, G. R., Steiner, T. The Weak Hydrogen Bond; Oxford University Press: Oxford, 1999.

4. Hobza, P., Špirko, V., Selzle, H.L., Schlag, E.W. J. Phys. Chem. A **102** (1998) 2501
5. Hobza, P., Havlas, Z. Chem. Rev. **100** (2000) 4253
6. Boys, S. F., Bernardi, F. Mol. Phys. **19** (1970) 553.
7. Simon, S., Duran, M., Danneberg, J. J. J. Chem. Phys. **105** (1996) 11024.
8. Hobza, P., Havlas, Z. Theor. Chem. Acc. **99** (1998) 372.
9. Foster, J.P., Weinhold, F. J. Am. Chem. Soc. **102** (1980) 7211.
10. Gaussian 98, Revision A.7, M. J. Frisch, G. W. Trucks, H. B. Schlegel, G. E. Scuseria, M. A. Robb, J. R. Cheeseman, V. G. Zakrzewski, J. A. Montgomery, Jr., R. E. Stratmann, J. C. Burant, S. Dapprich, J. M. Millam, A. D. Daniels, K. N. Kudin, M. C. Strain, O. Farkas, J. Tomasi, V. Barone, M. Cossi, R. Cammi, B. Mennucci, C. Pomelli, C. Adamo, S. Clifford, J. Ochterski, G. A. Petersson, P. Y. Ayala, Q. Cui, K. Morokuma, D. K. Malick, A. D. Rabuck, K. Raghavachari, J. B. Foresman, J. Cioslowski, J. V. Ortiz, A. G. Baboul, B. B. Stefanov, G. Liu, A. Liashenko, P. Piskorz, I. Komaromi, R. Gomperts, R. L. Martin, D. J. Fox, T. Keith, M. A. Al-Laham, C. Y. Peng, A. Nanayakkara, C. Gonzalez, M. Challacombe, P. M. W. Gill, B. Johnson, W. Chen, M. W. Wong, J. L. Andres, C. Gonzalez, M. Head-Gordon, E. S. Replogle, and J. A. Pople, Gaussian, Inc., Pittsburgh PA, 1998.

Hydrophobic Solvation in Liquid Water Via Car–Parrinello Molecular Dynamics: Progress and First Results

Barbara Kirchner[1,2] and Dominik Marx[1]

[1] Lehrstuhl für Theoretische Chemie
   Ruhr–Universität, D-44780 Bochum, Germany
[2] *present address*:
   Organisch–Chemisches Institut
   Universität Zürich , Winterthurerstr. 190
   CH-8057 Zürich, Switzerland

Abstract. Solvation of apolar substances in liquid water plays a crucial role in "wet chemistry" as well as in biochemistry due to the relation to hydrophobicity effects. However, knowledge of the microscopic details of structure and dynamics at such solute-water "interfaces" is rather limited. The object of this study is a single hydrogen radical in water at ambient conditions. This system was chosen, because it is known experimentally that the diffusion of H in water is much faster than the self-diffusion of water itself and only marginally slower than the fast Grotthuss diffusion of H^+ in water. This paradox awaits an explanation, in particular since pioneering classical molecular dynamics simulations predict a clathrate-like cage and do not confirm fast diffusion, probably because of insufficiently accurate model potentials. This limitation can be circumvented by using Car–Parrinello molecular dynamics, where the interactions are derived from density functional electronic structure calculations.

1   General Introduction

The final goal of this project is to investigate hydrophobic solvation effects in aqueous solution. As the most simplistic example we start with a single hydrogen radical dissolved in liquid water at ambient conditions. This system has been studied under several aspects in literature, for instance experiments of different isotopes (H, D, Mu) have been undertaken in order to determine the hyperfine coupling constants and to measure the effect of mass on particle diffusion [1]. Here, it is found that such an atom undergoes a surprisingly fast diffusion and in particular its lightest "isotope", Mu, has a diffusion coefficient larger than that of proton diffusion! The diffusion mechanism is not understood at all but must be related to hydrophobicity since He and Ne also feature quite fast diffusion (surmounting the self diffusion of water itself!). The main structural question to answer in this context is how a "hydrophobic species" — here a bare hydrogen radical — gets embedded in liquid water, i.e. how the atom is solvated by the surrounding water molecules in the three-dimensional hydrogen bonded network. A more technical issue is related to the fact that this is an open shell system (one unpaired spin) concerning its electronic structure.

Within the framework of traditional computational chemistry solvation effects are usually treated by one of two possible models: Either the cluster approximation or the reaction field method. In the cluster approximation, a group of molecules is extracted from a classical simulation trajectory and treated as a super molecule in a regular, static electronic structure calculation. In the reaction field method, the solvent is represented by embedding the reacting molecules (and possibly also a few solvent molecules) within a dielectric continuum. These types of methods always present problems when dealing with highly associated liquids. In such liquids there are strong and in particular directional bonding interactions between the solvent molecules, which competes with the preferences due to the solute. It was recently shown [2] that both of the methods mentioned above are particularly difficult to use in conjunction with liquid water. An alternative method to treat solvent effects in such liquids is provided by the Car–Parrinello molecular dynamics simulation (CPMD) scheme, see Ref. [3] for a comprehensive review. CPMD combines classical molecular dynamics simulation techniques with electroic structure calculations on the fly. It is possible with CPMD to simulate solvent effects in which hydrophobic particles are surrounded by solvent molecules, and to directly extract properties of interest during the simulation itself, since electronic structure information is available at all times. This makes the ab initio [2,4–8] investigation of hydrated entities in the liquid phase (at certain values of temperature and pressure) feasible. The situation of an hydrogen radical dissolved in water leaves space for two-fold speculation. One possibility is that the atom sits between the water molecules and takes part in the hydrogen bonding network. The other is a scenery in which the atom could be scavenged in a cage build up through the hydrogen bonding. The latter scenery would resemble a crystalline like situation of a solid state cluster similar to the well-known class of clathrates e.g. Cl_2*6H_2O. As the hydrogen radical is an apolar substance the question of hydrophobic hydration arises, in particular whether the liquid in the neighbourhood of the solute will feature an enhanced or decreased structure. Several studies have discussed this issue controversially and thus no clear picture has been drawn [9]. A third point to discuss is the diffusional mechanism of the hydrogen radical in water in comparison to the self-diffusion of water and the diffusional motion of the formed cavity. The diffusion of the hydrogen radical is observed to be pretty fast, much faster than for instance the one of simple ions and only slightly slower than proton diffusion. One possible explanation might be that the motion of the hydrogen radical is determined by the formation and breaking of hydrogen bonds, i.e. by the hydrogen bonding dynamics. Another idea found in the literature is jump diffusion from one cavity to a neighbouring one via "channels".

1.1   Previous Investigations

Although the system seems very simple there have not been many detailed microscopic simulational explorations in literature, the few existing are: [10–12]. But these simulational studies are all based on model potentials, which are classified by the authors themselves as "very simple" model potentials and which have been criticized with regard to their incorrect description of the particular interactions. The used potential are all — without any exception — pairwise additive and at most semi empirical. In addition, two different potential models lead to vastly different results concerning the structure around the solute [12].

The picture drawn from [10,11] supports a clathrate-like hydrated hydrogen radical, i.e. the hydrogen radical sits in a cavity consisting of about 18 water molecules. This picture is based on the difference between the location of the first peaks of the oxygen–hydrogen radical and the oxygen–oxygen radial pair correlation functions. Recent experiments to determine the hyperfine interaction of atomic hydrogen in liquid water and ice do confirm these results indirectly [13], but with a very crude model derived from surface tension arguments, i.e. a macroscopic quantity is applied to a clearly microscopic situation. However, it is found in the study [12] that another more reliable pair potential, which was explicitly fitted to experiment in previous studies, leads to an oxygen–hydrogen radical peak much closer to the oxygen–oxygen peak. Taken seriously, this would contradict the previous interpretation [10,11]. The qualitative difference concerning the relative location of the oxygen–hydrogen radical peak is discussed, but without resolving this issue.

Not much has been published concerning the influence of the hydrogen radical on the water structure close to the impurity, which is surprising as these studies are focused on the structure of the solvated hydrogen radical. In [10] it is found that the oxygen–oxygen radial pair correlation function for pure water differs little from that of the water oxygens close to the hydrogen radical. [10]

In both [10,11] it has been impossible to calculate the diffusion coefficients without large uncertainties. The authors find similar diffusion constants for both the hydrogen radical and the water molecules, which clearly contradicts experimental findings [1]; the authors themselves call their calculation unreliable in this respect. Nevertheless they advocate that the diffusion of the hydrogen radical is limited by the breaking and reformation of solvent hydrogen bonds in the case. If this is so, they predict that there should be no isotope effect. This is clearly not the case as shown more recently [1]: a pronounced difference in diffusion constants is found for light atoms in water. This, however, is a delicate issue since the de Broglie wavelength and thus the spatial requirement of the solute atoms depends on the mass so that isotope effects could result indirectly. Still, to look at the hydrogen bonding dynamics as the main source for what is going on in the system seems to be reasonable. Interestingly, the diffusion of neon is also quite fast, indeed it is faster than the self-diffusion of water [14], which was interpreted in terms of fast motion through a network of channels in liquid water [14]. The diffusion coefficients of the hydrogen isotopes H, D, Mu, and He are even larger than that of Ne and are found to follow an inverse cube root dependency on the mass [1], whereas mass-independence was seen only for Kr, Xe, and Rn. This clearly signals non-Stokesian diffusional behaviour for light atoms in water.

2   Simulational details

The preparation of an initial condition close to the equilibrium is quite essential for CPMD simulations, as CPMD is very time consuming. This "pre-equilibration" has been done using traditional molecular dynamics simulations with published model potentials. The necessary system size of one hydrogen radical plus 63 H_2O molecules used in [10,11] lies in the range of nowadays accessible system sizes for CPMD. In Ref. [12], i.e. one of the classical MD studies, it was shown that very similar results were obtained with 108 H_2O molecules, which suggests that a sort of convergence has been obtained. After the traditional MD simulation preliminary CPMD calculations were performed on the CRAY T3E of the MPG at RZG in Garching using 64–128 nodes. From this 6ps trajectory we took a configuration to start on the CRAY in Stuttgart.

As simulation box we took a simple cubic one. The lattice parameter — in this case the box-length — was chosen to be 12.4Å. The cutoff for the plane wave basis is 70Ryd. As the system contains one unpaired electron, the multiplicity is 2. We used for this open shell problem the local spin density (LSD) approximation. The functional BLYP and non conserving pseudo potentials were used. To be able to accelerate the simulation, we took the mass of a deuterium for each hydrogen atom in the water molecules including the single hydrogen radical. Rescaling from D to H results is done using the experimental conversion factor of 1.22 [15].

2.1   Equilibrium: Nosé–Hover Thermostats

To force the system to a state near the equilibrium, we used the method of Nosé and Hoover. For details concerning the method we refer to the original papers [16–18]. The target temperature for the Nosé–Hoover thermostats has been chosen to be 330K in order to allow for fast diffusion. We set the maximum number of steps to 1800, the fictitious electron mass to 1100a.u., the time step to 8a.u. ($\approx$0.2fs) and stored the wavefunction and coordinates in a restartable file every 500th step. In total the Nosé–Hoover runs had a length of 3.6ps.

2.2   Simulation: Microcanonical MD

During the course of this simulation (hereafter called *NVE*) again a temperature decrease was observed. To avoid this behaviour we changed from a microcanonical (*NVE*) to a canonical (*NVT*) ensemble, but using very gentle thermostats. This series we call *NVT*. In total the *NVE* run is 2.5 ps and the *NVT* is 4.1 ps so far. Whereas the average temperature of the *NVE* is 307K, the average temperature of the *NVT* amounts to 308K.

2.3   Technical details

Taken into account that the results are stored in the restart files sometimes more and less often, 5000 time steps with a MD time step of 8a.u., i.e. approximately 1ps, take about 30 hours on the CRAY-T3E using 128 processing elements with about 35MB peak memory each, i.e. the total RAM requirement is 4.5 Gigabytes.

A trajectory length of 10ps is the minimum to obtain good statistics concerning structural quantities, and a reliable estimate of diffusion coefficients requires *at least* twice that length. As one file necessary to restart jobs amounts to about 2GB, the disc space of 20GB is required. In summary the following resources have been used for the proposed project so far: **time:** 8 of 18 months,**CPU hours:** 49878 of 128000 hours on 128 nodes, **nodes:** 128, **memory per node:** 35MB, **disc space:** 20GB.

3   Results

We define a molecular cage which surrounds the hydrogen radical as **cavity** by collecting the neighbour molecules from a geometrical criteria to calculate the center of mass given by these neighbour molecules. If the distance is smaller than a certain amount a molecule is chosen as neighbour. Next it is tested whether the angle — built from hydrogen radical–oxygen vector (from the next nearest water molecule) to hydrogen radical–oxygen vector from the water with the smaller distances — is smaller than a certain limit. With this choice we wanted to make sure that rather molecules are included which have a free "sight" onto the hydrogen radical than molecules which are hidden behind other water molecules, but which are maybe closer. We call this neighbour molecules H_2O_{in} and the other water molecules H_2O_{bulk}. The given results are preliminary results with still nonsufficient statistics.

3.1   Structure

Description of the structure can be initiated by the radial pair correlation function $g(r)$. Figure 1 shows the radial pair correlation function for oxygen atoms of the water molecules, for the hydrogen radical and the oxygen atoms of the water molecules and for the hydrogen radical and the hydrogen atoms of the water molecules.

The maximum of the O–O function is at shorter distances than the one for the hydrogen radical function, i.e. in average the water molecules are closer to

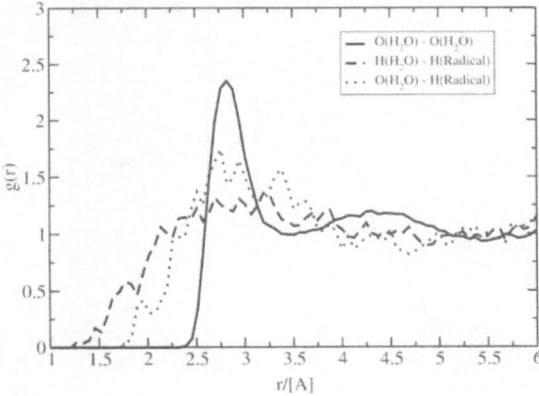

Fig. 1. Radial pair correlation function g(r) against the distance r in Ångström of the Car–Parrinello trajectory.

each other than to the hydrogen radical. Nevertheless the hydrogen radical peaks are very broad, which shows that closer distances than the closest oxygen distance are sometimes possible for certain configurations. Comparing the two hydrogen radical functions, one finds that the peak of the hydrogen–hydrogen radical function starts earlier to increase and has its maximum at locations before the oxygen–hydrogen radical function. This means that in average the hydrogen atom of the water molecules are pointing towards the hydrogen radical. This feature is of course in accordance with electronic structure calculations [23] where tests show the intermolecular structure of the most important attractive interaction to be when the hydrogen radical is coordinated via a hydrogen atom of the water molecule rather than via the lone pair of the oxygen atom.

In comparison with the previous empirical molecular dynamics simulations [10] (which we reproduced ourselves by classical MD in order to check these results) we find the maximum of both the functions obtained with the CPMD method much earlier than the one obtained with the classical MD program using a crude pair potential, see Figure 2. This, however, is in agreement with the results as obtained in Ref. [12] using the much more reliable hydrogen radical potential fitted to experiment with the maximum for the hydrogen radical–oxygen function at 3.0Å and 2.9Å for the hydrogen radical–hydrogen function. Again the relation to the intermolecular potential is the origin for such a behaviour. Both, our CPMD results and the ones obtained by fitting to experiment, agree in that the hydrogen atoms are closer to the hydrogen radical than the oxygen atoms.

To better understand the structure we also calculated the static orientation of the O–H bond with respect to the O–hydrogen radical vector. We separated between cavity-molecules (in) and bulk-molecules. Figure 3

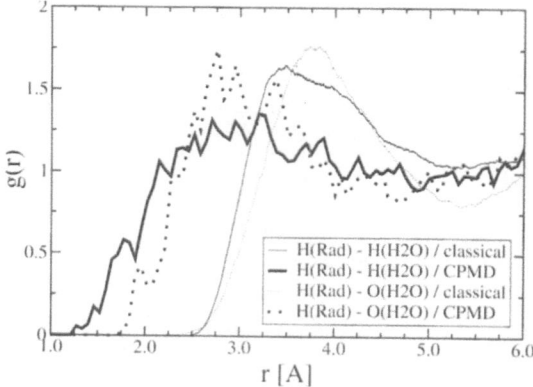

Fig. 2. Radial pair correlation function g(r) against the distance r in Ångström of the Car–Parrinello trajectory compared to the classical simulation for the hydrogen radical with the atoms of the water molecule [10].

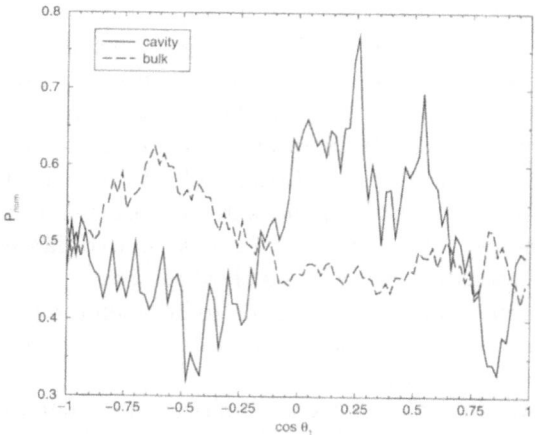

Fig. 3. Distribution function for the angle cosines describing the orientation of the water O–H bonds with respect to the hydrogen radical oxygen vector. Solid line: cavity water. Dashed line: bulk water.

plots the normalized distribution function against the angle cosines. For the cavity water molecules a very pronounced feature can be seen between $\cos \alpha = 0.1$ ($\alpha = 84°$) and $\cos \alpha = 0.6$ ($\alpha = 53°$). This means, that most often the symmetry axis of the water molecule almost points towards the hydrogen radical. (The other possibility would be that the hydrogen radical, the oxygen atom and one hydrogen atom of the water molecule lie on a straight line.) More detailed analysis of the cavity structure is under way.

3.2   Electronic Structure

Additional insights into the electronic structure of the clathrate cage and the state of the solvated hydrogen radical can be obtained by calculating the electron localization function (ELF). This function has been introduced by Becke and Edgecombe [19] to circumvent the problem of unitary orbital transformation for getting information on localized electronic groups. The ELF itself is a dimensionless localization index calibrated with respect to the uniform-density electron gas as reference. The ELF values lie between zero and one, with one corresponding to perfect localization and the ELF value of $\frac{1}{2}$ corresponding to electron-gas-like pair probability. As this study has been undertaken in the unrestricted (spin polarized) density functional theory framework, a new formula for the unrestricted case had to be implemented in the CPMD code. This formula was derived by Kohout and Savin [20].

$$
\mathrm{ELF}_{\mathrm{open\ shell}} = \left[1 + \left(\frac{\tau - \frac{1}{8}\frac{(\nabla\rho_\alpha)^2}{\rho_\alpha} - \frac{1}{8}\frac{(\nabla\rho_\beta)^2}{\rho_\beta}}{2^{\frac{2}{3}} c_F (\rho_\alpha^{\frac{5}{3}} + \rho_\beta^{\frac{5}{3}})} \right)^2 \right]^{-1}
$$

For depicting this function we choose two configurations, where the hydrogen radical once is located at a short distance 1.74Å to the next oxygen atom and once at a far distance of 3.00Å. We cut out the cavity together with the hydrogen radical to calculate ELF.

In Figure 4 the results are depicted. First, one should note, that every water molecule shows the well-known features of the lone pair — by choosing one water molecule and looking at the shaded sausage-like region behind the oxygen atoms. Also electrons are highly located at the hydrogen atoms. On the left hand side in Figure 4 ELF for the first cluster is depicted. It can be seen, that for the closest water molecule — the one with the 1.74Å distance — the hydrogen radical is coordinated sideways, not letting only the hydrogen atoms of the water molecule point towards the hydrogen radical, but to the same amount also the oxygen atom. The water molecule a little bit above in the middle is coordinated via the hydrogen atoms. On the right hand side the second cluster is depicted. Here the hydrogen radical is coordinated via the O–H bond of the water molecule. Whereas the first configuration is very rare the latter is very common.

For the hydrogen radical, it is impressive in both pictures, how much more space the higher value of ELF in the region of the hydrogen radical occupies compared to the regions around the hydrogen atoms of the water molecules, i.e. the green sphere around the hydrogen is much bigger in the case of the hydrogen radical than in the case of the water hydrogen atom. This shows in both cases a higher locality of electrons in a wider range of space.

The difference in the two pictures concern the shape of the basins around the hydrogen radical: in the first cluster the surface has a shape like a disc, whereas in the second cluster the surface has a more spherical shape. This

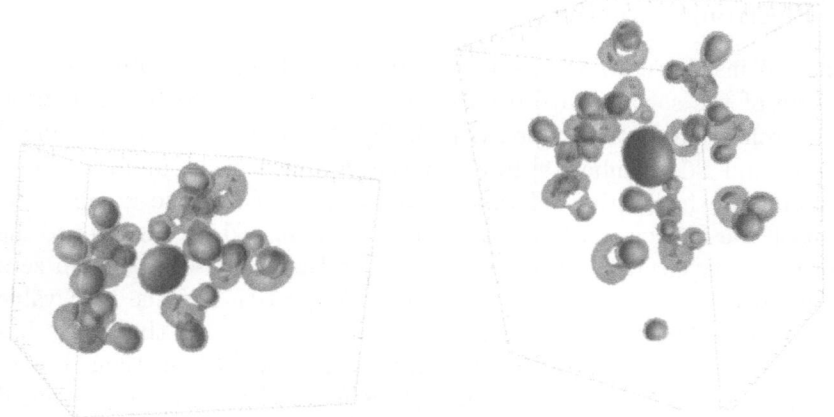

Fig. 4. The electron localization function with the values of 0.89 (green) and 0.8 (blue). On the right side a cluster in which the distance to the next oxygen amounts 1.74Å, on the left side an example with 3.00Å is depicted.

shows that our calculation gives different polarizations and thus qualitatively the right intermolecular forces.

3.3   Diffusion

Figure 5 displays the mean square displacement plotted against the time of the *NVT*. One realizes first that the curve of the hydrogen radical and the cavity increases much more than the one for water. The motion of the vector between the hydrogen radical and the cavity gives more or less a straight line. More information is given in Table 1, where the diffusion constants of the different entities and the different simulations together with the experimental one are listed. As these are very preliminary results based on an insufficient

Table 1. Diffusion constants in $Å^2 ps^{-1}$. The values for the heavy atom are directly obtained by the simulation, these are multiplied by a factor of 1.22 to give the light atom values. And are listed behind the light atom values in brackets. Note, that for 64 H_2O[21] a value of 0.28 and for 32 H_2O[22]a value of 0.13 has been given.

particle	Exp.[1]		*NVT*		*NVE*	
	298K		308K		307K	
H(D)	0.7	(0.5)	2.6	(2.1)	1.0	(0.8)
$H_2O_{all}(D_2O)$	0.24		0.27	(0.22)	0.20	(0.16)
$H_2O_{bulk}(D_2O)$			0.31	(0.25)	0.23	(0.19)
$H_2O_{in}(D_2O)$			0.10	(0.08)	−0.05	(−0.04)
Cavity(C_{heavy})			1.8	(1.4)	1.2	(1.0)

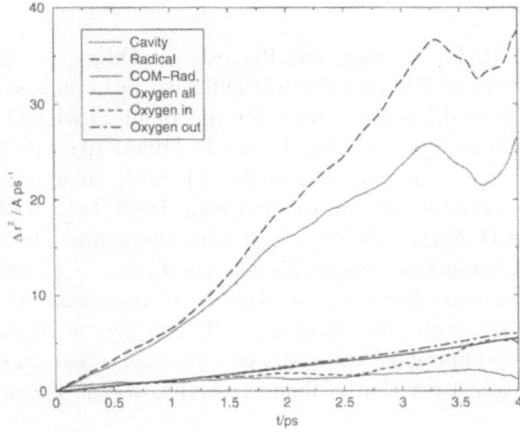

Fig. 5. The mean square displacement plotted again the time.

sampling time a comparison with the experiment is difficult. One can only infer that the values are in the same range as the experimental ones. This is also found when comparing the pure water values, where the 64 water system [21] has a by a factor of two higher value than the 32 molecules system [22]. However the main feature is that the hydrogen radical and the cavity move much faster than the water molecules themselves. The water molecules in the cavity are much slower than the water molecules outside the cavity. These results indicate a picture where the hydrogen radical moves via structural diffusion, i.e. the hydrogen radical moves together with the surrounding, but slowly exchanging water molecules, which are connected to the hydrogen radical by attractive forces. An indirect proof for this is the coordination via the hydrogen atoms. Electronic structure calculations [23] show that this geometry is a minimum on the potential surface of one hydrogen atom with one water molecule.

4   Conclusion and outlook

In conclusion, our CPMD simulations are the first to show a *faster* diffusion of a small hydrophobic molecule in water. This outstanding result has probably to do with a *structural diffusion mechanism*.

Notes and Comments. In order to arrive at hard conclusions, the trajectory has to be continued and a more detailed analysis has to be carried out. Barbara Kirchner would like to thank the HLRS for providing this project with calculation time. She also thanks the Schweizerischer Nationalfonds and the Novartis foundation for financial support.

References

1. E. Roduner, P. L. W. Tregenna-Piggott, H. Dilger, K. Ehrensberger, and M. Senba: Effect of Mass on Particle Diffusion in Liquids studied by Electron Spin Exchange and Chemical Reaction of Muonium with Oxygen in Aqueous Solution. J. Chem. Soc. Faraday Trans. **91** (1995) 1935–1940

2. D. Marx, M. Sprik, and M. Parrinello: Ab initio molecular dynamics of ion solvation. The case of Be^{2+} in water. Chem. Phys. Lett. **273** (1997) 360–366

3. J. Hutter and D. Marx: Ab Initio Molecular Dynamics: Theory and Implementation. In J. Grotendorst, editor, Modern Methods and Algorithms of Quantum Chemistry. John von Neumann Institute for Computing. Jülich (2000) 301–450

4. A. Curioni, M. Sprik, W. Andreoni, H. Schiffer, J. Hutter, and M. Parrinello: Density Functional theory-based molecular dynamics simulation of acid-catalyzed chemical reaction in liquid trioxane. J. Am. Chem. Soc. **119** (1997) 7218–7229

5. K. Doclo and U. Röthlisberger: Ab intio molecular dynamics simulation of the gas-phase reaction of hydroxyl radical with nitrogen dioxide radical. Chem. Phys. Lett. **297** (1998) 205–210

6. E. J. Meijer and M. Sprik: Ab inito molecular dynamics study of the reaction of water with formaldehyde in sulfuric acid solution. J. Am. Chem. Soc. **120** (1998) 6345–6355

7. U. Röthlisberger, M. Sprik, and M. L. Klein: Living polymers. Ab initio molecular dynamics study of the initiation step in the polymerization of isoprene induced by ethyl lithium. J. Chem. Soc. Faraday Trans. **94** (1998) 501–508

8. M. Mohr, D. Marx, M. Parrinello and H. Zipse: Solvation of Radical Cations in Water — Reactive or Unreactive Solvation? Chem. Eur. J. **6** (2000) 4009–4015

9. W. Blokzijl and J. B. F. N. Engberts. Hydrophobic Effects. Opinions and Facts. Angew. Chem. Int. Ed. Engl. **32** (1993) 1545–1579

10. J. S. Tse and M. L. Klein: Are hydrogen atoms solvated by water molecules? J. Phys. Chem. **87** (1983) 5055–5057

11. B. De Raedt, M. Sprik, and M. L. Klein: Computer simulation of muonium in water. J. Chem. Phys. **80** (1984) 5719-5724

12. H. D. Gai and B. C. Garrett: Path integral calculations of the free energies of hydration of hydrogen isotopes (H, D, and Mu). J. Chem. Phys. **98** (1994) 9642–9648

13. E. Roduner, P. W. Percival, P. Han, and D. M. Bartels: Isotope and temperature effects on the hyperfine interaction of atomic hydrogen in liquid water and in ice. J. Chem. Phys. **102** (1995) 5989–5997

14. M. Holz, R. Haselmeier, R. K. Mazitov, and H. Weingärtner: Self-Diffusion of Neon in Water by ^{21}Ne NMR. J. Am. Chem. Soc. **116** (1994) 801–802

15. E. H. Hardy, A. Zygar, M. D. Zeidler, M. Holz and F. D. Sacher: Isotope effect on the translational and rotational motion in liquid water and ammonia J. Chem. Phys. **114** (2001) 3174–3181

16. S. Nosé: A unified formulation of the constant temperature molecular-dynamics methods. J. Chem. Phys. **81** (1984) 511–519

17. S. Nosé: A molecular-dynamics method for simulations in the canonical ensemble. Mol. Phys. **52** (1984) 255–268

18. W. G. Hoover. Canonical dynamics - equilibrium phase-space distribution. Phys. Rev. A **31** (1985) 1695–1697

19. A. D. Becke and K. E. Edgecombe: A simple measure of electron localization in atomic and molecular systems. J. Chem. Phys. **92** (1990) 5397–5403
20. M. Kohout and A. Savin: Atomic shell structure and electron numbers. Int. J. Quant. Chem. **60** (1996) 876–882
21. P. L. Silvestrelli and M. Parrinello: Structural, electronic, and bonding properties of liquid water from first principles. J. Chem. Phys. **111** (1999) 3572–3580
22. M. Sprik, J. Hutter, and M. Parrinello: Ab initio molecular dynamics simulation of liquid water: Comparison of three gradient-corrected density functionals. J. Chem. Phys. **105** (1996) 1142–1152
23. J. Z. Larese and Q. M. Zhang: Layer-by-layer melting of argon films on graphite — a neutron-diffraction study. Phys. Rev. Lett. **64** (1990) 922–925

Ab initio Molecular Dynamics Simulation of Hydrogen Fluoride at Several Thermodynamic States

Markus Kreitmeir[1], Jens Jørgen Mortensen[2], Helmut Bertagnolli[1], and Michele Parrinello[3]

[1] Inst. f. Physikal. Chem., Univ. Stuttgart, Pfaffenwaldring 55,
   D-70569 Stuttgart, Germany
[2] Dep. of Physics, Techn. Univ. of Denmark, Building 307, DK-2800 Lyngby,
   Denmark
[3] Max-Planck Institut f. Festkörperforschung, Heisenbergstr. 1,
   D-70569 Stuttgart, Germany

Abstract. Liquid hydrogen fluoride is a simple but interesting system for studies on the influence of hydrogen bonds on physical properties. Several attempts have been made to describe the microscopic structure by means of classical simulation methods. But only an *ab initio* simulation method was able to reproduce the structure satisfactorily. Now, new experimental data on the structure is available for a broad range of thermodynamic states. For these, *ab initio* simulations were performed to obtain deeper insight into the microscopic structure. Car-Parrinello molecular dynamics were used to account for the highly polarizable system. A technical description of the simulations and first interpretations of the results are given.

1 Introduction

Advances in computer technology have changed scientific research quite profoundly. The ability to perform an almost incredible huge amount of calculations in decreasing time allows to tackle more and more relevant physical problems via computer simulations. Computer simulations are in between theory and experiment, but, by nature, have a stronger affinity to experimental work. One *performs* computer simulations just as one performs experiments, using computers instead of laboratories.

There are several good reasons for computer experiments and just a few of them will be mentioned here. They allow the substitution of expensive experiments e.g. the simulation of crash tests carried out by the automobile industries. Simulations can also be a tool for probing theories and thereby assist theoretical advances.

Moreover, simulations make data accessible, which is hard to obtain by experiments. This is the case concerning systems under extreme conditions of temperature and pressure. But instead of only extending the range of classical experiments, simulations are able to enhance the insight into the system. Molecular dynamics (MD) simulations provide data which cannot be obtained by experiments. An example is the application of simulation techniques in order to deepen our knowledge of hydrogen bonds.

During the last decades the importance of hydrogen bonds with respect to the structure of matter has become more and more obvious. The intramolecular shape of biomolecules, such as proteins or other macromolecules like the DNA, are strongly influenced by hydrogen bonds, but furthermore also the intermolecular structure is affected. For instance the properties of water and alcohols are strongly determined by hydrogen bonds. As hydrogen fluoride (HF) forms the strongest hydrogen bonds known, it has become a popular model system for investigations on this field of interest.

Solid hydrogen fluoride is constituted of parallel oriented zig-zag chains and it is apparent from its properties that these chains, to a certain extend, also persist in the liquid phase. To describe the properties of liquid hydrogen fluoride, many simulations have been carried out, see e.g. [1–6]. The classical simulations using pair potentials [1–3,5] are able to describe thermodynamic properties, but, in general, fail to depict the structure of the liquid. This is in contrast to the *ab initio* molecular dynamics simulation carried out by Röthlisberger and Parrinello [6], which was constrained to the liquid at ambient conditions.

A description of the structure of fluids can be given by the atom pair correlation function $g_{\alpha\beta}(r)$. For any central atom of kind α the function $g_{\alpha\beta}(r)$ gives the averaged radial probability for finding an atom of kind β at the distance r. For binary fluids as hydrogen fluoride, standard diffraction experiments yield only the total atom pair correlation function $G(r)$, which is the weighted sum of individual atom correlation functions. Hence, details of the structure cannot be derived from the function $G(r)$.

The benefit of the function $G(r)$ for MD simulations is that the quality of the structure resulting from simulations can be evaluated by comparison with data obtained by diffraction experiments. The total atom pair correlation function from diffraction experiments as well as the ones resulting from a classical (taken from [5]) simulation using pair potentials and an *ab initio* simulation (taken from [6]) are shown in Figure 1. Throughout this article, the intramolecular part of the $G(r)$ is omitted to emphasize the intermolecular part. As can easily be seen the classical simulation fails to reproduce the benchmark set by the diffraction experiment, whereas the *ab initio* MD simulation shows good agreement and describes the structure satisfactorily.

We assume that the underlying reason for this is the omission of polarizability of the classical MD approaches (the potential described by P. Jedlovszky et al [4] is, to our knowledge, the only exception). In our opinion it is essential to include this property for the performance of simulations of hydrogen fluoride, and as *ab intio* MD introduces polarizability immanently, we have chosen the MD method first introduced by R. Car and M. Parrinello [8] for our work.

Due to its toxicity and its aggressive behavior, liquid hydrogen fluoride has not been investigated by diffraction experiments until 1985 [7], when

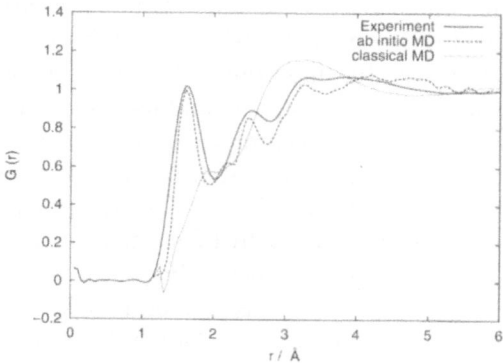

Fig. 1. Comparison of total atom pair correlation functions of hydrogen fluoride at 293 K (taken from [5–7]) obtained with different methods, see text beneath for details.

Deraman and coworkers carried out neutron scattering experiments on the liquid at ambient conditions. So, our knowledge of the structure of the system under interest was constrained to a single thermodynamic state. Now, the total atom pair correlation function of hydrogen fluoride is determined for a broad range of thermodynamic states by diffraction experiments recently carried out by Pfleiderer et al [9].

The experiments point out that there are significant changes in the average number of hydrogen bonds per molecule depending on the temperature. Remarkable changes in the $G(r)$ in comparison to ambient conditions are observed at the phase points given in Tab. 1, where states I and II belong to the same number density C. At state I hydrogen fluoride is liquid, whereas at states II and III it forms a supercritical fluid.

In order to obtain a better understanding of the structural changes indicated, MD simulations of the thermodynamic states given by Tab. 1 using the Car-Parrinello method were performed.

Table 1. Phase point, temperature T, number density C, and pressure p of hydrogen fluoride taken from [9]. Diffraction experiments were performed at these conditions.

phase point	T/K	C/nm^{-3}	p/bar
I	373	24.0	12.0
II	473	24.0	319.0
III	473	19.5	166.0

2   Theory

This section introduces a few basic concepts, so the reader may be able to tackle the contents of the following sections. It is definitely not the intention of this section to give an introduction to *ab initio* MD, this would be far beyond the scope of this article. A good summary of the Car-Parrinello molecular dynamics method is given, for example, by D. K. Remler and P. A. Madden [10].

The charged nuclei in an *ab initio* MD are moved according to Newton's equations of motion due to the forces which act on them. These forces arise, besides the interaction of the ions, from the electrons. The electronic wave functions are calculated using the density functional formalism of Kohn and Sham. The Kohn-Sham orbitals are expanded in a plane-wave basis set f_G^{PW}

$$f_G^{PW}(r) = \frac{1}{\Omega} e^{iG \cdot r} \tag{1}$$

Where the G are compatible with the imposed periodic boundary conditions and Ω is the volume of the simulation cell. The number of plane waves is controlled by an energy cut-off in the following way

$$\frac{1}{2} G^2 \leq E_{cut} \tag{2}$$

The size of the basis set, e.g. the number of basis functions N_{PW}, can be approximated by the fixed (respectively chosen) plane wave cut-off energy

$$N_{PW} \approx \frac{\Omega}{6\pi^2} E_{cut}^{3/2} \tag{3}$$

In other words, the number of plane wave basis functions depends on the cut-off energy as well as the volume of the simulation cell, e.g. the number of simulated particles.

In the region of the cores, the electronic wave function is rapidly oscillating. To decrease the number of basis functions needed for an accurate description of the whole system (regions near and far from nuclei), one uses pseudopotentials to describe the electron density near the nuclei [10]. As the electronic wave function is delocalized, only the system's electron density can be directly deduced from it. But it is possible to obtain localized orbitals for the electrons by means of a transformation. These so called localized Wannier functions [11,12] meet better the chemists idea of orbitals.

3   Technical Details of the Simulation

The sample consisted of 54 hydrogen fluoride molecules which were arranged in a body centered cubic (bcc) lattice in the initial setup. This improbable structure for hydrogen fluoride was chosen to speed up equilibration. All MD simulations were performed at constant energy in a periodically repeated cubic cell. The volume of the cell (see Tab. 2) was chosen to represent the experimental density (taken from [13,14]). The time step was 0.121 fs and a fictitious mass for the electronic degrees of freedom of $\mu = 800$ a.u. was chosen. Pseudopotentials taken from [6] were used for hydrogen and fluoride, and a plane-wave cut-off energy of $E_{cut} = 60$ Ry was chosen. The resulting number of plane-wave basis functions (N_{PW}) is given by Tab. 2.

All calculations were performed with the program CPMD [15] executed on the Cray T3E at the HLRS. The computational requirements are scaling with the size of the problem, e.g. the number of electrons, with an exponent between 2 and 3. The batch queue for 128 nodes and a maximum of 12h computation time turned out to have a good turn-around cycle for our purposes, so we were able to accomplish tasks without waiting periods. For smooth execution of the program, the complex coefficients of the plane wave basis functions as well as their derivatives have to be kept in memory. For example, for 216 electronic states (the number of molecules used multiplied by the number of states per molecule) and 59589 plane wave functions a total of more than 400 Megabyte is needed for storage.

Table 2. Phase point, temperature T, length of the edge of the cubic simulation cell a, sampling time t_s, CPU time t_1 per MD step, and CPU time t_2 for MD step including calculation of Wannier functions.

phase point	T / K	a / Å	t_s / ps	N_{PW}	t_1 / s	t_2 / s
I	373	13.1037	3.3	59589	8.2	31.4
II	473	13.1037	5.8	59589	8.2	31.4
III	473	14.0428	2.1	73483	23.0	48.4

The preparation of the system was carried out by heating up the system to an average temperature of 600 K, which was held for roughly 0.5 ps thus the system was completely losing memory of the initial configuration. Afterwards cooling was started by rescaling the velocities of the nuclei with a factor of 0.9998 per MD step until the desired temperature (see Tab. 2) was reached.

After an initial equilibration time of roughly 1 ps per state, the system was sampled for several picoseconds (see Tab. 2). For each state, localized Wannier functions were calculated for a simulation period of approximately 1 ps. Tab. 2 gives an overview of the CPU time needed per MD step as well as the time needed per MD step including the calculation of the localized Wannier functions.

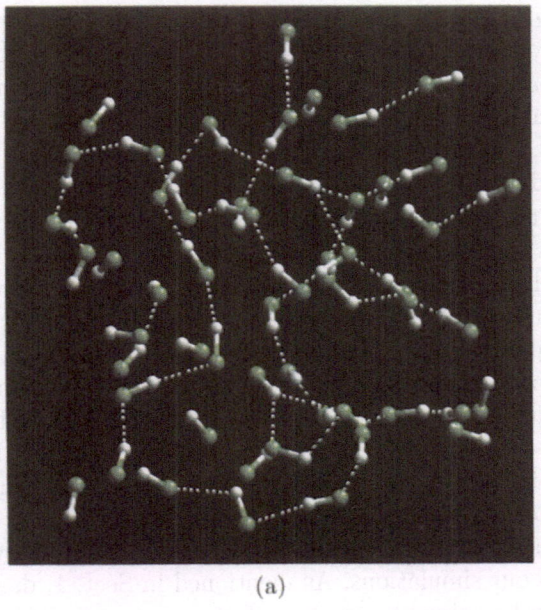

(a)

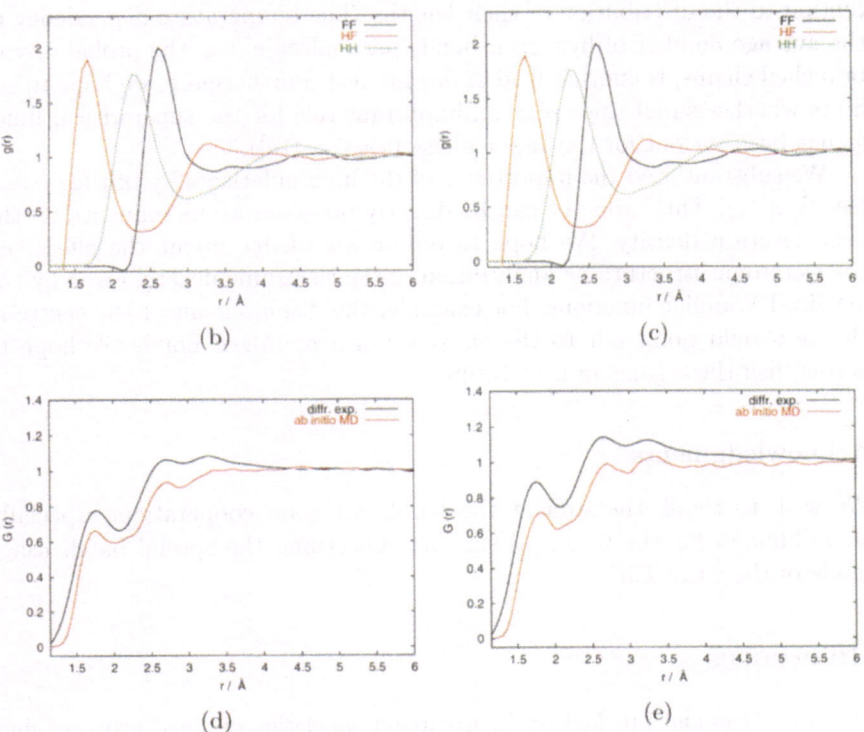

(b)   (c)

(d)   (e)

Fig. 2. (a) Instantaneous configuration of liquid HF. (b) Comparison of $G(r)$ for state I. (c) Atom pair correlation functions $g_{HH}(r)$, $g_{HF}(r)$ and $g_{FF}(r)$ for state I. (d), and (e) Comparison of $G(r)$ for state II and III respectively.

4   Results and Outlook

Fig. 2 shows a typical configuration of liquid hydrogen fluoride taken from our simulation data. The predominating element of the structure, the winding zig-zag chains, can easily be seen. This characteristic feature can be observed at all thermodynamic states investigated.

The total atom pair correlation functions $G(r)$ calculated from our MD data is shown in Fig. 2 for the thermodynamic state I. For comparison, the $G(r)$ obtained by neutron diffraction taken from [9] is shown. In fact, the total atom pair correlation functions show good agreement. In contrast to diffraction experiments, individual atom pair correlation functions are easily obtained from simulations. Fig. 2 also shows the individual atom pair correlation functions $g_{HH}(r)$, $g_{HF}(r)$, and $g_{FF}(r)$ for state I.

Moreover, the total atom pair correlation functions for the states II and III are given in Fig. 2. Again, the comparison to the results obtained by diffraction experiments shows good agreement.

The results presented here are only a first, simple analysis of the data obtained from our simulations. As mentioned in Sect. 1, deeper insight into the system can be gained by further efforts.

Therefore we will analyze the length of the HF chains with special attention to the distribution of their lengths. The temperature dependency of the average number of hydrogen bonds per molecule, e.g. the probability of branched chains, is another field of our interest. Furthermore, we hope to get hints whether small rings play an important role for the supercritical fluid, as has been set out for the vapor phase (see e.g. [16]).

We substantiated the importance of the high polarizability of the system (see Sect. 1). This property can be directly observed by its influence on the local electron density. We hope to obtain knowledge about the effects on the electronic structure by analyzation of the electronic density given by the localized Wannier functions. For example, the displacement of the centre of charge should point out to the existence of a hydrogen bond. We hope to accomplish these goals in near future.

Acknowledgments

We wish to thank the staff at the HLRS for good cooperation, especially H. Pöhlmann for the technical support concerning the special batch usage mode of the Cray T3E.

References

1. M.L. Klein and I.R. McDonald, Computer simulation of liquid hydrogen fluoride, *J. Chem. Phys.*, **71**, (1979), 298–308.
2. M.E. Cournoyer and W.L. Jorgensen, An improved molecular potential function for simulations of liquid hydrogen fluoride, *Mol. Phys.*, **51**, (1983), 119–132.

3. P. Jedlovszky and R. Vallauri, Structural Properties of liquid HF: a computer simulation investigation, *Mol. Phys.*, **93**, (1998), 15–24.

4. P. Jedlovszky and R. Vallauri, Computer simulations of liquid HF by a newly developed polarizable potential model, *J. Chem. Phys.*, **107**, (1997), 10166 – 10176.

5. R.G. Della Valle and D. Gazillo, Towards an effective potential for the monomer, dimer, hexamer, solid and liquid forms of hydrogen fluoride, *Phys. Rev. B*, **59**, (1999), 13699–13706.

6. U. Röthlisberger and M. Parrinello, *Ab initio* molecular dynamics simulation of liquid hydrogen fluoride, *J. Chem. Phys.*, **106**, (1997), 4658–4664.

7. M. Deraman, J.C. Dore and J.G. Powles, Structural studies of liquid hydrogen fluoride by neutron diffraction, *Mol. Phys.*, **55**, (1985), 1351–1367.

8. R. Car and M. Parrinello, Unified approach for molecular dynamics and density-functional theory, *Phys. Rev. Lett.*, **55**, (1985), 2471–2474.

9. T. Pfleiderer, I. Waldner, H. Bertagnolli, K. Tödheide and H.E. Fischer, The structure of liquid and supercritical hydrogen fluoride from neutron scattering, *J. Phys. Chem*, **113**, (2000), 3690 – 3696.

10. D.K. Remler and P.A. Madden, Molecular dynamics without effective potentials via the Car-Parrinello approach, *Mol. Phys.*, **70**, (1990), 921–966.

11. N. Marzari and D. Vanderbilt, Maximally localized generalized Wannier functions for composite energy bands, *Phys. Rev. B*, **56**(20), (1997), 12847.

12. P. L. Silvestrelli, N. Marzari, D. Vanderbilt and M. Parrinello, Maximally-localized Wannier functions for disordered systems: Application to amorphous silicon, *Solid State Com.*, **107**(1), (1998), 7.

13. E.U. Franck and W. Spalhoff, Fluorwasserstoff I. Spezifische Wärme, Dampfdruck und Dichte bis zu 300° C und 300 at, *Z. Elektrochem.*, **61**, (1957), 348–357.

14. E.U. Franck, G. Wiegand and R. Gerhardt, The Density of Hydrogen Fluoride at High Pressures to 973 K and 200 MPa, *J. Supercrit. Fluids*, **15**, (1999), 127.

15. J. Hutter, A. Alavi, T. Deutsch, M. Bernasconi, S. Goedecker, D. Marx, M. Tuckerman and M. Parrinello, *CPMD*, MPI für Festkörperforschung and IBM Zürich Research Laboratory (1995-1999).

16. M.A. Suhm, HF Vapor, *Ber. Bunsenges. Phys. Chem.*, **99**, (1999), 1159 – 1167.

Quantum Chemical Calculations of Transition Metal Complexes

Jan Frunzke and Gernot Frenking

Fachbereich Chemie, Phillipps-Universität Marburg, Hans Meerwein Straße, 35032 Marburg

1 Introduction

Transition metal complexes show a wide variety of chemical reactions. To gain insight into the bonding situation of these complexes and the transition states involved in these reactions is not only crucial for understanding the underlying principles, but even more for finding new reaction pathways or optimising reaction conditions in chemical industry. Where experiments fail to obtain the needful results, modern quantum chemical approaches can be utilised to investigate chemical systems and predict their properties. This is a challenging task for computational chemists and the necessary calculations, particularly at high levels of theory, are demanding in computational resources. Our research focuses on quantum chemical calculations of transition metal compounds using ab initio methods and density functional theory. The goal of our investigations is the exact calculation of bond energies and activation barriers, which are very difficult to obtain experimentally. In the course of our investigation we found out that coupled cluster calculation at the CCSD(T) level in conjunction with quasirelativistic small-core effective core potentials and valence basis functions of DZ+P quality give very accurate results. We are studying now the strength of various transition metal ligand bonds and the reaction profiles of important transition metal compounds in order to make predictions for new experiments. Coupled cluster calculations are computationally extremely demanding and thus, the access to supercomputers is absolutely necessary for our research. Calculations of this quality make it possible for us to predict accurate data about chemical reactions and bond energies, which can be considered as true top research on a worldwide level. The following chapters give an overview about the research of our group using computational resources of the HLR Stuttgart.

2 The Nature of the Chemical Bond Between a Transition Metal and a Group-13 Element: Structure and Bonding of Transition Metal Complexes with Terminal Group-13 Diyl Ligands ER (E = B - Tl; R = Cp, N(SiH₃)₂, Ph, Me)

The results of this study give a comprehensive picture about the bonding situation in transition metal complexes with terminal group-13 diyl ligands. The equilibrium geometries and first TM-ER bond dissociation energies of 35 transition metal complexes with group-13 ligand atoms [(CO)$_4$Fe-ECp], [(CO)$_4$Fe-EN(SiH$_3$)$_2$], [(CO)$_5$W-EN(SiH$_3$)$_2$], [(CO)$_4$Fe-EPh], and [TM(ECH$_3$)$_4$] (TM = Ni, Pd, Pt; E = B, Al, Ga, In, Tl) have been calculated at the DFT level of theory using gradient-corrected exchange and correlation functionals (BP86). The bonding situation in the complexes was examined with the help of the NBO[3] and CDA[2] partitioning schemes.

Table 1 shows the calculated bond lengths and bond angles and the theoretically predicted Fe-E bond dissociation energies of the axial and equatorial isomers of [(CO)$_4$Fe-ECp]. The calculations predict that the [(CO)$_4$Fe-ECp] complexes with an axial position of the ECp ligand are slightly more stable than the equatorial forms, except for E = Tl, where the equatorial form is 0.4 kcal/mol lower in energy than the axial form. The energy differences between axial and equatorial isomers are for all complexes very small (< 1 kcal/mol). The calculations show that the bond energies of the TM-ER bonds are rather high and follow in all cases the trend B > Al > Ga ~ In > Tl. The TM-ER bonds where R is a poor π-donor (Ph, CH$_3$) are shorter and stronger than those where R is a π-donating group (Cp, N(SiH$_3$)$_2$).

Table 1. Calculated bond lengths in Å and energies in [kcal/mol] of (CO)$_4$-Fe-ECp at BP86//II. Experimental values are given in parentheses

E	isomer	Sym.	Fe-E	E-C$_{Cp}$	Fe-CO$_{ax}$	Fe-CO$_{eq}$	∠E-Fe-CO$_{ax}$	∠E-Fe-CO$_{eq}$	ΔE^a	D$_e^b$	D$_o^b$
B	ax	C$_1$	1.962(2.010)^c	1.830-1.838 (1.811-1.817)^c	1.788(1.793)^c	1.765 (1.774,1786)^c	179.6	84.7^f	0.0	77.99	75.02
B	eq minimum	no energy	-	-	-	-	-	-	-	-	-
Al	ax	C$_1$	2.242(2.231)^d	2.240-2.243 (2.140-2.153)^d	1.768 (1.796)^d	1.772 (1.768)^d	179.6	85.2^f	0.0	53.12	51.35
Al	eq	C$_1$	2.233	2.254-2.256	1.778	1.772	79.5^f	123.8^f	+0.8	52.36	50.49
Ga	ax	C$_1$	2.330 (2.273)^e	2.355-2.356 (2.226)^e	1.755 (1.781)^e	1.782 (1.789)^e	180.0	87.7^f	0.0	32.89	31.62
Ga	eq	C$_1$	2.334	2.381-2.379	1.789	1.769	84.8^f	121.4^f	+0.5	32.41	31.01
In	ax	C$_s$	2.465	2.478-2.480	1.752	1.783	180.0	87.8^f	0.0	33.86	31.69
In	eq	C$_s$	2.470	2.498-2.501	1.789	1.767	84.9^f	121.8^f	+0.6	33.30	32.03
Tl	ax	C$_1$	2.580	2.432-3.056	1.748	1.789	179.6	88.6^f	0.0	16.69	15.81
Tl	eq	C$_1$	2.607	2.508-2.958	1.796	1.765	87.5, 87.9	118.8, 121.0	-0.4	17.11	16.13

^{a} Energy difference between axial and equatorial isomers [kcal/mol]. ^{b} Dissociation energy of the Fe-E bond [kcal/mol]. ^{c} Reference [7]. ^{d} Reference [8]. ^{e} Reference [9]. ^{f} Average value of slightly different angles.

Table 2. NBO data of $(CO)_4$-Fe-ECp at BP86/II[a]

E	isomer	q[Fe(CO)₄]	q(Fe)	q(E)	q(C₅H₅)	$p_x(E)$[b]	$p_y(E)$[b]	$p_z(E)$	P(Fe-E)	P(Fe-CO)[c]
				Complexes $(CO)_4$*-Fe-ECp*						
B	ax	-0.51	-0.56	0.32	0.19	0.51	0.51	0.71	0.48	0.70
B	eq	no energy minimum								
Al	ax	-0.67	-0.58	1.18	-0.51	0.29	0.29	0.26	0.48	0.69
Al	eq	-0.60	-0.59	1.12	-0.52	0.33	0.23	0.26	0.44	0.65
Ga	ax	-0.46	-0.51	0.96	-0.50	0.27	0.27	0.21	0.49	0.77
Ga	eq	-0.37	-0.55	0.88	-0.51	0.29	0.22	0.21	0.41	0.69
In	ax	-0.53	-0.49	1.06	-0.53	0.25	0.25	0.19	0.48	0.78
In	eq	-0.44	-0.54	0.98	-0.54	0.20	0.19	0.27	0.40	0.69
Tl	ax	-0.41	-0.45	0.89	-0.48	0.20	0.18	0.14	0.39	0.67
Tl	eq	-0.30	-0.51	0.81	-0.51	-	-	-	0.32	0.73
				Free ligands ECp[d]						
B	ax			0.05	-0.05	0.31	0.31	0.69		
B	eq	no energy minimum								
Al	ax			0.59	-0.59	0.14	0.14	0.26		
Al	eq			0.59	-0.59	0.14	0.14	0.25		
Ga	ax			0.59	-0.59	0.15	0.15	0.18		
Ga	eq			0.59	-0.59	0.15	0.15	0.17		
In	ax			0.61	-0.61	0.15	0.15	0.18		
In	eq			0.61	-0.61	0.15	0.15	0.17		
Tl	ax			0.64	-0.64	0.16	0.14	0.10		
Tl	eq			0.63	-0.63	0.15	0.16	0.10		

[a] Partial charges q, p-orbital population, Wiberg bond indices P.
[b] $p(\pi)$ AO of atom E.
[c] CO_{ax} trans to ECp; CO_{eq} in case of the equatorial isomer.
[d] Calculated using the frozen geometries in the complexes.

Table 2 shows the NBO results of the complexes. The Fe atom carries in all complexes a rather large negative charge, while the partial charge at the ligand atoms E is positive and very high for E = Al - Tl. The calculated charge distribution suggests that the Fe-E bond has a strong ionic character.

Table 3 gives the NBO and CDA results which show the changes that take place in the electronic structure when the Fe-E bond is formed. The population of the degenerate $p(\pi)$ AOs at atom E given by the NBO method increases, while the positive atomic charge at E increases in the complexes. Thus, there is a significant reorganization of the σ and π charges at atom E which leads to comparatively small changes in the total charge. The calculated $\Delta q_\sigma(E)$ values suggest that there is a substantial σ-donation Fe←ECp which is clearly higher than the π-backdonation Fe→ECp. The same result is given by the CDA calculations (Table 3).

Table 4 shows the optimized bond lengths and bond angles and the calculated Fe-E bond energies of the iron complexes $[(CO)_4Fe-EN(SiH_3)_2]$. The theoretically predicted interatomic distances and bond energies of the tungsten complexes $[(CO)_5W-EN(SiH_3)_2]$ are shown in Table 5.

The results for the boron complex are intriguing, since only the equatorial isomer was found to be a minimum on the potential energy surface. This is opposite to the Cp complex where only the axial isomer is a stable species (Table 1). The compound $[(CO)_4Fe-BN(SiMe_3)_2]$ has been synthesized, but the

Table 3. NBO and CDA data of the changes in the metal and ligand population of $(CO)_4$-Fe-ECp at BP86/II[a]

E	isomer	$\Delta q(E)$	$\Delta q_\pi(E)$	$\Delta q_\sigma(E)$	$\Delta q(C_5H_5)$	$d(CpE \rightarrow Fe)$	$b(CpE \leftarrow Fe)$
B	ax	+0.27	-0.40	+0.67	+0.24	0.598	0.277
B	eq		no energy minimum				
Al	ax	+0.59	-0.30	+0.89	+0.08	0.401	0.188
Al	eq	+0.53	-0.28	+0.81	+0.07	0.364	0.164
Ga	ax	+0.37	-0.24	+0.61	+0.09	0.413	0.039
Ga	eq	+0.29	-0.21	+0.50	+0.08	0.399	0.051
In	ax	+0.45	-0.19	+0.64	+0.06	0.361	0.187
In	eq	+0.35	-0.09	+0.44	+0.07	0.342	0.153
Tl	ax	+0.25	-0.08	+0.33	+0.16	0.238	0.102
Tl	eq	+0.18	-	-	+0.12	0.240	0.063

[a] Change in partial charges Δq; difference of the π-population Δq_π, and the σ-charges Δq_σ of the atom E, change
in the partial charges $\Delta q(C_5H_5)$ of the C_5H_5 fragments between the complex and the free ligand, positive values
indicate that the electronic charge decreases, negative values indicate that the electronic charge increases, charge
donation d and backdonation b.

Table 4. Calculated bond lengths in [Å] and energies in [kcal/mol] of $(CO)_4$-Fe-$EN(SiH_3)_2$.

E	isomer	Sym.	Fe-E	E-N	Fe-CO$_{ax}$	Fe-CO$_{eq}$	∠E-Fe-CO$_{ax}$	∠E-Fe-CO$_{eq}$	ΔE^a	D_e^b	D_o^b
B	ax	no energy minimum	-	-	-	-	-	-	-	-	-
B	eq	C_1	1.818	1.382	1.777	1.790	77.6	125.7	-	85.83	83.09
Al	ax	C_s	2.211	1.802	1.771	1.777	177.4	86.1[c]	0.0	51.84	50.23
Al	eq	C_s	2.199	1.810	1.782	1.777	80.2	124.2	-1.1	52.98	51.18
Ga	ax	C_s	2.274	1.876	1.764	1.783	178.7	87.9[c]	0.0	39.68	38.20
Ga	eq	C_s	2.273	1.887	1.791	1.774	84.5	122.7	+0.2	39.53	37.87
In	ax	C_s	2.420	2.023	1.758	1.783	179.1	88.0[c]	0.0	38.92	37.49
In	eq	C_s	2.422	2.034	1.791	1.770	84.5	123.5	+0.7	38.20	36.65
Tl	ax	C_s	2.530	2.123	1.752	1.789	179.5	89.2[c]	0.0	25.35	24.17
Tl	eq	C_s	2.540	2.144	1.796	1.769	87.5	121.4	+0.4	24.95	22.47

[a] Energy difference between axial and equatorial isomers [kcal/mol]
[b] Dissociation energy of the Fe-E bond [kcal/mol]
[c] Average value of slightly different angles.

authors did not report the structure of the complex[10]. A recent theoretical study of the parent compound $[(CO)_4Fe\text{-}ENH_2]$ at the DFT level predicted that the axial isomer is a minimum on the potential energy surface, and that it is lower in energy than the equatorial form[14].

The $[(CO)_4Fe\text{-}EN(SiH_3)_2]$ compounds have generally stronger TM-E bonds than the $[(CO)_5W\text{-}EN(SiH_3)_2]$ compounds. The difference in the bond energy becomes smaller for the heavier group-13 elements and finally vanishes for the thallium complexes where the Fe-Tl and W-Tl bond strengths are nearly the same. For both series of $[TM\text{-}EN(SiH_3)_2]$ holds, however, that the trend of the TM-E bond energies is the same as for the TM-ECp complexes, i.e. B > Al > Ga ~ In > Tl.

Table 5. Calculated bond lengths in [Å] and dissociation energies in [kcal/mol] of $(CO)_5$-W-EN$(SiH_3)_2$. Experimental values are given in parentheses.

E	Sym.	W-E	E-N	W-CO$_{trans}$	W-CO$_{cis}$	∠E-W-CO$_{trans}$	∠E-W-CO$_{cis}$^{a}	D_e^{b}	D_o^{b}
B	C_2	2.125	1.383	2.078	2.059	180.0	88.3	75.08	73.4
		(2.152)^c	(1.339)^c						
Al	C_{2v}	2.530	1.816	2.028	2.057	180.0	89.1	44.40	43.4
Ga	C_{2v}	2.551	1.893	2.022	2.058	180.0	89.6	36.69	35.7
In	C_{2v}	2.679	2.033	2.015	2.057	180.0	89.9	36.46	35.5
Tl	C_{2v}	2.792	2.134	2.004	2.057	180.0	89.7	25.24	24.5

^{a} Average value of slightly different angles. ^{b} Dissociation energy of the W-E bond [kcal/mol]. ^{c} Reference [10]

Table 6. NBO data of $(CO)_4$-Fe-EN$(SiH_3)_2$ at BP86/II^a.

E	isomer	q[Fe(CO)$_4$]	q(Fe)	q(E)	q(N)	p$_x$(E)^b	p$_y$(E)^c	p$_z$(E)	p$_z$(N)^f	P(Fe-E)	P(Fe-CO)^d
					Complexes $(CO)_4$-Fe-EN$(SiH_3)_2$						
B	ax	no energy minimum									
B	eq	-0.31	-0.58	0.59	-1.49	0.39	0.48	0.61	1.67	0.65	0.62
Al	ax	-0.64	-0.60	1.31	-1.79	0.26	0.23	0.20	1.79	0.53	0.67
Al	eq	-0.57	-0.63	1.23	-1.79	0.20	0.30	0.21	1.78	0.51	0.63
Ga	ax	-0.53	-0.56	1.14	-1.73	0.24	0.20	0.19	1.79	0.53	0.71
Ga	eq	-0.43	-0.60	1.06	-1.73	0.17	0.25	0.20	1.78	0.50	0.65
In	ax	-0.57	-0.53	1.21	-1.71	0.21	0.17	0.16	1.79	0.50	0.73
In	eq	-0.48	-0.58	1.13	-1.71	0.15	0.23	0.17	1.78	0.47	0.65
Tl	ax	-0.46	-0.48	1.07	-1.66	0.18	0.14	0.13	1.79	0.44	0.79
Tl	eq	-0.36	-0.54	1.00	-1.67	0.13	0.17	0.13	1.78	0.40	0.69
					Free ligands EN$(SiH_3)_2$^{e}						
B	ax	no energy minimum									
B	eq			0.51	-1.62	0.21	0.06	0.57	1.72		
Al	ax			0.79	-1.83	0.11	0.02	0.23	1.77		
Al	eq			0.79	-1.83	0.11	0.02	0.23	1.77		
Ga	ax			0.76	-1.78	0.11	0.02	0.23	1.77		
Ga	eq			0.77	-1.78	0.11	0.02	0.23	1.77		
In	ax			0.78	-1.75	0.10	0.01	0.22	1.76		
In	eq			0.78	-1.75	0.10	0.01	0.22	1.76		
Tl	ax			0.79	-1.72	0.10	0.01	0.20	1.76		
Tl	eq			0.77	-1.72	0.10	0.01	0.19	1.77		

^{a} Partial charges q, p-orbital population and Wiberg bond indices P.
^{b} $p(\pi)$ AO of atom E which is perpendicular to the Si-N-Si plane.
^{c} $p(\pi)$ AO of atom E which is in the Si-N-Si plane.
^{d} CO$_{ax}$ trans to EN$(SiH_3)_2$; CO$_{eq}$ in case of the equatorial isomer.
^{e} Calculated using the frozen geometries in the complexes.
^{f} Population of the nitrogen lone-pair orbital.

Tables 6 and 7 show the results of the NBO partitioning scheme for the aminoborylene complexes. The Fe-E and W-E bonds have a strong ionic character. This becomes obvious from the calculated charge distribution, which gives large positive charges for the group-13 elements E and large negative charges for the transition metals. The in-plane $p(\pi)$ AO of E is nearly empty in the free ligand, but it becomes equally or even higher occupied than the out-of-plane $p(\pi)$ in the complex. Both $p(\pi)$ AOs of E become higher populated in the complexes, which shows that the Fe→E π-backdonation is quite effective.

Table 7. NBO data of $(CO)_5$-W-EN$(SiH_3)_2$ at BP86/II[a].

E	$q[W(CO)_5]$	$q(W)$	$q(E)$	$q(N)$	$p_x(E)$[b]	$p_y(E)$[c]	$p_z(E)$	$p_\pi(N)$[f]	$P(W\text{-}E)$	$P(W\text{-}CO)$[d]
			Complex (CO)$_5$-W-EN(SiH$_3$)$_2$							
B	-0.36	-0.87	0.69	-1.50	0.30	0.31	0.66	1.78	0.82	0.75
Al	-0.61	-0.99	1.30	-1.81	0.17	0.11	0.23	1.78	0.56	0.84
Ga	-0.54	-0.96	1.20	-1.75	0.16	0.10	0.21	1.78	0.53	0.87
In	-0.60	-0.93	1.27	-1.73	0.14	0.08	0.17	1.78	0.50	0.89
Tl	-0.50	-0.86	1.15	-1.68	0.12	0.07	0.14	1.78	0.44	0.94
			Free ligand EN(SiH$_3$)$_2$[e]							
B			0.51	-1.62	0.21	0.06	0.57	1.72		
Al			0.79	-1.83	0.11	0.02	0.22	1.77		
Ga			0.77	-1.78	0.11	0.02	0.23	1.77		
In			0.78	-1.75	0.10	0.01	0.22	1.76		
Tl			0.77	-1.72	0.10	0.01	0.19	1.76		

[a] Partial charges q, p-orbital population and Wiberg bond indices P.
[b] $p(\pi)$ AO of atom E which is perpendicular to the Si-N-Si plane.
[c] $p(\pi)$ AO of atom E which is in the Si-N-Si plane.
[d] CO_{ax} trans to EN$(SiH_3)_2$; CO_{eq} in case of the equatorial isomer.
[e] Calculated using the frozen geometries in the complexes.
[f] population of the nitrogen lone-pair orbital.

Table 8. NBO and CDA data of the changes in the metal and ligand population of $(CO)_4$-Fe-EN$(SiH_3)_2$ at BP86/II[a]

E	isomer	$\Delta q(E)$	$\Delta q_\pi(E)$	$\Delta q_\sigma(E)$	$\Delta q_\pi(N)$	$\Delta q(N(SiH_3)_2)$	$d(E{\to}Fe)$	$b(E{\leftarrow}Fe)$
B	ax		*No energy minimum*					
B	eq	+0.08	-0.60	+0.68	+0.05	+0.23	0.583	0.476
Al	ax	+0.52	-0.36	+0.88	-0.02	+0.12	0.419	0.237
Al	eq	+0.44	-0.37	+0.81	-0.01	+0.13	0.368	0.208
Ga	ax	+0.38	-0.31	+0.69	-0.02	+0.15	0.393	0.128
Ga	eq	+0.29	-0.29	+0.58	-0.01	+0.14	0.356	0.145
In	ax	+0.43	-0.27	+0.70	-0.03	+0.14	0.214	0.165
In	eq	+0.35	-0.27	+0.62	-0.02	+0.13	0.211	0.175
Tl	ax	+0.28	-0.21	+0.49	-0.03	+0.18	0.243	0.079
Tl	eq	+0.23	-0.19	+0.42	-0.01	+0.13	0.222	0.061

[a] Change in partial charges Δq; difference of the π-population Δq_π, and the σ-charges Δq_σ of the atom E, change
in the partial charges $\Delta q(N(SiH_3)_2)$ of the $N(SiH_3)_2$ fragments between the complex and the free ligand, positive values
indicate that the electronic charge decreases, negative values indicate that the electronic charge increases and charge
donation d and backdonation b.

Tables 8 and 9 show the differences in the charge distribution of the fragments when the TM-E bond is formed. Note that the change in the partial charge at the boron atom of $[(CO)_4Fe\text{-}BN(SiH_3)_2]$ ($\Delta q(E) = +0.08$ e) is smaller than the alteration in the partial charge of the aminosilyl group ($\Delta q(N(SiH_3)_2) = +0.23$ e). The NBO analysis suggests that the TM$\leftarrow$EN$(SiH_3)_2$ σ-donation is very high for all elements E, and that the largest σ-donation is found for E = Al. The CDA results indicate that the

Table 9. NBO and CDA data of the changes in the metal and ligand population of $(CO)_5$-W-EN$(SiH_3)_2$ at BP86/II[a]

E	$\Delta q(E)$	$\Delta q_\pi(E)$	$\Delta q_\sigma(E)$	$\Delta q_\pi(N)$	$\Delta q(N(SiH_3)_2)$	$d(E \rightarrow W)$	$b(E \leftarrow W)$
B	+0.18	-0.34	+0.52	-0.06	+0.18	0.750	0.329
Al	+0.51	-0.15	+0.66	-0.01	+0.10	0.566	0.287
Ga	+0.43	-0.13	+0.56	-0.01	+0.10	0.509	0.258
In	+0.49	-0.11	+0.60	-0.02	+0.11	0.309	0.186
Tl	+0.38	-0.08	+0.46	-0.02	+0.12	0.432	0.155

[a] Change in partial charges Δq; difference of the π-population Δq_π, and the σ-charges Δq_σ of the atom E, change
in the partial charges $\Delta q(N(SiH_3)_2)$ of the $N(SiH_3)_2$ fragments between the complex and the free ligand, positive values
indicate that the electronic charge decreases, negative values indicate that the electronic charge increases and charge
donation d and backdonation b.

Table 10. Calculated bond lengths in [Å] and dissociation energies in [kcal/mol] of $(CO)_4$-Fe-EPh at BP86/II. Experimental values are given in parentheses.

E	isomer	Sym	Fe-E	E-C_{Ph}	Fe-CO_{ax}	Fe-CO_{eq}	$\angle$E-Fe-CO_{ax}	$\angle$E-Fe-CO_{eq}	ΔE^a	$D_e{}^b$	$D_o{}^b$
B	ax	C_s	1.800	1.515	1.824	1.774	179.3	85.1^d	0.0	102.77	99.79
B	eq	C_{2v}	1.797	1.519	1.778	1.794	76.0	126.5	+0.4	102.38	99.27
Al	ax	C_s	2.206	1.952	1.777	1.775	179.4	85.2^d	0.0	63.51	61.54
Al	eq	C_{2v}	2.199	1.955	1.782	1.777	78.4	125.4	-0.1	63.60	61.53
Ga	ax	C_s	2.263 (2.225)^c	1.983 (1.943)^c	1.771 (1.766)^c	1.780 (1.764)^c	179.6	86.8^d	0.0	55.03	53.16
Ga	eq	C_1	2.264	1.988	1.787	1.774	81.9	124.6	+1.7	53.33	51.36
In	ax	C_s	2.411	2.134	1.764	1.780	179.7	86.9^d	0.0	53.24	51.48
In	eq	C_1	2.414	2.141	1.787	1.770	82.0	124.9	+2.0	51.29	49.43
Tl	ax	C_s	2.506	2.216	1.760	1.784	179.8	88.0^d	0.0	42.52	40.95
Tl	eq	C_1	2.515	2.229	1.790	1.768	84.7	123.7^d	+2.5	40.04	38.36

[a] Energy difference between axial and equatorial isomers [kcal/mol]. [b] Dissociation energy of the Fe-E bond [kcal/mol]. [c] Reference [13]. [d] Average value of slightly different angles.

σ-donation becomes less for the heavier group-13 elements. Both methods agree, however, that the σ-donation is larger than the π-backdonation.

The TM complexes $[(CO)_n TM$-$ER]$ discussed so far had substituents R at the group-13 atom which are strong π-donors. The phenyl group in $[(CO)_4Fe$-$EPh]$ should provide only weak π-stabilization of the out-of-plane $p(\pi)$ of E through conjugation with the π-electron of the phenyl ring, and the in-plane $p(\pi)$ AO should not receive any stabilization by the phenyl ring. Electronic stabilization of the electron deficient group-13 atoms should mainly come from Fe$\rightarrow$E π-backdonation.

Table 10 shows the most important calculated bond lengths and bond angles. The experimental values for the complex $[(CO)_4Fe$-$EPh^*]$ are also given[13].

The calculated Fe-EPh bond dissociation energies are also significantly higher than the Fe-ECp and [Fe-EN$(SiH_3)_2$] compounds. Even the thallium

Table 11. NBO data of $(CO)_4$-Fe-EPh at BP86/II^a.

E	isomer	$q[Fe(CO)_4]$	$q(Fe)$	$q(E)$	$q(C_6H_5)$	$p_x(E)^b$	$p_y(E)^c$	$p_z(E)$	$p_\pi(C_6H_5)$	$P(Fe\text{-}E)$	$P(Fe\text{-}CO)^d$
						Complexes $(CO)_4$-Fe-EPh					
B	ax	-0.36	-0.59	0.65	-0.29	0.41	0.34	0.65	5.88	0.76	0.57
B	eq	-0.39	-0.91	0.66	-0.27	0.31	0.41	0.66	5.84	0.64	0.53
Al	ax	-0.73	-0.60	1.27	-0.54	0.23	0.20	0.24	5.93	0.51	0.65
Al	eq	-0.66	-0.62	1.20	-0.54	0.15	0.30	0.27	5.90	0.50	0.63
Ga	ax	-0.63	-0.56	1.12	-0.49	0.21	0.18	0.24	5.94	0.52	0.68
Ga	eq	-0.54	-0.59	1.05	-0.51	0.13	0.27	0.26	5.94	0.51	0.64
In	ax	-0.67	-0.53	1.16	-0.49	0.18	0.16	0.23	5.95	0.49	0.70
In	eq	-0.58	-0.56	1.08	-0.50	0.11	0.26	0.25	5.90	0.48	0.66
Tl	ax	-0.60	-0.50	1.04	-0.44	0.15	0.14	0.19	5.96	0.44	0.74
Tl	eq	-0.54	-0.59	0.98	-0.44	0.10	0.19	0.21	6.03	0.42	0.66
						Free ligands EPh^e					
B	ax			0.45	-0.45	0.11	0.01	0.69	5.85		
B	eq			0.45	-0.45	0.11	0.01	0.69	5.87		
Al	ax			0.73	-0.73	0.05	0.00	0.34	5.93		
Al	eq			0.73	-0.73	0.05	0.00	0.34	5.94		
Ga	ax			0.70	-0.70	0.05	0.00	0.36	5.95		
Ga	eq			0.70	-0.70	0.05	0.00	0.36	5.94		
In	ax			0.70	-0.70	0.05	0.00	0.36	5.95		
In	eq			0.70	-0.70	0.06	0.00	0.35	5.94		
Tl	ax			0.68	-0.68	0.04	0.00	0.36	5.95		
Tl	eq			0.68	-0.68	0.04	0.00	0.35	5.96		

^{a}Partial charges q, p-orbital population and Wiberg bond indices P.
$^b p(\pi)$ AO of atom E which is perpendicular to the phenyl plane.
$^c p(\pi)$ AO of atom E which is in the phenyl plane.
^{d} CO_{ax} trans to EPh; CO_{eq} in case of the equatorial isomer.
^{e} Calculated using the frozen geometries in the complexes.

Table 12. NBO and CDA data of the changes in the metal and ligand population of $(CO)_4$-Fe-EPh at BP86/II^a.

E	isomer	$\Delta q(E)$	$\Delta q_\pi(E)$	$\Delta q_\sigma(E)$	$\Delta q(C_6H_5)$	$\Delta q_\pi(C_6H_5)$	$\Delta q_\sigma(C_6H_5)$	$d(PhE{\rightarrow}Fe)$	$b(PhE{\leftarrow}Fe)$
					Complexes $(CO)_4$-Fe-EPh				
B	ax	+0.20	-0.63	+0.83	+0.16	-0.03	+0.19	0.516	0.473
B	eq	+0.21	-0.60	+0.81	+0.18	+0.03	+0.15	0.453	0.526
Al	ax	+0.54	-0.38	+0.92	+0.19	+0.00	+0.19	0.453	0.345
Al	eq	+0.47	-0.40	+0.87	+0.19	+0.04	+0.15	0.399	0.335
Ga	ax	+0.42	-0.34	+0.76	+0.21	+0.01	+0.20	0.383	0.264
Ga	eq	+0.35	-0.35	+0.70	+0.19	+0.00	+0.19	0.343	0.283
In	ax	+0.46	-0.29	+0.75	+0.21	+0.00	+0.21	0.388	0.277
In	eq	+0.38	-0.31	+0.69	+0.20	+0.04	+0.16	0.333	0.277
Tl	ax	+0.36	-0.25	+0.61	+0.24	-0.01	+0.25	0.273	0.180
Tl	eq	+0.30	-0.25	+0.55	+0.24	-0.07	+0.31	0.236	0.161

^{a} Change in partial charges Δq; difference of the π-population Δq_π and the σ-charges Δq_σ of the atom E, change
in the partial charges $\Delta q(C_6H_5)$ of the C_6H_5 fragments between the complex and the free ligand, positive values
indicate that the electronic charge decreases, negative values indicate that the electronic charge increases, charge
donation d and backdonation b.

Table 13. Calculated bond lengths in [Å] and energies in [kcal/mol] of TM(ECH₃)₄ at BP86/II and B3LYP/II. Experimental values are given in parentheses.

E	Sym.	TM-E BP86	TM-E B3LYP	E-C BP86	E-C B3LYP	D_e^a BP86	D_e^a B3LYP	D_o^a BP86	D_o^a B3LYP
Complexes Ni(ECH₃)₄.
B	T_d	1.771	1.764	1.553	1.548	91.36	83.80	87.26	79.70
Al	T_d	2.153	2.142	1.995	1.987	61.58	55.64	59.18	53.24
Ga	T_d	2.214	2.210	2.047	2.037	49.62	43.15	47.13	40.66
		(2.170)[b]		(2.014)[b]					
In	T_d	2.347	2.341	2.190	2.182	51.07	45.44	49.07	43.44
		(2.310)[c]		(2.195)[c]					
Tl	T_d	2.451	2.452	2.291	2.281	-[e]	28.39	-[e]	27.00
Complexes Pd(ECH₃)₄.
B	T_d	1.923	1.923	1.552	1.547	76.19	67.46	72.49	63.76
Al	T_d	2.295	2.295	1.994	1.987	52.86	46.02	50.46	43.62
Ga	T_d	2.355	2.362	2.051	2.040	40.34	33.43	38.24	31.33
In	T_d	2.466	2.468	2.192	2.185	43.71	37.42	41.70	35.41
Tl	T_d	2.603	2.622	2.307	2.293	26.81	19.88	25.40	18.47
Complexes Pt(ECH₃)₄.
B	T_d	1.931	1.928	1.551	1.546	89.34	82.68	85.64	78.98
Al	T_d	2.302	2.298	1.993	1.983	62.80	57.32	60.31	54.83
Ga	T_d	2.347	2.345	2.040	2.028	48.89	43.34	46.70	41.15
In	T_d	2.467	2.464	2.180	2.171	51.85	46.79	49.85	44.79
Tl	T_d	2.576	2.578	2.286	2.272	32.07	26.20	28.58	24.71

[a] First dissociation energy of TM-E bond [kcal/mol]. [b] Reference [11]. [c] Reference [1]. [d] calculated with ZPE corrections from B3LYP/II. [e] No SCF convergence.

complex [(CO)₄Fe-TlPh] has a rather high bond energy $D_e = 42.5$ kcal/mol and thus, may be a promising target for experimental studies. The ordering of the bond strength is the same as for the other compounds, i.e. B > Al > Ga ~ In > Tl.

Table 11 shows the results of the NBO analysis. Like in the other complexes with TM-E bonds there is strong Coulombic attraction between the positively charged atom E and the negatively charged iron atom. Table 12 gives the changes in the charge distribution of the fragments after the Fe-E bond is formed. The calculated data support the conclusion that the Fe→EPh π-backdonation in [(CO)₄Fe-EPh] is stronger than in the other group-13 complexes. The calculated Fe→EPh π-backdonation is only slightly higher than the Fe→EN(SiH₃)₂ π-backdonation (Table 8), although the p(π) AOs of E in free EPh are nearly empty (Table 11), while they are partly occupied in EN(SiH₃)₂ (Table 6). The Fe←EPh σ-donation is also the largest among the investigated group-13 complexes. Part of the Fe←EPh σ-donation comes from the phenyl group, which donates between 0.15–0.31 e towards E (Table 12).

Table 14. NBO data of $TM(ECH_3)_4$ at BP86/II and B3LYP/II[a].

E	$q[TM(ER)_3]$		$q(TM)$		$q(E)$		$q(CH_3)$		$p_x(E)$[b]		$p_y(E)$		$p_z(E)$		$P(TM-E)$	
	BP86	B3LYP	BP86	B3LYP	BP86	B3LYP	BP86	B3LYP	BP86	B3LYP	BP86	B3LYP	BP86	B3LYP	BP86	B3LYP
	Complexes Ni(ECH₃)₄															
B	$-$[d] 0.04		$-$[d] 0.16		$-$[d] 0.31		$-$[d] -0.35		$-$[d] 0.39		$-$[d] 0.39		$-$[d] 0.73		$-$[d] 0.56	
Al	-0.11	-0.10	-0.42	-0.48	0.71	0.76	-0.61	-0.64	0.32	0.31	0.32	0.31	0.34	0.33	0.55	0.52
Ga	-0.07	-0.07	-0.24	-0.30	0.62	0.67	-0.56	-0.60	0.28	0.26	0.28	0.26	0.35	0.34	0.55	0.51
In	-0.11	-0.14	-0.37	-0.44	0.66	0.72	-0.57	-0.61	0.29	0.29	0.29	0.29	0.31	0.30	0.56	0.53
Tl	-0.06	-0.08	-0.23	-0.30	0.58	0.64	-0.52	-0.57	0.24	0.21	0.24	0.21	0.31	0.30	0.56	0.52
	Complexes Pd(ECH₃)₄															
B	0.06	0.02	0.24	0.20	0.27	0.30	-0.33	-0.35	0.40	0.38	0.40	0.38	0.73	0.73	0.61	0.60
Al	-0.13	-0.10	-0.49	-0.50	0.74	0.77	-0.62	-0.64	0.28	0.26	0.28	0.26	0.35	0.34	0.46	0.44
Ga	-0.10	-0.11	-0.37	-0.40	0.66	0.70	-0.67	-0.60	0.25	0.23	0.25	0.23	0.36	0.35	0.46	0.43
In	-0.14	-0.18	-0.56	-0.59	0.72	0.77	-0.58	-0.62	0.25	0.23	0.25	0.23	0.31	0.30	0.47	0.45
Tl	-0.11	-0.13	-0.43	-0.46	0.64	0.69	-0.53	-0.58	0.19	0.17	0.19	0.17	0.31	0.30	0.45	0.42
	Complexes Pt(ECH₃)₄															
B	0.09	0.05	0.37	0.35	0.23	0.24	-0.32	-0.33	0.43	0.41	0.43	0.41	0.77	0.78	0.69	0.68
Al	-0.17	-0.16	-0.69	-0.69	0.77	0.80	-0.60	-0.63	0.29	0.29	0.29	0.29	0.36	0.36	0.49	0.48
Ga	-0.15	-0.15	-0.58	-0.60	0.69	0.72	-0.54	-0.58	0.26	0.25	0.26	0.25	0.36	0.36	0.49	0.47
In	-0.20	-0.16	-0.78	-0.81	0.75	0.79	-0.56	-0.59	0.26	0.25	0.26	0.25	0.30	0.30	0.49	0.48
Tl	-0.17	-0.13	-0.68	-0.69	0.67	0.71	-0.50	-0.54	0.21	0.20	0.21	0.20	0.30	0.29	0.47	0.46
	Free Ligand ECH₃[c]															
B					0.47	0.48	-0.47	-0.48	0.03	0.03	0.03	0.03	0.70	0.71		
Al					0.73	0.74	-0.73	-0.74	0.01	0.01	0.01	0.01	0.37	0.37		
Ga					0.69	0.70	0.69	-0.70	0.01	0.01	0.01	0.01	0.40	0.40		
In					0.69	0.71	-0.69	-0.71	0.01	0.01	0.01	0.01	0.39	0.39		
Tl					0.66	0.68	-0.66	-0.68	0.01	0.00	0.01	0.00	0.40	0.39		

[a] Partial charges q, p-orbital population, Wiberg bond indices P.
[b] $p(\pi)$ AO of atom E which is perpendicular to the TM-E-C axis.
[c] Calculated using the frozen geometries in the complexes; identical value have been found in the free ligands for all complexes of Ni, Pd and Pt.
[d] No SCF convergence

The compounds $[Ni(ECH_3)_4]$, $[Pd(ECH_3)_4]$, and $[Pt(ECH_3)_4]$ are different from the other complexes with group-13 ligands which have been discussed so far, because they are *homoleptic* complexes of *late* transition metals. The BP86/II[4] calculations of $[TM(ECH_3)_4]$ encountered severe convergence problems, which could not be resolved in a few cases. We decided to carry out calculations at B3LYP/II, which did not suffer from the same SCF problems. We report the results for $[TM(ECH_3)_4]$ at both levels of theory, which makes it also possible to see the alterations of the results when different methods are used.

The calculations show that the first TM-ECH₃ bond dissociation energies are rather high. The theoretical bond energies at B3LYP/II are always 5 - 8 kcal/mol higher than the BP86/II values (Table 13). The calculated first TM-CO bond dissociation energies at the CCSD(T)/II level of theory are D_o = 22.3 kcal/mol for $[Ni(CO)_4]$, D_o = 7.5 kcal/mol for $[Pd(CO)_4]$, and D_o = 10.9 kcal/mol for $[Pt(CO)_4]$. The trend of the TM-ECH₃ bond energies shows for the transition metals the order Ni $\sim$ Pt > Pd, while for the group-13 elements the same trend as for the other complexes is found B > Al > Ga $\sim$ In > Tl.

The results of the NBO partitioning scheme are shown in Table 14. The BP86 and B3LYP values are not very different from each other. The $p(\pi)$ orbitals of the atoms E, which are nearly empty in the free ligand ECH₃, become significantly populated in the complexes. This is a strong indication that there is substantial TM→ECH₃ π-backdonation in the molecules.

Table 15. NBO and CDA data of TM(ECH₃)₄ at BP86/II and B3LYP/II^a.

E	Δq(E)		Δqπ(E)		Δqσ(E)		Δq (CH₃)		d(H₃CE→TM)		b(H₃CE←TM)	
	BP86	B3LYP	BP86	B3LYP	BP86	B3LYP	BP86	B3LYP	BP86	B3LYP	BP86	B3LYP
							Complexes Ni(ECH₃)₄					
3	-^{b}	-0.17	-^{b}	-0.72	-^{b}	+0.55	-^{b}	+0.13	0.665	0.670	0.505	0.482
\l	-0.02	+0.02	-0.62	-0.60	+0.60	+0.62	+0.12	+0.10	0.554	0.597	0.501	0.461
ia	-0.07	-0.03	-0.54	-0.50	+0.47	+0.47	+0.13	+0.10	0.586	0.604	0.446	0.410
n	-0.03	+0.01	-0.56	-0.56	+0.59	+0.57	+0.12	+0.10	0.513	0.526	0.418	0.384
[1	-0.08	-0.04	-0.46	-0.42	+0.54	+0.46	+0.14	+0.10	0.454	0.470	0.404	0.368
							Complexes Pd(ECH₃)₄					
3	-0.20	-0.18	-0.74	-0.70	+0.54	+0.52	+0.14	+0.13	0.675	0.680	0.575	0.551
\l	+0.01	+0.03	-0.54	-0.50	+0.55	+0.53	+0.11	+0.10	0.551	0.563	0.561	0.552
ia	-0.03	+0.00	-0.48	-0.44	+0.45	+0.44	+0.02	+0.10	0.536	0.541	0.468	0.429
n	+0.03	+0.06	-0.48	-0.44	+0.51	+0.50	+0.11	+0.11	0.425	0.432	0.389	0.364
[1	-0.02	+0.01	-0.36	-0.34	+0.34	+0.35	+0.13	+0.10	0.413	0.525	0.368	0.329
							Complexes Pt(ECH₃)₄					
3	-0.24	-0.24	-0.80	-0.76	+0.56	+0.52	+0.15	+0.15	0.645	0.650	0.633	0.610
\l	+0.04	+0.06	-0.56	-0.56	+0.60	+0.62	+0.13	+0.11	0.619	0.639	0.553	0.551
ia	0.00	+0.02	-0.50	-0.48	+0.50	+0.50	+0.15	+0.12	0.571	0.576	0.496	0.494
n	+0.06	+0.08	-0.50	-0.48	+0.56	+0.56	+0.13	+0.09	0.480	0.488	0.401	0.402
[1	+0.01	+0.03	-0.40	-0.40	+0.41	+0.43	+0.16	+0.14	0.487	0.508	0.365	0.361

^{a} Change partial charges q, difference of the π- population and the σ -charges of the atom E between the complexes and the free ligands Δq_π and Δq_σ; positive values indicate that the electronic charge decrease, negative values indicate that the electronic charge increase and charge donation d and backdonation b.

^{b} No SCF convergence

The CDA results support the picture of strong TM→ECH₃ π-backdonation, although the TM←ECH₃ σ-donation is still larger. However, the sum of the TM-E donor-acceptor interactions leads to a degree of covalent bonding which is less than a single bond. It follows that the high bond dissociation energies are largely caused by charge attraction.

The results presented here show that a discussion about the question whether the TM-ER bonds should be considered as single bonds or triple bonds is meaningless. The TM←ER π-backdonation may indeed become as important as the TM→ER σ-donation, but the sum of the covalent interactions do not even give a bond order 1. This is different to the bonding situation in TM carbyne complexes, where the bond order for the L_nTM-CR bond is between 1.7–2.1 for Fischer-type carbyne complexes, and even between 2.3–2.5 for Schrock-type carbyne complexes[12]. Thus, [(CO)₄FeGaAr*] is not a "ferrogallyne"[13]. However, drawing a Lewis structure of the molecule with a Fe-Ga single bond is also not an appropriate representation of the bonding situation[6]. The problem lies in the weakness of the graphical representation of the bond in terms of electron-pair bonding. A triple-bond notation would be appropriate if only the covalent contributions to the Fe-EAr* bond shall be sketched, which are, however, only a minor part of the iron-gallium bonding interactions.

References

1. Uhl, W.; Pohlmann, M.; wartchow, R., *Angew. Chem.* **1998**,110, 1007; *Angew. Chem. Int. Ed.* **1998**, 37, 961.
2. Dapprich S., Frenking G., *J. Phys. Chem.* **1995**, 99, 9352.
3. Reed A.E., Curtiss L.A., Weinhold F., *Chem. Rev.* **1988**, 88, 899.
4. Frenking G., Antes I., Böhme M., Dapprich S., Ehlers A.W., Jonas J., Neuhaus A., Otto M., Stegmann R., Veldkamp A., Vyboishchikow S.F., In: Lipkowitz K.B., Boyd D.B. (eds) *Reviews in computational chemistry*, **1996**, Vol. 8, VCH, New York, pp 63-144.
5. Werner H.-J., Knowles P.J., Universität Stuttgart and University of Birmingham.
6. Cotton F.A., Feng X., *Organometallics* **1998**, 17, 128.
7. Cowley, A.H.; Lomeli, V.; Voigt, A., *J. Am. Chem. Soc.* **1998**, 120, 6401.
8. Weiss, J.; Stetzkamp, D.; Nuber, B.; Fischer, R.A.; Boehme, C.; Frenking, G., *Angew. Chem.* **1997**, 109, 95; *Angew. Chem. Int. Ed. 1997*, 36, 70.
9. Jutzi, P.; Neumann, B.; Reumann, G.; Stammler, H.G., *Organometallics* **1998**, 17, 1305.
10. Braunschweig, H.; Kollmann, C.; Englert, U., *Angew. Chem. Int. Ed.* **1998**, 37, 3179.
11. Uhl, W.; Benter, M.; Melle, S.; Saak, W.; Frenking, G.; Uddin, J., *Organometallics* **1999**, 18. 3778.
12. Vyboishchikov, S.F.; Frenking, G., *Chem. Eur. J.* **1998**, 4, 1439.
13. Su, J.; Li, X.-W.; Crittendon, R.C.; Campana, C.F.; Robinson, G.H., *Organometallics* **1997**, 16, 4511.
14. (a) Ehlers, A.W.; Baerends, E.J.; Bickelhaupt, F.M.; Radius, U., *Chem. Eur. J.* **1989**, 4, 210. (b) Radius, U.; Bickelhaupt, F.M.; Ehlers, A.W.; Goldberg, N.; Hoffmann, R., *Inorg. Chem.* **1998**, 37, 1080. (c) Bickelhaupt, F.M.; Radius, U.; Ehlers, A.W.; Hoffmann, R.; Baerends, E.J., *New J. Chem.* **1998**,1.

Computer Simulation of Protein Unfolding

Andreea Daniela Gruia, Stefan Fischer, and Jeremy C. Smith

Lehrstuhl für Biocomputing, IWR, Universität Heidelberg, Im Neuenheimer Feld 368, D-69120 Heidelberg, Germany

Abstract. Salt bridges are frequently present in globular proteins and therefore they have been extensively studied in order to understand their role in protein stability. Here salt bridge breaking is found to hinder unfolding in nonequilibrium molecular dynamics simulations performed to investigate the kinetic stability of a small protein, Staphylococcal nuclease (SNase) with C-terminal truncation (SNaseΔ). The simulations were performed in explicit solvent at 300 K, 400 K, 430 K and 450 K. The surface salt bridge between arginine 105 and glutamate 135 was found to form a kinetic trap for unfolding. The salt bridge was stable in the simulations over 1 ns at physiological temperature and over 500 ps at 450 K. When the potential function was modified so as to weaken the electrostatic interaction between arginine 105 and glutamate 135, the salt bridge broke at 430 K and the protein started to unfold. These results suggest that breaking of this particular salt bridge presents a significant barrier to the unfolding of SNaseΔ. The use of the Cray T3E supercomputer has enabled this process to be simulated with statistical significance.

1 Introduction

The genome sequencing projects have stimulated further interest in understanding the interactions that drive protein folding. Molecular dynamics (MD) simulation techniques represent a powerful tool in analyzing the mechanisms through which the proteins adopt an unique three-dimensional structure by providing atomic details of the folding process. The major limitation of MD techniques is that the accessible simulation time scales are in the range of tens of nanoseconds, while proteins fold in milliseconds to seconds. To overcome this problem, unfolding simulations are preferred for the study of conformational changes in proteins because they are performed under extreme conditions that accelerate the denaturation events to the MD accessible time scale. Therefore, proteins are perturbed by using high temperatures, pressure, solvent or pH changes in order to induce denaturation [1-3]. Parallellisation of MD simulation programs and their implementation on parallel supercomputers represent an efficient strategy that will contribute in the near future to solution of the protein folding problems.

In the last decades there has been an increasing interest in the contribution of electrostatic interactions to protein folding and stability. As a result of many experimental and theoretical studies that examined the effect of electrostatic interactions on protein structure, it has been suggested that hydrogen-bonding patterns and salt bridges may play an important role in fold specificity [4,5]. However, it is difficult to generalize on the role of buried or surface salt bridges in protein stability.

Of particular interest are the solvent-exposed salt bridges that are ubiquitous in globular proteins. Their contribution to protein folding stability is unclear. Experimental data supported by theoretical approaches estimate the contribution of surface salt bridges to protein folding stability to be small, i.e. in the range of 0.4-1.5 kcal/mol [6-8]. In contrast, Anderson et al. [9] have shown that a single surface salt bridge contributes 3-5 kcal/mol to the folding free energy of T4 lysozyme. Moreover, Spek et al. [10] have found that salt bridges stabilize the GCN4 leucine zipper by 1.7 kcal/mol. Experimental data concerning thermophilic and hyperthermophilic proteins suggest that the large number of surface salt bridges plays a key role in stabilizing the native structure of the proteins against thermal denaturation [11-13].

Staphylococcal nuclease (SNase) was chosen as a model for molecular dynamics simulations. It is a small protein of 149 residues (Fig.1) that has no disulphide bridges.

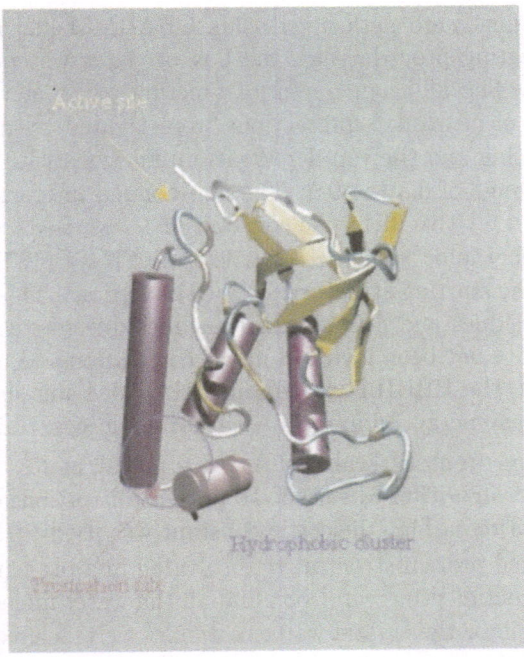

Fig. 1. Ribbon representation of the three dimensional structure of native Staphylococcal nuclease from 1STN.pdb [16] created using VMD program [21]. The truncation site at residue 136 is shown. The position of the hydrophobic cluster that stabilizes the C-terminal end of the protein is indicated in the blue circle. This consists of residues Gly 109, Pro 42, Tyr 54, Lys 134, Leu 137, Ile 139 and Trp 140.

Various techniques such as circular dichroism, fluorescence, neutron scattering, nuclear magnetic resonance, etc have been used to characterize its physical properties and unfolding/folding behavior [14-16]. A truncated construct of Staphylococcal nuclease (SNaseΔ) lacking the 13 C-terminal residues has been shown to be disordered, but compact under physiological conditions [17,18]. This structure misses a hydrophobic cluster that has been suggested to anchor the last helix of the structure (helix $\alpha 3$) to the rest of the protein, thus stabilizing the C-terminal end of the native fold [19,20].

In the present work nonequilibrium molecular dynamics simulations of the wild-type folded truncated construct are performed in explicit solvent to investigate the unfolding transition. High temperatures are used to help overcome the energy barriers of the transition process. We have found that one particular salt bridge between arginine 105 and glutamate 135 contributes significantly to trapping the truncated construct in a metastable folded state.

2   Methods

All the simulations were performed using CHARMM [22] in the NPT ensemble. The heating procedure for the box of water was performed in the NVE ensemble. The all-atom potential function parameter set PARAM22 was used. The electrostatic interactions were treated with a force switch smoothing function and the van der Waals interactions with a shift function with a cut off range of 6.5 to 10 Å. The non-bonded interaction list was cut off at 11 Å and a relative dielectric constant of 1 was used.

The crystallographic structure of wild type SNase (1STN.pdb)[16] was used to create the starting structure for the simulations. This structure lacks 5 N-terminal residues and 8 C-terminal residues due to weak electron density and they have not been included in the calculations. In order to add the hydrogen bonds, the HBUILD routine in CHARMM was used. The resulting structure was energy minimized with harmonic constraints (with force constants ranging from 10 kcal/mol·Å^2 to 0.1 kcal/mol·Å^2). For this, the Steepest Descent algorithm was used. To prevent distorsions of the structure during minimization, a high dielectric constant was used.

The minimized protein structure was solvated in a pre-equilibrated 55 Å-sided truncated octahedron water box and all the water molecules that overlapped the protein or the crystal water molecules were deleted. Nine chloride ions were added to preserve the electro-neutrality of the system. The system was run at 300 K to equilibrate for ~100 ps. Weak harmonic constraints (force constant of 0.1 kcal/mol·Å^2) were applied on the protein in order to prevent denaturation. At the end of the equilibration procedure, the last 5 residues from the C-terminal end were deleted and 17 water molecules were added to replace them. The system was then run at 300 K for a second equilibration for ~ 30 ps.

One 1-nanosecond simulation was performed at 300 K without any constraints. The system was then heated to 400 K, 430 K and 450 K for 50 ps, 80 ps and 70 ps, respectively. The heating was performed from 300 K in small increments of 10 K every 10 ps in the NPT ensemble to prevent denaturation due to temperature increase. For the heating procedures the coordinates and velocities after the first 100 ps production run at 300 K were used. One production run was performed at 400 K for 550 ps and one at 450 K for 200 ps. Six production simulations (200 ps each) were performed at 430 K, assigning random starting velocities.

In a final set of simulations, the strength of the salt bridge between arginine 105 and glutamate 135 was weakened by neutralizing their net charges (+1 and -1 for arginine and glutamate, respectively, in the standard charge model). Although their net charges were removed, the atomic charges were not completely turned off in order to avoid undesired hydrophobic behavior. Six simulations (200 ps each) were performed at 430 K by neutralizing the charges immediately after 80 ps of heating.

All computations were done on 64 processors, every 100 ps of production run requiring $\sim$ 7 CPU hours per node.

3   Results and Discussion

3.1   Salt bridge persists at 300 K and 400 K

The stability of the native-like conformation of the truncated construct was initially probed under physiological conditions at 300 K in aqueous solution in one simulation of length of 1 ns. Of particular interest for the analysis is helix $\alpha 3$ which is the element of secondary structure closest to the C-terminal end. In Fig. 2A are plotted the root mean square deviation (RMSD) of the entire protein (residues 6-136, black) and of the helix $\alpha 3$ (residues 121-135, red), calculated with respect to the crystal structure. The protein remained very close to the crystal structure average over the entire length of the trajectory with an RMSD of $\sim$1.0 Å on. The $\alpha 3$ helix showed no major perturbation with an RMSD of $\sim$0.7 Å, indicating that is stable throughout the 1 ns simulation.

Graphical examination of the trajectory suggests that the stability of the native structure might be due to a salt bridge formed between arginine 105 and glutamate 135 (Fig.3). This salt bridge might anchor the a3 helix to the body of the protein. To test the stability of this particular salt bridge, a simulation was performed at 400 K. The RMSD of the $\alpha 3$ helix is $\sim$1.7 Å, larger than at 300 K. The salt bridge was stable for 500 ps, further confirming that the barrier to breaking it is high.

3.2   Salt bridge breaks at 450 K

Fig. 2B shows the RMSD of the entire protein and of the $\alpha 3$ helix with respect to the crystal structure calculated from the 200 ps simulation at 450

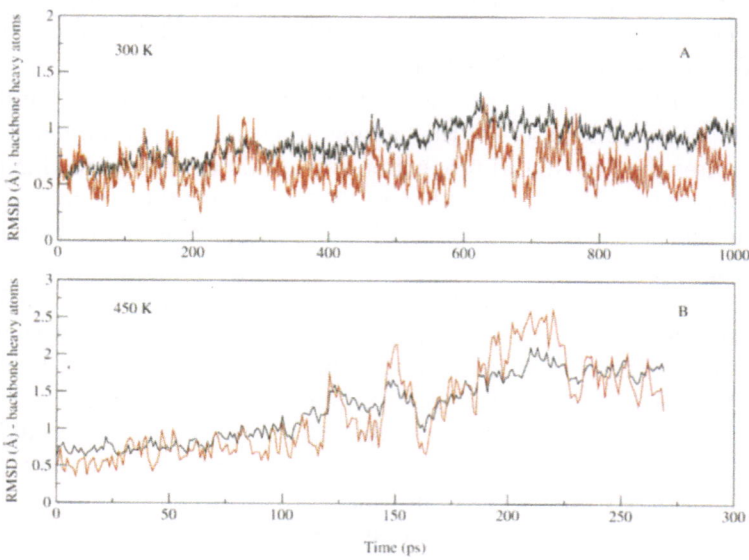

Fig. 2. RMSD of the entire protein (black) and of the α3 helix (red) with respect to the crystal structure at A) 300 K and B) 450 K. The first 50 ps in plot A and the first 70 ps in plot B represent heating. The RMSD of the α3 helix was calculated by first orienting and superimposing the structure in each frame of the trajectory on the crystal structure.

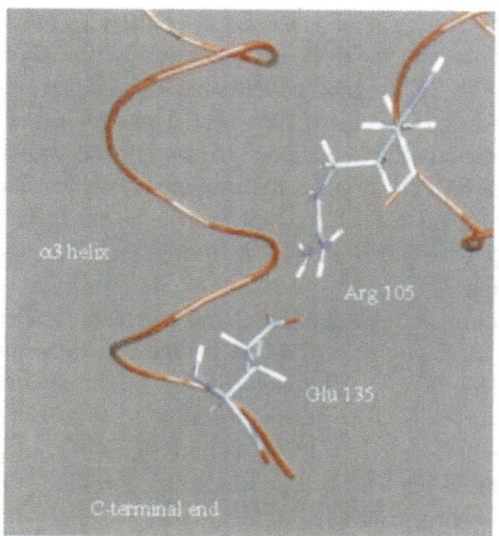

Fig. 3. The salt bridge between Arg 105 and Glu 135 anchors helix α3 to the rest of the protein. The picture was created using the VMD program [21].

K (the first 70 ps in the plot represent heating). The salt bridge exhibited
"breathing" motions with short life times ($\sim$ 10 ps) in closed conformations,
suggesting that high temperatures destabilize it. After 180 ps the salt bridge
broke and subsequently the α3 helix started to move away from the rest of the
protein, beginnnig to unfold. The slow, but steady increase in the total RMSD
(black curve) of the protein indicates that the protein undergoes a structural
change. This change in conformation is mainly due to the movement of α3
helix away from the protein.

3.3 Salt bridge effect at 430 K

To probe the statistical significance of the salt bridge effect we chose another
temperature (430 K) at which we estimated the salt bridge to be stable on
a short time scale ($\sim$200 ps) and performed 6 simulations. Results of these
simulations are shown in Fig.4.

In Fig. 4A the RMSD of the entire protein averaged over the 6 trajectories with standard charges on Arg 105 and Glu 135 (black curve) and of the
entire protein averaged over the 6 trajectories with the 105 - 135 salt bridge
charges neutralized (red curve) are plotted (first 80 ps represent heating).

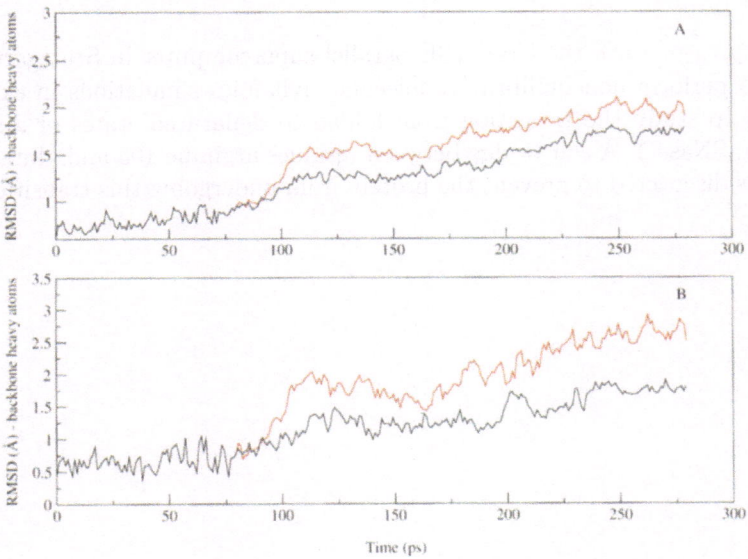

Fig. 4. A) RMSD of the entire protein at 430 K averaged over the six trajectories
with standard charges on Arg 105 and Glu 135 (black) and with neutralized charges
(red); B) RMSD of the α3 helix at 430 K averaged over the six trajectories with
standard charges (black) and with neutralized charges (red). The first 80 ps in all
plots represent heating.

The electrostatic interaction between the two residues was weakened immediately after the heating procedure at 80 ps. This results in a sharp increase in the RMSD of the entire protein when compared to the simulation with standard charges. The values of the RMSD reached in the end by the protein in both sets of simulations were around 2 Å, which are similar to the values obtained at the end of the 400 K and 450 K simulations. This suggests that the unfolding transition from the folded state of SNaseΔ also took place at 430 K, but the removal of the salt bridge accelerated the process.

The effect of the salt bridge is also noticeable in Fig. 4B. In this figure, the RMSD of the α3 helix averaged over the six trajectories with standard charges on Arg 105 and Glu 135 (black curve) and with neutralized charges (red curve) are plotted. As it can be seen, the RMSD of the α3 helix in the simulations with neutralized charges exhibits the same sharp increase due to salt bridge breaking.

Calculated RMSDs of all the elements of secondary structure in SNaseΔ showed that the sharp increase in the RMSD of the entire protein on charge neutralization is due entirely to the RMSD of the α3 helix. This is supported by the fact that the α3 helix contributes to all the major peaks in the RMSD of the entire protein (data not shown).

4 Conclusions

In the present work the Cray T3E parallel supercomputer in Stuttgart was used to perform nonequilibrium molecular dynamics simulations in explicit solvent to study the transition from folded to denatured states of a small protein, SNaseΔ. A salt bridge between residues arginine 105 and glutamate 135 was discovered to prevent the protein from undergoing this transition on

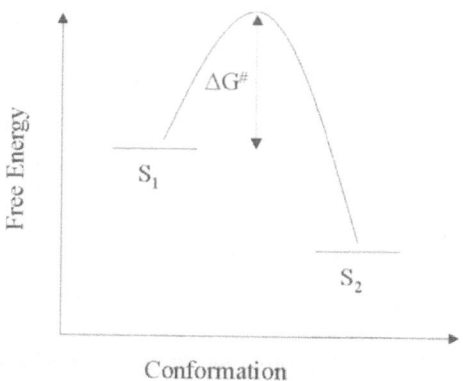

Fig. 5. Schematic energy diagram to illustrate the transition form the high-energy folded state (S1) to the unfolded experimental state (S2) of SNaseΔ. $\Delta G^{\#}$ is the activation energy for the transition.

the nanosecond time scale under physiological conditions. When the strength of this salt bridge is weakened, the salt bridge breaks and the protein starts to unfold rapidly. These results suggest that the truncated native structure protein is locked in a metastable state. Six trajectories performed at 430 K indicate that this salt bridge effect is statistically significant. Therefore, we suggest that breaking of this particular salt bridge is a significant initial barrier to the unfolding transition. This transition barrier is schematically drawn in Fig. 5.

In summary, then, the present work indicates that surface salt bridges may be important in protein folding/unfolding by stabilizing metastable states and thereby introducing rate-limiting steps in folding/unfolding transitions.

Acknowledgements

All the simulations were performed on a Cray T3E supercomputer at the Rechenzentrum Universität Stuttgart. The work was supported by the Protein Folding Grant (ID No.11755) offered by Höchleistungsrechenzentrum Stuttgart which is gratefully acknowledged. We wish to thank Dr. Heinz Pöhlmann and Mr. Bogdan Costescu for the excellent technical support they provided. We are grateful to Dr. Franci Merzel for his help with building the truncated octahedron.

References

[1]   Daggett, V., Levitt, M.: Protein unfolding pathways explored through molecular dynamics simulation; (1993) J. Mol. Biol. **232**, 600-619.

[2]   Alonso, D.O., Daggett, V.: Molecular dynamics simulations of protein unfolding and limited refolding : characterization of partially unfolded states of ubiquitin in 60% methanol in water; (1995) J. Mol. Biol. **247**, 501-520.

[3]   Lazaridis, T., Karplus, M.: "New view" of protein folding reconciled with the old through multiple unfolding simulations; (1997) Science **278**, 1928-1931.

[4]   Dill, K.A.: Dominant forces in protein folding; (1990) Biochemistry **29**, 7133-7155.

[5]   Sharp, K.A., Honig, B.: Electrostatic interactions in macromolecules: theory and applications; (1990) Annu. Rev. Biophys. Chem. **19**, 301-332.

[6]   Horovitz, A., Serrano, L., Avron, B., Bycroft, M., Fersht, A.R.: Strength and cooperativity of contributions of surface salt bridges to protein stability; (1990) J. Mol. Biol. **216**, 1031-1044.

[7]   Serrano, L., Horovitz, A., Avron, B., Bycroft, M., Fersht, A.R.: Estimating the contribution of engineered surface electrostatic interactions to protein stability by using double-mutant cycles; (1990) Biochemistry **29**, 9343-9352.

[8]   Sun, D.P., Sauer, U., Nicholson, H., Matthews, B.W.: Contributions of engineered surface salt bridges to the stability of T4 lysozyme determined by directed mutagenesis; (1991) Biochemistry **30**, 7142-7153.

[9]   Anderson, D.E., Becktel, W.J., Dahlquist, F.W.: pH-induced denaturation of proteins: a single salt bridge contributes 3-5 kcal/mol to the free energy of folding of T4 lysozyme; (1990) Biochemistry **29**, 2403-2408.

[10] Spek, E.J., Bui, A.H., Lu, M., Kallenbach, N.R.: Surface salt bridges stabilize the GCN4 leucine zipper; (1998) Protein Sci. **7**, 2431-2437.

[11] Musafia, B., Buchner, V., Arad, D.: Complex salt bridges in proteins: statistical analysis of structure and function; (1995) J. Mol. Biol. **254**, 761-770.

[12] Yip, W.C., Stillman, T.J., Britton, K.L., Artymiuk, P.J., Baker, P.J., Sedelnivoka, S.E., Engel, P.C., Pasquo, A., Chiaraluce, R., Consalvi, V., Scandurra, R., Rice, D.W.: The structure of Pyrococcus furiosus glutamate dehydrogenase reveals a key role for ion-pair networks in maintaining enzyme stability at extreme temperatures; (1995) Structure **3**, 1147-1158.

[13] Perutz, M., Raidt, H.: Stereochemical basis of heat stability in bacterial ferrodoxins and in haemoglobin A2; (1975) Nature **255**, 256-259.

[14] Serpersu, E.H., Shortle, D., Mildvan, A.S.: Kinetic and magnetic resonance studies of active-site mutants of staphylococcal nuclease: factors contributing to catalysis; (1987) Biochemistry **26**, 1289-1300.

[15] Loll, P., Lattman, E.E.: The crystal structure of the ternary complex of staphylococcal nuclease, Ca2+ and the inhibitor pdTp refined at 1.65 Å; (1989) Proteins: Struct., Funct., Genet. **5**, 183-201.

[16] Hynes, T.R., Fox, R.O.: The crystal structure of Staphylococcal nuclease refined at 1.7 Å resolution; (1991) Proteins: Struct., Funct., Genet. **10**, 92-105.

[17] Shortle, D., Meeker, A.K.: Residual structure in large fragments of staphylococcal nuclease: effects of amino acid substitutions; (1989) Biochemistry **28**, 936-944.

[18] Flanagan, J.M., Kataoka, M., Shortle, D., Engelman, D.M.: Truncated staphylococcal nuclease is compact, but disordered; (1992) Proc. Natl. Acad. Sci. USA **89**, 748-752.

[19] Yin, J., Jing, G: Tryptophan 140 is important, but serine 141 is essential for the formation of the integrated conformation of staphylococcal nuclease; (2000) J. Biochem. (Tokyo) **128**, 113-119.

[20] Li, Y., Jing, G.: Double point mutant F34W/W140F of staphylococcal nuclease is in a motlen globule state, but highly competent to fold into a functional conformation; (2000) J. Biochem. (Tokyo) **128**, 739-744.

[21] Humphrey, W., Dalke, A., Schulten, K.: VMD: visual molecular dynamics; (1996) J. Molec. Graphics **14**, 33-38.

[22] Brooks, B.R., Bruccoleri, R.E., Olafson, B.D., States, D.J., Swaminathan, S., Karplus, M.: CHARMM: A program for macromolecular energy, minimization and dynamics calculation (1983) J. Comp. Chem **4**, 187-217.

Computational Fluid Dynamics

Prof. Dr.-Ing. S. Wagner

Institut für Aero- und Gasdynamik, Universität Stuttgart
Pfaffenwaldring 21, 70550 Stuttgart

Many complex mechanisms in fluid flow can only be investigated by solving nonlinear partial differential equations. The solution of these equations is usually only possible by numerical methods that require computers, usually high performance computers. One typical example is laminar-turbulent transition. In this case, the unsteady, three-dimensional Navier-Stokes equations are solved by so- called direct numerical simulations (DNS) which means that all the complicated, unsteady phenomena are resolved by the numerical procedure, no model, as for instance turbulence models, is used. Many contributions to this subject were already published within the series *High Performance Computing in Science and Engineering*.

Gmelin et al. used DNS to actively control disturbances in a Blasius Boundary Layer in order to delay laminar-turbulent transition which help to reduce for instance drag, noise or flight mechanical instabilities. One approach to solve the problem is the superposition of anti-phase disturbances which might fail in case of high disturbance amplitudes. Gmelin et al. developed a new, very effective control algorithm with the aid of DNS to postpone transition in a two-dimensional boundary layer. Approximately 700.000 grid points in a plane parallel to the flow direction and normal to the flat plate and up to 25 Fourier modes in spanwise direction had to be used to solve the problem. The NEC SX-5 with 16 processors and 5 GB RAM was used. The code reached a performance of up to 1.6 GFLOPS on one processor at a vector operation ratio of 99.4 per cent.

Manhart et al. used DNS to study the behaviour of a boundary layer with an adverse pressure gradient and compared the results to statistical simulations (Reynolds Averaged Navier-Stokes, RANS, equations) with two-equation turbulence models. Approximately 36 million grid points contained the grid. 256 processors were used on the Cray T3E-900 with a performance of 70 MFLOPS per processor and an almost linear speed-up to more than 200 processors. An interesting result is that two-equation models have difficulties if the adverse pressure gradient persists over a long streamwise distance. A second peak was observed in the turbulent kinetic energy in the outer boundary layer which was not depicted by the two-equation turbulence model.

Göz et al. continued their simulation of bidisperse bubbly gas-liquid flows by a parallel finite- difference/front tracking method using the IBM RS/600 SP of the Scientific Supercomputing Center Karlsruhe. They discovered that systems with large deformable and small spherical bubbles show some novel features concerning both the motion of the bubbles and the induced liquid turbulence. This deviates remarkably from simulations with spherical bubbles only.

In order to reduce the computational effort of DNS to some extent large eddy simulation (LES) is used to simulate flows. In this case the large eddies of turbulence are treated as with DNS whereas the small eddies are approximated by a model, e.g. by the Smagorinsky subgrid-scale model used in the contribution by M. Schmid et al. They calculated the flow around a sphere at a Reynolds number of 50.000 and got good agreement of time averaged results with experimental data. Even vortex pairing as one possible consequence of the Kelvin-Helmholtz instability of the shear layer was predicted in good agreement with experiments. 64 processors of the CRAY T3E had to be used to perform the simulation.

Flows with chemical reaction is another example where high performance computing is mandatory. Gerlinger et al. used assumed probability density functions (pdf) to incorporate effects of temperature and species fluctuations on chemical reaction rates. They investigated supersonic combustion including temperature and species fluctuations. During simulations of two-dimensional flows on a NEC SX-5 they achieved an average performance of 874 MFLOPS on a single processor. For three-dimensional flows parallel computing will be mandatory.

Quite similar to the case of simulating chemical reaction is the problem of simulating 3-D magneto- hydrodynamic flow, especially in galaxies. Krause et al. used the NIRVANA code to simulate a hydrodynamic jet with parameters matched to the radio galaxy Cygnus A. 85 per cent of the user time the code was in vector mode with a vector operation ratio of 99.3 per cent. On a single processor of the NEC SX-5 they achieved a performance of 1 GFLOPS. For the computation of a complete simulation Krause et al. estimate 20.000 hours of CPU time. Even with 10 processors, the computation would take 80 days.

A highly complicated problem is two-phase flow that demands also huge computer resources. Hase et al. used a three-dimensional solver for the Navier-Stokes equations for incompressible two-phase flows on the basis of a Finite-Volume method with direct numerical simulation. The code is used to predict the behaviour of spherical and deformed droplets in a gas flow. The simulations were performed on the Cray T3E/512-900 using 16 CPU's. Typical computation time was 34305 seconds per CPU.

Giese et al. investigated also two-phase flow, especially in a pipe. The commercial code CFX-4.2 was used for these investigations that solve the Navier-Stokes equations for two-phase flows. The simulations were run on a NEC SX-4 requiring a memory of 1GB. Since the segregated solver of the version CFX-4 is sensitive in cases of strong phase interactions Giese et al. hope the new version CFX-5 with a coupled solver will overcome this problem.

Whereas the studies discussed so far investigated flow problems of more fundamental character the next three ones treat more practical engineering problems. In the first one, the influence of leakage on the main flow of a turbine is examined using the Navier-Stokes code ITSM3D. The studies show that leakage flow not only introduces mixing losses but can also dominate the secondary flow and can cause severe losses. 3.2 million grid points are necessary to resolve these secondary effects. On a 667 MHZ single processor Alpha-Dec Workstation 23 days of CPU time would have been necessary to perform one unsteady simulation. Using 4 processors on the NEC-SX4/SX5 a planned unsteady simulation is expected to be performed within two CPU days.

Another problem in turbomachinery is flow instability like "rotating stall" that was investigated by Ginter et al. for a one-stage axial water compressor with 30 runner and 30 stator blades. They used their computer program FENFLOSS that solves the RANS equations for an incompressible fluid. The simulation was carried out on a Hitachi SR 8000 computer.

Finally, Euler and Navier-Stokes solutions were used by Neef et al. to study flapping flight of birds. They used the FLOWer code of DLR (Deutsches Zentrum für Luft- und Raumfahrt) on one processor of NEC-SX5 and got an insight into the time dependent formation of tip vortices, the obtained thrust and drag. The obtained performance was 2GFLOPS with a vector operation ratio of more then 99 per cent.

Only 40 per cent of CFD papers submitted could be selected for publication.

DNS of Active Control of Disturbances in a Blasius Boundary Layer

C. Gmelin, U. Rist, and S. Wagner

Institut für Aerodynamik und Gasdynamik,
Universität Stuttgart, Pfaffenwaldring 21, D-70550 Stuttgart, Germany

Abstract. Many approaches with the objective to actively delay the laminar-turbulent transition in boundary layers are currently under investigation. These approaches, which are mostly based on the superposition of anti-phase disturbances fail in cases where high (nonlinear) disturbance amplitudes occur. One possible solution to overcome this problem is the direct feedback of instantaneous flow signals from the wall. In our case the spanwise vorticity (ω_z) on the wall is sensed, multiplied by a certain factor A and prescribed as a new boundary condition at the wall with some time delay Δt. This procedure (called ω_z-**control**) yields a robust algorithm which is less influenced by nonlinearities than other processes based on the linear superposition of disturbances (waves). The method was developed and evaluated using both linear stability theory and a three-dimensional spatial DNS code solving the complete Navier-Stokes equations.

1 Introduction

The most popular approach controlling transition is the superposition of disturbances with opposite phase to the existing waves. First attempts have been published by Milling [1], Liepmann et al. [2][3] and Kozlov et al. [4]. Until now this strategy has been realized many more times both experimentally [5] and numerically [6]. For disturbances with small (linear) amplitude a reduction in amplitude of up to 90% even in experiments is achievable. In contrast to their excellent performance in early transition stages these approaches don't work in a satisfactory manner in cases where high amplitudes occur due to nonlinear effects superposing disturbance and control wave. Moreover, the generation of large control waves which are necessary to cancel the initial wave with the aid of a suction/blowing slot sometimes causes very high velocities in the vicinity of these actuators, an effect which favors nonlinearities furthermore. These arguments make clear that there is a need for a smooth, robust control algorithm which is almost independent of the amplitude of the initial disturbance.

Several attempts have been made to control turbulent flows. Control via affection of the vorticity flux at the wall has been proposed by Koumoutsakos [7][8] whereas Choi, Moin & Kim [9] observed a damping effect on turbulent flows feeding back the instantaneous wall-normal velocity at a certain distance from the wall to the boundary. They report the establishing of a "virtual wall", i.e. a plane that has approximately no through-flow halfway between the detection plane and the wall and therefore a drag reduction of 25% and a strong reduction of turbulent flow structures. In some aspects our approach is very similar to their idea of feedback of instantaneous signals but in most cases flow data of the whole flow field is not available. In our case this problem is handled by sensing the spanwise vorticity at the wall, present as wall shear stress and easily measurable by hot film sensors or cavity hot wires [10], for example. These signals are multiplied with a certain factor and prescribed as wall-normal velocity v at the wall (for a detailed description see below in section 4).

An approach complementary to our LST investigations (section 5) has been applied by Joshi et al. [11] for plane Poiseulle flow using the spanwise shear at the wall as sensor variable and blowing/suction for actuation, as well. However, they converted the problem into a control theoretical one and determined the effect of the feedback control by the position of the zeros and poles of the system. Furthermore, they obtained an optimal sensor location relative to the actuator similar to our most effective phase shift between sensing and actuation. Another contribution concerning feedback control, again in planar Poiseulle flow was published by Hu et al. [12]. Analogical to our approach they modified the Orr-Sommerfeld equation to get some information about the stability of the controlled flow system. In contrast to the actuation via blowing/suction at the wall used in our investigations they modulated the wall temperature periodically to alter the viscosity of the fluid and therefore to stabilize the flow.

In our paper we use Direct Numerical Simulations (DNS) and Linear Stability Theory (LST) to explore the concept of ω_z-control and to evaluate its effects on the disturbances involved in the laminar-turbulent transition in a Blasius boundary layer. The DNS method and a discussion of results in the linear regime are presented in section 2 and 4, investigations using LST in section 5, active control of nonlinear disturbances in section 6 and a summary is given in section 7.

2   Numerical Method

All simulations were performed in a rectangular integration domain with the spatial DNS-code developed by Konzelmann, Rist and Kloker [13][14][15].

The flow is split into a steady 2D-part (Blasius base flow) and an unsteady 3D-part. The x-(streamwise) and y-(wall-normal) directions are discretized with finite differences of fourth-order accuracy and in the spanwise direction z a spectral Fourier representation is applied. Time integration is performed by the classical fourth-order Runge-Kutta scheme. The utilized variables are normalized with $\tilde{U}_\infty = 30\frac{m}{s}$, $\tilde{\nu} = 1.5 \cdot 10^{-5}\frac{m}{s^2}$ and $\tilde{L} = 0.05m$ ($\tilde{}$ denotes dimensional variables):

$$x = \frac{\tilde{x}}{\tilde{L}} \qquad y = \frac{\tilde{y}}{\tilde{L}} \qquad z = \frac{\tilde{z}}{\tilde{L}} \qquad t = \tilde{t} \cdot \frac{\tilde{U}_\infty}{\tilde{L}}$$

$$u = \frac{\tilde{u}}{\tilde{U}_\infty} \qquad v = \frac{\tilde{v}}{\tilde{U}_\infty} \qquad w = \frac{\tilde{w}}{\tilde{U}_\infty} \qquad Re = \frac{\tilde{U}_\infty \tilde{L}}{\tilde{\nu}} = 10^5 \quad ,$$

where u, v and w are the components of the unsteady velocity disturbances. This leads to the dimensionless frequency $\beta = 2\pi \tilde{f} \tilde{L}/\tilde{U}_\infty$, where $\tilde{f}$ is the frequency in $[Hz]$ and the dimensionless spanwise vorticity $\omega_z = \partial u/\partial y - \partial v/\partial x$. The Reynolds numbers formed with the displacement thickness are $Re_{\delta_1} \approx 500$ at the inflow boundary and $Re_{\delta_1} \approx 1300$ at the outflow boundary.

3   Computational aspects

Clearly the present work aims at gaining insight into new laminar-flow control strategies through numerical simulation. The necessary numerical tools have been developed and optimised for efficient use of the available resources earlier [16]. Due to their speed and the available large memory of the supercomputers of the HLRS either more computations were possible within a given time frame or specific cases could be analysed more thoroughly. Even our computations of eigenvalues in the framework of linear stability analysis profited from the computational advantages of the supercomputers compared to workstations. The spanwise Fourier ansatz principally reduces the 3-D problem in physical space to a set of $(K+1)$ complex 2-D problems in Fourier space thus enabling a largely parallel computation in Fourier space. However, the modes are coupled by the nonlinear convective terms of the vorticity transport equations and are transformed to physical space ("pseudospectral method" with de-aliasing procedure) for the calculation of the nonlinear vorticity terms, which in turn are parallelized in streamwise direction.

The uniform equidistant grid in x- and y-direction contains maximal 2882×241 points using up to 25 Fourier modes in spanwise direction. The problem has been computed on the hww-supercomputer, the NEC SX-5 (16 processors, 32 GB RAM) using approximately 5 GB RAM. The code reaches a performance of up to 1.6 GFLOPS at a vector operation ratio of 99.4% resulting in a computation time of ≈ 2.0 μs per gridpoint, time step and spanwise harmonic mode and a memory requirement of about 350 Byte per point and spanwise mode. About 88% of the overall time was spent in parallel execution and only 12% of the total computation time for I/O, system-calls and serial program parts. On the other hand, the investigations concerning linear stability theory (chapter 5) were conducted on another hww supercomputer, the NEC-SX-4 running serial using approximately 20MB RAM. Single eigenvalues have been calculated interactively.

The amount of accumulated data is approximately 2 GB per run but this value strongly depends on the case, the number of stored timesteps, and the resolution. Therefore it is very important to have the opportunity to store and to access the data rapidly. A very significant point in this context is the direct connection between the hww-supercomputers and the hww-fileserver, the hwwfs1 enabling a very fast exchange of data.

4   Control Mechanism

To actively damp disturbances in boundary layers we use the feedback of instantaneous signals of the spanwise vorticity fluctuations measured at the wall. These signals are multiplied by an amplitude factor $|A|$ and are pre-scribed as a v-boundary condition after a time delay Δt at the wall which is necessary to produce the phase shift Φ shown in Fig.1.

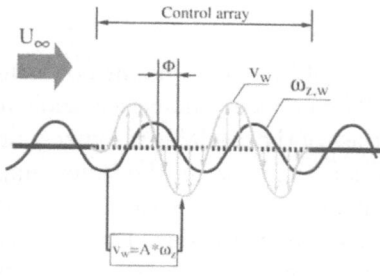

Fig. 1. Ilustration of the ω_z-control principle

The effects of changing $|A|$ and Δt resp. Φ will be studied in section 5 using linear stability theory. Here, results of DNS are analyzed in order to show how the method works in the linear case. For a small-amplitude 2D Tollmien-Schlichting (TS-) wave results of three simulations are compared with each other in Fig.2. Compared to the uncontrolled case there is a strong damping effect of the ω_z-control on the disturbance amplitude in Fig.2 (a). Because of the small amplitudes a linear superposition of waves is valid here, i.e. the controlled wave (———) in Fig.2 (a) can also be viewed as the addition of the uncontrolled wave (·······) with a control wave which arises when previously extracted v_w (index w denotes wall quantities) signals are prescribed in an otherwise undisturbed flow (- - - -).

However, this approach has to be clearly distinguished from the "classical" wave superposition principle [5], where an anti-phase disturbance of the same frequency and wave number is superposed to the initial perturbation. In the "classical" case, frequency and wave number of both disturbances are the same. Looking at the wave numbers of the present example in Fig.2 (b) one can clearly see, that this is not the case here using the ω_z-control, because the wave number α_R and therefore the phase velocity c_{ph} of the controlled wave is strongly different from the uncontrolled case.

A deeper insight into the acting mechanisms can be obtained by looking at the spatial linear 2D energy balance equation. This equation is derived from the 2D Navier-Stokes equations with the aid of a parallel flow assumption and a wave approach for the disturbances [17] [18]. Flow properties are split into steady mean and fluctuating quantities. Velocities in streamwise (x), wall normal (y) and spanwise (z) direction are $u = U + u'$, $v = V + v'$, $w = w'$, $\omega_Z = \Omega_z + \omega'_z$, respectively. Overlines denote the average over one period of

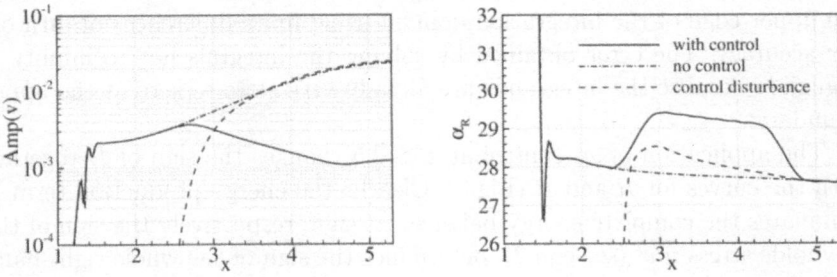

Fig. 2. v-Amplitude (a) and wave number (b) of undisturbed linear wave, controlled wave and control disturbance. Control parameters: $|A| = 7.5 \cdot 10^{-5}$, $\Delta t = 0 \Rightarrow \Phi = 0$, control array from $x = 2.4 \ldots 4.8$. Control disturbance is obtained by an extra simulation prescribing the wall signal of the run with control at the wall without initial disturbance. Due to the presence of linear waves addition of modes is valid.

time.

$$E = \frac{d}{dx} \int_0^\infty \frac{1}{2} U (\overline{u'^2} + \overline{v'^2}) dy = R + D + P +$$

$$\underbrace{- \nu \frac{d}{dx} \int_0^\infty \overline{v' \omega'_z} \, dy - \int_0^\infty (\overline{u'^2} - \overline{v'^2}) \frac{\partial U}{\partial x} \, dy + \overline{v'_w p'_w}}_{\text{small}} \qquad (1)$$

E is the spatial rate of increase of the fluctuation energy flux. $E > 0$ indicates a growth of disturbances whereas $E < 0$ means a weakening of disturbances. The terms which form the major part of the right hand side of equation (1) are

– the energy production term, where $\overline{u'v'}$ is the averaged Reynolds stress

$$R = \int_0^\infty -\overline{u'v'} \left(\underbrace{\frac{\partial U}{\partial y}}_{>0} + \underbrace{\frac{\partial V}{\partial x}}_{\to 0} \right) dy \quad , \qquad (2)$$

– the dissipation

$$D = -\nu \int_0^\infty \overline{\omega'_z{}^2} \, dy \quad , \qquad (3)$$

– and the pressure term

$$P = -\frac{1}{\rho} \frac{d}{dx} \int_0^\infty \overline{p'u'} \, dy \quad . \qquad (4)$$

All integrals are solved by integrating the flow quantities from the wall to the upper edge of the integration domain using finite differences of fifth order accuracy. The error obtained by solving the integrals not to infinity is negligible because the integrands are already very close to zero at the upper boundary.

The application of ω_z-control at $x > 2.5$ changes the sign of E together with the curves for R and P (Fig.3). Clearly, the energy production term R dominates the complete energy balance. Its sign, respectively the sign of the Reynolds stress $\overline{u'v'}$ (see eqn.2), determines the sign of the whole right-hand side of equation (1) and therefore the attenuation or growth of the regarded disturbance ($\overline{u'v'} > 0 \Rightarrow R < 0 \Rightarrow E < 0 \Rightarrow$ reduction of amplitude and vice versa). The first of the two remaining terms, the dissipation term D has always a damping effect whereas the pressure term P always tends to counteract the production term R.

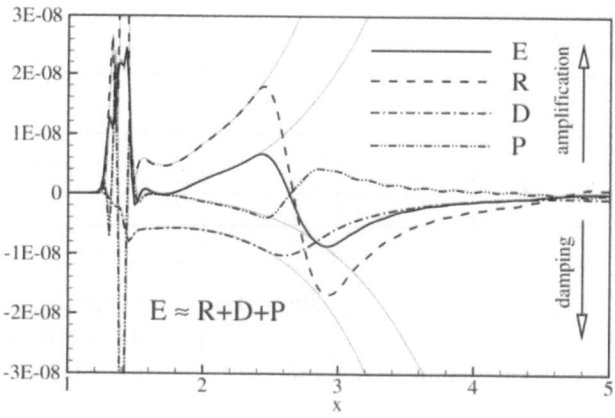

Fig. 3. Streamwise distribution of the main energy-balance terms based on eqn.(1), dotted lines: uncontrolled case for reference.

The change of sign of the Reynolds stress $\overline{u'v'}$ when control is applied is not caused by different u' or v' amplitudes but by its strong sensitivity to the phase difference $\Delta\Theta = |\Theta(u') - \Theta(v')|$ around $\Delta\Theta(y) = \frac{\pi}{2}$ [19] which is altered by the non-zero v_w.

5   Linear Stability Theory

To get an overview of the damping capabilities of the present concept and to optimize the parameters for further simulations, investigations using linear stability theory (LST) were performed. Therefore, the boundary conditions at the wall for the Orr-Sommerfeld (and Squire equation) had to be changed (index w denotes wall properties) to

$$v'_w = A \cdot \omega'_{z,w} \tag{5}$$

$$v'_w = A \cdot \left(\frac{\partial u'}{\partial y} - \frac{\partial v'}{\partial x} \right)$$

$$(1 + iA\alpha) \cdot v'_w - A \cdot \left(\frac{\partial}{\partial y} \right)_w u' = 0 \tag{6}$$

$$\text{with} \quad A = |A| \cdot e^{i\Phi}$$

where $|A|$ is the amplitude factor and Φ is the phase difference between v_w and $\omega_{z,w}$ similar to the time delay used in the simulations. Due to the ability to express ω_z in terms of u and v (eqn. 6) the discretized system remains a homogeneous eigenvalue problem which can be solved in the same way as the original Orr-Sommerfeld equation.

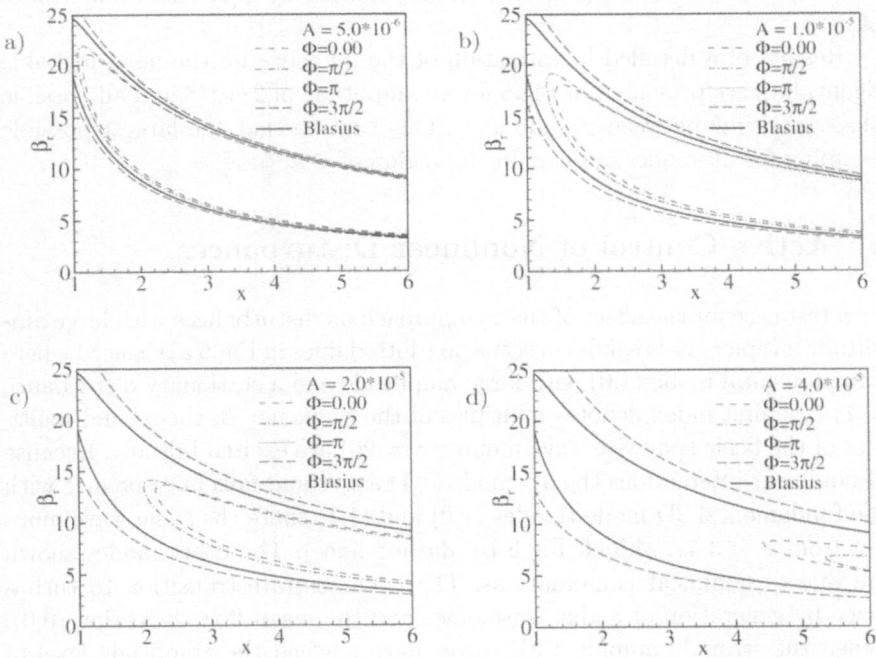

Fig. 4. Curves of zero amplification ($\alpha_i = 0$) for 2D-TS-modes with different control parameters. (a): $|A| = 5 \cdot 10^{-6}$, (b): $|A| = 1 \cdot 10^{-5}$, (c): $|A| = 2 \cdot 10^{-5}$, (d): $|A| = 4 \cdot 10^{-5}$.

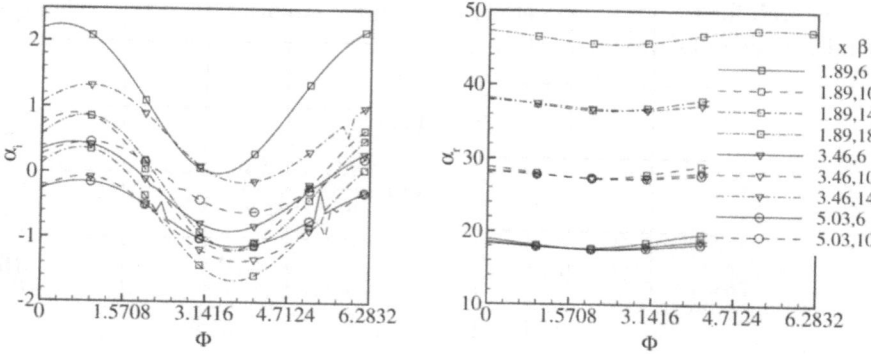

Fig. 5. Wave number α_R and amplification rate α_I from LST with ω_z-control $(v_w = |A| \cdot e^{i\Phi} \cdot \omega_{z,w})$ applied. $|A| = 2 \cdot 10^{-5}$ (Note that for the spatial approach $\alpha_i < 0$ means amplification.)

A strong damping effect and a significant reduction of the unstable area in the stability diagram (Fig.4) is already caused by very small amplitudes $|A|$.

Results of a detailed investigation of the influence on the most unstable eigenvalues are presented in Fig.5 for an amplitude of $2 \cdot 10^{-5}$ and all possible phase angles Φ between v_w and $\omega_{z,w}$. One can see that the largest possible damping for all modes appears in the region of $\Phi \approx \frac{\pi}{4} \dots \frac{\pi}{2}$.

6   Active Control of Nonlinear Disturbances

As a test case for the effect of the ω_z-approach on disturbances with large amplitude a typical K-breakdown scenario (dotted lines in Fig.6 a)) is used where a fundamental mode (1,0) with large amplitude and a stationary disturbance (0,1) (the first index denotes multiples of the frequency β, the second multiples of the basic spanwise wave number $\gamma = 20$) are excited initially. Because of nonlinear interactions the 3D-mode (1,1) arises and falls in resonance with the fundamental 2D-mode (modes (1,0) and (1,1) share the same wave number from $x \approx 3.4 \dots 4.0$ cf. Fig.6 b), dashed lines). The other modes shown are due to nonlinear combinations. They demonstrate transition to turbulence by generation of higher harmonics and the mean flow distortion (0,0). When the strongly amplified 3D-waves have reached the amplitude level of the fundamental mode, saturation sets in and transition to turbulence takes place (dashed lines). Applying ω_z-control to this scenario two main control effects can be distinguished: direct damping of nonlinear disturbances and the affection of the resonant behaviour. The first is comparable to the linear case where ω_z-control was shown to be able to directly damp TS-disturbances.

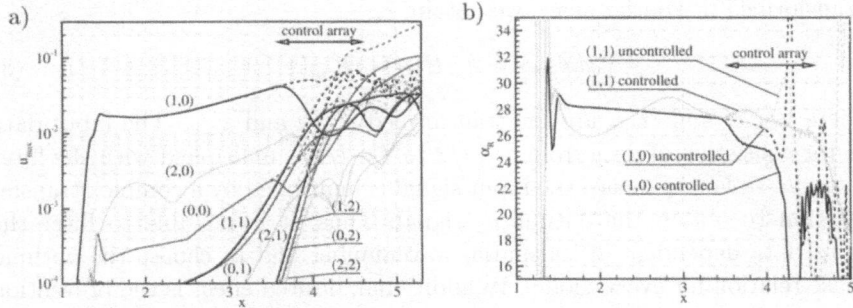

Fig. 6. K-breakdown, u_{max}-amplitudes vs. x.   Only 2D-modes controlled   ($|A| = 2 \cdot 10^{-4}$, $\Phi \approx \frac{\pi}{2}$). a) u_{max}-amplitudes, dashed lines: uncontrolled case, grey lines: higher harmonics. b) wavenumbers

In Fig. 6 only 2D modes, i.e. (1,0) and its higher harmonics were actively controlled with a phase Φ of approximately $\frac{\pi}{2}$ applied by a fixed time delay between $\omega_{z,w}$ and v_w (control array from $x = 3.5$ to $x = 4.6$). Despite the strongly nonlinear regime in this case the amplitudes of the 2D modes (1,0) and (2,0) are strongly decreased.

The damping of the 3D-modes is now due to the second effect mentioned above: resonance in 2D boundary layers is accompanied by phase synchronization of the resonant (1,1) to the fundamental (1,0) mode (i.e. both waves have the same phase speed $c = \frac{\beta}{\alpha_{\text{fundamental}}}$). Investigations using LST predict apart from changed amplification rates strongly altered wave numbers of the controlled mode (Fig.5). This effect leads to a decoupling of the resonant modes and can therefore suppress resonance. Fig.6 b) shows the wave numbers of the most important modes of the simulation mentioned above. Before the direct attenuation via ω_z-control can take effect the wave number of mode (1,0) is shifted to lower values (i.e. the wave is accelerated), the resonant mode (1,1) is not synchronized any more and resonance between (1,0) and (1,1) is prevented.

Looking at Fig. 5 one can observe that the optimal phase shift between ω_z and v is more or less independant of the frequency. Thus controlling with a fixed time delay between sensor and actuator signal yields to a different, non optimal control phase for every frequency. To obtain the desired Phase for every occuring frequency resp. wave number, we applied a spatial *FIR-Filter* of length l to the input data to treat every wavenumber in the proper way:

$$v_w(x,t) = |A| \sum_{x'=x-l/2}^{x+l/2+1} \underbrace{h(x'-x+\frac{l}{2}+1)}_{i(\text{Fig. 7})} \cdot \omega_{z,w}(x',t) \qquad (7)$$

Transformed in Fourier space we obtain:

$$V_w(x, \alpha) = A \cdot H(\alpha) \cdot \Omega_{z,w}(x, \alpha) \qquad (8)$$

where V_w, H and $\Omega_{z,w}$ are the transformed v_w, h and $\omega_{z,w}$. The input data vector consisting of data from $x - l/2$ to $x + l/2$ is multiplied with the filter vector h. In Fourier space the input signal is multiplied by a complex transfer function to obtain the output v_w-signal. Thus, it is possible to filter the input data dependent of its spatial wavenumber and to choose the optimal phase relation for every mode. An additional, desired effect is the prevention of instabilities, which might be introduced unintentionally by the actuator response to the flow field. Without the use of the present filter, above a certain amplitude level $|A|$ of approximately $2.5 \cdot 10^{-4}$ the occurence of high-frequency instabilities can superseed the active attenuation of instabilities and produce large amplification of all modes by nonlinear interaction.

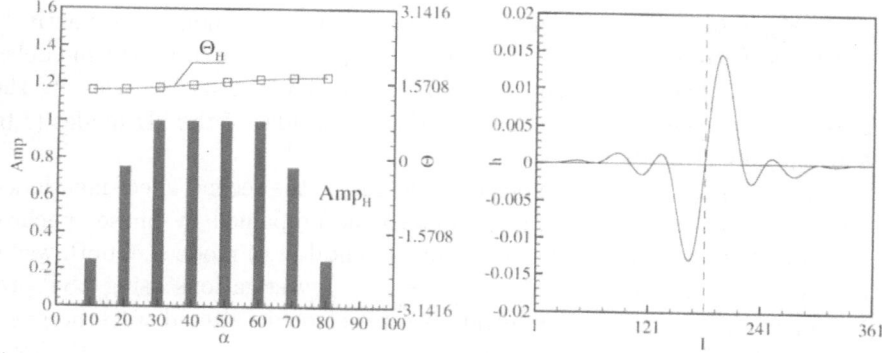

Fig. 7. Transfer function and filter coefficients of the spatial FIR-filter used to stabilize the ω_z-control.

Looking at further investigations applying ω_z-control in combination with a spatial filter in late nonlinear stages to both 2D and 3D modes shows that an amplitude reduction of more than one order of magnitude is possible. Fig.8 shows a simulation with control of the 2D (..,0) *and* 3D (..,1) modes where the control array extends from $x = 3.5$ to $x = 5.0$. Compared to previous simulations a further reduction of the disturbances is observed and larger control amplitudes are possible without the danger of undesirable high-frequency instabilities. The control amplitude is turned on via a spatial ramp function shown in Fig.8. With such an arrangement it is possible to prevent the occurence of transitional structures such as Λ-vortices or high shear layers when active control is applied only two wavelengths prior to their first appearence in the uncontrolled case (Fig. 9). As with other unsteady control

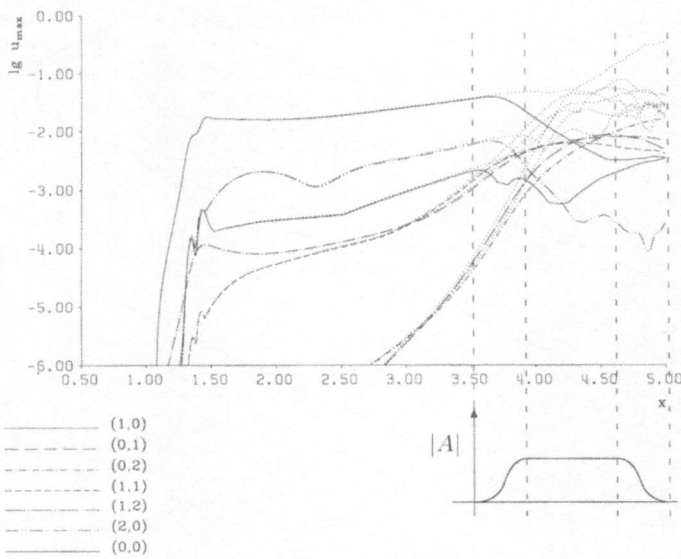

Fig. 8. K-breakdown, u_{max}-amplitudes vs. x.   2D- and 3D-modes controlled, application of a spatial FIR-filter ($|A| = 3 \cdot 10^{-4}$). Dotted lines: uncontrolled case. Small picture: Amplification factor $|A|$ vs. x.

strategies ω_z-control exhibits one shortcoming: stationary modes are not affected directly. Only by damping the fluctuating parts of the disturbance the nonlinear generation of these modes can be inhibited.

7   Summary

With the aid of Direct Numerical Simulations (DNS) it was possible to develop a simple, yet effective control algorithm to actively control the laminar-turbulent transition occurring in a 2D boundary layer. It combines two main effects: the direct attenuation caused by a change of the energy properties and a reduced resonance according to an altered phase velocity of the involved modes. Calculations using Linear Stability Theory (LST) show a strong dependence of the resulting wave number and amplification rate on the chosen amplitude and phase difference between ω'_z (sensed) and v'_{wall} (stimulated).

It is shown that this approach works very well even close to transition where the boundary layer instabilities have reached a highly nonlinear stage. Further investigations have to show how far transition can be shifted downstream and whether a complete relaminarisation of the flow is possible using this approach.

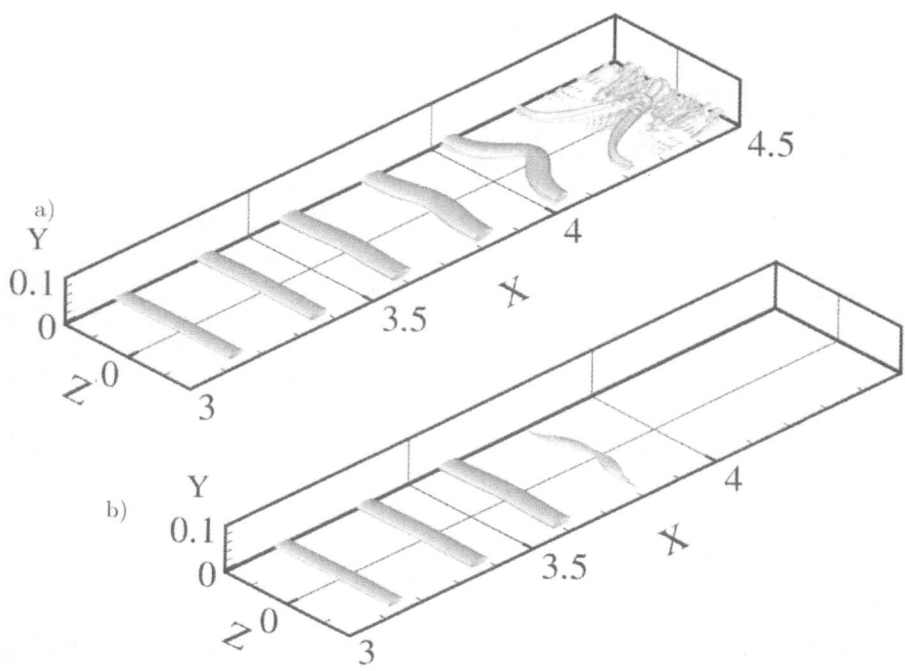

Fig. 9. Instantaneous vortex structures made visible by the λ_2-Criterium [20] for the uncontrolled case (a) and the controlled case (b), same case as Fig. 8

Acknowledgements

The support of this research work by the Deutsche Forschungsgemeinschaft DFG is gratefully acknowledged.

References

1. R. W. Milling. Tollmien-Schlichting wave cancelation. *Phys. Fluids*, 24:979–981, 1981.
2. H. W. Liepmann, G. L. Brown, and D. M. Nosenchuck. Control of laminar-instability waves using a new technique. *J. Fluid Mech.*, 118:187–200, 1982.
3. H. W. Liepmann and D. M. Nosenchuck. Active control of laminar-turbulent transition. *J. Fluid Mech.*, 118:201–204, 1982.
4. V. V. Kozlov and V. Y. Levchenko. Laminar-turbulent transition control by localized disturbances. In H. W. Liepmann and R. Narasimha, editors, *Turbulence Managment and Relaminarisation*. Springer Verlag, Berlin, Heidelberg, 1987. IUTAM-Symposium, Bangalore, India, 1987.
5. M. Baumann and W. Nitsche. Investigations of active control of Tollmien-Schlichting waves on a wing. In R.A.W.M. Henkes and J.L. van Ingen, editors,

Transitional Boundary Layers in Aeronautics, volume 46, pages 89–98. KNAW, Amsterdam, North Holland, 1996.

6. E. Laurien and L. Kleiser. Numerical simulation of boundary-layer transition and transition control. *J. Fluid Mech.*, 199:403–440, 1989.

7. P. Koumoutsakos. Active control of vortex - wall interactions. *Phys. Fluids*, 9, 1997.

8. P. Koumoutsakos, T. R. Bewley, E. P. Hammond, and P. Moin. Feedback algorithms for turbulence control - some recent developments. AIAA 97-2008, 1997.

9. H. Choi, P. Moin, and J. Kim. Active turbulence control for drag reduction in wall-bounded flows. *J. Fluid Mech.*, 262:75–110, 1994.

10. M. Baumann, D. Sturzebecher, and W. Nitsche. On active control of boundary layer instabilities on a wing. In H.J. Heinemann W. Nitsche and R.Hilbig, editors, *Notes on Numerical Fluid Mechanics II*, volume 72, pages 22–29. Vieweg-Verlag, Braunschweig, 1998.

11. S. S.Joshi, J. L. Speyer, and J. Kim. A system theory approach to the feedback stabilization of infinitesimal and finite-amplitude disturbances in plane poiseulle flow. *J. Fluid Mech.*, 332:157–184, 1997.

12. H. H. Hu and H. H. Bau. Feedback control to delay or advance linear loss of stability in planar Poiseulle flow. In *Proc. R. Soc. London*, volume 447, pages 299–312, 1994.

13. M. Kloker. *Direkte Numerische Simulation des laminar-turbulenten Strömungsumschlages in einer stark verzögerten Grenzschicht*. Dissertation, Universität Stuttgart, 1993.

14. U. Konzelmann. *Numerische Untersuchungen zur räumlichen Entwicklung dreidimensionaler Wellenpakete in einer Plattengrenzschicht*. Dissertation, Universität Stuttgart, 1990.

15. U. Rist and H. Fasel. Direct numerical simulation of controlled transition in a flat-plate boundary layer. *J. Fluid Mech.*, 298:211–248, 1995.

16. P. Wassermann, M. Kloker, and S. Wagner. DNS of laminar-turbulent transition in a 3-D aerodynamics boundary-layer flow. In E. Krause and W. Jäger, editors, *High performance Computing in Science and Engineering*. Springer, 2000.

17. F. R. Hama, D. R. Williams, and H. Fasel. Flow field and energy balance according to the spatial linear stability theory of the Blasius boundary layer. In *Laminar-Turbulent Transition*. Springer-Verlag, 1980. IUTAM Symposium Stuttgart/Germany, 1979.

18. F. R. Hama and S. de la Veaux. Energy balance equations in the spatial stability analysis of boundary layers with and without parallel-flow approximation. Princeton University, Princeton, New Jersey, unpublished, 1980.

19. C. Gmelin and U. Rist. Active Control of Laminar-Turbulent Transition Using Instantaneous Vorticity Signals at the Wall. *Phys. Fluids*, pages 513–519, 2/2000.

20. J. Jeong and F. Hussain. On the identification of a vortex. *J. Fluid Mech.*, 285:69–94, 1995.

Statistical Analysis of a Turbulent Adverse Pressure Gradient Boundary Layer

M. Manhart[1] and T. Hüttl[1]

[1]) Lehrstuhl für Fluidmechanik, Technische Universität München
Boltzmannstr. 15, D-85748 Garching, Germany

Abstract. A direct numerical simulation (DNS) of an adverse pressure gradient boundary layer has been conducted which corresponds to an experiment of Watmuff [17]. The simulation extends longer into the adverse pressure gradient region than earlier simulations did. The results are compared to statistical simulations with two-equation turbulence models. The use of a high performance computer allows for evaluation of detailed flow physics. The results show that the changes in turbulence structure due to the pressure gradient are important for the balance of the turbulent kinetic energy. Statistical turbulence models have to account for these subtle changes in order to be successfull under adverse pressure gradient conditions.

1   Introduction

Zero pressure gradient turbulent boundary layers are the subject of a great number of theoretical, experimental and numerical investigations, which have established and similarity solutions for a wide Reynolds number range (Coles [1], Fernolz and Finley [2]). Compared to that, investigations of turbulent boundary layers with a superimposed pressure gradient in streamwise direction are quite rare. Generally, the experimental access to turbulent boundary layers decelerated by adverse pressure gradients is exacerbated by their tendency to exhibit three-dimensional mean flows and a stronger sensitivity to initial conditions. The prediction of turbulent adverse pressure gradient boundary layers with Reynolds averaged simulations (RANS) often fails because presently available turbulence models cannot account for the changes in turbulence structure with respect to the zero pressure gradient case.

Direct numerical simulation (DNS) has become a reliable tool to investigate the detailed physics of turbulent flows. This includes budgets of the mean momentum and Reynolds stress balance, whose knowledge is a key in further improvement of turbulence models for RANS. DNS studies of adverse pressure gradient boundary layers can be found in Spalart and Watmuff [15], Na and Moin [11,12], Skote et al. [14], Manhart [9,7] and Hüttl et al. [4]. The present study is a continuation of the latter works. It is designed according to an experiment of Watmuff [17] at a Reynolds number of $Re_\theta = 670$ which is accessible to DNS. It consists of an accelerated region which provides a well defined turbulence structure followed by a decelerated region (adverse pressure gradient) which is in the focus of our investigation.

It has been shown (Hüttl et al. [4]), that the difficulties of RANS grow with streamwise length of the adverse pressure gradient zone. Therefore, we decided to do a DNS which extends for a longer streamwise distance into the adverse pressure gradient region than the prior works. In order to test turbulence models, RANS computations have been performed in Hüttl et al. [4] with a boundary layer code. Two turbulence models have been used: The standard $k-\omega$-model (Wilcox [18]) and Menter's SST model [10]. Both models can be integrated to the wall without using wall functions. The computational domain and the boundary conditions are chosen the same as in the DNS.

2   Numerical method

2.1   The basic scheme

Our DNS code MGLET is based on a finite volume formulation of the Navier-Stokes equations on a staggered Cartesian non-equidistant grid. The spatial discretization is of second order (central) for the convective and diffusive terms. For the time advancement of the momentum equations an explicit second-order time step (leapfrog with time-lagged diffusion term) is used.

The Poisson equation for the pressure is solved by an iterative point-wise velocity-pressure iteration like that described in Hirt et al. [3]. It can be used as a single-grid iteration or as a smoother in a multigrid cycle. The maximum divergence is chosen in order to keep the maximal velocity error below $\Delta u_{max} \leq 10^{-5} u_\infty$ (according to the relation $div_{max} = \Delta u_{max}/\Delta x_{min} \cdot l/u_\infty$). Here, u_∞ and l are the characteristic velocity and length scales, respectively.

2.2   Zonal grid algorithm

The refinement for the local grids is done by dividing one coarse grid cell into 8 fine grid cells. The coarse and the fine grid are arranged in an overlapping way, so that the coarse grid is defined globally (global grid) and the fine grid is defined only locally (zonal grid). Each second cell-face of the local fine grid lies exactly on a coarse grid cell-face. The overlapping of the grids allows for flexible handling of grid refinement.

In our approach the coarse-grid and the fine-grid solutions are fully coupled [6,8]. We use averaging over four cell faces for the velocities and averaging over 8 grid cells for the pressure restriction. The solution on the coarse-grid level in the non-overlapping region serves as a boundary condition for the fine grid. The interpolation of the coarse grid variables to the fine grid boundary points is done by a first order interpolation, in order to provide conservation of mass and momentum fluxes. If the interpolation were not conservative, the turbulent fluctuations would be strongly damped near the grid interface [6]. For solving the Poisson equation on both levels, we use the pressure correction on the coarse grid as a new pressure estimate for the fine grid in a multigrid cycle.

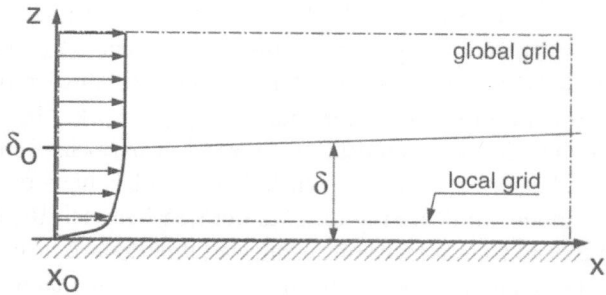

Fig. 1. Configuration and geometry of the zonal grid simulation of an adverse pressure gradient boundary layer.

2.3   Configuration and boundary conditions

The geometry used for the simulations of turbulent boundary layers is sketched in Figure 1. The streamwise, spanwise and wall-normal directions are denoted by x, y and z, respectively. In the whole computational box with dimensions l_x, l_y and l_z the global grid is defined. In addition to that, a locally refined grid is used between the wall and the wall-normal position z_{lg}. Manhart [6] showed for zonal DNS of channel flow, that z_{lg} should preferably lie in the logarithmic region of the mean streamwise velocity profile.

At the inlet plane, time dependent boundary conditions are needed to prescribe turbulent velocity fluctuations. They are generated by taking fluctuations from a position $x_{bc} = 10\delta_0$ downstream and superposing them onto a mean velocity profile, for which a time-averaged velocity profile of a zero-pressure gradient boundary layer simulation of Spalart [16] is taken that matches the Reynolds number desired. This procedure gives similar results than the method proposed by Richter et al. [13] for LES of boundary layers and the method of Lund et al. [5] applied to a zero-pressure gradient boundary layer.

In the exit plane and at the top surface, a Neumann condition for the velocities and a Dirichlet condition for the pressure is used. The velocity derivatives normal to the boundary are set to zero. At the outflow boundary the pressure is set to zero. Near the exit plane, a negative effect of this simple formulation on the flow quantities is visible up to about two boundary layer thicknesses upstream. At the top surface the pressure distribution in streamwise direction is set to realise the pressure gradient on the flow. It has been derived from Bernoulli's equation using the experimental data of Watmuff [17].

At the wall, impermeability and no-slip conditions are realized and in spanwise direction periodic boundary conditions are used.

Table 1. Parameters of the grid and the computational domain.

direction	streamwise (x)	spanwise (y)	wall normal (z)
length l_i	$0.983m$	$0.0614m$	$0.072m$
$l_i/\delta_{0,ref}$	81.92	5.12	6.0
$n_{i,coarse}$	1024	128	144
$n_{i,zonal}$	2048	256	32
parallel blocks	32	8	1
$\Delta x_{i,coarse}/\delta_{0,ref}$	0.08	0.04	0.01...0.107
$\Delta x_{i,zonal}/\delta_{0,ref}$	0.04	0.02	0.005
$\Delta x_{i,zonal}^+$	12	6	1.5

Table 2. CPU-time requirements of the DNS.

	coarse grid	coarse + zonal grid
$n_{tot} = n_x \cdot n_y \cdot n_z$	$20.1 \cdot 10^6$	$35.7 \cdot 10^6$
timesteps	214 250	122 000
$\Delta t\, u_0/\delta_{0,ref}$	0.01	0.005
$t_{cpu}/\Delta t$	1317 CPU-sec	2575 CPU-sec
processors	128	256

2.4   Computational requirements and flow domain

The computational requirements of direct numerical simulations are extremely high. Therefore, the simulations have been performed on the high performance computer Cray T3E-900, a massively parallel computer (HLRS computing center, Stuttgart/Germany). Considerable efforts have been made to optimize the code. For single processor runs of the code MGLET, the minimum requirement of core memory is 12 words per grid point. Parallel computations have additional requirements in order to keep neighbouring information [8]. About 70 MFLOPS per processor can be achieved and the required CPU-time per time step and grid point lies between $2.5 \cdot 10^{-5}$ and $4.0 \cdot 10^{-5}$ seconds. The elapsed time per time step depends on the number of processors used. If the problem size is large enough, parallelisation gives a linear speed-up to more than 200 processors on the Cray T3E-900 [8].

The experimental domain of Watmuff's experiment can be divided into a region of favourable-pressure-gradient from $x = 0.2m$ to $0.6m$ and an adverse-pressure gradient region from $x = 0.6m$ on. Experimental data is available from $x = 0.2m$ to $2.0m$ (at $0.05m$ intervals). The core memory of the high performance computer limits the length computational domain $0.29m \leq x \leq 1.273m$, for which a total number of $n_{tot} = 35.7 \cdot 10^6$ grid points is required. Table 1 shows details of the DNS flow domain lengths l_i, the number of grid points n_i of the coarse and zonal grid and the mesh sizes Δx_i. Δx_i^+ is normalized with $u_{\tau,max}$ at $x \approx 0.6m$ and ν. Parameters of the CPU-time requirements of the DNS are shown in Table 2.

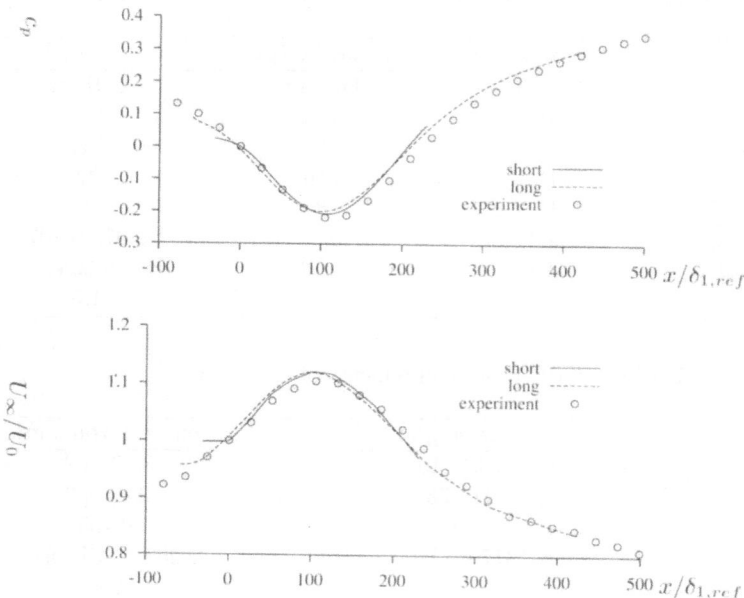

Fig. 2. Development of the mean wall pressure coefficient c_p (top) and free stream velocity U_∞ (bottom). Symbols: Watmuff [17].

The reference position of the Simulation ($x = 0.0$) corresponds to the position $x = 0.4m$ in the experiment which is in the region of a favourable pressure gradient. The inflow plane of the DNS lies $10\delta_0$ boundary layer thicknesses upstream of that position in order to let the turbulence structure develop. The reference length and velocity are the displacement thickness $\delta_{1,ref}$ and the velocity at the boundary layer edge U_0 at the reference position.

3   Results

3.1   Global quantities

The global quantities like mean pressure coefficient c_p, mean freestream velocity U_∞, mean wall friction coefficient c_f and displacement thickness δ_1 are good indicators if a simulation meets the experimental values. Often they are the only quantities which are required for an engineering point of view. The quantities are compared between the present DNS ('long'), and a former DNS with a shorter domain ('short', Manhart [9,7]) and the experiment of Watmuff [17]. The results of the RANS models are shown for c_f and δ_1, respectively.

The pressure coefficient $c_p = (p - p_0)/(1/2\rho u_0^2)$ at the wall initially is decreasing (accelerated flow). After $x/\delta_0 = 15$ the adverse pressure gradient

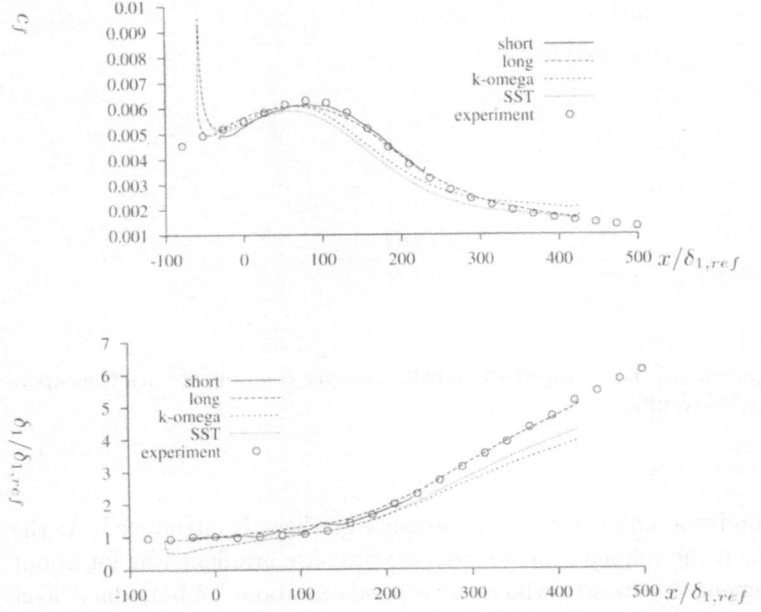

Fig. 3. Development of the mean wall friction coefficient c_f (top) and displacement thickness $\delta_1/\delta_{1,ref}$ (bottom). Symbols: Watmuff [17].

which decelerates the flow sets in (Figure 2). The favourable pressure gradient leads to an increasing wall friction and to a delay in the growth of the boundary layer (Figure 3). The adverse pressure gradient leads to a falling skin friction and a faster growth of the displacement thickness. The RANS are able to reproduce the behaviour of the skin friction coefficient at least qualitatively. A full quantitative prediction of c_f, however, was not achieved by the RANS, especially in the later part of the adverse pressure gradient region. This could partly be due to scaling problems. Note, that the reference length $\delta_{1,ref}$ is the displacement thickness at $x = 0.0$. In the RANS, δ_1 is collapsing shortly after the inflow plane. Therefore, we used the DNS scaling for the RANS results, because identical boundary conditions are used in both DNS and RANS, respectively. If we had used δ_1 from the actual RANS run, then the accordance between RANS and DNS (or experiment) would have been even worse.

3.2 Statistics of first and second order

In Figure 4, we show the mean streamwise velocity component at two streamwise positions. The first is located in the favourable pressure gradient region ($x/\delta_{1,ref} = 53$) shortly after the reference position, the second is far in the adverse pressure gradient region ($x/\delta_{1,ref} = 368$). At the first position all results collapse more or less onto one curve. That shows the feasability of

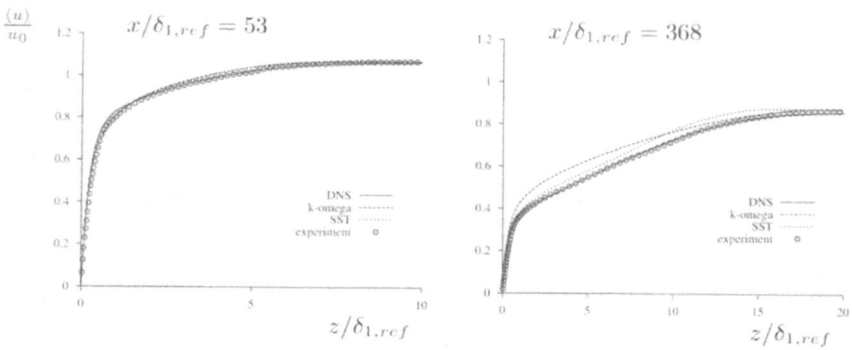

Fig. 4. Comparison of the averaged streamwise velocity component with the experiment and RANS-results

the inflow condition and the way, the pressure gradient is introduced. At the second position the influence of the adverse pressure gradient was for about 200 displacement thicknesses which corresponds to about 32 boundary layer thicknesses. One can see the typical shape of an adverse pressure gradient velocity profile. The accordance between DNS and experiment is perfect, but the RANS deviate from this typical shape.

The profiles of the turbulent kinetic energy $k = 1/2 \left(\langle u'^2 \rangle + \langle v'^2 \rangle + \langle w'^2 \rangle \right)$ show the reason for the poor performance of the RANS models. Already at the first position they are not able to reproduce the high peak near the wall which appears in the DNS and the experiment. The shape of k is more complicated at the second position. The turbulent kinetic energy shows two peaks. One at the wall which is the remainder of the near-wall peak upstream and a second one at $z/\delta_{1,ref} \approx 7$ which is developed in the adverse pressure gradient region. Both, DNS and experiment, show this double peak feature, the DNS at little smaller values than the experiment. The RANS only show one broad peak.

3.3 Mean momentum balance

In order to shed more light into the detailed flow physics, the momentum balance has been computed from the statistical quantities. The momentum balance is expressed by the Reynolds equation

$$\frac{\partial U_i}{\partial t} = -\frac{\partial}{\partial x_j} U_i U_j - \frac{1}{\rho}\frac{\partial P}{\partial x_i} - \frac{\partial}{\partial x_j}\overline{u_i' u_j'} + \nu\frac{\partial^2 U_i}{\partial x_j^2}. \tag{1}$$

Here, U_i and P are mean quantities, i.e. averaged in time and homogeneous y-direction, and u_i' are the local and temporal deviation from the mean quantities (fluctuations) and $\overline{(.)}$ means averaging over time and the homogeneous

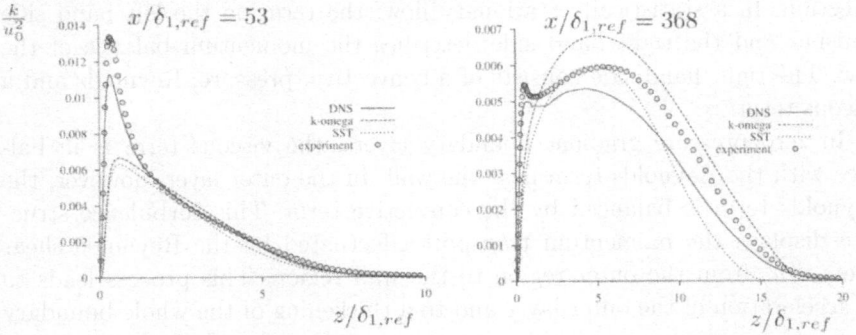

Fig. 5. Comparison of the turbulent kinetic energy with the experiment and RANS-results

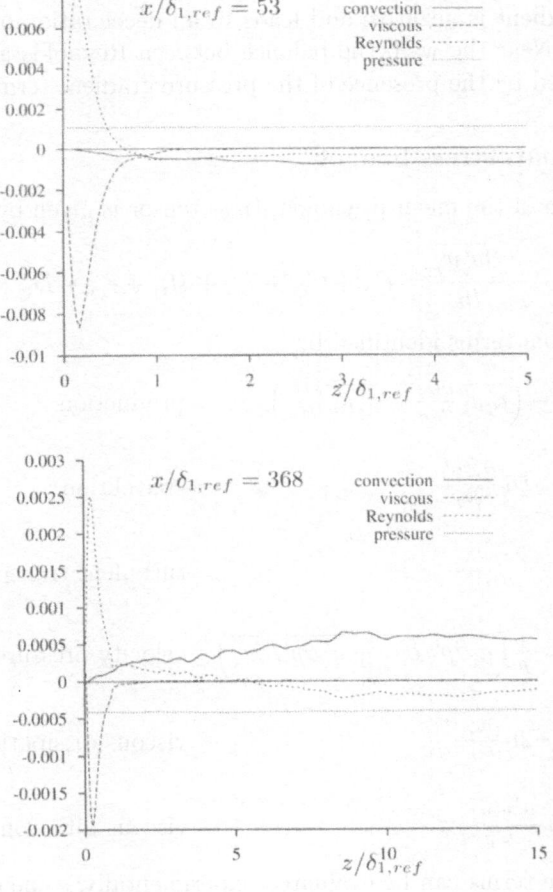

Fig. 6. Mean Momentum balance normalized by $U_0^2/\delta_{1,ref}$

direction. In a statistically stationary flow, the term on the left hand side vanishes and the right hand side describes the momentum balance of the flow. The right hand side consists of a convective, pressure, Reynolds and a viscous term.

In zero pressure gradient boundary layers, the viscous term is in balance with the Reynolds term near the wall. In the outer layer, however, the Reynolds term is balanced by the convective term. This turbulence structure displays the momentum transport effectuated by the Reynolds shear stress $\overline{u_i' u_j'}$ from the outer region to the wall region. This process leads to an acceleration of the outer layer and to a thickening of the whole boundary layer.

At the first position in the favourable pressure gradient zone (Figure 6), these relations are only altered by the addition of a constant pressure term throughout the layer. This term adds to the balance of the Reynolds- and viscous term near the wall and leads to an acceleration of the flow in the outer layer (visible as negative convective term) because there it is larger than the Reynolds term. In the adverse pressure gradient zone (second position) the pressure gradient is negative and leads to an deceleration of the flow in the outer layer. Near the wall, the balance between Reynolds and viscous term is only altered by the presence of the pressure gradient term.

3.4   Reynolds stress budget

The equation of the mean Reynolds stress tensor is given by

$$\frac{\partial \overline{u_i' u_j'}}{\partial t} = P_{ij} + C_{ij} + T_{ij} + \Pi_{ij} + \epsilon_{ij} + D_{ij}. \tag{2}$$

with individual terms identified by

$$P_{ij} = -\left(\overline{u_i' u_k'} \frac{\partial U_j}{\partial x_k} + \overline{u_j' u_k'} \frac{\partial U_i}{\partial x_k} \right) \qquad \text{production}$$

$$C_{ij} = -U_k \frac{\partial \overline{u_i' u_j'}}{\partial x_k} \qquad \text{convektion}$$

$$T_{ij} = -\frac{\partial \overline{u_i' u_j' u_k'}}{\partial x_k} \qquad \text{turbulent transport}$$

$$\Pi_{ij} = -\frac{1}{\rho} \left(\overline{u_i' \partial p' / \partial x_j} + \overline{u_j' \partial p' / \partial x_i} \right) \quad \text{velocity pressure gradient} \tag{3}$$

$$\epsilon_{ij} = -2\nu \overline{\frac{\partial u_i'}{\partial x_k} \frac{\partial u_j'}{\partial x_k}} \qquad \text{viscous dissipation}$$

$$D_{ij} = \nu \frac{\partial^2 \overline{u_i' u_j'}}{\partial x_k^2} \qquad \text{viscous diffusion}$$

Some of these terms can be evaluated experimentally, some easily and some very hardly (e.g. dissipation). But especially the pressure gradient term is

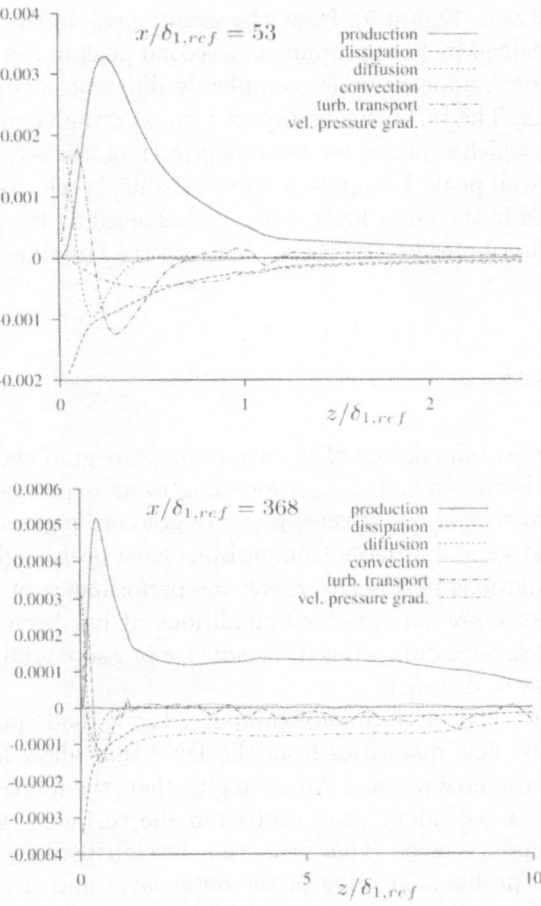

Fig. 7. Reynolds stress budgets

out of reach for wind tunnel experiments. Profiles of all the terms for the longitudinal Reynolds stress $\overline{u'u'}$ are shown in Figure 7 for the two different streamwise positions. In the favourable pressure gradient zone the budget resembles that of a zero pressure gradient boundary layer. The longitudinal Reynolds stress is produced mainly in the buffer layer (which corresponds here to $z/\delta_{1,ref} \approx 0.3$). These fluctuations are brought to the wall by turbulent transport and viscous diffusion which immediately at the wall balances the dissipation. The velocity pressure gradient term is negative throughout the whole layer which means that this term distributes kinetic energy from the longitudinal component to the other two velocity components.

In the adverse pressure gradient region, the production term has decreased by a factor of about 6 compared to its peak in the favourable pressure gradient zone. This goes hand in hand with a decrease of the turbulent kinetic energy by a factor of two (Figure 5). Now, the second peak in the turbulent kinetic energy is explained by the upcoming of a second peak in the production term. This second peak appears under completely different conditions as the first near wall peak. The turbulent transport term and the viscous diffusion term is small here, which explains the relative growth of this secondary peak compared to the wall peak. Dissipation, albeit visible, is relatively small, so that the production in the outer layer is mainly balanced by the pressure gradient term which distributes kinetic energy between the individual Reynolds stress components.

4 Conclusions

A direct numerical simulation of an adverse pressure gradient turbulent boundary layer has been conducted corresponding to an experiment. This simulation reaches further in the adverse pressure gradient region than earlier simulations of the same flow case. Comparisons have been made with Reynolds averaged simulations in order to check the performance of turbulence models under adverse pressure gradient conditions. It has been shown that two-equation models have difficulties if the adverse pressure gradient persists over a long streamwise distance.

The use of a high performance computer has it made possible to extract statistics of the flow quantities from the DNS that allow for a detailed investigation of the flow physics. According to that, the following observations can be made. A secondary peak evolves in the turbulent kinetic energy in the outer boundary layer. This peak can develop because (i) the shear of the mean flow profile is growing in the outer layer and (ii) the mechanisms which damp the production near the wall (turbulent transport, viscous diffusion) are not present there. In order to be successful in decelerated boundary layers, a turbulence model has to take into account these subtle changes in turbulence structure.

Acknowledgment. We gratefully acknowledge the support of the High Performance Computing Centre Stuttgart (HLRS).

References

1. D. Coles. The turbulent boundary layer in a compressible fluid. In *Report R-403-PR*. The Rand Corporation, Santa Monica, CA, 1962.
2. H.H. Fernholz and P.J. Finley. The incompressible zero-pressure-gradient turbulent boundary layer: an assessment of the data. *Prog. Aerospace Sci.*, 32:245–311, 1996.
3. C.W. Hirt, B.D. Nichols, and N.C. Romero. Sola – a numerical solution algorithm for transient fluid flows. In *Los Alamos Sci. Lab.*, Los Alamos, 1975.

4. T.J. Hüttl, G. Deng, M. Manhart, and R. Friedrich. Testing turbulence models by comparison with DNS data of adverse-pressure-gradient boundary layer flow. to appear in: High Performance Computing in Science and Engineering '00, 2000.

5. T.S. Lund, X. Wu, and K.D. Squires. Generation of turbulent inflow data for spatially-developing boundary layer simulations. *J. Comp. Phys*, 140:233–258, 1998.

6. M. Manhart. Zonal direct numerical simulation of turbulent plane channel flow. In R. Friedrich and P. Bontoux, editors, *Computation and visualization of three-dimensional vortical and turbulent flows. Proceedings of the Fifth CNRS/DFG Workshop on Numerical Flow Simulation*, volume 64 of *Notes on Numerical Fluid Mechanics*. Vieweg Verlag, 1998.

7. M. Manhart. Direct numerical simulation of an adverse pressure gradient turbulent boundary layer on high performance computers. In E. Krause and W. Jäger, editors, *High Performance Computing in Science and Engineering '99*, pages 315–326, Berlin, Heidelberg, New York, 1999. Springer.

8. M. Manhart. Direct numerical simulation of turbulent boundary layers on high performance computers. In E. Krause and W. Jaeger, editors, *High performance Computing in Science and Engineering 1998*. Springer Verlag, 1999.

9. M. Manhart. Using zonal grids for direct numerical simulation of turbulent boundary layers with pressure gradients. In W. Nitsche, H.-J. Heinemann, and R. Hilbig, editors, *Vol. 72, Notes on numerical fluid mechanics*, pages 299–306. Vieweg-Verlag, Braunschweig, 1999.

10. F. R. Menter. Zonal two-equation k-ω turbulence models for aerodynamic flows. In *AIAA 24th Fluid Dynamics Conf., AIAA Paper 93-2906*, 1993.

11. Y. Na and P. Moin. Direct numerical simulation of turbulent boundary layers with adverse pressure gradient and separation. Report No. TF-68, Thermosciences Division, Department of mechanical engineering, Stanford University, 1996.

12. Y. Na and P. Moin. Direct numerical simulation of a separated turbulent boundary layer. *J. Fluid Mech.*, 370:175–201, 1998.

13. K. Richter, R. Friedrich, and L. Schmitt. Large-eddy simulation of turbulent wall boundary layers with pressure gradient. In *6th Symposium on Turbulent Shear Flows, Toulouse*, pages 22/3/1–22/3/7, 1987.

14. M. Skote, D.S. Henningson, and R.A.W.M. Henkes. Direct numerical simulation of self-similar turbulent boundary layers in adverse pressure gradients. *Flow, Turbulence and Combustion*, 60:47–85, 1998.

15. P. R. Spalart and J. H. Watmuff. Experimental and numerical study of a turbulent boundary layer with pressure gradients. *J. Fluid Mech.*, 249:337–371, 1993.

16. P.R. Spalart. Direct simulation of a turbulent boundary layer up to $R_\theta = 1410$. *J. Fluid Mech.*, 187:61–98, 1988.

17. J.H. Watmuff. An experimental investigation of a low Reynolds number turbulent boundary layer subject to an adverse pressure gradient. In *Ann. Res. Briefs*, pages 37–49. Center for Turbulent Research, 1989.

18. D. C. Wilcox. *Turbulence Modeling for CFD*. DCW Industries, Inc., La Canada, CA, 1993.

Simulation of Bidisperse Bubbly Gas-Liquid Flows by a Parallel Finite-Difference/ Front-Tracking Method

Manfred F. Göz[1], Bernard Bunner[2], Martin Sommerfeld[1], and Grétar Tryggvason[3]

[1] Institut für Verfahrenstechnik, Fachbereich Ingenieurwissenschaften, Martin-Luther-Universität Halle-Wittenberg, D-06099 Halle, Germany
[2] Microfluidics and Biotechnology Group, Coventor, 625 Mount Auburn Street, Cambridge MA 02138, USA
[3] Department of Mechanical Engineering, Worcester Polytechnic Institute, 100 Institute Road, Worcester, MA 01609, USA

Abstract. Three-dimensional direct numerical simulations of bidisperse bubbly gas-liquid flows in a triply-periodic cubic domain are carried out. The numerical method utilizes a finite difference scheme for the solution of the Navier-Stokes equations for the flow field and a front tracking method for the resolution and movement of the deformable bubble interfaces. These numerical experiments aim at evaluating the dependency of bubble interactions, bubble rise velocities, and bubble-induced liquid turbulence on parameters like bubble size, interface deformability, bubble size distribution, and gas volume fraction in order to gain insight into these dispersed multiphase flows. The temporal evolutions of bubble velocities and liquid turbulent kinetic energies in various bidisperse bubble systems with spherical and deformable bubbles are presented for comparison. Also given are some details about the requirements and performance of these simulations on a parallel computer platform.

1 Motivation

Bubbly gas-liquid flows occur in a variety of natural phenomena as well as industrial processes. Notably, bubble columns find wide application in chemical process engineering due to their excellent heat and mass transfer characteristics. Their main feature is that the flow behaviour is determined by the interaction of the bubbles. As it is virtually impossible to account for all the details of these interactions taking place on a microscale within the scope of a large-scale calculation, models have to be developed which integrate the microscopic properties of a complex two- or multiphase flow into a mesoscopic description. The proper formulation of such models on the mesoscale in turn determines the success of macroscopic flow computations.

In bubble columns with an initially still liquid, all motion is induced solely by bubbles introduced into the column through a membrane at the bottom and rising due to buoyancy. Recent experimental [1,2] and numerical [3–5] results point towards a strong dependency of the flow behaviour on bubble size and gas volume fraction (or 'voidage'). Both the motion of bubbles in a swarm

and the bubble-induced liquid turbulence are different from those associated
with a single bubble due to bubble interactions and resulting fluctuations.
While it is easily understandable that bubble size and thus deformability
as well as bubble number density affect the frequency and strength of these
interactions, it is not yet clear how to quantify such dependencies.

Moreover, the bubble size distribution in real flows is not monomodal but
polymodal, due to the lack of control of the bubble size at the injectors and
to coalescence and breakup. This has an additional effect on mutual bubble
interactions and the resulting swarm behaviour due to different rise velocities,
shapes and wakes of bubbles of different sizes. A continuous size distribution
is usually observed in experiments. However, this distribution often has two
distinct peaks, hence it is justified to focus on bidisperse systems.

As these important effects have not been investigated systematically, there
is a lack of understanding of the fundamental mechanisms leading to experimentally observed global flow patterns in bubbly gas-liquid flows, hampering
the development of practical models. While it is impossible at present to
compute the turbulent flow in an entire bubble column using direct numerical simulations, DNS allows to gain detailed insight into the micro- and
mesoscopic flow field properties for a wide range of parameters. Here this approach is used to study the flow of gas bubbles in a liquid and its dependence
on bubble size, size distribution, interface deformability, and void fraction.

2 Numerical aspects

The three-dimensional Navier-Stokes equations, formulated for a single flow
field with variable density and viscosity, are solved utilizing a finite difference method combined with a front tracking algorithm allowing for a fully
deformable interface between a bubble and the ambient liquid [3]. At each
time step the density field is reconstructed from the new front positions, and
the viscosity values are then adjusted according to the updated density levels.
For computational reasons, the discontinuities in density and viscosity at the
interfaces are smoothed over 2–3 grid points.

The Navier-Stokes equations, augmented by an interfacial source term to
account for the stress boundary conditions due to surface tension, are solved
by a second-order accurate projection method, using centered differences on a
fixed, staggered grid. The incompressibility condition leads to a non-separable
elliptic equation for the pressure which is solved by a multigrid Poisson solver.

In order to accurately determine the surface tension at the interface between gas and liquid, each front is resolved by a set of marker points connected
by triangular elements. The marker points are advected with a velocity interpolated from the grid velocity. Conversely, the density, viscosity and surface
tension are determined from the position of the front and distributed onto the
grid. Curvature and surface tension are calculated using a local polynomial
fit to the front.

In the present calculations, periodic boundary conditions in all three spatial directions are prescribed, so that the computational domain represents a small section of the flow field in the middle of a bubble column. In order to keep the influence of system size effects to a minimum, the number of bubbles should not be too small. On the other hand, it cannot be chosen arbitrarily large because of the high computational cost of the simulations.

The large amount of memory and CPU time required for the simulations, as well as the nature of the problem, where a large number of bubbles are distributed approximately uniformly in a large domain, suggest the use of parallel computers. The computational domain – a cube in our case – can easily be decomposed into equisized subdomains, which are assigned to different processors and are thus worked on in parallel. Since the physical variables like the velocity and density fields live on a regular, fixed grid, this gives perfect load balance of all grid-based calculations, i.e., a perfectly balanced Navier- Stokes solver which accounts for most of the computational cost. For our simulations we employed between 8 and 18 processors, depending on problem size and memory requirements.

Dealing with the fronts is more complicated since they move across the boundaries of the subdomains. A dynamic master-slave approach was developed to parallelize the fronts. For each front, the processor corresponding to the subdomain where most of the front is located is called 'master' while the processors or subdomains supporting the rest of the front are called 'slaves'. Several times within each time step, the 'master' collects the parts of the fronts that belong to the 'slaves', performs some calculations, such as front restructuring or curvature calculation, and sends the updated data back to the 'slaves'. Since each front consists of a linked list of points and elements, the approach can also be described as a linked list of linked lists. It ensures that all processors have the same information.

The overhead due to data communication between processors (using the MPI library) is small, typically less than 10% of the total CPU time. The memory overhead may be higher because a considerable amount of front data is duplicated on different processors. However, this memory overhead is usually less than 20% of the total memory and does not represent a serious constraint on modern distributed-memory parallel computers.

3   Simulation results

Various bidisperse systems with spherical and/or ellipsoidal bubbles are under investigation. The chosen void fractions are 2, 6, and 12%; the number of bubbles ranges from 16 to 72. Some parameters of the simulations are summarized in Table 1. In a first series of simulations – bd2, bd6a, bd6b, bd12–, we chose an equal number of small and large bubbles with volume ratio 1:2. In addition to the void fraction and the number of bubbles, buoyancy-driven bubbly flows are governed by two dimensionless parameters. One is the Eötvös number, $Eo = \rho_l g d^2 / \sigma$, which can be viewed as a dimensionless

area, and the other is the Galileo number, $Ga = \rho_l^2 g d^3 / \mu_l^2$, which is similar
to a Reynolds number squared, where the velocity scale is based on the acceleration of gravity g and the bubble diameter d. The Galileo number of the
small and large spherical bubbles is given by 600 and 1200, respectively. The
surface tension σ of the interface between the two fluids was chosen such that
$Eo = 0.63$ and $Eo = 1.00$ for the small and large bubbles, respectively. This
implies that all bubbles remain approximately spherical, as can be seen on
the left side of fig. 1, with the maximum amount of deformation experienced
by the large bubbles being about 5%. The 6% case was run twice, but with
a different number of bubbles, to explore system size effects. Time-averaged
results for these simulations were presented previously [4]. Here, we will show
time-dependent quantities.

In a second series consisting of the simulations named bd2a, bd6c, and
bd12a, the diameter of the large bubbles was set to double the diameter
of the small bubbles and there are 8 times more small bubbles than large
bubbles, so that the volumes occupied by the small and the large bubbles are

Table 1. Parameters of the simulations: Void fraction and number of small and
large bubbles, grid size, memory in MB occupied by the grid and front variables
per processor, cost (defined as CPU days per second real time and per processor)
and total number of CPU days per processor.

run	voidage	bubbles	size	grid	front	cost	days
bd2	2%	10 + 10	192^3	177	67	1.44	86.7
bd6a	6%	8 + 8	128^3	54	29	1.22	73.1
bd6b	6%	29 + 29	192^3	177	101	2.21	120.0
bd12	12%	9 + 9	104^3	30	27	1.45	87.4
bd2a	2%	32 + 4	256^3	210	236	1.77	123.7
bd6c	6%	40 + 5	192^3	81	196	1.25	100.4
bd12a	12%	64 + 8	192^3	81	219	2.04	183.5

Table 2. Breakdown of the major computational tasks, in percents: solving elliptic
equations, interpolating between front and fixed grids, performing front calculations, front and grid communications, Input/Output.

run	elliptic	interfacing	front calc.	front comm.	grid comm.	I/O
bd2	67.2	13.3	8.1	3.5	1.8	0.75
bd6a	41.2	25.9	15.9	7.8	2.4	1.11
bd6b	42.8	23.8	16.8	6.3	1.1	1.50
bd12	47.3	21.0	17.2	6.3	3.6	1.07
bd2a	56.5	16.3	10.5	6.0	1.5	1.25
bd6c	47.9	20.5	14.4	8.1	2.2	1.51
bd12a	39.0	24.7	16.7	9.3	1.8	2.20

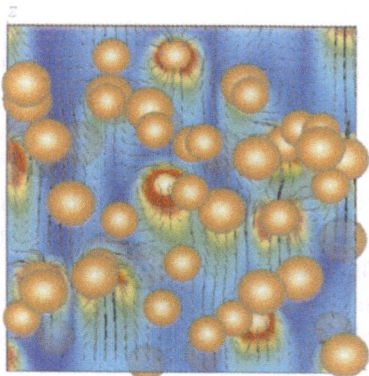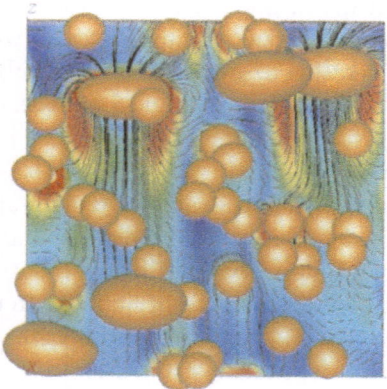

Fig. 1. Instantaneous bubble configurations together with plane sections of the streamlines and contour of the vorticity. Left: bd6b system, with 29 small and 29 large bubbles. Right: bd6c system, with 40 small and 5 large bubbles.

the same. The surface tension was selected such that the Eötvös numbers of the small and large bubbles are, respectively, 1.0 and 4.0. Hence, while the small bubbles (with $Ga = 900$) remain nearly spherical, the large bubbles ($Ga = 7200$) deform into ellipsoids whose shape fluctuates as they interact with their neighbours (fig. 1, right).

In each case, bubbles and liquid are released from rest. From the experience of simulations of monodisperse systems with spherical bubbles [3], it is expected that after an initial transient the system behaviour tends towards a statistical steady state, where averaged quantities like the average bubble rise velocity become approximately constant. This may not be true in systems containing deformable bubbles [3,5]. The resulting bubble velocities are scaled as a Reynolds number based on the mean diameter as length scale.

3.1 Systems of spherical bubbles with volume ratio 1:2

For each system of a given void fraction, the rise Reynolds numbers of all bubbles are collected in figure 2. It is seen that the bubbles move unsteadily but fluctuate about a mean value (both within their respective size class and with respect to the average motion of all bubbles). The mean values decrease with increasing void fraction, while the fluctuations increase. The latter can be clearly seen not only from the fluctuation ranges, but also from the increasing overlapping of the lower and upper bands of curves representing the rise velocities of small and large bubbles, respectively. It is natural that bubble interactions and thus velocity fluctuations increase in strength and frequency as the system becomes denser, i.e. as the bubbles get closer. This is not only an effect of voidage but also of the number of bubbles, as is obvious from the plots for the two 6% systems (see the discussion in [4]).

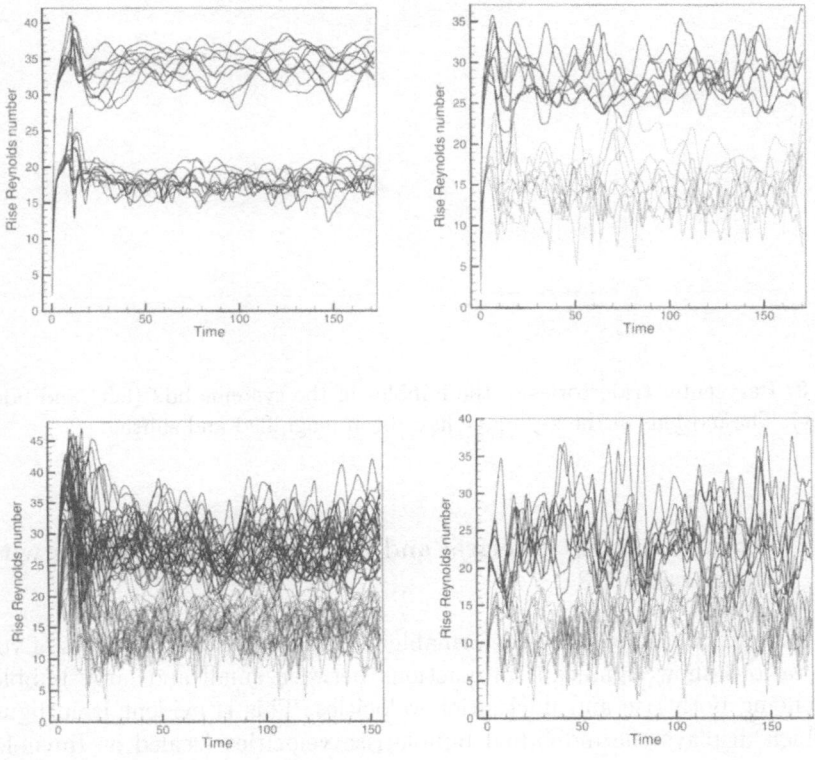

Fig. 2. Individual rise Reynolds numbers vs. time of all bubbles in the systems bd2 (upper left), bd6a (upper right), bd6b (lower left), and bd12 (lower right).

Another interesting observation is the fact, that the 2% system and the large 6% system (bd6b) settle rather quickly towards a statistical steady state, while the small 6% system (bd6a) and much more so the 12% system show considerable oscillations in the averaged values (not shown). Nevertheless, also in the last two systems the oscillations appear to be centered about approximately constant long-time averages.

The overall movements of the bubbles are reflected in their barycenter trajectories in the three coordinate planes (fig. 3). In contrast to the other systems like bd6b (fig. 3, right), system bd2 (fig. 3, left) shows a pronounced anisotropy between the two horizontal directions, including extended drifts first in the positive and then in the negative y-direction during bubble rise.

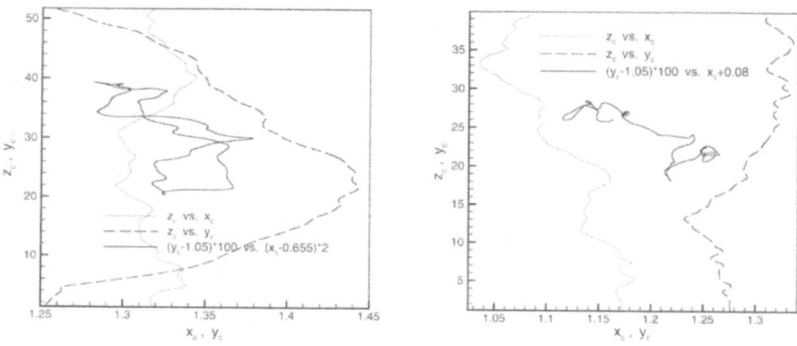

Fig. 3. Barycenter trajectories of the bubbles in the systems bd2 (left) and bd6b (right). The motions in the x-y plane have been magnified and shifted.

3.2   Systems of small spherical and large ellipsoidal bubbles with volume ratio 1:8

The systems containing large deformable and small spherical bubbles of volume ratio 8 show significant interactions between small and large bubbles influencing both rise and fluctuation velocities. This is evident from figure 4, which displays the individual bubble rise velocities (scaled as Reynolds numbers) for the systems with 2% and 12% void fraction. The rise velocities of the bubbles of each size class oscillate strongly about their respective mean values, but stay within a certain fluctuation range except for temporary outbursts indicating strong bubble interactions. The vertical velocity fluctuations of the small bubbles are relatively large compared to those of the large bubbles. The horizontal fluctuations of the spherical bubbles remain small, while those of the ellipsoidal bubbles are large due to their unsteady shape deformation and rocking motion. In addition, unlike the systems with volume ratio 1:2, the horizontal fluctuations of the large bubbles are of much greater amplitude in the 2% system than in the higher void fraction systems, because the large bubbles have more room to perform sideways motion.

In contrast to the 2% system with spherical bubbles of volume ratio 2 (bd2), the system bd2a shows no significant anisotropy in the two horizontal directions, as may be concluded from the barycenter trajectories (fig. 5). The rocking motion of the large bubbles induces more pronounced oscillations in the trajectories than could be observed in the spherical bubble systems.

The 6% system, bd6c, settles fairly quickly towards a statistical steady state (not shown). The other two systems, however, show some unexpected behaviour. After the initial acceleration phase, in which the small and large bubbles rise mostly independently of each other, an intermediate phase of strong interaction between the two size classes occurs in the 12% system (fig. 4, right). While these outbursts abate, a steady drift in the velocities takes

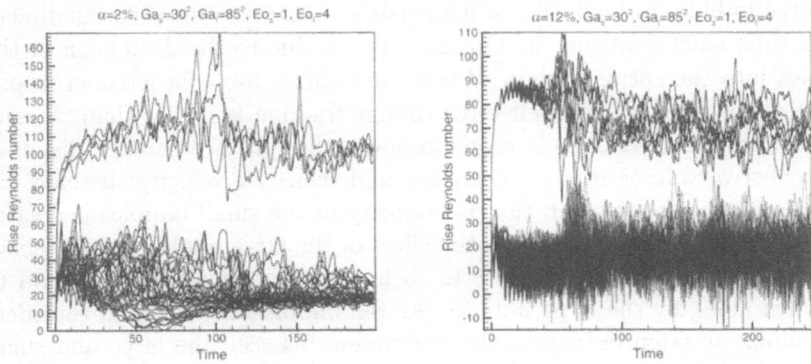

Fig. 4. Individual rise Reynolds numbers vs. time of all bubbles in the systems bd2a (left) and bd12a (right).

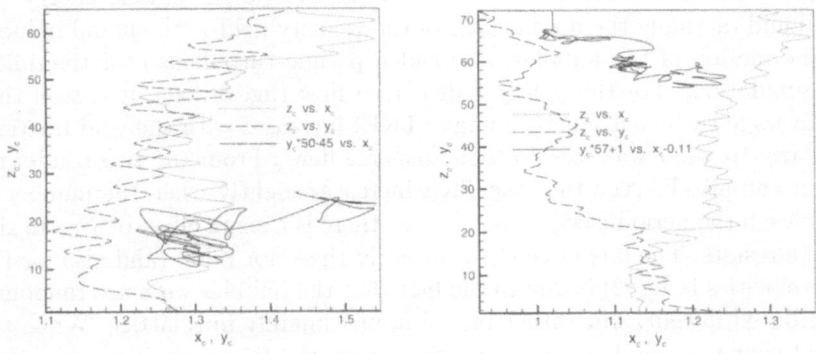

Fig. 5. Barycenter trajectories in the systems bd2a (left) and bd6c (right).

over, which accelerates the small bubbles and decelerates the large bubbles. Apparently the large bubbles pull the small ones along with them in this very dense system, but are also hindered by the presence of the many small and slow bubbles. Even after a long simulation time, this process appears not to have completed.

The unexpected behaviour of the 2% system bd2a is probably mostly caused by an initial nonhomogeneous bubble configuration. By mistake in the initialization of the bubbles' position, the bubbles were placed in a corner of the computational domain instead of being distributed over the whole domain. After the simulation starts and bubbles are released, it takes some time until the bubbles disperse throughout the domain. Unlike the other results, where the bubbles experience a quick acceleration from rest and the average rise velocity quickly reaches a steady-state value, the rise velocity for the large bubbles in bd2a (fig. 4, left) shows a slow rise until the nondimensional time reaches about 70. This slow rise is due to the dispersion of the bubbles into the entire domain. When the bubbles are concentrated in one corner of the domain, their effective volume fraction is higher than 2% and their hindered rise velocity is therefore lower than at 2%. As they disperse, their effective volume fraction decreases and their rise velocity therefore increases. On the other hand, the rise velocity of the small bubbles decreases as time increases. This illustrates the effect of the large bubbles on the small ones. When the bubbles are close to each other, the small bubbles tend to be pulled along by the large bubbles. As the bubbles disperse and the effective volume fraction decreases, the interaction between the large and small bubbles decreases and this "pulling" effect decreases too.

3.3   Turbulent kinetic energy of the liquid

Figure 6 shows the evolution in time of the "turbulent kinetic energy" in the liquid (actually the fluctuations of the velocity field in the liquid induced by the motion of the bubbles, also called pseudo-turbulence) for the different simulations. For the lightly bidispersed flow (fig. 6, left), it is seen that the average turbulent kinetic energy (TKE) increases with the void fraction, similarly to what was seen in monodisperse flows. From the two results for 6%, it can also be seen that the TKE increases slightly when the number of bubbles in the periodic cell increases, i.e., there is a small effect of system size on the results. The large overshoot at early times for bd6b (and also for the rise velocities in fig. 2) is due to the fact that the bubbles were not randomly distributed initially but rather placed approximately in a lattice. While the initial configuration has a strong influence on the transient, it has no effect on the steady-state, average results.

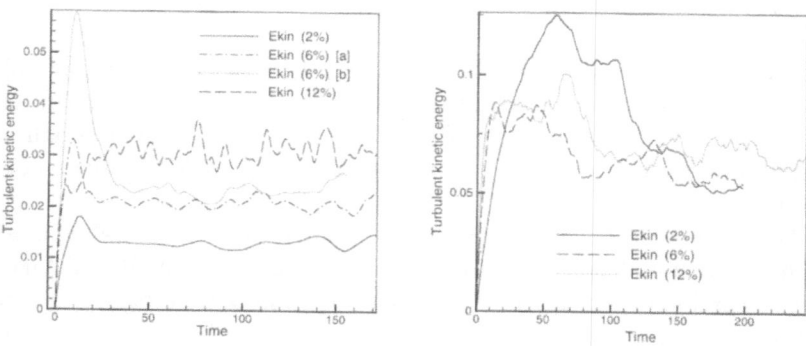

Fig. 6. Liquid turbulent kinetic energy vs. time in the bidisperse systems with spherical bubbles only (left) and with deformable large bubbles (right).

The results are quite different for the simulations with 1:8 volume ratio (fig. 6, right). Firstly, the values of the TKE are larger by factors of 8 to 2. Secondly, it appears that the TKE is about the same for 2%, 6% and 12%. For the 2% case, a large initial peak is observed, likely because of the non-homogeneous initial positioning of the bubbles mentioned above. If we neglect this transient phase and consider only the results for $t > 100$, when the bubbles have dispersed throughout the entire domain, it appears that the TKE for 2% is within 20% of the corresponding values for 6% and 12%.

We note that a previous study of monodisperse deformable bubbly flows showed similar trends in the evolution of the turbulent kinetic energy with void fraction [3]. This suggests that the flow field induced in the liquid by the bubbles is determined mainly by the few large and deformable bubbles, and not by the many small and spherical bubbles.

4   Conclusions

Direct numerical simulations with a parallel finite difference/front tracking scheme are used to study the evolution of bidisperse gas bubble swarms rising through a liquid initially at rest. The aim is to gain deeper insight into the micro- and mesoscopic behaviour of bubbly flows and support their modelling, e.g. in the framework of an Euler-Lagrange approach. Various bidisperse bubble systems are under consideration and are compared against each other and against monodisperse systems with the same void fractions. The bidisperse systems with spherical bubbles of the small volume ratio 2 between large and small bubbles behave similarly to the corresponding monodisperse systems. In contrast, the systems with large deformable and small spherical bubbles of volume ratio 8 show some novel features concerning both the motion of the bubbles and the induced liquid turbulence.

Acknowledgements

The authors gratefully acknowledge the generous grant of computer time on the IBM RS/6000 SP by the Scientific Supercomputing Center Karlsruhe, as well as travel grants by the DAAD and the US National Science Foundation (grant INT- 9726759) within the Foreign Exchange Visitor programme. MFG and MS are supported by the Deutsche Forschungsgemeinschaft.

References

1. Schlüter M., Räbiger N. (1998) Bubble swarm velocity in two phase flows. In: Proc. of the ASME Heat Transfer Division, HTD-Vol. 361, Volume 5, Anaheim.
2. Bröder D., Laín S., Sommerfeld M. (2000) Experimental studies of the hydrodynamics in a bubble column. In: Proc. 5th German-Japanese Symposium on Bubble Columns (Dresden, May 2000), TU Bergakademie Freiberg, pp. 125–130.

3. Bunner, B.: Numerical simulation of gas-liquid bubbly flows. Ph.D. thesis, The University of Michigan, Ann Arbor 2000.
4. Göz, M.F., Bunner, B., Sommerfeld, M., Tryggvason, G.: Simulation of bubbly gas-liquid flows by a parallel finite difference/front tracking method. In: E. Krause & W. Jäger (eds.), High Performance Computing in Science and Engineering 2000 (Springer 2001), pp. 326–337.
5. Göz, M.F., Bunner, B., Sommerfeld, M., Tryggvason, G.: Direct numerical simulation of bidisperse bubble swarms. Proc. of the 4th Int. Congress on Multiphase Flow ICMF-2001 (New Orleans, May 2001).

Vortex Shedding in the Turbulent Wake of a Sphere at Subcritical Reynolds Number

M. Schmid, V. Bakić, and M. Perić

Fluid Dynamics and Ship Theory Section, Technical University of
Hamburg-Harburg, Lämmersieth 90, D–22305 Hamburg, Germany

1 Introduction

The vortex shedding frequency of large coherent structures in turbulent wakes
is weakly influenced by the shape of the body or viscous effects [3]; it is of
the same order for many bodies if the wake width is used for scaling. On
the other hand the instability frequency of the separated shear layer depends
upon the boundary layer thickness before separation from the body and the
Reynolds number. The wake of the flow around a sphere at Re=50 000 is
turbulent due to the instability of the shear layer. It is examined using large-
eddy simulation and visualization experiments. The primary instability is of
Kelvin-Helmholtz type, which causes the shear layer roll-up. The instability
depends on the Reynolds number and is mainly influenced by viscous effects,
thus only simulation offers the possibility to predict the flow correctly.

2 Theory

The mathematical model for the considered fluid flow are the Navier-Stokes
equations, which consist of the continuity equation,

$$\frac{\partial}{\partial t} \int_V \rho \, dV + \int_S \rho \underline{v} \cdot \underline{n} \, dS = 0 \tag{1}$$

and the momentum equation:

$$\frac{\partial}{\partial t} \int_V \rho \underline{v} \, dV + \int_S \rho \underline{v} \, \underline{v} \cdot \underline{n} \, dS = \int_S \underline{\underline{T}} \cdot \underline{n} \, dS . \tag{2}$$

The fluid is assumed to be incompressible and the stress tensor $\underline{\underline{T}}$ contains
viscous and pressure contributions according to Stoke's law for Newtonian
fluids, while volume forces are neglected.

The filtering of the Navier-Stokes equations in space provides the equations used for large-eddy simulation. The approach is described most easily
by the filtering of a function $f(x)$,

$$\overline{f}(\underline{x}) = \int_V G(\underline{x} - \underline{x}') f(\underline{x}') d\underline{x}' , \tag{3}$$

where $G(x)$ represents the filter function. In the filtered Navier-Stokes equations, all additional terms created by the filtering (which stem from the non-linear convective fluxes) are dumped into an additional subgrid stress tensor $\underline{T}^{\text{sgs}}$, which is modeled assuming isotropic structure of turbulence (Boussinesq approximation). For the sake of simplicity the elements of the subgrid stress tensor will be expressed using the Einstein summation convention

$$T_{ij}^{\text{sgs}} = 2\mu_t \overline{D}_{ij} + \frac{1}{3}T_{kk}^{\text{sgs}}\delta_{ij}, \tag{4}$$

where the contribution of normal stresses T_{ii}^{sgs} on the right hand size is linked to the pressure and is without dynamic effect. The turbulent viscosity is computed using the Smagorinsky model [7], where eddy viscosity is a function of the filter size $\triangle$, the filtered deformation rate $\overline{D}_{ij} = \frac{1}{2}(\partial \overline{u}_i/\partial x_j + \partial \overline{u}_j/\partial x_i)$, and the dimensionless wall distance n^+, which reduces the subgrid stresses in the vicinity of solid walls [1]:

$$\mu_t = \rho \left[\triangle C_s \left(1 - e^{\left(-n^+/25\right)^3}\right)^{\frac{1}{2}} \right]^2 \sqrt{2\overline{D}_{ij}\overline{D}_{ij}}, \tag{5}$$

The Smagorinsky constant is chosen as $C_s = 0.1$, which is most often used. The filter size depends upon the local mesh size and the cubic root of the control volume is used as an approximation for it. Details about the underlying numerical method of second-order accuracy can be found in [4].

3   Computational Details

The flow around a sphere of diameter D held by a stick (for experimental realization purpose) of diameter $d = 0.1D$ is simulated with the dimensionless time step size $\Delta t\, U/D = 0.025$ on a grid with 1.86×10^6 CVs at Re=50 000. The grid is refined locally at the surface of the sphere, along the stick and in regions of large variable changes, as indicated by previous test simulations.

A large amount of CPU time was spent to optimize the grid, because the computation of 1000 time steps at least was necessary for each test simulation. For the time averaged results already shown in [5] time integration was carried out over 30 000 time steps (particles in the flow would pass within this time through a domain of the size of sphere more than 300 times). Second order discretization methods are used (midpoint rule integral approximation, linear interpolation and central differences), except in regions far from the sphere and wake, where a small portion of a first order upwind approximation is blended with the second order approximation in order to avoid oscillations on a coarse grid (there is no turbulence there and the flow is basically potential).

The method of domain decomposition is used for parallelization: the model is subdivided into pieces of equal size, and the variables are exchanged across interfaces after each iteration. The efficiency of parallelization, depends upon the problem size and the number of used processors; for the simulation of the flow around the sphere 64 processors on CRAY T3E have been used. For the same number of processors the total efficiency was determinded around 90% [6].

3.1 Results

Time averaged results like the drag coefficient or Reynolds stress profiles of the simulation are in good agreement with the available measurements [5]. Here we concentrate on the study of vortex shedding and shear layer instability. Only instantaneous results in one plane are shown here and compared to the experiment.

The visualization of the flow is realized in a water tank by towing the sphere and injecting dye through the surface of the sphere. At the same Reynolds number as in the computation, the laminar boundary layer separation near the equator could be observed. Like in a mixing layer vorticity is high in the shear layer due to the large velocity gradient. The layer rolls up forming vortices short distance after separation. The fluid from the higher-velocity region outside the wake enters regions of low velocity, thus creating vortices; their size depends upon the Reynolds number (the observed size for Re=5 000 [6] is larger than for Re=50 000).

The dimensionless frequency is usually called Strouhal number St $= \frac{U}{Df}$, which is made dimensionless by the undisturbed velocity U and sphere diameter D. For convenience St is called frequency f_{KH} here. There are two possibilities to measure the frequency of the shear layer instability: Kim & Durbin [2] measured the velocity in the shear layer with a hot wire probe. Their data doesn't show a unique power law dependence for frequency on Reynolds number: for $10^3 <$ Re $< 10^4$ frequency scales approximately with Re$^{0.75}$ while at higher Reynolds numbers the exponent is closer to 0.66. Another possibility of determing the frequency is used here: the vortices were counted over a finite time period, which might be less accurate. Experiments were carried out in the range 22 000 < Re < 175 000 and the fit through 7 points gives a power law:

$$f_{KH} = 0.0039\mathrm{Re}^{0.695}. \tag{6}$$

In the visualization of the simulated pressure field, vortices are indicated by low pressure in Fig. 1. The counting of vortices over a dimensionless time period of $T = tU/D = 6.25$, where t is the simulation time, U is the undisturbed velocity and D is the sphere diameter, gives $f_{KH} = 3.7$. The value found in the visualization experiment is $f_{KH} = 7.3$ where vortices are counted during $T = 11.3$. This value agrees quite well with data presented in [2].

For the flow around a circular cylinder the Kelvin-Helmholtz frequency is usually normalized by the shedding or Karmán frequency f_K of the large vortices in the wake. It is found [8] that the frequency scales for Re$< 10^5$

with a power law

$$\frac{f_{KH}}{f_K} = 0.0235 Re^{0.67}, \tag{7}$$

where the exponent and the constant is found by analyzing experimental data. The author's analysis indicates a higher exponent for $Re > 10^5$:

$$\frac{f_{KH}}{f_K} \sim Re^{0.7}. \tag{8}$$

Assuming the same law for the shear layer instability applies to the flow around sphere gives $f_{KH} = 6.28$ based on $f_K = 0.19$, which is computed form the time history of forces in the simulation. This value is relatively close to the result of analysis of experimental observations, suggesting that indeed the phenomena in the shear layer of sphere and cylinder are similar. This is due to the fact that the thickness of the shear layer is small compared to the radius of curvature in circumferential direction.

Two reasons can be responsible for the difference between simulation and experiment: the first is related to the way of calculating the frequency by counting the vortices. A more accurate way to get the frequency is the Fourier transformation of velocity as a function of time at monitoring points in the shear layer (similar to putting a hot-wire probe into the shear layer). The second reason lies in the resolution of the separated shear layer in the computation. Pressure plots at each time slice similar to those in Fig. 1 show the separation of low pressure regions from the stationary pressure minimum before the equator, but not all off them lead to the formation of a large eddy. It is suspected that the numerical resolution isn't high enough to resolve every detail of the shear layer and some of the weaker eddies disappear due to numerical errors. The simulated frequency is actually almost exactly half of that found in the experiment. In Fig. 2 velocity vectors in one plane are shown: each arrow represents one CV, and small eddies are represented by the minimum of four arrows. Even smaller eddies are not resolved in a LES; the numerical grid must be refined to capture all structures, but that would lead to a DNS and an excessive storage and computing time. Comparison of vortices from simulation (Fig.2) and from experiment (Fig. 3) also show that simulation produces larger vortices at a lower frequency.

In both the experiment and LES vortex pairing is observed. For simplification the process is described as two-dimensional: the roll up of the shear layer leads to folding and breakup of a continuous vortex sheet into discrete vortices, which is related to the shear layer instability frequency f_{KH}. The presence of subharmonics leads to the pairing process [9], which is the sequential amalgamation of these vortices into discrete vortices forming larger vortical structures. According to these rules vortex pairing is expected and observed in both experiment (Fig. 3) and simulation (Fig.2).

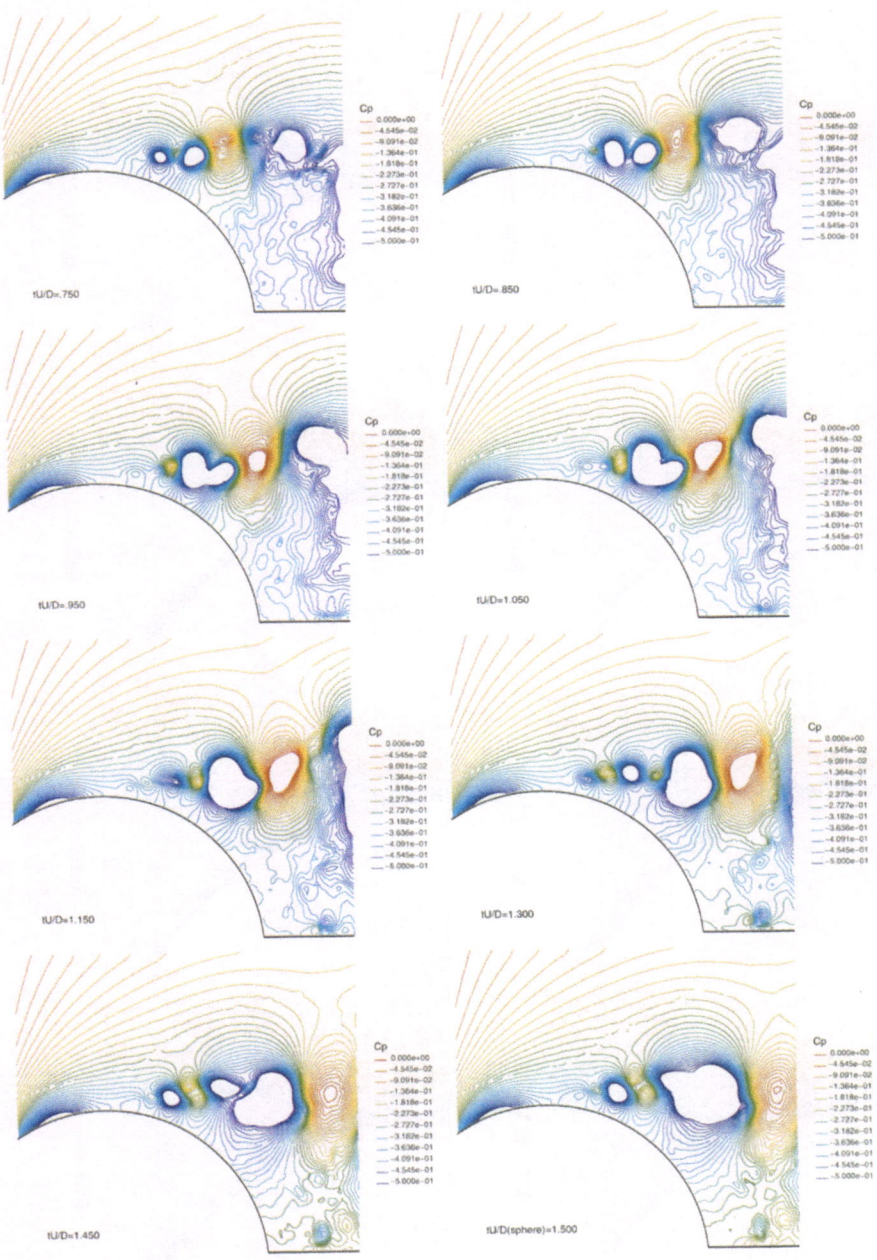

Fig. 1. Contours of pressure coefficient C_p in a plane of the flow around a sphere. Low pressure region indicate vortices, two pairings are observed in discontinuous time slices.

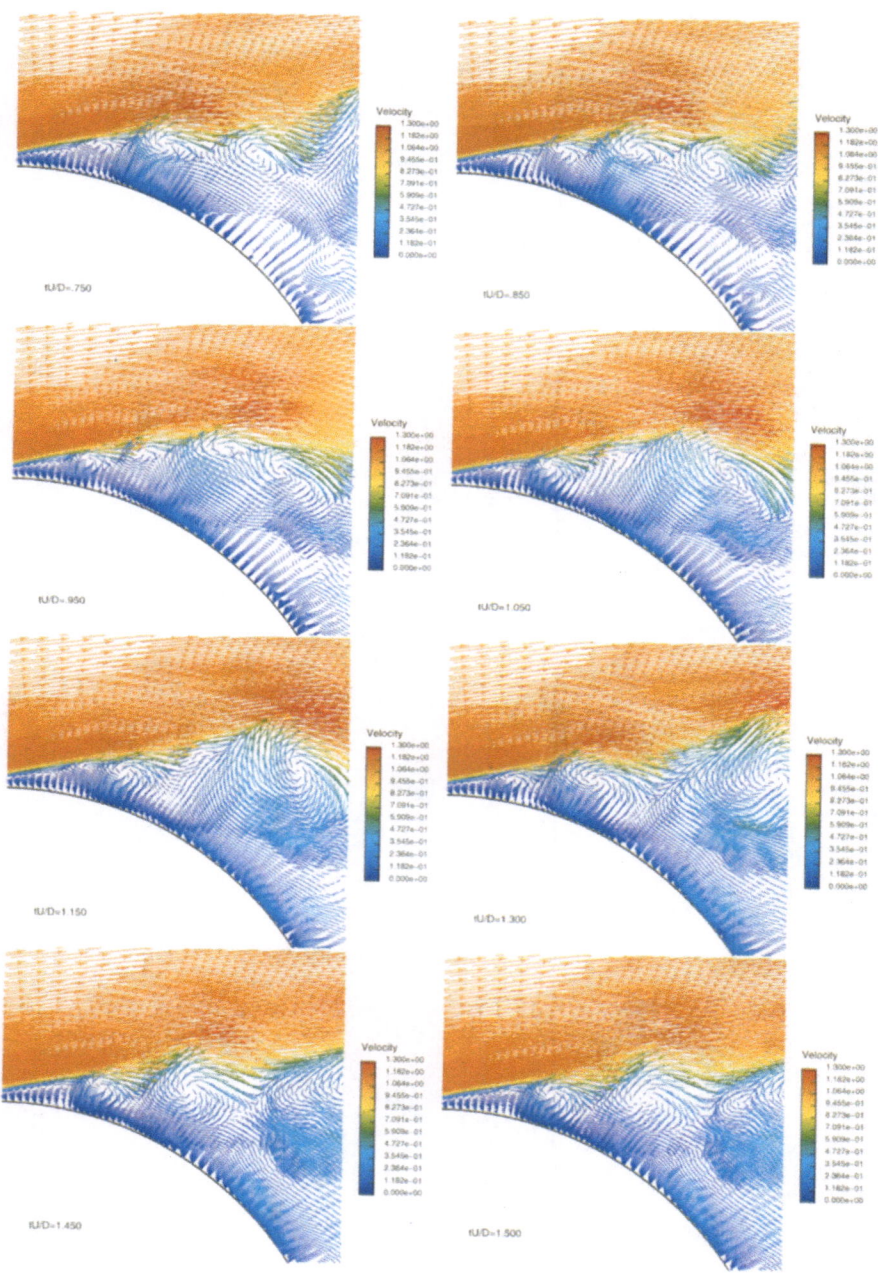

Fig. 2. Velocity vectors in one plane at same time slices as in Fig. 1.

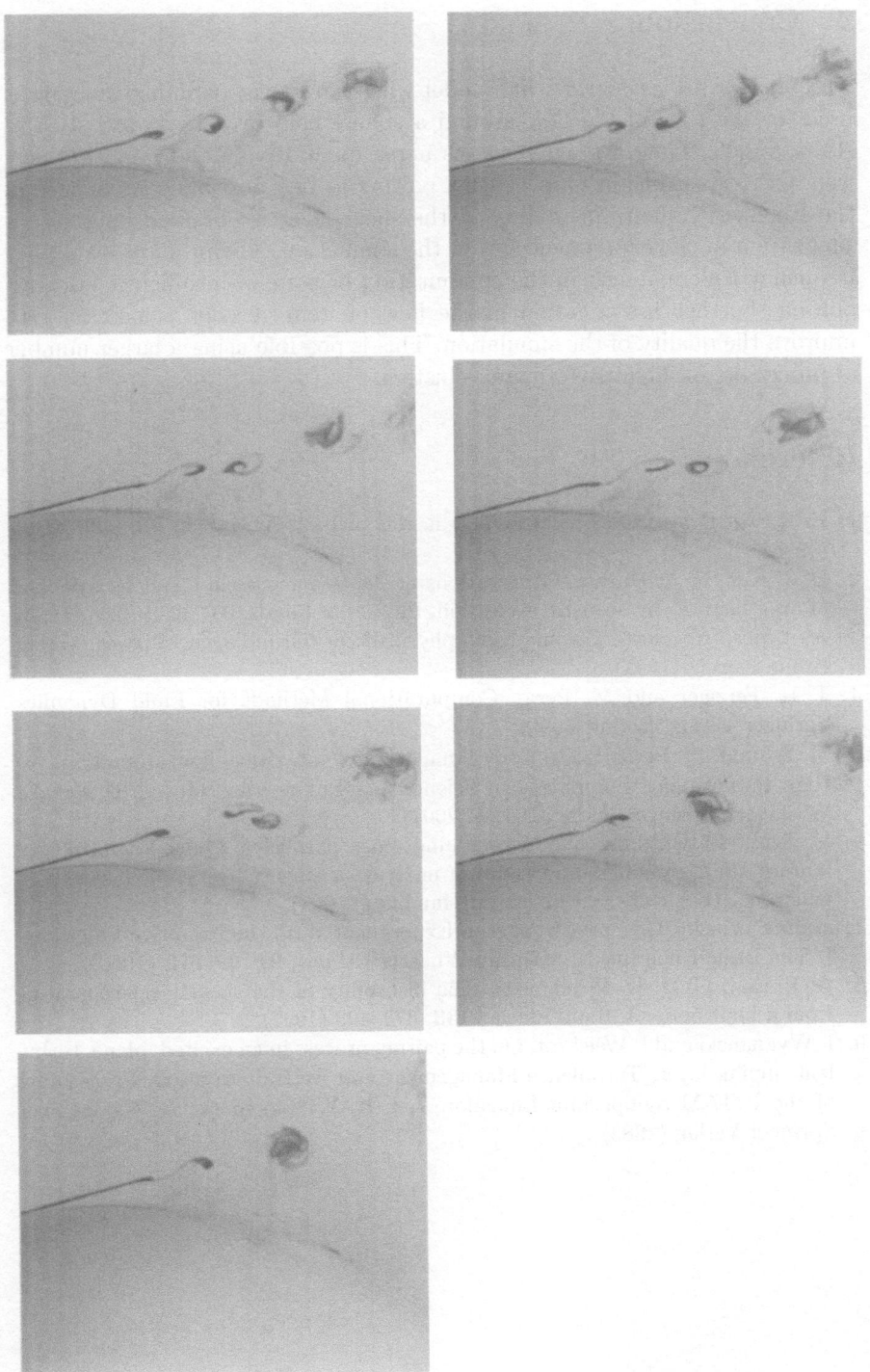

Fig. 3. Visualization of vortex pairing (sequence starts top left and finishes bottom right) in the water tank through dye injection shows the coalescence of two vortices forming a larger one.

4   Conclusions

The Smagorinsky subgrid-scale model with van-Driest damping function is used for the LES of the flow around a sphere held by a backward stick at $Re = 50\,000$. Time averaged results agree qualitatively and quantitatively well with experimental data. Vortex pairing as one possible consequence of the Kelvin-Helmholtz instability of the shear layer is observed both by visualization in the experiment and in the simulation. The primary instability frequency isn't matched in the computation because of insufficient grid resolution. Further investigation of the flow on refined grids is necessary to improve the quality of the simulation. This is possible using a larger number of processors on high performance clusters.

References

1. E. R. van Driest, On turbulent flow near a wall, J. Aero. Sci., **23**, 1007–1011, (1956).
2. H. J. Kim, P. A. Durbin, Observations of the frequencies in a sphere wake and of drag increase by acoustic excitation. Physics of Fluids, **31**, 3260–3265 (1988).
3. A. Leder, Abgelöste Strömungen: physikalische Grundlagen, Vieweg Verlag, Braunschweig (1992).
4. J. H. Ferziger and M. Perić, Computational Methods for Fluid Dynamics, Springer Verlag, Berlin (1996).
5. M. Schmid, M. Perić, Large Eddy Simulation of subcritical flow around Sphere, High Performance Computing in Science and Engineering '00 (ed. E. Krause, W. Jäger), Springer Verlag, Berlin (2001).
6. V. Seidl, Entwicklung und Anwendung eines parallelen Finite-Volumen Verfahrens zur Strömungssimulation auf unstrukturierten Gittern mit lokaler Verfeinerung, Dissertation, Universität Hamburg (1997).
7. J. Smagorinsky, General Circulation Experiment With the Primitive Equations: I. The Basic Experiment, Monthly Wheater Review, **91**, 99–164, (1963).
8. A. Prasad, C. H. K. Williamson, The instability of the shear layer separating from a bluff body. J. Fluid Mech., **333**, 375–402 (1997).
9. I. Wygnanski and I. Weisbrot, On the pairing process in an excited, plane, turbulent mixing layer, Turbulence Management and Relaminarisation, Proceedings of the IUTAM Symposium Bangalore (ed. H. W. Liepmann, R. Narashima), Springer Verlag (1988).

Assumed PDF Modeling with Detailed Chemistry

Peter Gerlinger and Manfred Aigner

Institut für Verbrennungstechnik der Luft- und Raumfahrt,
Universität Stuttgart, Pfaffenwaldring. 38-40, 70569 Stuttgart, Germany

Abstract. Turbulent fluctuations exert a significant influence on chemical production rates. If finite-rate chemistry is employed the use of probability density
functions (pdf) allows to account for turbulence chemistry interaction. In this paper an assumed pdf approach incorporates the effects of temperature and species
fluctuations on chemical reaction rates. The pdf's assumed are a clipped Gaussian
distribution for temperture and a multivariate β-pdf for an arbitrary number of different species. Finite-rate chemistry is usually associated with large discrepancies in
chemical time scales. Therefore implicit or at least point implicit numerical schemes
are required for time integration. Thus the pdf-equations and pdf influenced source
terms are discretized by backward Euler formulations. Results show that the high
numerical stability of the employed LU-SGS algorithm is maintained. A detailed
investigation of the performance of the scheme is given. This includes a comparison
between NEC SX-4 and NEC SX-5.

1 Introduction

The simulation of high speed combustion is a major challenge for numerical
schemes as well as for modelling. Due to extremely high flow velocities fluid
mechanical time scales and combustion time scales may be in the same order
of magnitude. Most combustion models used at lower Mach numbers may not
be applied for this reason. In addition often lifted flames occur that requires
a correct simulation of ignition delay times. Fortunately hydrogen is the fuel
prefered in supersonic combustion. The corresponding reaction mechanism
may already be described by 20 reactions and 9 different species. A still more
demanding problem is turbulence chemistry interaction in case of finite-rate
chemistry. There are two kinds of pdf approaches that may be used to calculate averaged species production rates: The evolution pdf-approach of Pope
[1] that usually employs a Monte Carlo solver and an assumed-pdf approach
[2–4]. While for the first one the form of the pdf may freely evolve, it is presumed in the latter one and completely defined knowing only the first couple
of moments. In moment methods the assumed pdf approach is exlusively used
to calculate mean chemical production terms. From a physical point of view
the Monte Caro pdf method is more accurate but requires substantially more
computer resources [5]. Additionally, methods for convergence acceleration
that are widely used for moment methods are not available for Monte Carlo
solvers. In the last years multigrid methods have been developed that may

be used for finite rate combustion without [6] and with presumed pdf modelling [7]. Thus assumed pdf methods are computationally more attractive for complex three dimensional simulations or combustion processes that require a large number of different species.

2   Governing Equations and Numerical Scheme

The simulation of supersonic combustion problems requires to solve the full compressible Navier-Stokes, turbulence and species transport equations as given in eq. (1). There are different problems encountered in doing this

- the set of governing equations is numerically stiff thus requiring an implicit solver,
- if multigrid is used for convergence acceleration the solver has to damp out efficiently high frequency error components,
- the solver should be able to deal with high cell aspect ratios that are common for high Reynolds number flows in near wall regions,
- the solver should be able to vectorize in all parts of the numerical scheme and achieve good performance even for block structured grids with small block sizes,
- due to the large number of transport equations to be solved this has to be done in a very efficient way,
- turbulence chemistry interaction has to be taken into account.

An efficient solution for these problems also depends on the kind of computer used. For the 2D problems that are investigated in this paper no parallization is used due to the limited demand in computer memory. A 3D version of the code has been adapted to highly parallel systems using MPI. In case of the 2D simulations presented here, all necessary Jacobians are kept in memory. This is an advantage because the LU algorithm requires Jacobians in both sweeps (lower (L) and upper (U) sweep). In the 3D-code a recalculation is performed. Thus most of the investigations of this paper deal with the vectorization of the implicit solver. Due to dependencies the implicit part is much harder to vectorize than explicit one. The two-dimensional set of equations to be solved is given by

$$\frac{\partial \mathbf{Q}}{\partial t} + \frac{\partial (\mathbf{F} - \mathbf{F}_\nu)}{\partial x} + \frac{\partial (\mathbf{G} - \mathbf{G}_\nu)}{\partial y} = \mathbf{S} \qquad (1)$$

where the conservative variable vector is

$$\mathbf{Q} = \left[\bar{\rho}, \bar{\rho}\tilde{u}, \bar{\rho}\tilde{v}, \bar{\rho}\tilde{E}, \bar{\rho}q, \bar{\rho}\omega, \bar{\rho}\sigma_e, \bar{\rho}\sigma_Y, \bar{\rho}\tilde{Y}_i \right]^T, \qquad i = 1, 2, ..., N_k - 1, \quad (2)$$

$\mathbf{F}$ and $\mathbf{G}$ are inviscid, and $\mathbf{F}_\nu$ and $\mathbf{G}_\nu$ are viscous fluxes in x- and y-direction, respectively. The source vector $\mathbf{S}$ results from turbulence and chemistry. The variables in eq. (2) are density $\bar{\rho}$, velocity components $\tilde{u}$ and $\tilde{v}$, the total specific energy $\widehat{E}$, the turbulence variables $q = \sqrt{k}$ (k = turbulent kinetic energy) and $\omega = \epsilon/k$ (ϵ = dissipation rate of k), the variance of energy, σ_e, the species mass fractions, $\tilde{Y_i}$, and the sum of their variances, σ_Y. N_k is the number of different species. The simulation of hydrogen combustion involves a 9-species (N_2, O_2, H_2, H_2O, OH, O, H, HO_2, and H_2O_2), 19-step reaction mechanism based on [8]. For turbulence closure a two-equation low-Reynolds-number q-ω turbulence model [9,10] is employed. Finally, the source vector appearing in eq. (1) is given by

$$\mathbf{S} = \left[0,0,0,0,S_q,S_\omega,S_{\sigma_e},S_{\sigma_Y},\overline{S}_i\right]^T, \quad i = 1,2,...,N_k-1, \quad (3)$$

where S_q and S_ω are source terms of the q-ω model [9], S_{σ_e} and S_{σ_Y} are source terms within the variance conservation equations, and $\overline{S}_i$ are time averaged source terms resulting from chemistry. The way $\overline{S}_i$ is calculated will be disscussed in Sect. 3. The LU-SGS (Lower-Upper Symmetric Gauss-Deidel) algorithm may be expresses symbolically by

$$(D+L+U)\,\Delta\mathbf{Q}^{n+1} = -\Delta t\,\mathbf{R}, \quad (4)$$

where D is a diaginal, L a lower, and U an upper triangular operator. Finally $\mathbf{R}$ is the discretized residual vector and $\Delta\mathbf{Q}$ the update during one time step. This system may be approximately factored and solved in two consecutive steps [11,12]:

Lower sweep: $\qquad\qquad (D+L)\,\Delta\bar{\mathbf{Q}} = -\Delta t\,\mathbf{R} \qquad\qquad\qquad (5)$

Upper sweep: $\qquad\qquad (D+U)\,\Delta\mathbf{Q}^{n+1} = D\,\Delta\bar{\mathbf{Q}} \qquad\qquad (6)$

and the solution is updated by $\mathbf{Q}^{n+1} = \mathbf{Q}^n + \Delta\mathbf{Q}^{n+1}$. Chemistry is treated in an implicit fashion and fully coupled with the fluid motion. The source and viscous Jacobians add to the diagonal D forming a matrix which has to be inverted directly at every grid point. Local time stepping is used to enhance convergence to a steady state. As to allow a complete vectorization of this algorithm the implicit part (left side of eqs. (5) and (6)) is fully vectorized too. This is done by sweeping along diagonal lines through the computational domain. The procedure is shown in Fig. 1 for the lower sweep. Because all volumes located on diagonals are computational independent it is possible to vectorize these time consuming subroutines. The disadvantage especially for 2D simulations is the short vector length which is given by the smaller number of volumes for both coordinate directions. Additionally much performance is lost at the beginning and at the end of each sweep due to still smaller vector lengths. A detailed study of this subject follows in Sect. 5. On the other hand most implicit algorithms encounter still worse problems

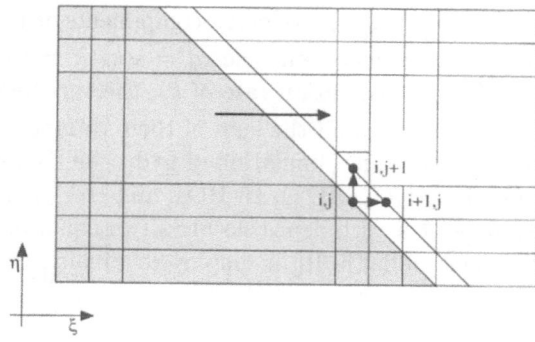

Fig. 1. Lower sweep through a 2D computational domain along diagonal lines.

concerning vectorization than the LU-SGS scheme does. At any grid point the numerical scheme requires the inversion of source term matrices that contribute to the diagonal operator D. This is done by a Gauss algorith that also has been vectorized along the diagonals of the computational domain. However, this very efficient procedure encounters performance reductions

- for short vector lengths that often occur in case of block structured complex geometries
- if multigrid techniques are used for convergence acceleration because at coarse grid levels the vector length automatically decreases.

Nevertheless a strong reduction in CPU time is achieved by the multigrid technique even in cases with combustion [6,7,10]. In the following section the assumed pdf approach is explained in more detail. Especially the formation of source Jacobians is highly complicated. These Jacobians are given in [7] as well as further details concerning the interaction of the pdf model with the numerical scheme and the multigrid method.

3   Assumed PDF Closure

A major problem in simulating turbulent reacting flows by use of averaged equations is the treatment of mean reaction rates. Reynolds decomposition (in mean and fluctuation) usually employed in moment methods may not be used for highly non-linear chemical production rates. A large number of higher order correlations would appear that can not be modeled. The instantaneous production rate of species i is given by

$$S_i = M_i \sum_{r=1}^{N_r} \left[(\nu_{i,r}'' - \nu_{i,r}') \left(k_{f_r} \prod_{l=1}^{N_k} c_l^{\nu_{l,r}'} - k_{b_r} \prod_{l=1}^{N_k} c_l^{\nu_{l,r}''} \right) \right], \qquad (7)$$

where k_{f_r} and k_{b_r} are forward and backward reaction rates of reaction r and c_l are species concentrations. It is well known that $\overline{S}_i \neq S_i(\overline{p}, \widetilde{T}, \widetilde{Y}_1, \widetilde{Y}_2, ..., \widetilde{Y}_{N_k})$

and both sides of this equation often differ orders in magnitude. If the joint pdf of all source term variables $P(\hat{\rho}, \hat{T}, \hat{Y}_1, ..., \hat{Y}_{N_k})$ is known, the time averaged production rates can be calculated by

$$\overline{S}_i = \int S_i(\hat{\rho}, \hat{T}, \hat{Y}_1, ..., \hat{Y}_{N_k}) \, P(\hat{T}, \hat{Y}_1, ..., \hat{Y}_{N_k}) \, d\hat{\rho} \, d\hat{T} d\hat{Y}_1 .... d\hat{Y}_{N_k} . \quad (8)$$

The integration is performed over all realizable values of the sample space variables $\hat{\rho}$, $\hat{T}$, and $\hat{Y}_i$. In case of an assumed pdf approach it is impossible to define the shape of a pdf that involves all these variables. Therefore statistical independence is assumed for temperature, density and species fluctuations [4]. In this case the joint pdf P can be written as the product of a temperature P_T and a composition pdf P_Y and a δ-function

$$P(\hat{\rho}, \hat{T}, \hat{c}_1, ..., \hat{c}_{N_k}) = P_T(\hat{T}) \, P_Y(\hat{Y}_1, ..., \hat{Y}_{N_k}) \, \delta(\hat{\rho} - \overline{\rho}) . \quad (9)$$

If the shapes of P_T and P_Y are known, the averaged chemical production rates may be calculated by eq. (8).

3.1 Gaussian PDF

For temperature distribution a Gaussian PDF is chosen [4,6]:

$$P_T(\hat{T}) = \frac{1}{\sqrt{2\pi\sigma_T}} \, exp\left[-\frac{(\hat{T} - \tilde{T})^2}{2\,\sigma_T}\right] , \qquad \sigma_T = \widetilde{T''^2} . \quad (10)$$

This pdf is completely defined if the first $(\tilde{T})$ and second moments (σ_T) are known. While the first moment is obtained from the energy equation an additional equation for σ_T has to be solved. In the present case an equation for the energy variance $\sigma_e = \widetilde{e''^2}$ is used

$$\frac{\partial}{\partial t}(\overline{\rho}\sigma_e) + \frac{\partial}{\partial x_j}(\overline{\rho}\tilde{u}_j\sigma_e) - \frac{\partial}{\partial x_j}\left[\left(\frac{\mu}{Pr} + \frac{\mu_t}{Pr_t}\right)\frac{\partial\sigma_e}{\partial x_j}\right] =$$

$$2\frac{\mu_t}{Pr_t}\left(\frac{\partial\tilde{e}}{\partial x_j}\right)^2 - 2\,C_e\overline{\rho}\sigma_e\omega - 2\,(\hat{\gamma} - 1)\,\overline{\rho}\,\sigma_e\frac{\partial\tilde{u}_j}{\partial x_j} . \quad (11)$$

Unclosed correlations are modeled with gradient type approximations [13,4,7], the laminar and turbulent Prandtl numbers, Pr and Pr_t, are 0.9 and 0.7, respectively. For futher details consult [7]. Using an averaged specific heat

$$\hat{c}_v = \hat{c}_v(\tilde{T}, \tilde{Y}_k) = \sum_{k=1}^{N_k} \tilde{Y}_k \frac{\int_0^{\tilde{T}} c_{v_k}(\tilde{T}) \, d\tilde{T}}{\tilde{T}} , \quad (12)$$

the temperature variance is recovered from the energy variance by

$$\widetilde{T''^2} \approx \widetilde{e''^2}/\hat{c}_v^2 \quad (13)$$

neglecting effects of species fluctuations on $\hat{c}_v$ and on the heat of formation. Due to the limited validity of the Arrhenius fuction in terms of temperature a clipped Gaussian distribution is usually employed [6] instead of the full one, given in eq. (10).

3.2 Multivariate Beta PDF

Much more demanding than a one scalar pdf is a composition pdf that has to describe statistics for an arbitrary number of gas species. In addition a significant presumption for such a multi-variate composition pdf is the possibility to obtain an analytical solution for the averaged production rates given by eq. (8). Such a β-pdf has been proposed by Girimaji [2,3]

$$P_Y(\hat{Y}_1, \hat{Y}_2, ..., \hat{Y}_{N_k}) = \frac{\Gamma\left(\sum_{m=1}^{N_k} \beta_m\right)}{\prod_{m=1}^{N_k} \Gamma(\beta_m)} \left[\delta\left(1 - \sum_{m=1}^{N_k} \hat{Y}_m\right) \prod_{m=1}^{N_k} \hat{Y}_m^{\beta_m - 1}\right] \quad (14)$$

where

$$\beta_m = \tilde{Y}_m B, \qquad B = \left[\frac{\sum_{m=1}^{N_k} \tilde{Y}_m\left(1 - \tilde{Y}_m\right)}{\sigma_Y} - 1\right], \qquad \sigma_Y = \sum_{m=1}^{N_k} \widetilde{Y_m''^2}. \quad (15)$$

This pdf is completely defined by its first $(\tilde{Y}_i)$ and second (σ_Y) moments. An important advantage of this pdf is that only one additional parameter, the sum of species variances σ_Y is needed. Thus only one additional transport equation has to be solved. This equation can be derived from the species conservation equations [2,4], and its modeled form is given by

$$\frac{\partial}{\partial t}(\bar{\rho}\sigma_Y) + \frac{\partial}{\partial x_j}(\bar{\rho}\tilde{u}_j\sigma_Y) - \frac{\partial}{\partial x_j}\left[\bar{\rho}(D + D_t)\frac{\partial\sigma_Y}{\partial x_j}\right] =$$
$$2\sum_{i=1}^{N_k} \bar{\rho}D_t \frac{\partial\tilde{Y}_i}{\partial x_j}\frac{\partial\tilde{Y}_i}{\partial x_j} - C_{\sigma_Y}\bar{\rho}\sigma_Y\omega + 2\sum_{i=1}^{N_k} \overline{Y_i'' S_i} \quad . \quad (16)$$

Again, unclosed correlations are modeled with gradient type approximations [4,13], D and D_t denote laminar and turbulent diffusion coefficients, and the model constant C_{σ_Y} equals 0.5. All terms on the right hand side represent source terms. The first one constitutes main production, the second one dissipation. The last term is calculated analytically with known pdf and needs no further modeling. This term causes a strong dissipation of σ_Y in the main reaction zone.

3.3 Averaged Production rates

With both pdfs defined and under presumpion of statistical independence the averaged production rate given in eq. (8) may now be calculated from

$$\overline{S}_i = \int S_i(\hat{T}, \hat{Y}_1, ..., \hat{Y}_{N_k}) P_T \cdot P_Y \, d\hat{T} d\hat{Y}_1 .... d\hat{Y}_{N_k} . \quad (17)$$

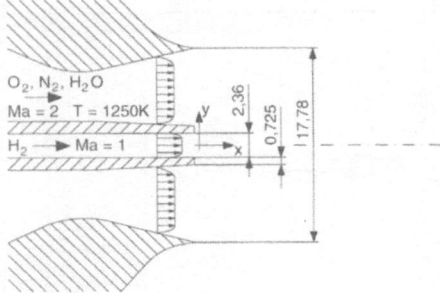

Fig. 2. Geometry (mm) for the Cheng et al. [14] combustion experiment.

Because the temperature is only involved in the reaction rates the corresponding integrals may be calculated separately from the remaining part of the source term. Mean reaction rates $\bar{k} = \int k\,P_T\,d\hat{T}$ have to be calculated numerically for any forward and backward reaction. As to avoid recalculation at any time step, these values are stored in look-up tabels depeding on $\tilde{T}$ and σ_T. These tabels are kept in memory during the computation.

4   Results and Discussion

There are only few available test cases concerning supersonic combustion where temperature and species fluctuations have been measured. On the other hand especially rms (root mean square) values are important for the validation of an assumed pdf model. The supersonic hydrogen-air diffusion flame of Cheng et al. [14] is investigated for program validation. A sketch of the axisymmetric burner geometry is given in Fig. 2. A stream of pure hydrogen is injected at sonic speed into the surrounding vitiated high speed air flow ($Y_{H_2O} = 0.1749$, $Y_{N_2} = 0.5802$, and $Y_{O_2} = 0.2449$) with a Mach number of 2 and a temperature of 1250 K. The high temperature is obtained by precombustion. In this case a lifted flame develops under ambient conditions. The calculation is performed using a 5 block grid with 120 x 24, 96 x 24, 120 x 80, 96 x 8 and 128 x 56 volumes. The grid is highly refined near solid walls and also covers a small region of the interior part of both tubes. This is necessary for an accurate simulation of the shock wave pattern at the tube outlets. Figure 3 shows calculated temperature contours. Ignition takes place about 3 cm downstream of the injector and a maximum temperature of about 2200 K is obtained. The corresponding rms values of temperature are plotted in Fig. 4. The main production of temperature variance in regions of large average temperature gradients. Highest values are obtained during ignition with values up to 440 K. Species and temperature profiles have been measured at different locations downstream of the injector. Results are plotted for the location $x = 51.74$ mm downstream of the injector. This location is

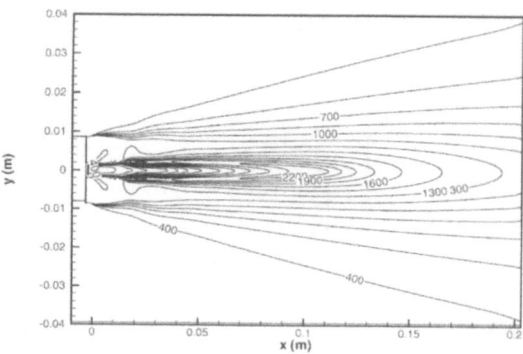

Fig. 3. Calculated temperature contours ($\Delta T = 150\,\mathrm{K}$) with assumed pdf closure.

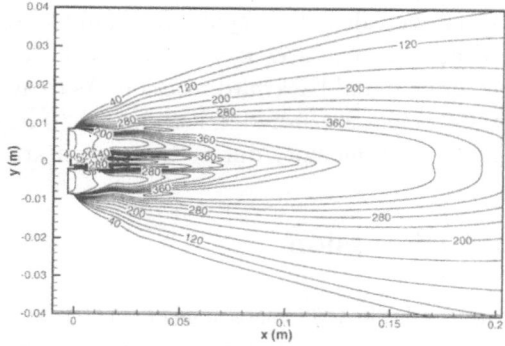

Fig. 4. Calculated rms of temperatur ($\Delta T = 40\,\mathrm{K}$) with assumed pdf closure.

shortly after ignition took place. Thus the flame still shows a ring-like structure that is typical for diffusion flames. The simulation is performed using a clipped Gaussian distribution for temperature and the Girimaji β-pdf for species composition. Numerical results are plotted by solid lines and experimental ones are given by symbols. The overall agreement between experiment and simulation is quite good if the complexity of the supersonic flow is taken into account . However there are still some minor problems with temperature fluctuations in the outer region. The greatest discrepancy between experiment and simulation is observed in the rms values of the OH molar fractions. A very important point for supersonic combustors is the ignition delay. For this freestream experiment the numerically predicted point of ignition agrees very well with the experimentally obtained location.

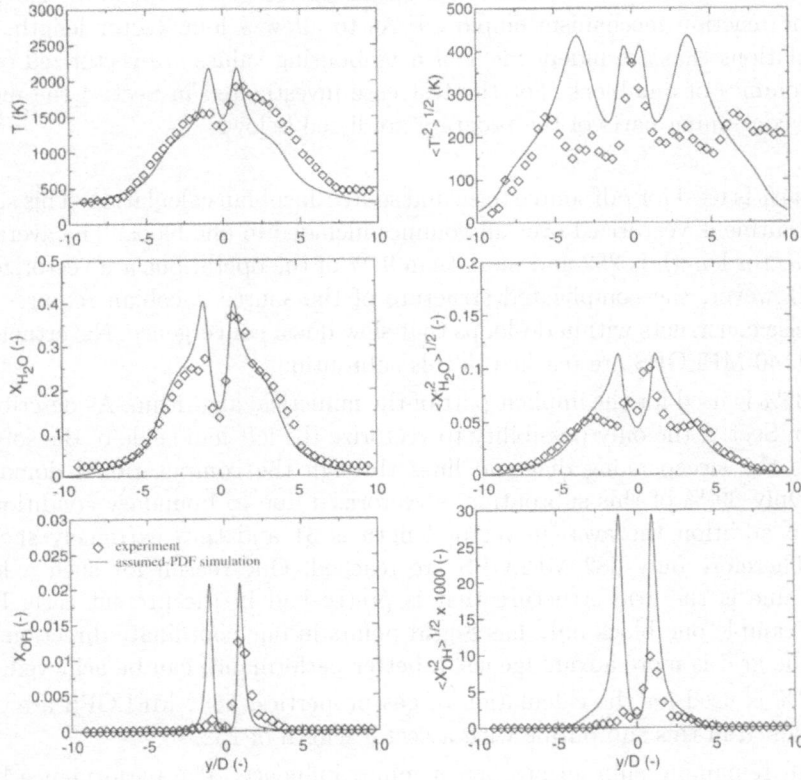

Fig. 5. Mean values (left) and rms (right) of temperature, H_2O, and OH molar fractions at $x = 51.74\,\mathrm{mm}$ downstream of the injector.

5   Performance

Due to the moderate memory requirements for 2D simulations all calculations are performed on a single processor of the NEC SX-5. For comparison the same simulations are also done on the older NEC SX-4. In the present case it is very disadvantageous that the vector length for some blocks is extremely short. Therefore performance data is also given for another similar test case with a more favourite mesh size. For simulation analysis the program FTRACE has been used on both machines, NEC SX-4 and NEC SX-5.

5.1   Simulation on the NEC SX-5

The assumed pdf modeling is very time consuming due to a complex source term calculation and complicated source term Jacobians. The average CPU time for a simulation with pdf modeling is roughly doubled in comparison to laminar chemistry. However these values also strongly depend on the size of the reaction mechanism employed. As to allow a long vector length, all simulations that are independent of neighbouring values are vectorized over all volumes of one block. For the test case investigated in Sect. 4 the most time consuming parts of the program are listed below:

- 44% is used for pdf source term and source Jacobian calculation. This subroutine is vectorized over all volumes included in one block. The average vector length is 252 and more than 97% of the operations are vectorized. However, the complicated structure of the source Jacobian requires if-else constructs within do-loops that slow down convergence. Nevertheless 1140 MFLOPS are reached in this subroutine.
- 36% is used for the implicit part of the numerical algorithm. As described in Sect. 2 the only possibility to vectorize the left hand side of the solver is the sweep along diagonal lines through the computational domain. Only 89 % of this subroutine is vectorized due to boundary conditions. In addition the average vector length is 31 and thus extremely short. Therefore only 382 MFLOPS are reached. One reason for such a low value is the grid structure that is pretty bad in the present case. For example one block only has 8 grid points in one coordinate direction. If the grid is more advantageous, a better performance can be achieved.
- 7% is used for the calculation of gas properties. 1852 MFLOPS are obtained in this subroutine with a vector length of 252.
- all remaining subroutines are of minor influence. The performance lies between those stated above.

The highest performance obtained is 3294 MFLOPS but the subroutine is unimportant. The average values of the total program are 874 MFLOPS and a vector length of 94. Because only 400 iterations have been analysed input and output still has some influence aggravating these results. If the complexity of the implicit scheme is considered the overall performance is quite good

5.2   Comparison NEC SX-5 against NEC SX-4

As described in the section before, most time consuming parts are the pdf subroutines, the impicit solver and the gas property calculation. In comparison to the NEC SX-4 the number of MFLOPS increases for the NEC SX-5 by a factor of 1.75 for the pdf calculations, 1.61 for the implicit solver, and 1.04 for the gas property calculation. The vectorized operation ration as well as the vector length is the same on both machines. For the total program the performance increased by a factor of 1.73. The achieved peak performance of 1556 MFLOPS on the NEC SX-4 was less than half the corresponding performance on the NEC SX-5.

5.3   Grid dependence of performance parameters

Due to the short vector length in the implicit part of the the numercal algorithm a strong dependence of performance parameters on the computational grid is expected. The test case investigated above was disadvantageous due to blocks with extremely small grid sizes in one coordinate direction. Therefore a second similar test case has been investigated on the NEC SX-5. The simulation is done with the same reaction mechanism and the same number of conservation equations. Thus the only difference between both calculations is the computational grid. Again a 5 block grid has been used with now 264×40, 216×56, 272×72, 136×72, and 176×64 volumes, respectively. This causes a shift in the importance of the different subroutines. While the improvement for the pdf subroutine is negligible the implicit solver achieves an improvement by a factor of 1.51. The average vector length is increased from 30.4 to 46.6. In total the improvement is not as strong as expected. With 955 MFLOPS the average performance is only increased by about 10%.

Acknowledgments

We wish to thank the *Deutsche Forschungsgemeinschaft (DFG)* for financial support of this work within the Collaborative Research Center SFB 259 at the University of Stuttgart.

References

1. Pope, S. B.: PDF Methods for Turbulent Reactive Flows. Prog. Energy Comb. Sci., **11**, (1985) 119–192
2. Girimaji, S. S.: A Simple Recipe for Modeling Reaction-Rates in Flows with Turbulent Combustion. AIAA paper 91-1792 (1991)
3. Girimaji, S. S.: Assumed β-pdf Model for Turbulent Mixing: Validation and Extension to Multiple Scalar Mixing", Comb. Sci. Techn., **78**, (1991) 177–196
4. Baurle, R. A., Alexopoulos, G. A., Hassan, H.A.: Modeling of Supersonic Combustion Using Assumed Probability Density Functions. J. Prop. Power **10** (1994) 777–786
5. Möbus, H., Gerlinger, P., Brüggemann, D.: Comparison of Eularian anf Lagrangian Monte Carlo PDF Methods for Turbulent Diffusion Flames. Comb. Flame **124** (2001) 519–534
6. Gerlinger, P., Stoll, P., Brüggemann, D.: An Implicit Multigrid Method for the Simulation of Chemically Reacting Flows. J. Comp. Phys. **146** (1998) 322–345
7. Gerlinger, P., Möbus, H., Brüggemann, D.: An Implicit Multigrid Method for Turbulent Combustion. J. Comp. Phys. **167** (2001) 247–276
8. Jachimowski, C. J.: An Analytical Study of the Hydrogen-Air Reaction Mechanism with Application to Scramjet Combustion. NASA TP 2791, (1988)
9. Coakley, T. J., Huang, P. G.: Turbulence Modeling for High Speed Flows. AIAA paper 92-0436 (1992)

10. Gerlinger, P., Brüggemann, D.: An Implicit Multigrid Scheme for the Compressible Navier-Stokes Equations with Low-Reynolds-Number Turbulence Closure. J. Fluids Eng. **120** (1998) 257–262
11. Jameson, A. , Yoon, S.: Lower-Upper Implicit Schemes with Multiple Grids for the Euler Equations. AIAA J. **25** (1987) 929–937
12. Shuen, J. S.: Upwind Differencing and LU Factorization for Chemical Non-Equilibrium Navier- Stokes Equations. J. Comp. Phys. **99** (1992) 233–250
13. Gaffney, R. L. Jr, White, J. A., and Girimaji, S. S.: Modeling Turbulent Chemistry Interactions Using Assumed PDF Method. AIAA Paper 92-3638 (1992)
14. Cheng, T. S., Wehrmeyer, J. A., Pitz, R.w., Jarrett, O. Jr., and Northam, G. B.: Raman Measurements of Mixing Modeling and Finite-Rate Chemistry in a Supersonic Hydrogen Air Diffusion Flame. Comb. Flame **99** (1994) 157-173

A 3D Hydrodynamic Simulation for the Cygnus A Jet as a Prototype for High Redshift Radio Galaxies

Martin Krause* and Max Camenzind

Landessternwarte Königstuhl,
69117 Heidelberg, Germany

Abstract. We report the 3D simulation of a hydrodynamic jet with parameters matched to the radio galaxy Cygnus A. For this simulation, a cylindrical grid is used in order to save computational resources. The jet is injected in pressure equilibrium into a King type cluster atmosphere with slight random modifications in order to break the symmetry. The jet is simulated as a bipolar outflow with the back-flows allowed to interact with each other. Inward motion of the shocked external medium is observed in the symmetry plane, which is clearly visible in the plot of the derived bremsstrahlung emission. Based on these numerical results, we attempt to explain the spiral-like x-ray structures observed in Cygnus A. We propose that they are fingers of included and expanding shocked external medium. These results are extrapolated to higher redshift radio sources, which are thought to reside in even denser environments. Here, we propose that the same mechanism, as observed in this simulation, could explain the large amounts of cool line emitting gas within the hot radio bubble. Mainly due to hardware problems at HLRS, we were not able to compute the model upto the desired extention. The results are therefore preliminary.

1 Astrophysical Introduction

1.1 High Redshift Radio Galaxies (HZRG)

Extragalactic jets emerge from the cores of bright elliptical galaxies [2,3]. This is the result of extensive observations on neighboring radio galaxies (see e.g. [3]). Radio galaxies have, however, been found upto redshifts of 5. Many characteristic properties of these progenitors are not known of in any detail. According to our present understanding of the formation of galaxies, present–day bright ellipticals are not the product of a late normal merging sequence of fainter galaxies. Indeed, new results obtained by the Hubble Space Telescope (HST) suggest that early radio galaxies, which still consist of individual gas clumps, are the progenitors of the bright ellipticals (in the sense of proto–CD galaxies). Super–massive Black Holes with masses upto a few billion solar masses have been found in the centers of these bright ellipticals. These dark masses are probably formed in the early evolution of the core region of these

---

* email: M.Krause@lsw.uni-heidelberg.de

galaxies at redshifts > 4. The Black Holes are thought to be the main drivers for the jets, which are the sources of the radio emission.

Radio galaxies and quasars, which are thought to be very similar to radio galaxies but seen at smaller viewing angles, with their jets therefore constitute an important diagnostic tool for examining the state of matter in the early Universe. These sources, in general, show four distinct components in their spectrum:

1. the infrared–optical–UV continuum of the central source (emission from the accretion disk and its surroundings);
2. the radio–optical continuum of the jets (synchrotron emission from relativistic particles in the jet);
3. the X–ray and γ continuum (inverse Compton emission from the inner jets);
4. narrow and broad emission lines from heated gas.

The analysis of these components will provide constraints on the emission mechanisms, the state of the thermal plasma, or contributions from stellar and non-stellar sources. This information is essential for the understanding of formation and evolution of galaxies and their active nuclei in the early Universe.

More than 150 radio galaxies are now detected at high redshift (> 2), half of them within the last three to four years [6]. The redshift record for radio galaxies is now above five [1]. Many of these sources are located in the center of young clusters of galaxies because of their large rotation measures. Their redshift corresponds to a cosmological lookback–time of at least 80% the age of the Universe. This observed redshift range also corresponds to the maximum in the redshift distribution of quasars. The spatial density of these objects was at that time about 300 times higher than today.

1.2   Interaction of HZRG Jets with the External Medium

Another observed property of these objects is important for our understanding of the early Universe. Distant radio galaxies often show huge haloes of ionized gas extending to 150 kiloparsecs and having a velocity dispersion of 700 to 1000 km/s. These emission line regions are quite clumpy, their luminosities reaching up to 10^{37} Watts. The kinematics turns out to be complex [12]. The observed characteristics show that the radio sources have quite a big impact on the emission line regions. The correlation between the extent of the radio structure with velocity dispersion and extent of the Lyman α emission is evidence for an interaction of the jet and the ionized matter. Such interactions result from time to time in a bending of the radio jet. Properties of the observed Lyman α line profiles imply gas masses of the haloes of about 10^9 solar masses. This gas might consist of 10^{12} clouds with a diameter of about the size of the solar system [12].

With the beginning of the operation of the Chandra x-ray satellite on August 19, 1999, spatially resolved images of HZRGs in the x-ray regime became available [6,9]. In the few cases observed so far, the extended x-ray emission is aligned with the radio axis and the spectrum is well fit by a thermal bremsstrahlung model with plasma temperatures of $\approx 10^8$ K. These sources also show aligned Lyman α emission. This is an unavoidable consequence when extragalactic jets encounter a dense medium. As we are going to show in detail in a future publication, bremsstrahlung and Lyman α emission arise together in the shocked extragalactic medium behind the bow-shock of a propagating jet. In this shock the intergalactic medium (IGM) is heated to $\approx 10^8$K. In this temperature regime cooling via bremsstrahlung is particularly effective. When the density is > 1 cm^{-3}, the shocked medium can cool in less than one million years to $\approx 10^5$ K. Since cooling also increases the density, which in turn increases the cooling rate, a thermal instability sets in. In regions where the gas has cooled to $< 10^6$ K, cooling by line emission becomes important and the gas is cooled further to $\approx 10^4$ K. Below this temperature, the cooling curve drops very fast.

1.3   The Special Case of Cygnus A

The radio galaxy Cygnus A can be considered – with some restrictions – as a prototype for HZRGs: It is located in a comparatively dense environment (n = $(10^{-2} - 10^{-1})$ cm^{-3}), and emits on a high power level at radio frequencies, comparable to much higher redshift sources [4,5]. The cooling time t_c against thermal Bremsstrahlung in the Cygnus A environment is considerably longer than in typical HZRG environments:

$$t_c = 1.6 \left(\frac{v_s}{100 \text{ km/s}} \right) \left(\frac{1 \text{ cm}^{-3}}{n_{ext}} \right) \text{Mio.yrs} .$$

Here v_s is the bow-shock velocity and n_{ext} is the number density of the undisturbed external medium. With a jet velocity of $v_{jet} \approx 0.4\, c$ (c: speed of light) [5], implying – according to momentum balance – a bow-shock velocity of:

$$v_s \approx \sqrt{\epsilon\eta}\, v_{jet} \approx 1000 \text{ km/s} ,$$

where $\eta = \rho_{jet}/\rho_{ext} \approx 10^{-3}$ is the density contrast between jet matter and external medium and $\epsilon = A_{jet}/A_{head} \approx 0.1 - 0.3$ is the relative increase of the jet head's area, a cooling time of $t_c \approx 320$ Mio. yrs is derived. This indicates that during the radio source lifetime of a few times 10^7 years, bremsstrahlung can radiate some 10^{45} erg/s ($\approx 10\%$ of the total energy content of the shocked external medium), which is sufficient to be well detected, but still much less than needed in order to initiate line emission. Therefore, cooling by thermal bremsstrahlung can be studied in this source without the effects of cooling by line emission. Even the back-reaction of the cooling on the hydrodynamics of the underlying flow is negligible. Therefore, it is justifiable to compute a pure hydrodynamic model and infer the emission afterwards from the computed hydrodynamic variables. Cygnus A has the further advantage that it is located relatively nearby (redshift $z = 0.0562$, corresponding to a luminosity distance of ≈ 800 million light years). Therefore, detailed x-ray observations with the Chandra observatory are available [14].

In the present contribution, we report a 3D hydrodynamic simulation of the jet in the Cygnus A radio galaxy. Our model includes a hot cluster atmosphere in hydrostatic equilibrium. Because we pay special attention to the behavior of the shocked cluster gas, we have computed the jet propagation in both directions from the galaxy center. This removes unphysical effects from the boundary conditions on the side of the jet inflow in virtually all previous jet simulations known to the authors. The symmetry is broken by a certain increase of the density at random points in the external medium, where the right-hand side is assigned to a higher probability and amount.

2   Numerics

2.1   The extended *NIRVANA* code (*NIRVANA_CP*)

For the computation in this contribution, the magneto-hydrodynamic (MHD) code *Nirvana_CP* was employed. The main part of this code (*NIRVANA*) was written by Udo Ziegler [15]. In that version, it solves the MHD equations in three dimensions (3D) for density ρ, velocity $\boldsymbol{v}$, internal energy e, and magnetic field $\boldsymbol{B}$:

$$\frac{\partial \rho}{\partial t} + \nabla \cdot (\rho \boldsymbol{v}) = 0 \tag{1}$$

$$\frac{\partial \rho \boldsymbol{v}}{\partial t} + \nabla \cdot (\rho \boldsymbol{v} \boldsymbol{v}) = -\nabla p + \frac{1}{4\pi} (\boldsymbol{B} \cdot \nabla) \boldsymbol{B} - \frac{1}{8\pi} \nabla \boldsymbol{B}^2 \tag{2}$$

$$\frac{\partial e}{\partial t} + \nabla \cdot (e \boldsymbol{v}) = -p \, \nabla \cdot \boldsymbol{v} \tag{3}$$

$$\frac{\partial \boldsymbol{B}}{\partial t} = \nabla \times (\boldsymbol{v} \times \boldsymbol{B}) \ . \tag{4}$$

NIRVANA can be characterized by the following properties:

1. explicit Eulerian time–stepping,
2. operator–splitting formalism for the advection part of the solver,
3. method of characteristics–constraint–transport algorithm to solve the induction equation and to compute the Lorentz forces [10];
4. artificial viscosity has been included to dissipate high–frequency noise and to allow for shock smearing in case the flow becomes supersonic.

The code was upgraded to *NIRVANA_C* and extensively tested by Markus Thiele [11] in order to take into account optically thin cooling. Consequently, (3) was replaced by

$$\frac{\partial e}{\partial t} + \nabla \cdot (e \boldsymbol{v}) = -p \, \nabla \cdot \boldsymbol{v} - \mathcal{K} \ . \tag{5}$$

There is also an additional equation due to generation and depletion of individual atomic species, with masses m_α, by ionization and recombination:

$$\frac{\partial \rho_\alpha}{\partial t} + \nabla \cdot (\rho_\alpha v) = k_\alpha \ . \tag{6}$$

$\mathcal{K}$ and k_α are computed by the following equations:

$$\mathcal{K} = -\sum_{\alpha=0}^{N} \sum_{\beta=0}^{N} \frac{\rho_\alpha}{m_\alpha} \frac{\rho_\beta}{m_\beta} \Lambda_{\alpha \beta} \tag{7}$$

$$k_\alpha = \sum_{\alpha=0}^{N} \left(m_\alpha \sum_{\beta=0}^{N} \left(\sum_{\gamma=\beta}^{N} \frac{\rho_\beta}{m_\beta} \frac{\rho_\gamma}{m_\gamma} \mathrm{IF}^+_{\alpha \beta \gamma} - \rho_\alpha \frac{\rho_\beta}{m_\beta} \mathrm{IF}^-_{\alpha \beta} \right) \right) \ , \tag{8}$$

where the Λ and IF functions summarize the details of atomic physics and are functions of temperature and, in general, also of the number densities ρ_α/m_α.

Recently, this code was further upgraded to *NIRVANA_CP* by Markus Thiele and one of the authors (Martin Krause). This upgrade consisted in vectorization and parallelization for use on a NEC SX-5 computer by OPEN_MP like methods provided by the standard NEC C compiler for this machine. The code was revised and fine tuned in order to achieve the maximum possible performance for the given hardware. At the time of writing, all parts of the code were working and tested. Nevertheless, fine tuning for optimization of the magnetic and cooling parts was still in progress.

2.2 Test Calculations

The reliability of the code was tested by computing test problems with the new code version on SX-5 and comparing the result to the output of the old – sufficiently validated – version run on a Linux PC. Since in the following we are concerned with only hydrodynamic calculations, we report here a hydrodynamic test problem with one species.

We have computed the one-dimensional Sod shock problem, which even has a semi-analytic solution [8]. We have used a grid of 4095 cells in order to get the maximum possible vectorization efficiency. This grid was initially divided into two areas. On the left-hand side, density and temperature were set to one, and on the right-hand side they were set to 0.125 and 0.8, respectively. The velocity was zero, everywhere, and the adiabatic index 1.4. Due to the ten times higher pressure on the left hand side, a shock wave develops. The computation was carried out both on a Linux PC and on the NEC SX-5 at the High Performance Computing Center in Stuttgart (HLRS). The result after 6000 timesteps and a physical time of $2.3 \ 10^{-5}$ is shown in Fig. 1 Differences between the two computations could be found only in the velocity in regions where it is close to zero. This velocity differences of the order $\Delta v \approx 10^{-12}$ should reflect the numerical truncation error. The test calculation also demonstrates the high suitability of *NIRVANA_CP* in the handling of shock waves.

As is shown in Table 1, we achieved a high degree of vectorization. 85% of the user time the code was in vector mode, the vector operation ratio was 99.3%, and with the one SX-5 processor we used here, we achieved a FLOP rate of 1.0 GFLOPS. This is the value we expected, given the maximum performance of 4 GFLOPS, which only can be achieved if the instructions are especially fine tuned. The high vectorization degree also was indicated by the compiler report: almost every significant loop was vectorized.

Table 1. Profile Information for Sod Test

****** Program Information ******			
Real Time (sec) :	12.189758	User Time (sec) :	8.827777
Sys Time (sec) :	0.628303	Vector Time (sec):	7.538145
Inst. Count :	306726730	V. Inst. Count :	110947808
V. Element Count :	27955444095	FLOP Count :	8572226294
MOPS :	3188.936809	MFLOPS :	971.051522
A.V. Length :	251.969323	V. Op. Ratio (%):	99.304546
Memory Size (MB):	48.000000	MIPS :	34.745636
I-Cache (sec) :	0.314695	O-Cache (sec) :	0.043142
Bank (sec) :	0.000112	:	

3   Simulation Setup

A cylindrical grid was used for the jet simulation (see Fig. 2). The computational domain is described by cylindrical coordinates: $Z \in [-6.9\,\text{kpc}, 6.9\,\text{kpc}]$, $R \in [0, 5.7\,\text{kpc}]$ and $\phi \in [0, 2\pi]$. 255, 107 and 73 grid points were used in the Z,R and ϕ directions, respectively. We have a resolution of 10 points per beam radius (ppb) in the R and Z direction and an R dependent resolution in the ϕ direction of 12 ppb at the jet boundary. This is a very poor resolution for a jet simulation. For example, we expect the jet to move 20% faster then in a converged simulation. However, for a study of the external medium it should be sufficient [7]. We connect the grid in the ϕ direction by a special kind of periodic boundary condition:

$$f(Z, R, 0) = f(Z, R, N_\phi^{\max} - 2) \tag{9}$$
$$f(Z, R, N_\phi^{\max} - 1) = f(Z, R, 1) \tag{10}$$
$$f(Z, R, N_\phi^{\max}) = f(Z, R, 2) \ . \tag{11}$$

Here $N_\phi^{\max}$ denotes the maximum cell index in the ϕ direction and f stands for every hydrodynamic variable. As usual with *NIRVANA*, two boundary cells

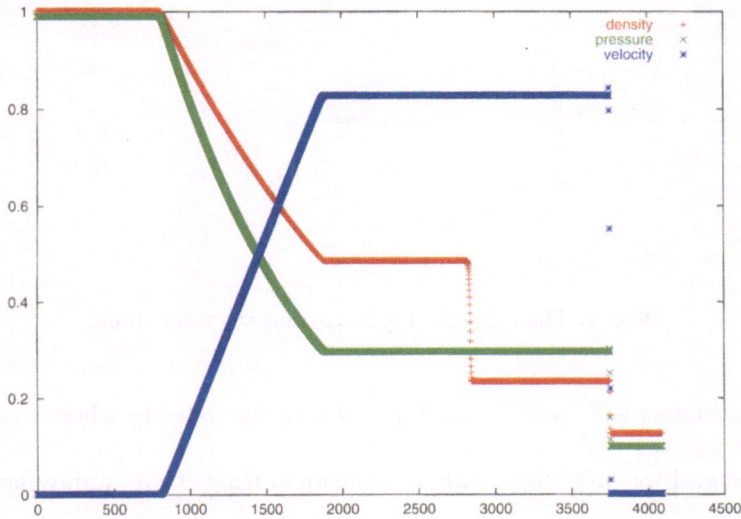

Fig. 1. Density, pressure and velocity for the Sod shock test problem after 6000 timesteps, corresponding to a physical time of $2.3\ 10^{-5}$. The output from the SX-5 computation is plotted, but the output from the PC would be indistinguishable on this graph. The units are normalized in order to fit in the graph, and on the horizontal axis, grid points are indicated.

are used on the upper side and one cell on the lower side. These conditions should implicate a smooth variation of the hydrodynamic variables across the ϕ boundary. Special care should also be applied to the boundary condition on the axis. In the present version of the code we still use conditions for rotational symmetry. But in the near future this also will be replaced by a connection boundary condition of the following type:

$$f(Z, 0, \phi) = f(Z, 2, (\phi + \pi)\ \mathrm{mod}(2\pi))\ , \tag{12}$$

with the required modifications for the vector variables due to antisymmetry and the staggered mesh. There are several advantages of the cylindrical grid:

1. Since the jet itself has cylindrical geometry, the required volume for the computational domain is reduced by $1 - \pi/4 \approx 22\%$ (for a grid of equal extension in R and positive Z direction).
2. The grid points are naturally concentrated in the area of the jet beam. This is preferable if no cooling of the external medium is taken into account or if the cooling length there is long.
3. There is one special axis (the Z axis) with much more points than the two other directions. This increases the performance for vectorization, since only inner loops – and therefore one of the directions – are vectorized.

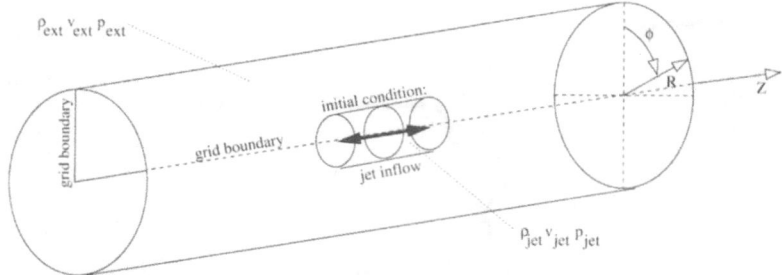

Fig. 2. The cylindrical grid used for the simulation.

This technique will make it finally possible to simulate the whole volume of a radio galaxy.

The grid was initialized with an isothermal King cluster atmosphere:

$$\rho(r) = \frac{\rho_0}{\left(1 + \left(\frac{r}{r_c}\right)^2\right)^{3\beta/2}} , \tag{13}$$

where $r = \sqrt{R^2 + Z^2}$ denotes the spherical radius, $\rho_0 = 1.2 \; 10^{-25}$ g/cm^3 the characteristic density, $\beta = 0.75$ and $r_c = 35$ kpc is the core radius. In order to break the bipolar and axial symmetry, this density profile was modified by random perturbations of the following kind:

1. With 10% probability the density was increased by a factor between 1 and 1.4.
2. With 5% probability the density was increased by a factor between 1 and 2, only if the cell was unmodified in the first process and the Z coordinate was positive.

The jet is injected in the middle of the grid in the region $R \in [0, 0.55 \text{ kpc}]$, $\phi \in [0, 2\pi]$, and the first two cells in the positive and negative Z direction, respectively. This region is treated as a further boundary and has the constant values: $\rho_{\text{jet}} = 6.68 \; 10^{-28}$ g/cm^3, $v_Z = \pm 0.4c$, where c denotes the speed of light, and the plus sign applies for the positive Z region and the minus sign for the negative one, and the pressure was set in order to match the external pressure at that position. This gives a slightly varying density contrast across the grid of $\eta = \rho_{\text{jet}}/\rho_{\text{ext}} \approx 7 \; 10^{-3}$ and an internal Mach number $M = 9.8$. The adiabatic index used here is $\gamma = 5/3$, for a nonrelativistic gas.

4 Results

Because of the small grid size used (288 MB), the vector registers could be filled only once each time an inner loop was encountered. This decreases the performance somewhat to the measured value of 753 MFLOPS. The result of the computation is shown in Figs. 4a & 4b. In the polar plot one can see the smooth variation of the density across the ϕ boundary. If one looks closely on the axis region, one can discover an artificial discontinuity there. This is due to the rotational symmetry boundary condition, and is to be replaced as discussed above. The development of the shocked external medium seems to be unimpressed by the troubles on the axis. The striking result of this simulation is that this material is completely forced to move inward in the vicinity of the "symmetry" plane. No deformation of the bow-shock outwards is detected. Although the contact discontinuity between the shocked external and jet medium seems to be quite stable, the interaction of the back-flows in the symmetry plane seems to pull shocked external medium into the radio cocoon. Here it can in principle be reaccreted by the active nucleus. This can, of course, not be explored with the present simulation. The yellow regions in Figs. 4a & 4b represent density values, that are a few times less then the external medium and about a hundred times more than the jet plasma. This indicates thinned out shocked external medium, since in a simulation without cooling, jet plasma can hardly get that dense. We therefore find finger-like structures of shocked external medium in between the cocoon of shocked jet plasma. Figure 3 shows a cut through such a region. It shows that in the fingers not only the density but also the pressure is enhanced. Because of this, the fingers expand adiabatically and become thinner as can be seen in Fig. 4a when following such a structure inwards. We also added a plot of the derived emission due to bremsstrahlung (Fig. 4d). Here, one can immediately see the enhancement at the edges. The adiabatic expansion of the above mentioned fingers suppresses their length somewhat compared to the density plot. But still they have an extention of ≈ 3 jet radii.

5 Discussion

The results of the previous section confirm the findings of [7]. We find the same x-ray cavity within the radio bubble and the enhancement behind the bow shock surface. However, the sheets discussed above constitute a two phase medium, which might explain the structures of strong x-ray emission within the radio bubble of Cygnus A (the yellow spiral-like structures in Fig. 4d). We are inconclusive about this interpretation, given the early state of the simulation. But, if this mechanism really works, it is likely to be also present in HZRG. There, the density of the external medium is considerably higher (see introduction). Therefore the shocked external medium, which enters the radio bubble, can cool very fast to temperatures, where it becomes

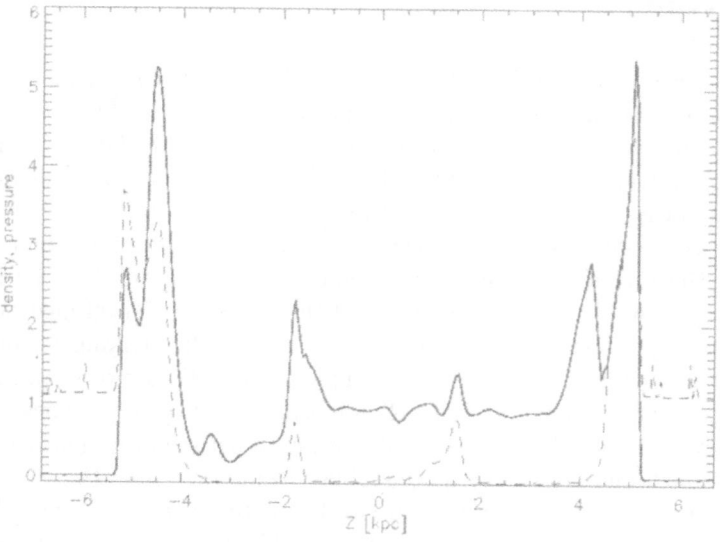

Fig. 3. Cut at $\phi = \pi/2$ and $R = 2.5$ kpc. The density (dashed line) is shown in units of 10^{-25} g/cm^3, the pressure (solid line) in units of 10^{-8} dyne/cm^2.

a strong Lyman α emitter. The cooled medium could then be heated by the UV radiation of the central active nucleus, and thus keep a high emissivity level for all the active phase of the system. The advantages of such a scenario would be:

1. New polarization data [13] indicate that in HZRG emission lines both, reflected light from the hidden active nucleus and direct emission, is roughly equally important. The proposed scenario also features both components.
2. Emission from regions where the nucleus can hardly illuminate is easily explained.
3. The mystery how dense cool line emitting gas comes into the radio bubble is explained.

This computation was intended as an input for a 10 times larger grid in R and Z directions. This is actually needed because the real radio galaxy Cygnus A has that extension. However, this was made impossible for now due to the recent severe hardware problems at HLRS. That problems damaged the data published here, so it cannot be used as input for a bigger grid. Netherthless, we think that the damage done to a few pixels spread throughout the grid does not inhibit the discussion of the so far produced data too much. For this computation 57.7 hours of CPU time and one processor of the NEC SX-5 were needed. We can now estimate the total CPU time T_{CPU} needed if one carries out this computation with a grid which grows with the jet: $T_{\mathrm{CPU}} = \int_0^{N_{\mathrm{ts}}^{\mathrm{max}}} \delta N_{\mathrm{cells}} dN_{\mathrm{ts}}$. Here, $\delta = 4.83\ 10^{-10}$ h/cell/timestep is the required CPU

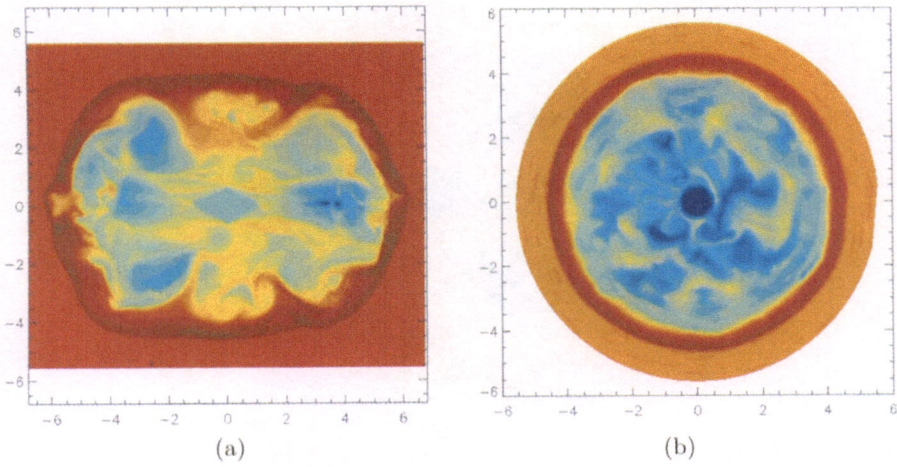

Fig. 4. (a) R-Z slice of the logarithmic density at a time of 0.66 Mio. yrs and $\phi = \pi/2$. As in b and c, red means high and blue means low. (b) R-ϕ slice of the logarithmic density at the same time as in a and $Z = -0.32$ kpc. Axis units are in kpc in a and b.

time per grid cell and timestep, $N_{cells} = 5.53 \ 10^{-4} N_{ts}^2$ cells/timestep^2 is the number of grid cells and N_{ts} is the timestep number. The values are measured in the present computation. It follows, that:

$$T_{CPU} \approx 8.9 \ 10^{-14} \cdot N_{ts}^3 \ \text{CPUhours} .\tag{14}$$

Therefore, for the computation of the full simulation, one needs about 20000 hours of CPU time. Even with 10 processors, the computation would take 80 days.

6   Future Work

With the hydrodynamic simulation discussed in this paper we could already enlighten some aspects of the shocked external medium in radio galaxies in a dense environment. However, we expect to get a considerably deeper understanding if magnetic fields, and especially, self consistent optically thin cooling is included. We expect to accomplish the optimization of these parts of *NIRVANA_CP* in the very near future. The progress of this project is much dependent on the further grant of CPU time since these computations turned out to be very time consuming.

Acknowledgments

This work was supported by the Deutsche Forschungsgemeinschaft (Sonder-forschungsbereich 437).

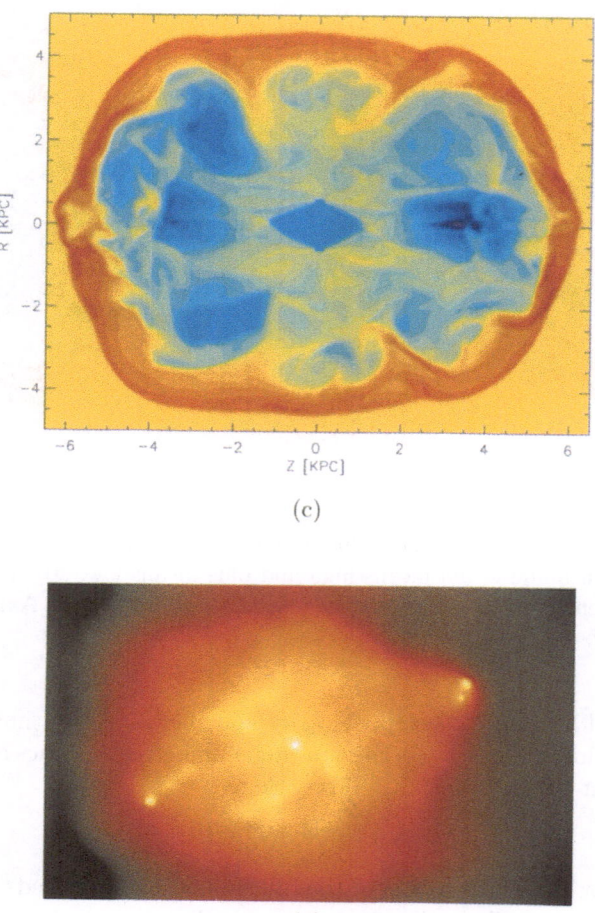

(c)

(d)

Fig. 4. (continued) (**c**) Emission due to bremsstrahlung of the same slice as in a. The shocked external material extents ≈ 3 jet radii into the cocoon. (**d**) X-ray image of the radio galaxy Cygnus A. The point sources the core and the two hot spots in the radio lobe. Credit: NASA/UMD/A.Wilson et al.

References

1. van Breugel et.al. 1999, in 'The Most Distant Radio Galaxies', eds: Röttgering, Best, Lehnert, p.49
2. Camenzind, M.: 1997, *Les noyaux actifs de galaxies*, Lecture Notes in Physics **m46**, Springer-Verlag (Heidelberg)
3. Camenzind, M.: 1999, in: The radio galaxy Messier 87, H.-J. Röser, K. Meisenheimer (eds.), Lecture Notes in Physics **530**, Springer-Verlag (Heidelberg), p. 252
4. Carilli, C.L., Perley, R.A., Harris, D.E.: 1994, Monthly Notices of the Royal Astronomical Society, **270**, 173

5. Carilli, C.L., Barthel, P.D.: 1996, The Astronomy and Astrophysics Review **7**, 1-54
6. Carilli et. al., 2000: in Gas & Galaxy Evolution, eds: Hibbard, Rupen & Gorkom
7. Clarke, D.A., Harris, D.E., Carilli, C.L.: 1997, Monthly Notices of the Royal Astronomical Society, **284**, 981
8. Courant & Friedrichs, 1948: *Supersonic flow and shock waves*, Interscience Publishers, New York
9. Fabian, A. C. Crawford, C. S., Ettori, S, Sanders, J. S.: 2001, Monthly Notices of the Royal Astronomical Society, **322**, L11
10. Hawley, J.F., Stone, J.M.: 1995, Comp. Phys. Comm. **89**, 127
11. Thiele, M.: 2000, Ph.D. thesis, University of Heidelberg, FRG
12. van Ojik, R., Röttgering, H.J.A., Miley, G.K., Hunstead, R.W.: 1997, Astronomy & Astrophysics **317**, 358
13. Vernet J., Fosbury R. A. E., Villar-Martýn M., Cohen M. H., Cimatti A., di Serego Alighieri S., 2001: Astronomy & Astrophysics, **366**, 7
14. Wilson, A.S., Young, A.J., Shopbell, P.L.: 2001 in: Proceedings of the Oxford Conference "Particles and Fields in Radio Galaxies", eds.: Laing, R.A. and Blundell, K.M., ASP Conference Series
15. Ziegler, U., Yorke, H.: 1997, Comp. Phys. Comm. **101**, 54

Parallel Computation of the Time Dependent Velocity Evolution for Strongly Deformed Droplets

M. Hase, M. Rieber, F. Graf, N. Roth and B. Weigand

Institute of Aerospace Thermodynamics, University of Stuttgart, Pfaffenwaldring 31, 70569 Stuttgart, Germany

Abstract. A fully three-dimensional numerical procedure has been used to predict the behavior of spherical and deformed droplets in a gas flow. The computational grid is moving with the droplet to minimize grid size and computation time. Numerical results of drag coefficients for spherical droplets show good agreement with literature data. The behavior of droplets with initially cylindrical or disk shapes has been compared with corresponding spherical droplets for different viscosities of the droplet liquid. For low viscosities the droplets are oscillating. For higher viscosities the initially strongly deformed droplets approach a spherical shape asymptotically. The influence of the strong initial deformation is shown. The simulation has been run on the Cray T3E/512-900 at the HLRS.

1 Introduction

The behavior of droplets moving with respect to the surrounding gas is very important for the description and modeling of sprays, heavily used in technical systems as for instance in automotive engines or gas turbines. In order to describe these processes the drag of the droplets or the development of the relative droplet velocity with time is of particular interest. In many cases the droplets are far away of being spherical. Strongly deformed droplets are observed during the disintegration of liquid sheets and ligaments into droplets. During the droplet transport they may be deformed due to the surrounding gas flow or due to collisions with a wall or other droplets.

The differences in the behavior between spherical and deformed droplets have been studied rarely in the past because of the strongly nonlinear effects in such processes. The present study presents a fully 3D approach to investigate these differences in detail.

2 Numerical Method

It is necessary to perform completely three-dimensional calculations to obtain reliable results on droplets in a surrounding gas flow. By including arbitrarily symmetry planes to the investigated problem, the vortex street behind the droplet cannot develop in a physical. This leads to wrong droplet velocities and wrong drag coefficients.

The problem under investigation has been attacked by the inhouse 3D CFD program FS3D (Free Surface-3D). The numerical program solves the Navier-Stokes equations for incompressible two-phase flows by using a Finite-Volume method with direct numerical simulation. The conservation equation of the liquid mass is solved using the Volume-of-Fluid method [1]. In this method an additional variable indicates if the fluid in a grid cell is liquid or gaseous. The computation of volume fluxes is based on a piecewise linear reconstruction of the interface (PLIC) [2]. Therefore the program allows the simulation of single droplets or small droplet groups, with strong deformations of the droplets [3-5]. The program FS3D has been parallelized using domain decomposition and the communication library MPI.

The two-phase flow field is described by the following equations for continuity and momentum

$$\nabla \cdot \mathbf{u} = 0, \tag{1}$$

$$\frac{\partial \rho}{\partial t} + \nabla \cdot (\rho \mathbf{u}) = 0, \tag{2}$$

$$\frac{\partial (\rho \mathbf{u})}{\partial t} + \nabla \cdot [(\rho \mathbf{u}) \otimes \mathbf{u}] = -\nabla p + \nabla \cdot \mu \left[\nabla \mathbf{u} + (\nabla \mathbf{u})^T \right] + \nabla \cdot \mathbf{T} \quad . \tag{3}$$

The last term in the momentum equation $\nabla \cdot \mathbf{T}$ accounts for surface tension effects. The capillary stress tensor $\mathbf{T}$ is defined only at the phase interface. The density and the viscosity are constant inside the gas and the liquid. Across the sharp interface separating liquid and gas the density and viscosity are computed with the liquid volume fraction f by

$$\rho(\mathbf{x}, t) = \rho_G + (\rho_L - \rho_G) f(\mathbf{x}, t) \tag{4}$$

$$\mu(\mathbf{x}, t) = \mu_G + (\mu_L - \mu_G) f(\mathbf{x}, t) \quad , \tag{5}$$

where the subscripts L and G refer to liquid and gas respectively. The advection of the volume fraction and thus of the discontinuity is governed by the transport equation

$$\frac{\partial f}{\partial t} + \nabla \cdot (\mathbf{u} f) = 0 \tag{6}$$

The simulations need to track the droplets in the gas stream for a longer time. In an inertial system the grid would be too large and the calculations would be therefore too expensive. To avoid these problems, the numerical code allows to move the grid with the center of mass of the droplet. The three-dimensional grid with about 500000 ($128 \times 64 \times 64$) control volumes used for the droplet simulations can be supposed to be a channel with quadratic cross-section and slip conditions at the side walls. The inlet of the channel has a spatially constant inlet velocity (x-direction), which equals the momentary mean velocity of the droplet in x-direction in an inertial reference system.

Hence the center of gravity of the droplet remains at the same x-position. To stabilize the droplet in y- and z-direction correction terms have been used, which are small in comparison to the other acceleration terms. The viscosity of the droplet liquid was chosen high enough to damp out possible resonance effects, which may result from these correction terms. At the outlet boundary a small but efficient damping zone was applied to avoid backflow into the system [6]. Using this combination of moving grid, stabilization in transverse directions and damping zone at the outlet boundary the droplet behavior can be simulated for long times using moderate grid sizes.

Typical simulation in the mentioned size takes place on 16 CPU's of the Cray T3E/512-900 with a domain of $32 \times 32 \times 32$ on each CPU. Depending on the viscosity the number of time steps differs between the simulations because of the explicit time discretization. An overview at the computational time needed for the simulations is given in the next section. In [7] the performance of the program FS3D has been analyzed in detail.

3   Results

The results shown here are for droplets with different initial shapes and different viscosities of the droplet liquid. The properties of the liquid were the same as for water, with density $\varrho_L = 1000 \, \text{kg/m}^3$ and surface tension $\sigma_L = 73 \cdot 10^{-3} \, \text{N/m}$ except for the viscosity which was n-times the viscosity of water $\mu_L = n \cdot \mu_{water} = n \cdot 1 \cdot 10^{-3} \, \text{kg/(ms)}$. The gaseous fluid was air with density $\varrho_G = 1.2 \, \text{kg/m}^3$ and viscosity $\mu_G = 18 \cdot 10^{-6} \, \text{kg/(ms)}$. In nearly all cases the volume of the liquid mass was the same as for a spherical droplet with the diameter of $1 \cdot 10^{-3} \, \text{m}$, the initial velocity of the droplet was $U_0 = 10 \, \text{m/s}$, and the droplet is accelerated by gravitational force in negative x-direction.

An impression of the flow field around a deformed droplet is given in Fig. 1. In Fig. 1a the velocity field around the droplet is shown and in Fig. 1b the vorticity field is presented. The damping zone starts at $x = 1.1 \cdot 10^{-2} \, \text{m}$, which is indicated by the vertical lines. On the right hand side of this line the flow field is damped to avoid backflow into the system.

Outside of the damping zone negligible influence on the flow field has been detected, as detailed studies have shown. The figure makes clear that the flow behind the droplet is fully three-dimensional and any assumptions on symmetries will cause wrong results.

First spherical droplets have been studied. From calculated temporal evolutions of the droplet velocity drag coefficients for different Reynolds numbers have been obtained. In these calculations the droplet size and the initial droplet velocity were different to the values given above to obtain a wide range of Reynolds numbers. The viscosity of the droplet liquid was chosen to be $\mu_L = 10 \cdot \mu_{water}$. A comparison between the numerical results and data from literature is given in Fig. 2. There is a good agreement between the calculations and the literature. It should be mentioned, that first calculations for droplets with a lower viscosity of the liquid show similar but slightly lower drag coefficients.

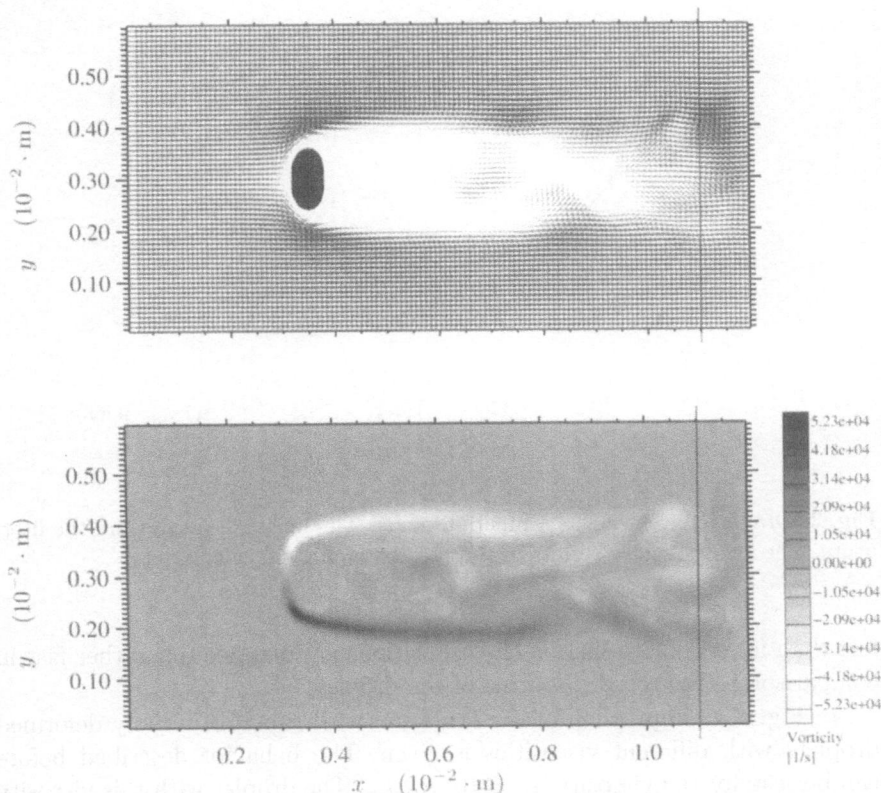

Fig. 1. Velocity field (Fig. 1a top) and vorticity field (Fig. 1b bottom) around a deformed droplet moving in air. Shown is a x,y-plane through the center of the droplet at the time $t = 0.00825$ s. The droplet had at the beginning of the calculations the shape of a circular disk with diameter $D_{d,0} = 1.805 \cdot 10^{-3}$ m. The viscosity of the droplet liquid is $\mu_L = 1000 \cdot \mu_{water}$.

In addition to initially spherical droplets two different initial shapes have been studied. One is a cylinder with diameter $D_{c,0} = 0.451 \cdot 10^{-3}$ m. The cylinder axis was set parallel to the direction of the initial velocity. The ratio of length to diameter has been chosen, that according to Rayleigh the cylinder will not disintegrate in two separate droplets [8]. The other initial shape is a circular disk with diameter $D_{d,0} = 1.805 \cdot 10^{-3}$ m. The direction of the initial velocity has been perpendicular to the flat side of the disk. The deformed droplets have the same volume as a spherical droplet with $D = 1 \cdot 10^{-3}$ m.

The evolution of the droplet shape with time depends on the viscosity of the droplet liquid. For low viscosity ($\mu_L = 10 \cdot \mu_{water}$) the droplet oscillates between prolate and oblate shape. This can be seen in Fig. 3. The initial velocity was $U_0 = 10$ m/s. This value is above the constant equilibrium velocity for a corresponding spherical droplet. For a prolate shape the deceleration is

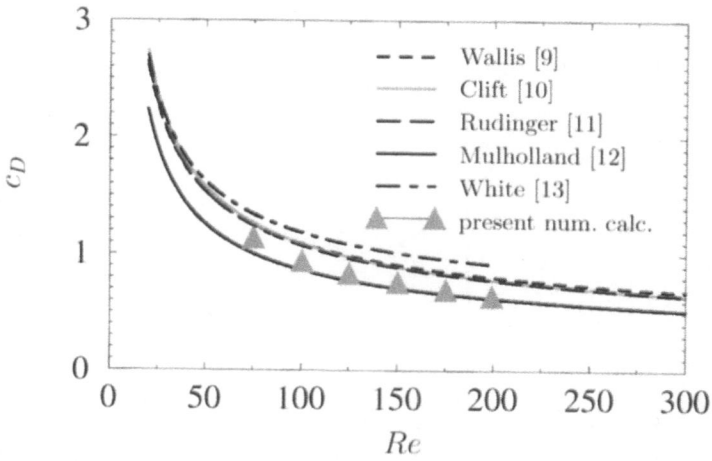

Fig. 2. Drag coefficient c_D as a function of Re. The lines are taken from the literature. The symbols show results of the present numerical calculations.

less than for an oblate shape. The oscillations are damped out rather fast in comparison to the relaxation time of the droplet.

In Fig. 4 an impression of the evolution with time for initially deformed droplets with different viscosities is given. The behavior described before can be seen for the viscosity $\mu_L = 10 \cdot \mu_{water}$. The droplet with this viscosity oscillates between prolate and oblate shape. The displayed droplet shapes are not the maximum deformation in the respective oscillation period. For the higher viscosity $\mu_L = 100 \cdot \mu_{water}$ the droplet is spherical between $t = 0.004\,s$ and $t = 0.006\,s$. The droplet with the viscosity $\mu_L = 1000 \cdot \mu_{water}$ is not oscillating. It approaches the spherical state asymptotically.

The evolution of the droplet velocity with time is shown in Fig. 5 for different initial droplet shapes and different viscosities of the droplet liquid. It has been found that for the medium viscosity $\mu_L = 100 \cdot \mu_{water}$ the droplets have a spherical shape after approximately 0.005 s for both initial shapes (cylindrical and disk). For the lower viscosity $\mu_L = 10 \cdot \mu_{water}$ and the higher viscosity $\mu_L = 1000 \cdot \mu_{water}$ it takes more than five times longer for the droplet to attain a spherical shape. In Fig. 5 the effect of the deformation on the droplet behavior can be seen. In the case shown here this effect is larger for higher viscosities. These results show that the drag coefficient changes with time not only due to the change of Reynolds number but also due to changing droplet shape.

In order to model the droplet behavior it is obviously important to know, when the deformed droplet reaches an approximately spherical shape and how large the influence of the deformation on the droplet velocity is. An indicator for sphericity is the droplet surface A of the deformed droplet compared with the surface A_0 of the corresponding spherical droplet. In Fig. 6 the ratio

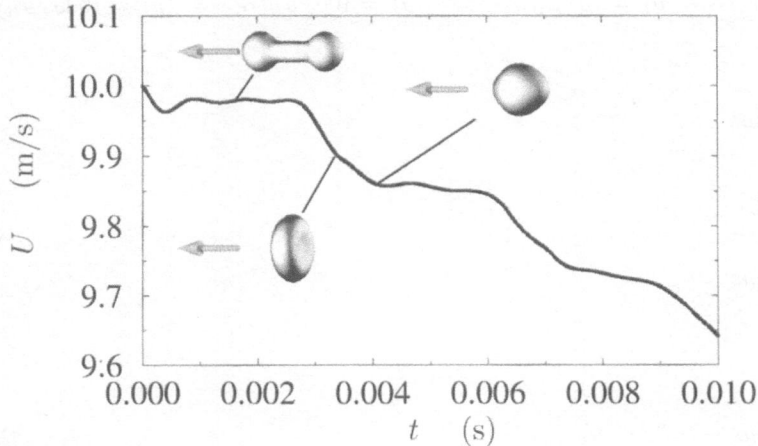

Fig. 3. Velocity U of a freely falling droplet in air as a function of time. Initially the droplet had a cylindrical shape falling parallel to the cylinder axis. In addition the shape of the oscillating droplet is shown at three different times. The arrows indicate the direction of the droplet velocity.

A/A_0 is plotted for three droplets with different viscosities of the droplet liquid. The time until a spherical shape is obtained depends on the viscosity of the liquid phase. For low viscosities the droplet oscillates. The lower the viscosity the longer the time before the oscillations are damped out. There exists an optimum value the asymptotical border case, where the droplet reaches the spherical shape fastest. Here this value seems to be approximately $\mu_L = 100 \cdot \mu_{water}$. For higher viscosities the droplet approaches the spherical shape without oscillations. Then the time until spherical shape is reached increases with increasing viscosity.

The computational time for the presented simulations differs due to the different viscosities of the liquid because of the varying stability restriction. For the lower viscosities $\mu_L = 10 \cdot \mu_{water}$ and $\mu_L = 100 \cdot \mu_{water}$ the CFL-condition [14] is the governing restriction. This leads to a similar number of time steps for the simulations with this viscosities. The number of time steps differs between 5823 time steps for the disk shape with $\mu_L = 100 \cdot \mu_{water}$ and 6203 time steps for the cylindrical shape with $\mu_L = 100 \cdot \mu_{water}$. The computation time is $13263\,s$ per CPU for the disk shape with $\mu_L = 100 \cdot \mu_{water}$ and $13956\,s$ for the cylindrical shape with $\mu_L = 100 \cdot \mu_{water}$. The variation of the number of time steps results of the velocity evolution for the different initial shapes.

For the high viscosity $\mu_L = 1000 \cdot \mu_{water}$ the momentum diffusion is the governing restriction [14]. For this case 17067 time steps for each shape of the initial deformation have been needed. The computational time for the simulations with the viscosity is $34305\,s$ per CPU.

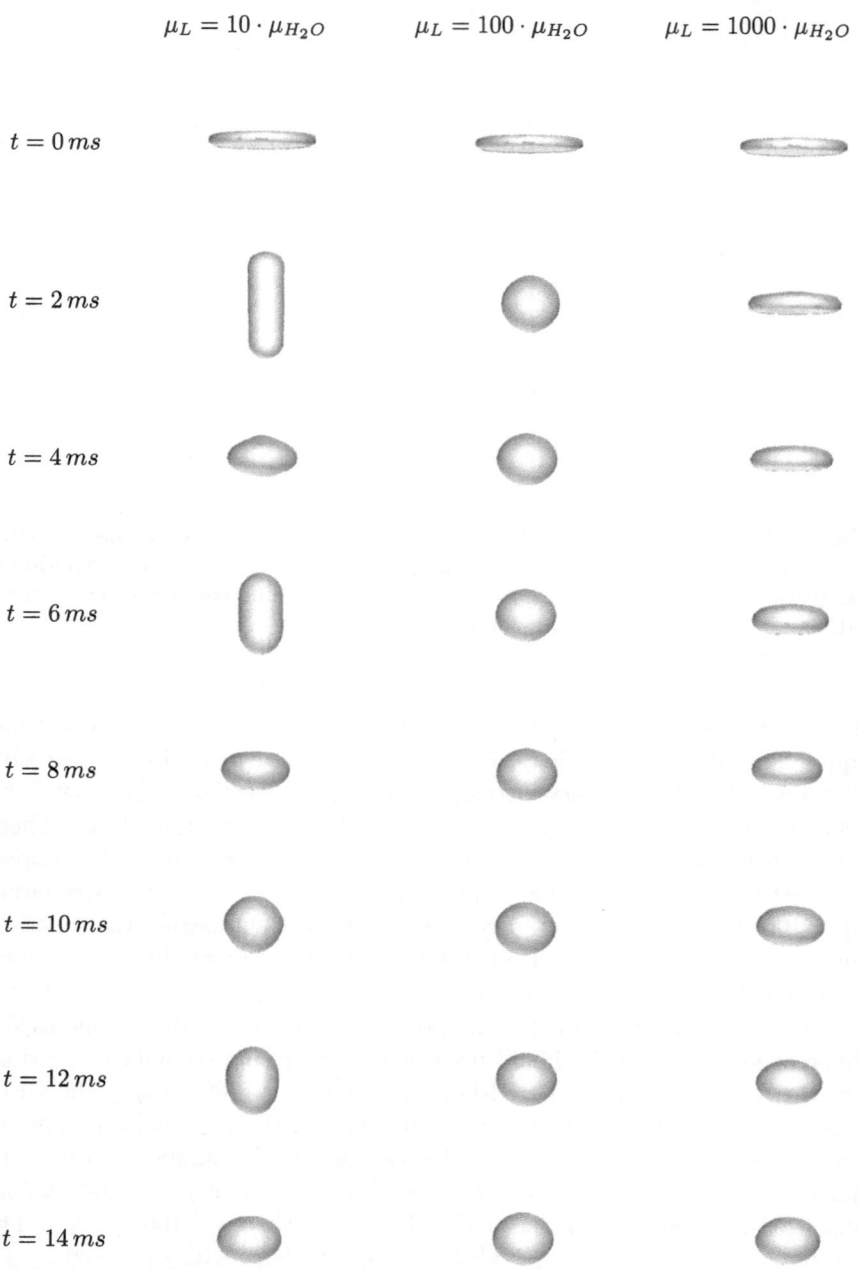

Fig. 4. Evolution of the droplet shape of an initially disk droplet as a function of time with three different liquid viscosities μ_L. The shape of the droplet with $\mu_L = 10 \cdot \mu_{H_2O}$ are not the maximum deformation in the oscillation periods. The direction of the initial velocity is perpendicular to the flat side of the disk.

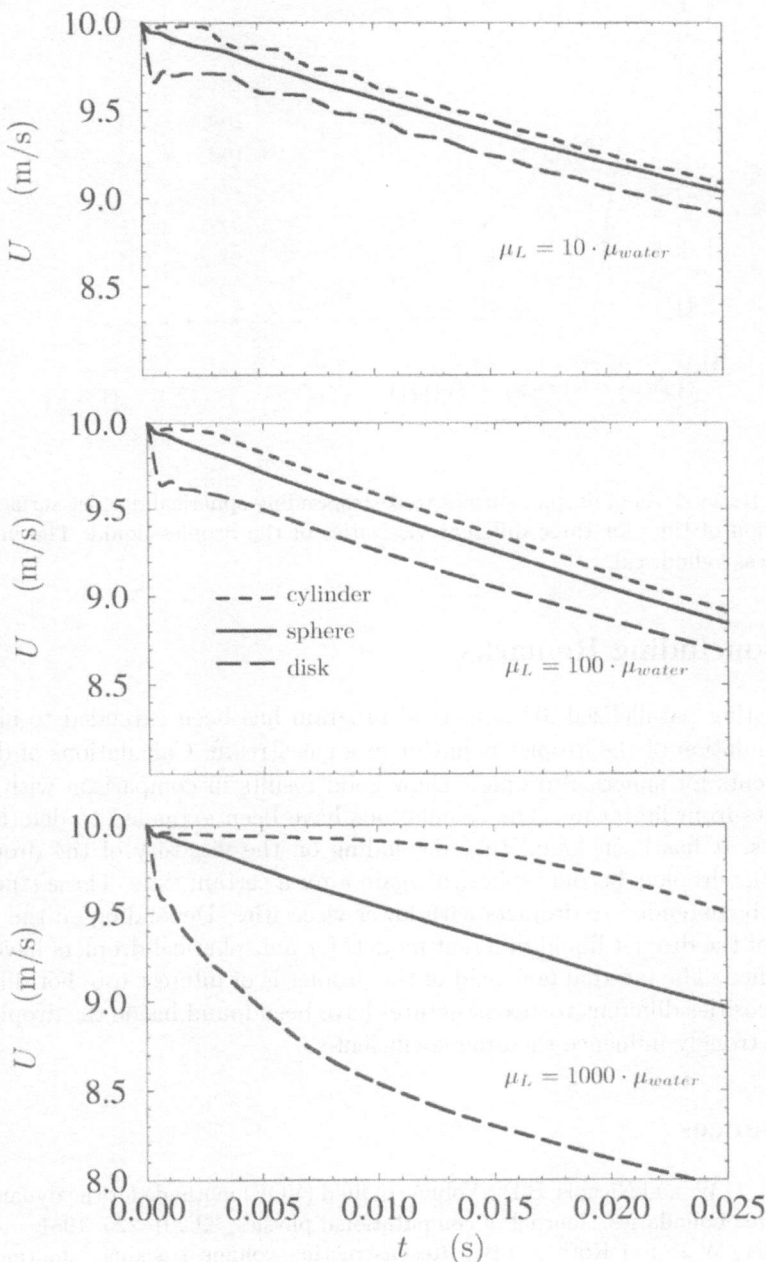

Fig. 5. Droplet velocity U as a function of time for three different initial droplet shapes and three different viscosities of the droplet liquid.

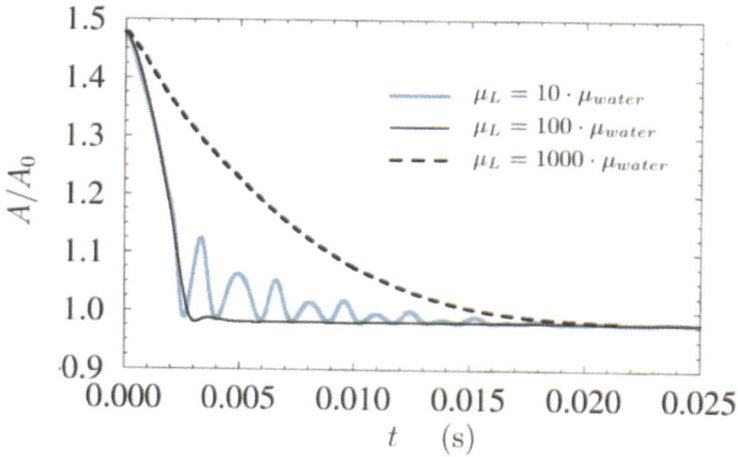

Fig. 6. Ratio A/A_0 of droplet surface to corresponding spherical droplet surface as a function of time for three different viscosities of the droplet liquid. The initial shape was cylindrical.

4   Concluding Remarks

An existing parallelized 3D-numerical program has been extended to allow the simulation of the droplet behavior in a gas stream. Calculations of drag coefficients for spherical droplets show good results in comparison with coefficients from literature. The calculations have been extended to deformed droplets. It has been found that depending on the viscosity of the droplet liquid the droplets become spherical again after a certain time. These studies should be extended to droplets with lower viscosities. Depending on the viscosity of the droplet liquid different models for nonspherical droplets have to be applied. The internal flow field of the droplet is of interest too. For different viscosities different vortex structures have been found inside the droplets, which strongly influence the drag coefficient.

References

1. Hirt, C.W. and Nichols, B.D.: Volume of fluid (VOF) methode for the dynamics of free boundaries. Journal of computational physics, 39:201–225, 1981.
2. Rider, W.J. and Kothe, D.B.: Reconstructing volume tracking. Journal of computational physics, 141:112–152, 1998.
3. Rieber, M. and Frohn, A.: Numerical simulation of splashing drops. In Proc. 14th Int. Conference on Liquid Atomization and Spray Systems, 1998.
4. Schelkle, M., Rieber, M. and Frohn, A.: Numerische Simulation von Tropfenkollisionen. Spektrum der Wissenschaft, pages 72–79, Januar 1999.
5. Rieber, M. and Frohn, A.: A numerical study on the mechanism of splashing. International Journal of Heat and Fluid Flow, June 1999.

6. Stanley, S. and Sarkar, S.: Simulations of Spatially Two-Dimensional Shear Layers and Jets. In Theoretical and Computational Fluid Dynamics. Springer-Verlag, 1997.

7. Rieber, M. and Frohn, A.: Parallel Computation of Interface Dynamics in Incompressible Two-Phase Flows. In Krause, E. and Jäger, W., editors, High-Performance Computing in Science and Engineering 99: Transactions of the High Performance Computing Center Stuttgart (HLRS), pages 241–252. Springer Verlag, 2000.

8. Rayleigh, F.R.S. and Lord, J.S.W.: On the instability of jets. Proc. London Math. Soc., 10:4–13, 1878.

9. Wallis, G.B.: One dimensional two-phase flows. McGraw-Hill, 1969.

10. Clift, R., Grace, J.R. and Weber, M.E.: Bubbles, Drops and Particles. Academic Press, 1978.

11. Rudinger, G.: Handbook of Power Technoloyg, volume 2, chapter Fundamentals of gas-particles flow. Elsevier Science Pub. Co., 1980.

12. Mulholland, J.A., Srivastava, R.K. and Wendt, J.O.L.: Influence of droplet spacing on drag coefficient in nonevaporating, monodisperse streams. AIAA Journal, 26:335–354, 1988.

13. White, F.M.: Viscous Fluid Flow. McGraw-Hill, 1974.

14. Roache, Patrick J.: Fundamentals of computational fluid dynamics. Albuquerque, NM : Hermosa Publishers Publ., 1998.

Simulation of Two-Phase Flow in Pipes

Tobias Giese and Eckart Laurien

Institute for Nuclear Technology and Energy Systems (IKE), University of
Stuttgart, Pfaffenwaldring 31, D-70550 Stuttgart, Germany;
E-Mail: giese@ike.uni-stuttgart.de, laurien@ike.uni-stuttgart.de

Abstract. A two-phase flow of water in a pipe is investigated experimentally and numerically. The simulation is performed with the two-fluid model which is a general approach for multiphase problems. To model phenomena like phase change and change of flow regime, adequate inter-phase exchange terms were implemented. The program CFX-4.2 was used with an enhanced two-fluid model for the analysis of complex pipe flow. The results show that high performance computing can be helpful for the investigation of applied flow problems.

1 Introduction

Two-phase flows in pipes are a matter of particular interest for the design of technical applications like power plants and chemical process plants. In contrast to single phase flow cases, the analysis of multiphase flow is difficult due to the modeling of the interaction of the phases in momentum, energy and turbulence. Phase change phenomena like boiling, cavitation and condensation may occur. Additionally, possible changes in flow regime of pipe flow i.e. from bubbly flow to stratified flow raise the complexity and prevent simple approaches for flow calculation. In the past, one dimensional approaches of pipe flow were used and one dimensional simulation codes have been developed. However, their calculations are still based on empirical data from experiments. To use the correlations, it is necessary to neglect the influence of flow regime or to define the character of the flow a priori, e.g. whether it is a bubbly flow or a stratified flow [1]. In a complex pipe system with phase interaction phenomena, it is not possible to get the necessary information about the flow regime in advance. To overcome the semi - empirical procedures, an approach with CFD seems to be promising. The two-fluid model is deemed to be a general approach for multiphase problems and is described in Sect. 3. The usage of CFD for problems with phase change phenomena makes it necessary to implement adequate models in the two-fluid description. For the flow case that was analysed experimentally, two models were developed: A model for the momentum interaction in different flow regimes and a model for the description of cavitation phenomena, see Sect. 4 and 5. To avoid that effects like secondary flows and gravity have to be neglected and to use the code for applied problems, a three dimensional approach for a complex pipe geometry is necessary. As a consequence, the computational effort is enormous and the usage of the HLRS supercomputers is necessary.

2   Analysed Flow Case and Experimental Results

The experiment was carried out by the authors at a test facility at Siemens, Erlangen in December 1999. Within the facility, a PVC pipe was installed which links a tank to a lower sited second tank. The height difference betweeen the tanks was 13m, the pipe diameter 0.1m and the overall length about 30m. Two flow cases were compared: A cold water (18°C) single phase flow and a warm water (99°C) two phase flow, both cases near atmospheric pressure (1 bar). The mass flux in the warm water flow case is reduced significantly in comparison to the cold water flow case. The difference can be explained by phase change phenomena such as boiling and cavitation. If the local pressure in the pipe drops below the saturation pressure, cavitation phenomena occur. In this two phase flow case, the prediction of mass flux is difficult due to the coupling between the phase change phenomena depending on the local static pressure in a pipe segment and the pressure loss in the entire pipe.

The first experiment was performed with cold water (18°C) which can be assumed to be a single phase flow case due to the lack of phase change phenomena. In the second case, the water in the upper tank is heated close to its saturation temperature ($\approx$ 99°C). This high temperature increases the saturation pressure of water and causes cavitation phenomena in low pressure regions of the pipe. Due to the fact that a two phase character of the flow increases the pressure loss, a decreased mass flux is expected.

Figure 1 is a schematic picture of the experimental setup. The instrumentation consists of a magnetic inductive flow sensor (MID) which was sited in a pipe segment with single phase flow character. Three local temperature and pressure measuring devices where placed at M1, M2 and M3. Additionally, at two locations of the pipe the volume fraction of steam is measured with a simple wire pair probe. To get an impression of the character of the flow, two windows (W1 and W2) allow the observation of the flow. The differences between the two flow cases are significant. In the single phase flow case (cold water) low pressure regions appear (0.3 bar) in certain sections of the pipe. These pressure values and the gained mass flux agree well with the values of a simple one dimensional analysis with the Bernoulli equation. The visual impression of the flow meets the picture of a single phase flow with a few air bubbles. In the flow case with water close to 99 °C, the pressure does not drop below the saturation pressure corresponding to the local temperature. The reason for this is the formation of steam which is visible in the windows of the pipe. A view on the jet leaving the pipe confirms that in a flow case with warm water, a gaseous phase is present. The main consequence of the different character of the flow is the reduced mass flux in the case with high temperature. In comparison with the low temperature case, the mass flux is reduced from 36kg/s to 23kg/s. This is a reduction of nearly 40% and can not be explained by measurement errors (mass flux: less that 5%). The reason for this reduction is the influence of the multiphase character of the flow e.g. the

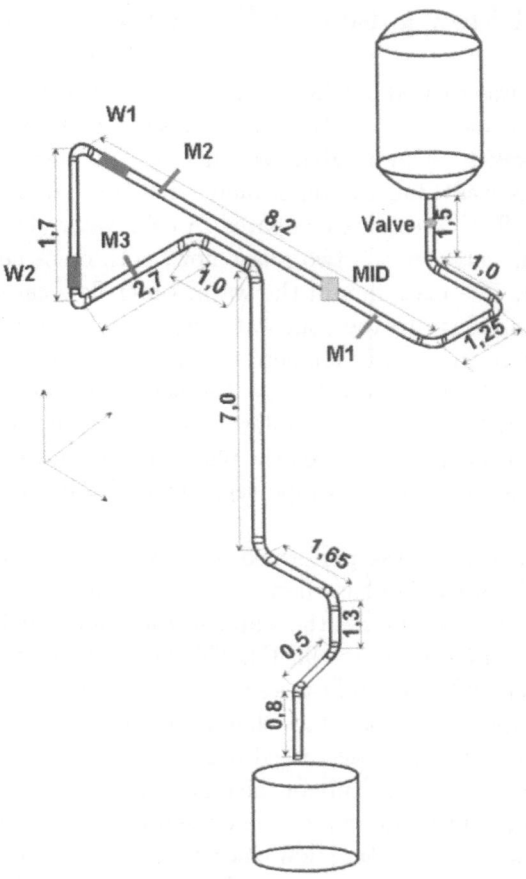

Fig. 1. Experimental Setup

production of steam by cavitation and, as a result, the increase of pressure loss in the pipe.

3   Two-Fluid Model

To model the two phase character of the flow, different approaches are possible. The Euler-Lagrange approach is adequate for the simulation of drops or bubbles in an embedding continuum, but it fails if complex flow regimes occur. To overcome this, the Euler-Euler approach is used. In this approach, both phases, liquid (index L) and steam (index G, gas), are regarded as two continua interpenetrating each other. With the help of two phase functions α_L defined unity in pure liquid and zero in pure gas, and α_G, defined unity in pure gas and zero in pure liquid, the basic equations for the average two-phase state quantities can be formulated [2]. The three dimensional mass, momentum and energy conservation equations are integrated for each phase. The mass conservation equation for the liquid phase

$$\frac{\partial}{\partial t}\left(\alpha_L \rho_L\right) + \nabla \cdot \left(\alpha_L \rho_L \tilde{\boldsymbol{u}}^L\right) = \Gamma_L^C \tag{1}$$

and for the gas phase

$$\frac{\partial}{\partial t}\left(\alpha_G \rho_G\right) + \nabla \cdot \left(\alpha_G \rho_G \tilde{\boldsymbol{u}}^G\right) = \Gamma_G^C \ , \tag{2}$$

the momentum conservation equations for the liquid phase

$$\frac{\partial}{\partial t}\left(\alpha_L \rho_L \tilde{u}_m^L\right) + \nabla \cdot \left(\alpha_L \rho_L \tilde{u}^L \tilde{\boldsymbol{u}}_m^L\right) = -\frac{\partial\left(\alpha_L \bar{p}\right)}{\partial x_m} + \nabla \cdot \left(\alpha_L \left(\underline{\tilde{\tau}}^L + \underline{\tau}^{Re\,L}\right)\right) \tag{3}$$
$$+ F_m^L + \alpha_L \rho_L g_m \beta \left(\tilde{T}^L - T_0\right)$$

and for the gas phase

$$\frac{\partial}{\partial t}\left(\alpha_G \rho_G \tilde{u}_m^G\right) + \nabla \cdot \left(\alpha_G \rho_G \tilde{u}^G \tilde{\boldsymbol{u}}_m^G\right) = -\frac{\partial\left(\alpha_G \bar{p}\right)}{\partial x_m} + \nabla \cdot \left(\alpha_G \underline{\tilde{\tau}}^G\right) \tag{4}$$
$$F_m^G + \alpha_G \rho_G g_m \beta \left(\tilde{T}^G - T_0\right)$$

the energy conservation equation for the liquid phase

$$\frac{\partial}{\partial t}\left(\alpha_L \rho_L \tilde{h}^L\right) + \nabla \cdot \left(\alpha_L \rho_L \tilde{h}^L \tilde{\boldsymbol{u}}_m^L\right) = \nabla \cdot \left(\alpha_L \left(\bar{\boldsymbol{q}}^L + \boldsymbol{q}^{Re\,L}\right)\right) \tag{5}$$
$$+ \nabla \cdot \left(\alpha_L \underline{\tilde{\tau}}^L \tilde{\boldsymbol{u}}_m^L\right)$$
$$+ c_{LG}^L \left(1/c_p^G \, \tilde{h}^G - 1/c_p^L \, \tilde{h}^L\right)$$
$$+ \Gamma_L^E$$

and for the gas phase

$$\frac{\partial}{\partial t}\left(\alpha_G \rho_G \tilde{h}^G\right) + \nabla \cdot \left(\alpha_G \rho_G \tilde{h}^G \tilde{\boldsymbol{u}}^G\right) = \nabla \cdot \left(\alpha_G \bar{\boldsymbol{q}}^G\right) \tag{6}$$
$$+ \nabla \cdot \left(\alpha_G \underline{\tilde{\tau}}^G \tilde{\boldsymbol{u}}^G\right)$$
$$+ c_{LG}^L \left(1/c_p^G \, \tilde{h}^G - 1/c_p^L \, \tilde{h}^L\right)$$
$$+ \Gamma_G^E$$

form the partial differential equation system that has to be solved.

In the equation for the liquid phase, the additional terms for turbulent fluxes $\tau^{Re\,L}$ and $\boldsymbol{q}^{Re\,L}$, describing the Reynolds-stresses and the turbulent heat fluxes, appear. The flow in the liquid phase is modelled as a turbulent flow with the k-ε-model.

The two-phase character of the flow can be found in the inter-phase exchange terms. The terms F_m^k model the momentum exchange and $c_{LG}^k \left(1/c_p^G \, \tilde{h}^G - 1/c_p^L \, \tilde{h}^L \right)$ the heat transfer between the phases (k=L,G). The heat exchange coefficient due to conduction can be modeled with a heat transfer coefficient c_{LG}^k [3]. If mass exchange between the phases occurs, it can be implemented into the code via the exchange terms Γ_k^C and Γ_k^E, describing mass and energy exchange due to phase change phenomena. The terms that model the inter-phase exchange can be summarized in the vector

$$ \boldsymbol{A}^k = \begin{bmatrix} \Gamma_k^C \\ F_1^k \\ F_2^k \\ F_3^k \\ c_{LG}^k \left(1/c_p^G \, \tilde{h}^G - 1/c_p^L \, \tilde{h}^L \right) + \Gamma_k^E \end{bmatrix} . \tag{7} $$

The inter-phase exchange terms in this approach exist for some of the flow cases, for example for dilute bubbly flow. But for many phenomena like boiling, cavitation and condensation and for different flow regime the terms are unknown.

4   Momentum Exchange

In the following, a momentum phase exchange model is discussed. The momentum interphase exchange terms, in Eq. (7), can be replaced by volume forces corresponding to the assumed physical effect in dependency on the flow regime. Figure 2 shows a flow situation with the liquid phase mainly in the lower part of the pipe and the gas phase mainly in the upper part of the pipe. For bubbly flow, the main interaction consists of the drag force of the bubbles moving relative to the liquid. With the assumption that the bubbles are spherical, the flow of water around a sphere can be used in analogy. The

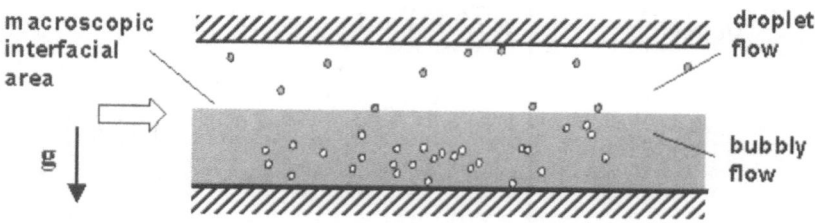

Fig. 2. Complex flow situation with bubbly and droplet flow

integration over all bubbles with diameter d_B in a control volume leads to a phase interaction force for bubbly flow

$$F_{\mathrm{B}} = \frac{3}{4} \frac{C_{\mathrm{D}}}{d_{\mathrm{B}}} \alpha_{\mathrm{G}} \rho_{\mathrm{L}} |\tilde{u}_{\mathrm{G}} - \tilde{u}_{\mathrm{L}}| (\tilde{u}_{\mathrm{G}} - \tilde{u}_{\mathrm{L}}) \ . \tag{8}$$

For droplet flow, the phase interaction force can be derived in almost the same manner. A single drop is moving in a gaseous surrounding. Summarizing all drops in a control volume this results in

$$F_{\mathrm{T}} = \frac{3}{4} \frac{C_{\mathrm{D}}}{d_{\mathrm{D}}} \alpha_{\mathrm{L}} \rho_{\mathrm{G}} |\tilde{u}_{\mathrm{G}} - \tilde{u}_{\mathrm{L}}| (\tilde{u}_{\mathrm{G}} - \tilde{u}_{\mathrm{L}}) \ . \tag{9}$$

The decision if the term for the bubbly flow or the term for the droplet flow is used in (3) and (5) can be made in dependency of the local volume fraction. Large values for α_L and small values for α_G indicate a bubbly flow. For values $\alpha_G < 0.25$ the bubbly flow correlation is used. For $\alpha_G > 0.75$, the droplet flow correlation is used. In between, an approach according to

$$F_{\mathrm{P}} = \left(\frac{\alpha_{\mathrm{G}} - 0.25}{0.5} \right)^n \cdot F_{\mathrm{D}} + \left(1 - \left(\frac{\alpha_{\mathrm{G}} - 0.25}{0.5} \right)^n \right) \cdot F_{\mathrm{B}} \ . \tag{10}$$

That means large values for α_L and small values for α_G indicate a bubble flow. Realistic values for the exponent n are a field of current research. Equation (10) is valid for all regions in Fig. 2 with the exception of the macroscopic interfacial area of stratified flow. In this area, a correlation for stratified flows according to Wang [4] is used

$$F_{\mathrm{Str}} = \frac{1}{2} A_{\mathrm{i}} f_{\mathrm{i}} \rho_{\mathrm{G}} |\tilde{u}_{\mathrm{G}} - \tilde{u}_{\mathrm{L}}| (\tilde{u}_{\mathrm{G}} - \tilde{u}_{\mathrm{L}}) \ . \tag{11}$$

A_{i} is the interfacial area and f_{i} an interfacial friction coefficient. The decision between (10) and (11) is based on an indicator for stratified layers. At the interfacial area of stratified flows, the gradient of the volume fraction has a maximum value. As a first step,

$$F^{\mathrm{K}} = F_{\mathrm{P}} \cdot \left(1 - \max \left(0; \min \left(\frac{|\nabla \alpha_{\mathrm{G}}|}{C_{\mathrm{Str}}}; 1 \right) \right) \right)$$
$$+ F_{\mathrm{Str}} \cdot \left(\left(0; \max \left(\frac{|\nabla \alpha_{\mathrm{G}}|}{C_{\mathrm{Str}}}; 1 \right) \right) \right) \tag{12}$$

with a parameter C_{Str} adapted to the grid density is used.

5   Cavitation Model

At IKE, attempts to analyse a two phase flow case in pipes lead to the development of a cavitation model for water with high temperature and, as a consequence, high saturation pressure [5]. The basic idea of the developed model is that the phase change in a cavitating flow is based on thermodynamic processes and not on bubble dynamics. As described in [6], heat transfer between a bubble and surrounding liquid can be used to quantify cavitation phenomena. Recent work extended this approach on phase change phenomena of drops in a gaseous surrounding to simulate flow situations shown in Fig. 2. To derive the inter phase mass transfer, the fact that the interfacial area of bubbles and drops is at saturation temperature T_{sat} is used. The saturation temperature depends on the local static pressure. The variation of the pressure, (assuming equal pressure for both phases) or the temperature of either steam or water leads to a heat transfer, which can be balanced with an inter-phase mass transfer. Assuming spherical bubbles and drops with diameters d_B and d_D, the heat transfer between the dispersed phase and its surrounding can be derived for both flow regimes. The inter-phase heat transfer coefficient α_B of a spherical bubble in motion with a relative velocity to the surrounding fluid can be approximated by

$$\alpha_B = \frac{Nu_B \cdot \lambda_L}{d_B} = \frac{\lambda_L}{d_B} \cdot \left(2 + 0.6 \cdot Re_B^{1/2} Pr_L^{1/3} \right) . \tag{13}$$

In this model, the Reynolds number Re_B and the Prandtl number Pr_L are used to predict the convective heat transfer according to the Ranz-Marshall correlation [3]. In equilibrium the heat flux $\Delta \dot{h}_B$ towards the interphase of one bubble can be used to predict the inter-phase mass transfer $\dot{m}_B$ at the bubble which is the basic value for the different source terms in the equation sets

$$\Delta \dot{h}_B = \Delta h_{LG} \cdot \dot{m}_B = \alpha_B \cdot (T_L - T_{sat}) \cdot \pi d_B^2 . \tag{14}$$

If λ_L is the thermal conductivity of the water and Δh_{LG} the evaporation enthalpy, this leads to the equation

$$\Gamma_B = (T_L - T_{sat}) \cdot \frac{6 \cdot \alpha_G \cdot \lambda_L}{d_B^2 \cdot \Delta h_{LG}} \cdot \left(2 + 0.6 \cdot Re_B^{1/2} Pr_L^{1/3} \right) \tag{15}$$

for the inter-phase mass transfer at the interfacial area of all bubbles per volume. If the liquid is superheated $(T_L > T_{sat})$, it is assumed that the transported heat (from the liquid to the interfacial area of the bubble) is used for further vaporization of the liquid phase. The Ranz-Marshall correlation used for the droplet region leads to the equation

$$\Gamma_D = (T_G - T_{sat}) \cdot \frac{6 \cdot (1 - \alpha_G) \cdot \lambda_G}{d_D^2 \cdot \Delta h_{LG}} \cdot \left(2 + 0.6 \cdot Re_D^{1/2} Pr_G^{1/3} \right) . \tag{16}$$

The decision which flow regime correlation, bubbly or droplet flow, is used for the prediction of inter-phase mass transfer is performed by the volume fraction of steam according to

$$\Gamma^C = \left(\frac{\alpha_G - 0.25}{0.5} \right)^n \cdot \Gamma_D + \left(1 - \left(\frac{\alpha_G - 0.25}{0.5} \right)^n \right) \cdot \Gamma_B . \tag{17}$$

The derivation of the inter-phase mass transfer based on bubbles and drops is misleading in the region of the "macroscopic" interfacial area of stratified flows. In this region, which can be identified by a large gradient of the phase fractions, a simple correlation for free surface flow is used. Due to the exchanged mass energy is transported between the phases. However, the thermal energy of the system has to be modified in the case of mass transport. The reason is that for the vaporization of liquid, the latent heat Δh_{LG} has to be removed from the system. To enable the calculations of different flow regimes, an approach

$$\Gamma_{\mathrm{L}}^{\mathrm{E}} = \alpha_{\mathrm{L}} \Delta h_{\mathrm{LG}} \Gamma_{\mathrm{L}}^{\mathrm{C}} \tag{18}$$
$$\Gamma_{\mathrm{G}}^{\mathrm{E}} = \alpha_{\mathrm{G}} \Delta h_{\mathrm{LG}} \Gamma_{\mathrm{G}}^{\mathrm{C}} \tag{19}$$

was implemented.

6   Simulation Code

To make a simulation in a complex geometry possible, the commercial code CFX-4.2 is used [7]. The code provides the basic two fluid equations and simple models for bubbly flow. However, for the simulations mentioned below, the terms for the momentum, mass and energy exchange were implemented by additional FORTRAN subroutines and the models of Sect. 3 and 4 were integrated. With these additional implementations, the simulation of the two phase flow in the experimental pipe is possible.

7   Simulation Results

To test the capabilities of the cavitation model, a parameter study of the two phase flow in a pipe bend was performed [5][8]. The influence of the bubble diameter and nucleation density on the development of steam in water near saturation temperature was studied. The simulations showed a clear influence of the secondary flows on the steam distribution in the bend, see Fig. 3. The pressure loss in the pipe bend showed a clear dependency on the cavitation intensity.

The application of the cavitation model and stratified flow model on the experimental pipe flow was performed. A numerical mesh of more than 280000 nodes was necessary to take effects like secondary flows in bends into account. These effects seem to have an important influence on the production and the location of the steam in the pipe. A plot of the volume fraction of the steam at the pipe wall of a certain section of the pipe is provided in Fig. 4. In the lower horizontal pipe section shown in Fig. 4, the flow regime converses from a bubbly flow to a stratified flow. The effect of gravity in this process is negligible in comparison to the effect of secondary flows in bends that transports the steam fraction to the inner radius of the bend. The mass

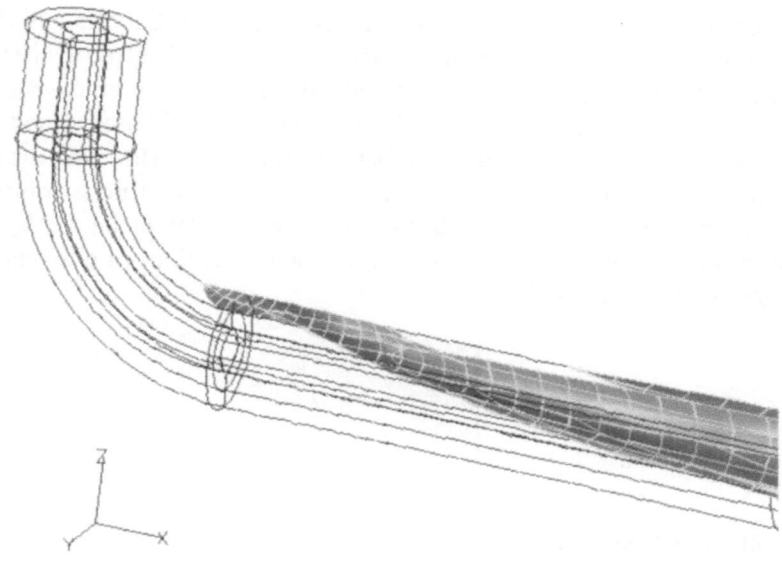

Fig. 3. Iso-surface of void fraction 5%

flux of the recent calculations is 25.5 kg/s in the two phase flow case and differs from the experimental result (23 kg/s). However, not all necessary model parameters have been correctly identified yet. An enhancement of the analysis to reach the experimental values is expected from a more realistic modeling of the stratified flow regime and a physical based relation instead of equation (10) for flow regimes between bubbly and droplet flow.

8   Computational Resources

The simulation was performed on a NEC SX-4 at the High Performance Computing Center Stuttgart. The computational effort is large due to the two-fluid approach and the additional set of Navier-Stokes equation for the second phase. Additionally, the segregated solver of the version CFX-4 is sensitive in cases of strong phase interaction. This requires small iteration steps. The new version CFX-5 with a coupled solver will overcome this problem [9]. The grid for the large geometry of the experiment requires a memory of 1GB. For future work, the usage of 4 or 8 processors parallel is planned which might be possible if a parallelized version of Software is available on NEC SX-5. The commercial part of the programme is partially vectorised and the authors vectorised their FORTRAN subroutines as far as possible.

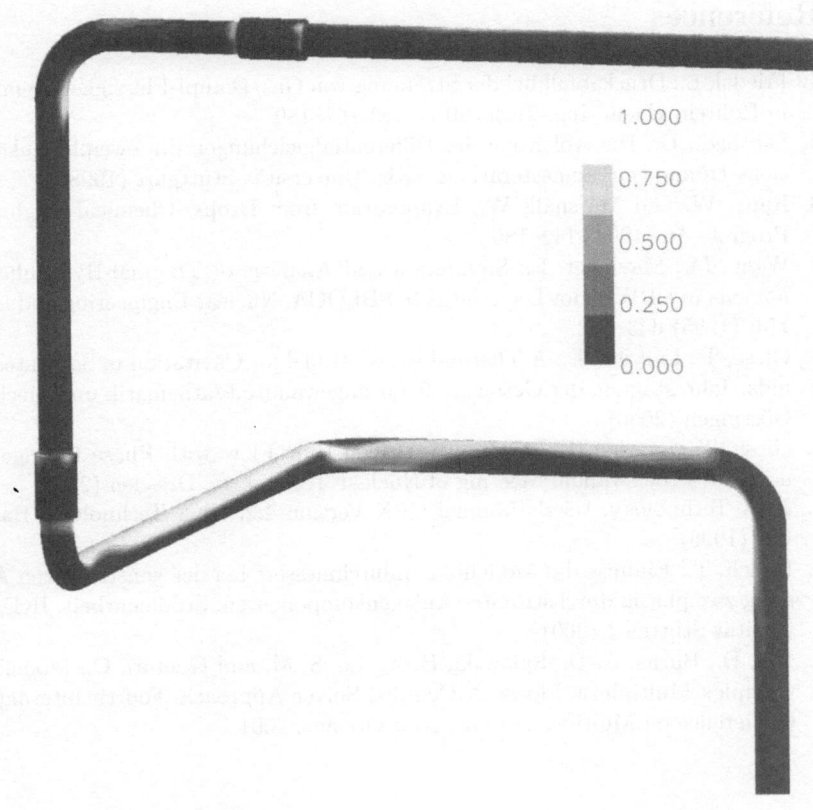

1.000

0.750

0.500

0.250

0.000

Fig. 4. Volume fraction of steam at pipe wall

9   Conclusion

It has been demonstrated that the investigated experimental setup can be analyzed using a commercial CFD code. New models have been implemented using the FORTRAN interface. In order to continue our model development, it will be necessary that, beside academic codes, industrial CFD codes are made available on supercomputers.

Acknowledgement

This study has been funded by the German Electricity Association (VDEW) and the Gemeinschaftskraftwerk Neckar GmbH (GKN)

References

1. Friedel, L.: Druckabfall bei der Strömung von Gas/Dampf-Flüssigkeitsgemischen in Rohren. Chem.-Ing.-Tech. **50** (1978) 167–180
2. Saptoadi, D.: Die Ableitung der Differentialgleichungen für Zweiphasenkonvektionsströmungen. Semesterarbeit. IKE, Universität Stuttgart (1998)
3. Ranz, W. and Marshall, W.: Evaporation from Drops. Chemical Engineering Progress **48** (1952) 142–180
4. Wang, M., Mayinger, F.: Simulation and Analysis of Thermal-Hydraulic Phenomena in a PWR Hot Leg related to SBLOCA. Nuclear Engineering and Design **155** (1995) 643–652
5. Giese, T., Laurien, E.: A Thermal Based Model for Cavitation in Saturated Liquids. Jahrestagung der Gesellschaft für angewandte Mathematik und Mechanik, Göttingen (2000)
6. Giese, T., Laurien, E.: A Gravity Driven Pipe Flow with Phase Change Phenomena. Proc. Annual Meeting of Nuclear Technology, Dresden (2001)
7. AEA Technology: User's Manual CFX Version 4.2. AEA Technology, Harwell, UK (1998)
8. Hirsch, T.: Einfluss des Modellblasendurchmessers bei der konstruktiven Auslegung zweiphasig durchströmter Anlagenkomponenten. Studienarbeit. IKE, Universität Stuttgart (2001)
9. Yin, D., Burns, A. D., Splawski, B. A., Lo, S. M. and Guetari, C.: Modeling of Complex Multiphase Flows: A Coupled Solver Approach, Fourth International Conference on Multiphase Flow, New Orleans, 2001

Computational Study of the Flow in an Axial Turbine with Emphasis on the Interaction of Labyrinth Seal Leakage Flow and Main Flow

Jan E. Anker, Jürgen F. Mayer, and Heinz Stetter

Institut für Thermische Strömungsmaschinen und Maschinenlaboratorium,
Universität Stuttgart, D-70550 Stuttgart, Germany

URL: http://www.itsm.uni-stuttgart.de

Abstract. This paper presents a numerical study of the flow in a 1.5 stage low-speed axial turbine with a straight labyrinth seal on the rotor blade and focuses on the interaction of the leakage flow with the main flow. The influence of the leakage flow on the flow field of the turbine is examined using the Navier-Stokes code ITSM3D. The impact of the re-entering leakage flow on the main flow in dependency of the leakage mass flow rate is studied at two different clearance heights of the labyrinth in the simulations. As demonstrated in this paper, leakage flow not only introduces mixing losses but can also dominate the secondary flow and induce severe losses. In agreement with the experimental data the computational results show that even at realistic clearance heights the leakage flow gives rise to negative incidence over a considerable part of the downstream stator which causes the flow to separate.

1 Introduction

Tip leakage flow can account for considerable parts of the total losses in performance of turbomachinery. The loss of performance due to leakage flow in turbomachinery with unshrouded blades has been intensively studied over the last decade. Much less work has been done on leakage flow over shrouded blades. The most obvious way to minimize losses caused by tip clearance flow in turbomachinery with shrouded blades is to reduce the leakage flow. Thus, up to now the research activities have concentrated on optimizing the sealing efficiency of labyrinth seals, such as the work of Stoff [18], Rhode et al. [13], Waschka [22], Rhode und Johnson [14] and Storteig [19]. In these references, isolated labyrinth seals and their sealing efficiency are considered. However, the influence of leakage flow on the main flow was not examined.

One of the first investigations on overall losses in turbomachinery with shrouded blades reported were performed by Kacker and Okapuu [8]. Later Denton [2] derived a theory for tip leakage losses of shrouded blades. These publications provide global correlations for losses caused by leakage flow, but they do not take into account that leakage flow effects not only result in entropy generation due to mixing but also can induce losses in subsequent stators due to the change of the flow field.

Wallis et al. [21] and Korschunov and Döhler [9] have studied the interaction of the leakage flow and the main flow in turbomachinery in detail. While the first paper concentrates on reducing aerodynamic losses by turning the leakage flow using bladelets on the shroud of the rotor in a four stage turbine, in the second paper the effect of an artificially produced jet (which simulates leakage flow) entering the main flow at different angles on a following linear cascade is studied.

In a project partly funded by European turbomachinery industry, Lehrstuhl für Dampf- und Gasturbinen (DGT), Ruhr-Universität Bochum and Institut für Thermische Strömungsmaschinen und Maschinenlaboratorium (ITSM), University of Stuttgart, are collaborating to study the influence of the leakage flow in a 1.5 stage axial turbine with a labyrinth seal on the periphery of the moving row experimentally and numerically. First experimental and numerical results from this project were published by Peters et al. [12]. While their paper concentrates on experimental results and how the tip leakage flow changes the secondary flow in the turbine, this paper focuses on the numerical modeling and on the leakage flow interaction directly behind the outlet gap of the labyrinth seal.

2   Test Rig and Measurement Data

The turbine considered in this paper is a 1.5 stage axial low-speed air turbine. It is operated in a test rig at the Ruhr-Universität Bochum where extensive steady-state measurements have been carried out [17]. The turbine stage consists of two identical stator blade rows and a rotor with a labyrinth seal on the shroud in between. The cylindrical profile of the turbine blades was taken from the tip-section of an industrial gas turbine rotor-blade. For the geometrical data of the turbine, the reader has to confer Table 1.

Figure 1a shows a meridional view of the turbine and its measurement planes. An azimuthal cut through the turbine with midspan velocity triangles and definitions of the flow angles are shown in Fig. 1b. The test rig is operated at peak efficiency condition (design condition) with a rotor speed of $n =$

Table 1. Test rig geometry and operational data

Number of blades (stator/rotor)	N_{st}/N_{rot}	37/36	Chord/pitch ratio (stator/rotor)	t/s	1.29/ 1.25
Chord length	s	0.163 m	Aspect ratio	h/s	1.043
Rotational speed	n	500 rpm	Hub/tip radius ratio	D_o/D_i	0.795
Stagger angle	β_{Bi}	37.5°	Blade height	h	0.17 m
Max. Reynolds number	Re	$4 \cdot 10^5$	Clearance height	s_{cl}	1 mm/ 3 mm
Flow coefficient	ϕ	0.35	Stage loading coefficient	ψ	2.4

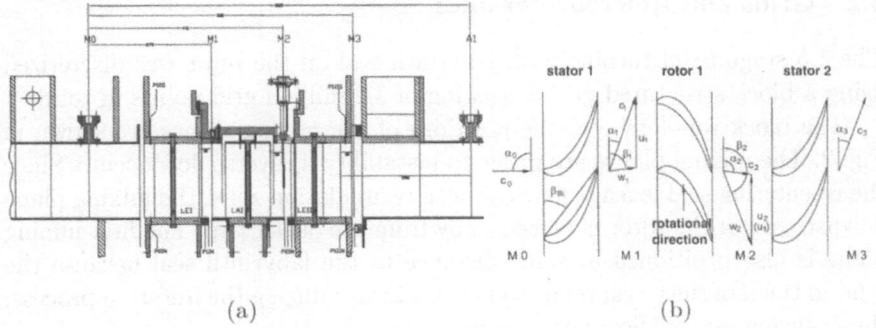

Fig. 1. Turbine geometry. **a** Measurement planes; **b** blading and angle definitions

500 rpm. At two different clearance heights $s = 1$ mm ($s/D = 0.07$ %) and $s = 3$ mm the flow fields in the planes M1, M2 and M3 were measured with a pneumatic 5-hole probe at defined inlet and outlet conditions in M0 and A1, respectively.

A detailed description of the test rig and the measurements carried out in Bochum can be found in [16].

3   Numerical Method

3.1   Flow Solver

The presented results of the interaction of leakage flow with the main flow were performed with the multistage Navier-Stokes solver ITSM3D. The solver has been developed at the Institute of Thermal Turbomachinery and Machinery Laboratory (ITSM) of the University of Stuttgart, see Jung et al. [7] and Merz [11].

The equations solved are the fully three-dimensional, unsteady Favre-averaged Navier-Stokes equations. The fluid is assumed to behave as a perfect gas. A modified algebraic Baldwin-Lomax model is used to describe the effect of turbulence. The solution method of the flow-solver is a cell-vertex central-difference finite volume scheme based on the work of Jameson [5].

ITSM3D is programmed in Fortran 77/90 and only some emulation and portability routines are written in C. The code was parallelized using MPI exploiting the multi-grid topology [6]. The steady state calculations presented in this paper were performed on the workstations of the institute of the authors and the results have already been published in [1].

Since the solver ITSM3D was originally developed for transonic flows and the flow in the turbine under consideration can be regarded as incompressible (Ma $= 0.05 - 0.15$) a preconditioning method based on the work of [20], [10] was implemented.

3.2   Grids and Reference Frames

The 1.5 stage axial turbine with labyrinth seal on the rotor was discretized using a block-structured grid consisting of 3.2 million grid points in total.

The block topology and the positions of the mixing planes are shown in Fig. 2. The mixing planes are prone to instability if reverse flow occurs. Since the re-entering seal leakage flow causes a recirculation zone, the mixing plane downstream of the rotor is placed away from the outlet gap. The first mixing plane is also positioned at some distance to the labyrinth seal because the fluid in the labyrinth seal tends to flow backward during the iteration process, thus causing reverse flow near the inlet gap.

The clearance height of the stator seals is down to 0.1 mm. Since the experimental data has shown the stator hub leakage flows to be neglectable, the stator seals were not modeled.

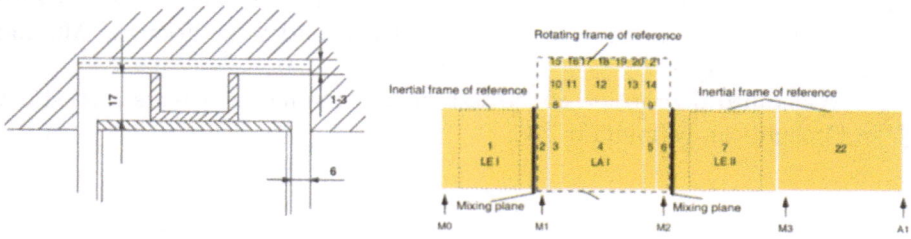

Fig. 2. Geometry and modeling of the labyrinth seal

3.3   Boundary Conditions

At the inlet and outlet, as well as at the mixing planes, a non-reflecting post-correction method based on the work of Giles [3] and Saxer [15] is applied to prevent spurious reflections from waves that leave the domain. The change in the time-dependence of the solution of the Navier-Stokes equations due to preconditioning is not taken into account in the boundary treatment yet. Since the development of appropriate boundary conditions is a subject of current research, the simulations run with the preconditioned scheme are therefore considered as preliminary.

At the inlet plane the measured radial distribution of total pressure and temperature in M0 and axial flow direction is specified. At the outlet boundary A1 the measured radial distribution of the static pressure is prescribed.

Solid surfaces are assumed to be adiabatic, and the no-slip condition is applied. Periodicity in the pitch-wise direction is ensured using phantom cells that keep copies of the periodic values such that the points on these boundaries can be treated like interior points.

4   Comparison between Experimental and Computational Results

To verify that the solver ITSM3D captures the important effects when leakage flow interacts with the main flow, the calculated circumferentially averaged velocities and yaw angles at the measurement planes M1 and M2 are compared to the data obtained experimentally in Bochum [17], see Figs. 3–5. The turbine was simulated with clearance heights of 1 mm and 3 mm. The calculations performed with and without preconditioning are identified by the labels "(PC)" and "(NonPC)", respectively.

As seen in Fig. 3, the calculated velocities and yaw angles in M1 for the clearance height of 1 mm are in good agreement with the measured ones.

The measured circumferentially averaged axial velocity in M2, depicted in Fig. 4, reveals that there is a reduced throughput in the hub region of the rotor, since the hubside passage vortex causes an overturning of the main flow. Apparently, the preconditioned solver captures the secondary flow effects better than the original unpreconditioned scheme does. However, the circumferential velocities are better predicted by the original solver. This may be caused by the inconsistent boundary treatment mentioned in the previous section.

Since the flow field in M1 is nearly independent of the clearance height, only the calculated averaged velocities in the plane M2 for the case of a clearance height of 3 mm are compared to experimental values in Fig. 5. The solution obtained by the preconditioned code deviates from the measurement in the midchannel. However in the important casing region, the calculated and measured values coincide. As in the case of 1 mm clearance height, in contrast to the original scheme the preconditioned scheme better resolves the secondary effects.

5   Visualization by Virtual Reality Techniques

With support from the visualization team of the HLRS it was possible to analyze the interaction of the tip leakage flow and the main flow in the CUBE of the Virtual Environments Lab of HLRS. In Fig. 8 different snapshots from our simulations visualized in the CUBE are shown.

By only looking at two-dimensional cuts at different locations in the turbine is very time consuming and requires a distinctive ability of imagination. The Virtual Reality based visualization proved to be a helpful tool for getting an overview over the simulation model, analyzing the flow field and detecting different flow phenomena.

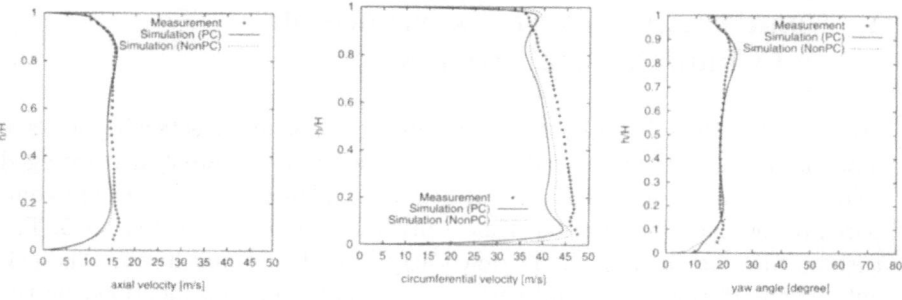

Fig. 3. Circumferentially averaged velocities and yaw angles over the relative channel height h/H at 1 mm clearance height in the measurement plane M1 (30 mm behind stator I)

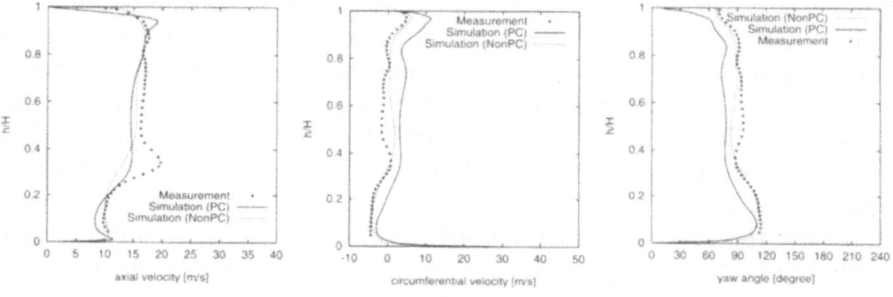

Fig. 4. Circumferentially averaged velocities and yaw angles over the relative channel height h/H at 1 mm clearance height in the measurement plane M2 (30 mm behind rotor I)

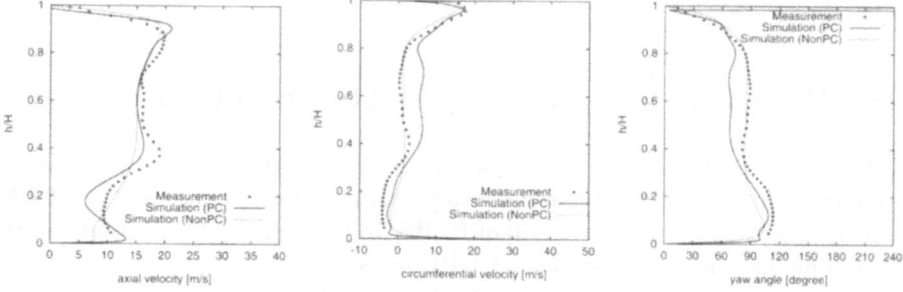

Fig. 5. Circumferentially averaged velocities and yaw angles over the relative channel height h/H at 3 mm clearance height in the measurement plane M2 (30 mm behind rotor I)

6   Unsteadiness and Inhomogeneity of the Tip Leakage Flow

The calculations showed that the mass flow rate through the labyrinth seal varies in the circumferential direction. In Fig. 6a lines of constant radial velocities in an axial plane in the middle of the outlet gap of the labyrinth seal are depicted. This graph is representative for other axial positions; it shows a variation of the radial velocity over the pitch. Correspondingly, the static pressure on the walls in the last cavity of the labyrinth seal varies in the circumferential direction (see Fig. 8b).

Compared to the variation of the radial velocities in the passage, the variation of the radial velocities in the outlet gap of the labyrinth seal is small. Since the velocity field of the wake is retarded, less impulse is necessary to deflect the flow in the wake than in the midchannel. Figure 6b, in which the lines of constant radial velocities are plotted over two pitches in a plane 11 mm behind the outlet gap of the labyrinth seal (and 11 mm ahead of M2), reveals the inhomogeneity of the leakage flow that can also be observed in Fig. 8c, where the re-entering leakage flow is visualized.

Since the labyrinth seal, the rotor and a region behind the outlet of the labyrinth seal have been modeled in a rotating frame of reference, the observed variations in the pressure and the velocities show that the tip leakage effect is an unsteady phenomenon. This result is confirmed by the work of Wallis et al. [21], where the flow in the labyrinth seal of the rotor of a four stage test rig was examined and pulsations in the velocities of the flow in the labyrinth seal were detected at a frequency corresponding to the rotational speed of the rotor.

7   Consequences of the Influence of Tip Leakage Flow

In Fig. 7 the calculated yaw angles in M2 (30 mm behind the trailing edge of the rotor) for a clearance height of 1 mm and 3 mm are shown. A yaw angle of $\alpha = 90°$ means an axial flow. Yaw angles of $\alpha < 90°$ cause a negative incidence of the following stator, whereas yaw angles of $\alpha > 90°$ lead to a positive incidence. From Fig. 7 it is clearly seen that the influence of the tip leakage flow in the case of a clearance height of 3 mm results in negative incidence on the upper 20 % of the stator. The reason for the reduction of the yaw angle with an increased clearance height is easily recognized from Fig. 8a: The streamlines visualized in the rotating frame of reference show that the flow over the labyrinth of the rotor is not turned properly.

Figure 8d shows the computed streamlines around the second stator at different radial positions: While the streamlines at the lower radial position show a correct flow angle, the upper streamlines reveal that the tip leakage flow leads to a separation and consequently to increased losses.

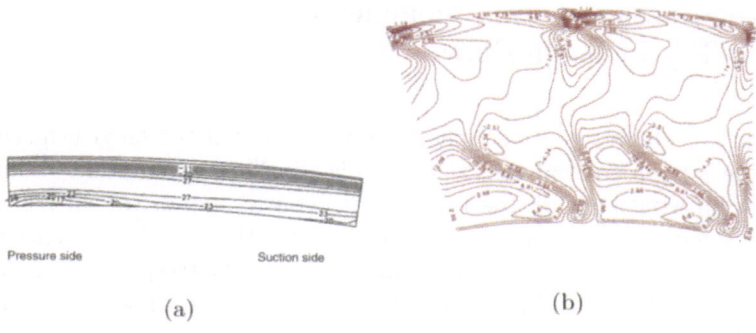

Pressure side   Suction side

(a)   (b)

Fig. 6. Radial velocities of the leakage flow. (a) Lines of constant radial velocity (in m/s) in the axial plane in the middle of the outlet gap of the labyrinth seal; (b) lines of constant radial velocity (in m/s) 11 mm behind the rotor and 11 mm upstream of M2 plotted over two pitches.

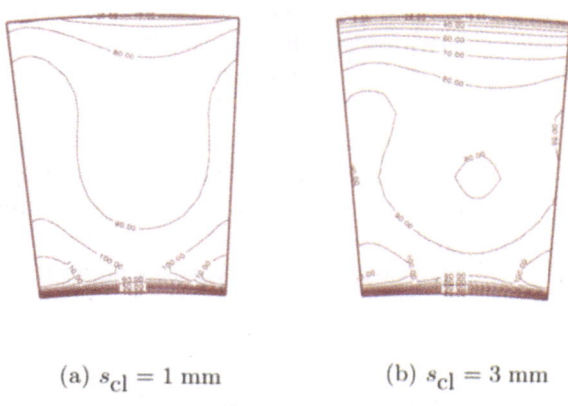

(a) $s_{cl} = 1$ mm   (b) $s_{cl} = 3$ mm

Fig. 7. Yaw angles in the measurement plane M2 (30 mm behind the trailing edge of the rotor) for two values of the clearance height s_{cl}. A yaw angle of $\alpha = 90°$ corresponds to the ideal case of axial flow.

8   Future Work and the Need for High Performance Computing

Until now only steady state calculations have been carried out. However, in the previous sections the nature of leakage flow has been proven to be unsteady. Since we want to avoid the use of mixing planes which eliminate parts of the inherently unsteady physics of turbine flow, in future work we will use the sliding-mesh approach and thereby include time-accurate simulations in our work.

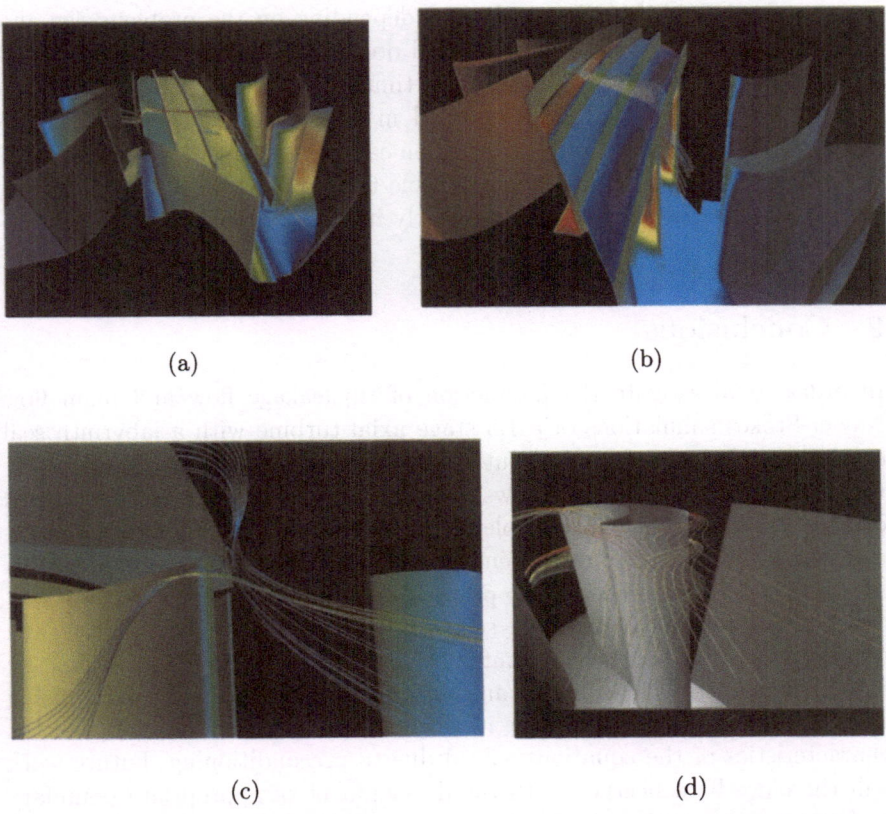

(a)   (b)

(c)   (d)

Fig. 8. Visualization of the leakage flow. (**a**) Undeflected streamlines in the labyrinth seal; (**b**) distribution of static pressure in the labyrinth seal; (**c**) the leakage flow re-entering the main flow behind the rotor; (**d**) calculated streamlines, showing a suction-sided incidence of the second stator.

For unsteady flow simulations a dual time-stepping scheme is implemented in the Navier-Stokes code ITSM3D. As discussed by Jung et al. [7], this scheme allows a faster convergence than using the explicit time-consistent multigrid scheme based on the work of He [4]. A detailed description of the flow solver ITSM3D for unsteady calculations and the time-inclination method for simulating turbomachinery with arbitrary pitch ratios can be found in [6].

In order to obtain a periodic solution using the flowfield of a preceding steady simulation as an initial condition, 7–10 blade passing periods are necessary. Realistically, 30–100 dual time steps per time step and 100 time steps for every blade passing period are needed. As shown in the previous sections 3.2 million grid points are necessary to resolve secondary flow effects. On a 667 MHz single processor Alpha-Dec Workstation $62.7\mu s/(\text{time step} \times \text{node})$ specific CPU time is used, which means that up to 23 days of CPU time in total are needed for performing one unsteady simulation. Using four processors on the NEC-SX4/SX5 platform – depending on the problem size and the decomposition of the computational domain – ITSM3D typically needs 1–$6\mu s$/(time step $\times$ node) specific CPU time. Assuming that $6\mu s$/(time step $\times$ node) specific CPU time are needed in general, on the NEC-SX4/SX5 platform the planned unsteady simulation can be performed within two CPU days. Thus, in order to achieve acceptable turn-around times when starting with the unsteady calculations, we clearly need High Performance Computing.

9   Conclusions

In order to investigate the interaction of tip leakage flow and main flow Navier-Stokes simulations of a 1.5 stage axial turbine with a labyrinth seal on the rotor have been carried out. Since the flow solver ITSM3D originally was developed for transonic flows and the flow in the turbine considered can be regarded as incompressible, a preconditioning technique was implemented. A comparison of measurement data with the solutions obtained with both the original scheme and the preconditioned scheme shows that the preconditioning scheme is resolving secondary effects better than the original scheme does. However, the circumferential velocities are better predicted by the original solver. This circumstance may result from the fact that by using the original boundary conditions, it was not accounted for the change of the characteristics of the equations solved due to preconditioning. Future work will therefore be concerned with the development of appropriate boundary conditions. Nevertheless, both schemes are capturing the important effects of the interaction of tip leakage flow and main flow.

In the simulations performed it was noted that the effect of the tip leakage flow is not uniform in the pitch-wise direction. At the suction side of the rotor wake, where less impulse is necessary to deflect the flow radially than in the midchannel, the entering leakage flow exhibits its greatest impact. Since the labyrinth seal, the rotor and a region behind the outlet were simulated in a rotating frame of reference, the circumferential inhomogeneity of leakage flow clearly shows that we have to deal with unsteady effects. Therefore, in future work unsteady measurements and simulations will be included.

In agreement with the experimental data, the computational results show that even at realistic clearance heights the leakage flow can give rise to suction-sided incidence of considerable parts of the downstream stator which causes the flow to separate. Thus, it can be concluded that leakage flow should be taken properly into account in the design or optimization process of turbomachinery.

Acknowledgements

The work was supported by the Forschungsvereinigung Verbrennungskraft-maschinen e.V. (FVV) which is gratefully acknowledged. Further, the authors are very thankful for the support given by the visualization group of HLRS visualizing the performed simulations in the CUBE.

References

1. Anker J.E.; Mayer J.F.; Stetter H., Computational Study of the Interaction of Labyrinth Seal Leakage Flow and Main Flow in an Axial Turbine, in: *4th European Conference on Turbomachinery, Fluid Dynamics and Thermodynamics*, edited by Bois G.; Decuypere R.; Martelli F., pp. 641–652, 2001
2. Denton J.D., Loss Mechanisms in Turbomachines, *Journal of Turbomachinery*, 115(4):621–656, October 1993
3. Giles M.B., Non–Reflecting Boundary Conditions for the Euler–Equations, TR–88–1, MIT Computational Fluid Dynamics Laboratory, 1988
4. He L., Time-Marching Calculations of Unsteady Flows, Blade Row Interaction and Flutter, *VKI-LS 1996-05*, 1996
5. Jameson A.; Schmidt W.; Turkel E., Numerical Solutions of the Euler Equations by Finite Volume Methods Using Runge Kutta Time–Stepping Schemes, *AIAA Paper 81-1259*, 1981
6. Jung A.R., *Berechnung der Stator-Rotor-Wechselwirkung in Turbomaschinen*, Dissertation, Universität Stuttgart, Institut für Thermische Strömungsmaschinen und Maschinenlaboratorium, 2000
7. Jung A.R.; Mayer J.F.; Stetter H., Unsteady Flow Simulation in an Axial Flow Turbine Using a Parallel Implicit Navier-Stokes Method, in: *High Performance Computing in Science and Engineering '98*, edited by Krause E.; Jäger W., Transactions of the High Performance Computing Center Stuttgart (HLRS) 1998, pp. 269–294, Springer-Verlag, 1999
8. Kacker S.C.; Okapuu U., A Mean Line Prediction Method for Axial Flow Turbine Efficiency, *ASME Journal of Engineering for Power*, 104(1):111–119, January 1982
9. Korschunov B.A.; Döhler S.W., Einfluss von Leckageströmungen an der Laufradspitze auf die aerodynamischen Charakteristiken des folgenden Leitgitters, *BWK – Brennstoff, Wärme, Kraft*, 48(7/8), 1996
10. Merkle C.L.; Sullivan J.A.; Buelow P.E.O.; Venkatswaran S., Computations of Flows with Arbitray Equations of State, *AIAA J.*, 36(4):515–521, 1998
11. Merz R., *Entwicklung eines Mehrgitterverfahrens zur numerischen Lösung der dreidimensionalen, kompressiblen Navier-Stokes-Gleichungen in mehrstufigen Turbomaschinen*, Dissertation, Universität Stuttgart, Institut für Thermische Strömungsmaschinen und Maschinenlaboratorium, 1998
12. Peters P.; Breisig V.; Giboni A.; Lerner C.; Pfost H., The Influence of the Clearance of Shrouded Rotor Blades on the Development of the Flowfield and Losses in the Subsequent Stator, *ASME Paper 2000-GT-478*, 2000
13. Rhode D.; Demko J.; Traigner U.; Morrison G.; Sobolik S., The Prediction of Incompressible Flow in Labyrinth Seals, *Trans. ASME, J. Fluids Engrg.*, 108:19–25, 1986
14. Rhode D.; Johnson J., Flow Visualisation and Leakage Measurements of Stepped Labyrinth Seals; Part 1: Annular Groove, *ASME Paper 96-GT-136*, 1996

15. Saxer A.P., *A Numerical Analysis of 3-D Inviscid Stator/Rotor Interactions Using Non-Reflecting Boundary Conditions*, PhD Thesis, MIT Gas Turbine Laboratory, 1992

16. Stetter H.; Pfost H.; Breisig V.; Anker J.E., Deckbandströmungseinfluß, *in: FVV-Heft R 504, Informationstagung Turbinen*, Forschungsvereinigung Verbrennungskraftmaschinen e. V., Frankfurt a. M., S. 125–154, 1999

17. Stetter H.; Pfost H.; Giboni A.; Anker J.E., Deckbandströmungseinfluß, *in: FVV-Heft R 509, Informationstagung Turbinen*, Forschungsvereinigung Verbrennungskraftmaschinen e. V., Frankfurt a. M., 2000

18. Stoff H., Incompressible Flow in a Labyrinth Seal, *J. Fluid Mech.*, 100:817–829, 1980

19. Storteig E., *Dynamic Characteristics and Leakage Performance of Liquid Annular Seals in Centrifugal Pumps*, Doktoravhandling (PhD Thesis), Institutt for Marint Maskineri, NTNU, Trondheim, Norway, 2000

20. Turkel E.; Radespiel R.; Kroll N., Assessment of Two Preconditioning Methods for Aerodynamic Problems, preprint submitted to Elsevier, 1997

21. Wallis A.M.; Denton J.D.; Demargne A.A.J., The Control of Shroud Leakage Flows to Reduce Aerodynamic Losses in an Low Aspect Ratio, Shrouded Axial Flow Turbine, *ASME Paper 2000-GT-475*, 2000

22. Waschka W.; Wittig S.; Kim S.; Scherer T., Heat Transfer and Leakage in High-Speed Rotating Stepped Labyrinth Seals, in: *AGARD-CP-527*, pp. 26.1–10, 1993

Numerical Simulation of Rotating Stall in an Axial Compressor

F. Ginter, A. Ruprecht, and E. Göde

Institute of Fluid Mechanics and Hydraulic Machinery, University of Stuttgart

Abstract. The flow instability "rotating stall" is simulated for a one-stage water model of an axial compressor with 30 runner and 30 stator blades. The domain of calculation of the entire 3d-model is splitted into 60 subdomains and simulated in parallel. The interaction between impeller and diffuser is taken into account by a coupling algorithm with overlapping grids. The simulation results are compared with experiments. The predicted angular speed of the single full span rotating stall cell and the temporal curve of the pressure in the interspace agree well with the experiment.

1 Introduction

The stable behaviour of pumping systems is in general a basic condition for a reasonable operation and high operating hours of the pump. Instabilities at part load can lead to reduction of efficiency, high dynamic load and cavitation up to the destruction of the pump. The operating range is mostly limited by the existence of instabilities. One of the well known instabilities of technical importance is rotating stall (RS).

Rotating stall incepts at part load and is associated with a complete change of the flow pattern. The flow field separates into one or more blocked regions and one or more sound regions. The blocked regions are characterized by extended backflow and are called rotating stall cell. After the onset of RS a new stable periodic flow state exists, which is signed by a constant rotating frequency of the RS cell, which is in the range of 50 % of the rotor frequency. Velocity and pressure distribution can be assumed to be stationary in a relative system, rotating with the speed of the cell.

Numerous experimental studies have been carried out since the seventies, to get knowledge about RS. Day et al., 1978, presented a basic work about the characteristic of axial compressors and Day and Cumpsty, 1978, discussed the flow pattern in the passages. An overview about RS can be found in Pampreen, 1993. A model of the cell propagation for fully developed RS is presented by Gyarmathy, 1996. This model is based on the momentum exchange between impeller and diffuser.

Due to the complexity of the flow phenomena and the effort of the calculation, only less publications about RS simulation can be found. Most of them use two- dimensional models or reduce the number of passages for three-dimensional calculations. Nishizawa and Takata, 1994, Outa et al., 1994, and

He, 1997, simulated RS in two-dimensional models. Ismail and He, 1997, calculate the inception of RS in a NASA Rotor 67 for 7 of 22 blades, based on a three-dimensional model. Hoying et al., 1999, simulate the inception of RS due to 8 blades, while introducing the tip clearance flow.

The goal of this project is the simulation of RS, to get better insight into the mechanism within the machine. Two-dimensional calculation of the investigated axial compressor have been carried out and are discussed in Saxer et al., 1999. Therefore the interest of the presented paper is focussed to the three-dimensional model of the entire machine. The investigated subject is fully developed RS with constant cell propagating speed, whereas the inception and onset of RS is not discussed.

2   Numerical methods

2.1   Basic equations

The basic equations are the time dependent Reynolds-averaged Navier-Stokes equations for an incompressible fluid. The Reynolds stresses are expressed by the Boussinesq' turbulent viscosity hypothesis.

For the diffuser the equations are solved in the absolute system, whereas the equations for the impeller are introduced as absolute velocities in the relative system. The momentum equations have the following form:

$$\frac{\partial U_i}{\partial t} + U_i \frac{\partial U_i}{\partial x_j} + \frac{1}{\rho}\frac{\partial p}{\partial x_i} - \left[(v + v_t)\left(\frac{\partial U_i}{\partial x_j} + \frac{\partial U_j}{\partial x_i}\right)\right] = \epsilon_{ijk}\Omega_j U_k \quad (1)$$

The angular velocity is $\Omega = 0$ for the absolute system. The momentum equation are solved together with the continuity equation, which is identical for both cases,

$$\frac{\partial U_i}{\partial x_i} = 0 \quad (2)$$

The turbulent viscosity is calculated by the mixing length model according to Prandtl. The usage of higher turbulence models is avoided, because the interest is focussed to fully developed RS. It is known, that fully developed RS is less influenced by friction, but dominated by advective forces, see Saxer et al., 1999. For the discussion of the inception mechanism friction might play a more important role, so that a more accurate prediction of turbulence and consequently higher order models may be necessary.

2.2   Simulation algorithm

The simulations are carried out by the program FENFLOSS, which has been developed at the Institute for more than a decade, see Ruprecht, 1989, and Ginter, 1997. The partial differential equations are solved by the Galerkin

Finite Element Method. The spatial discretization of the domain is performed by 8 node brick elements for three-dimensional problems. For the velocity components tri-linear approximation is applied. The pressure is assumed to be constant within the element. For advection dominated flow a Petrov-Galerkin formulation with skewed upwind orientated weighting functions is applied. The time discretization is done by a three-level fully implicit finite difference approximation of 2nd order.

For the solution of the momentum equation and continuity equation a segregated solution algorithm is used, see Ruprecht et al., 1999. Each momentum equation is handled independently. The linear equation systems are solved by the BICGSTAB2 algorithm with an incomplete LU decomposition for preconditioning. The pressure is treated by a modified UZAWA type pressure correction scheme. After the solution of the momentum and continuity equations the turbulence viscosity is predicted by algebraic relations, based on the velocity field. The whole procedure is carried out in a global iteration until convergence is obtained. For unsteady simulations the global iteration has to be carried out in each time step.

2.3   Coupling conditions

Because of the different flow patterns in the passages during RS state, the entire machine with 30 rotor blades and 30 stator blades has to be modelled. For the investigated problem reasonable simulation time can only be achieved by using high performance computation with massively parallel architecture, see also Heitele et al., 1999. The model is separated into two components, the impeller and the diffuser. Each passage of the components will be treated by one processor. The entire domain of simulation will be separated into 60 subdomains and distributed to exactly the same number of processors. The communication between the different subdomains is achieved by overlapping grids.

At the frontier between the single subdomains two different coupling conditions exist. The interfaces inside of one component are independent from time, so that the partner of the data exchange does not change during the simulation. The coordination can be fixed at the beginning of the simulation. The interface between impeller and diffuser is characterized by the motion of the grid against the other, due to the rotation of the impeller. In the general case non-matching grids with overlapping parts are allowed, see Fig. 1. The position of the nodes which are introduced into the communication is changing at each time step and must be predicted again.

The exchange of data between the two components of the machine is based on dynamical boundary conditions, which will be renewed at each global iteration. Downstream, the values of the nodes will be transformed from rotor to stator as Dirichlet boundary condition. Upstream, the fluxes will be transferred from stator to rotor as a surface integral. More details about the coupling algorithm can be found in Ruprecht et al., 1999.

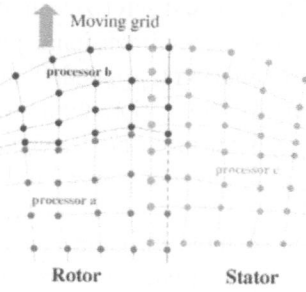

Fig. 1. Impeller and diffuser with non-matching coupling.

3   Simulated compressor and numerical model

The investigated axial compressor is a water model, which is located at the Laboratory of Turbomachines at the ETH Zurich. The axial compressor consists of four stages. The impeller and the diffuser are composed of 30 blades each . The diameter at the hub is 299 mm, the diameter at the tip is 380 mm, so that the hub/tip ratio is 0.82. The rotational speed is 120 min^{-1}. The design point is located at a flow coefficient of φ=0.396 related to the radius in the centre of the passage.

The numerical model consists of a fully three-dimensional model for a one-stage compressor with 30 impeller and 30 diffuser blades. The clearance flow between impeller blade tip and casing as well as between diffuser blade tip and hub do not influence significantly the flow characteristic of fully developed RS, see Harada, 1985. Therefore, the gaps have not been modelled. The FE-discretization is shown in Fig. 2. The blades are located in the middle of a subdomain.

An unrolled midspan cut of the domain of simulation is shown in Fig. 3. The domain in front of the stage and behind the stage is elongated to avoid

Fig. 2. FE discretization of impeller and diffuser blades.

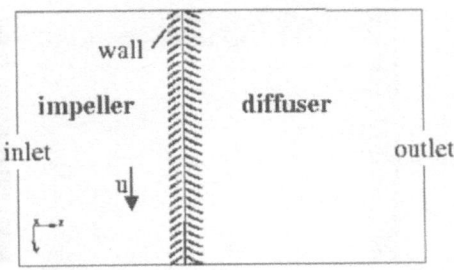

Fig. 3. Unrolled midspan cut of domain of calculation and boundary condition.

influences of the boundary conditions to the simulated flow field near the passages.

At the inlet the axial and the circumferential velocity components are assumed to be constant around the circumference. At the outlet the pressure is set to be 0. Impeller and diffuser blades, hub and casing are associated with wall boundary conditions. At the elongation from hub and casing to inlet and outlet symmetry boundary conditions are set to avoid changes of the flow angle of the incoming flow due to wall boundary conditions. Impeller and diffuser are combined by coupling conditions, which are described in chapter 2.3.

No disturbance is applied at the inlet, so that the interaction between impeller and diffuser is the only unsteady boundary condition. RS develops as an unstable flow state in the domain of calculation. For one revolution of the impeller 270 time steps are necessary. One time step corresponds to an angular turning of $\Delta\varphi=1.33^{o}$.

The grid structure of the single subdomains of impeller and diffuser are identical. The subdomains of the impeller are discretized by about 7300 elements, those of the diffuser by about 7000 elements. The entire domain of calculation is distributed into about 450000 elements.

4   Simulation results

The angular frequency Ω_{RS} of the cell and the blockage factor λ are characteristic values for RS. The angular frequency Ω_{RS} is located between 0 and the rotor frequency Ω_R and is defined in ratio to the rotor frequency Ω_R. The blockage factor λ describes the part of the blocked region,

$$\lambda = \frac{blocked\ region}{entire\ region} \tag{3}$$

The blocked region is bounded by a pressure peak and a pressure valley in the interspace between impeller and diffuser.

In the interspace a spot point is defined to show the temporal curve of the physical quantities near casing. In cylinder sections near the middle of

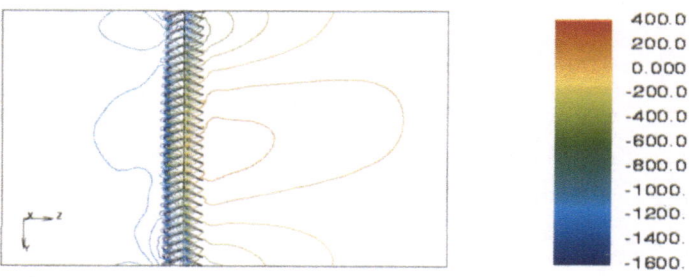

Fig. 4. Pressure distribution near hub at t=11 T$_R$.

Fig. 5. Pressure distribution near casing at t=11 T$_R$.

the passage and near casing cutting lines are defined in front and behind the impeller. The z-axis is chosen to be the rotating axis so that the z-component of the velocity corresponds to the transport component. The simulation was carried out on a Hitachi SR8000.

The flow field is simulated for the operating point Q/Q$_0$=0.63 (φ=0.25). For this operating point experimental results exist, so that a comparison is possible. The pressure difference between inlet and outlet amounts to about ΔP=1000 Pa. The experiment gives an average pressure difference of ΔP=950 Pa, so that experiment and calculation agree quite well.

The pressure difference against the outlet is plotted in Fig. 7. for the region in the interspace for a time range of t=8 T$_R$ to t=16 T$_R$. The RS cell is bounded by a pressure peak and a pressure valley. For the chosen operation condition, the computation shows one single stall cell of the full span type. This agrees well with the experiment.

In Fig. 8 the measured c_p-value of the pressure is plotted. The comparison between the temporal curves shows good agreement between simulation and experiment. The blockage factor due to the pressure maximum and minimum is predicted to λ=0.45. The experimentally found value varies between λ=0.34 and λ=0.4. The rotational speed of the RS-cell is simulated to Ω_{RS}/Ω_R=0.51. This agrees well with the experimental value of Ω_{RS}/Ω_R=0.54. The difference between pressure peak and pressure valley

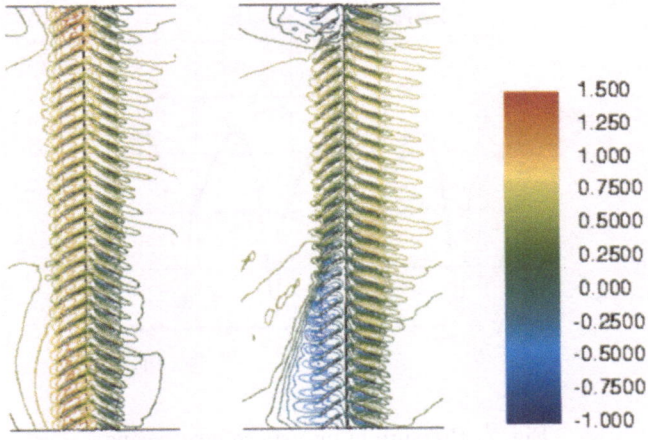

Fig. 6. Distribution of v_z at t=11 T_R, near hub (left) and near casing (right).

is simulated to ΔP=1650 Pa and is compared to the experimentally deter-mined value of ΔP=1830 Pa slightly lower. One reason for this difference may be the relatively coarse grid, so that a reduction of this difference can be expected with a grid refinement.

To get an insight into the flow structure the simulated results will be discussed based on two unrolled cylinder cuts at the time step t=11 T_R. The cylinder cuts are located 10 % of the blade height away from the hub and 10 % of the blade height away from the casing. The pressure distribution is shown in Figs. 4 and 5 as difference to the outlet pressure in Pa. In both plots the unequal pressure distribution around the circumference can be detected. In front of the impeller a region with low pressure is built, which indicates the existence of one full span RS-cell. Behind the diffuser higher pressure can be found. The distribution between hub and casing shows only slight differences.

The transport velocity v_z is shown in Fig. 6 for both cylinder cuts near hub and casing. The unequal distribution of the single passages can be recognized with one RS cell and separation of sound and blocked region. While through-flow with high velocity in the RS-cell is determined near hub, the region near casing is characterized by marked backflow. This leads to strong three- di-mensional flow pattern in the impeller. In the diffuser passages throughflow without backflow is predicted in both cases. The region with sound flow is characterized by throughflow in impeller and diffuser passages and shows low differences between hub and casing.

For time step t=11 T_R cuts are set in unrolled cylinder cuts near hub, in the middle of the passage and near casing. The cuts are compared for position in front of the impeller and in the interspace. For the cut in front of the impeller v_z is plotted in Fig. 9. The transport velocity shows a sig-nificant split between sound flow and RS cell. For the sound flow v_z shows

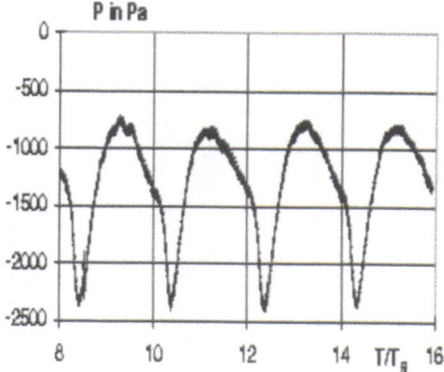

Fig. 7. Pressure in interspace near casing

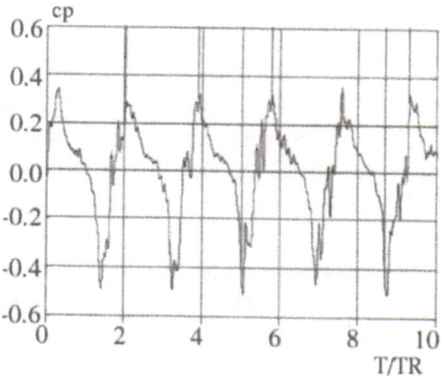

Fig. 8. Experimentally determined pressure near casing

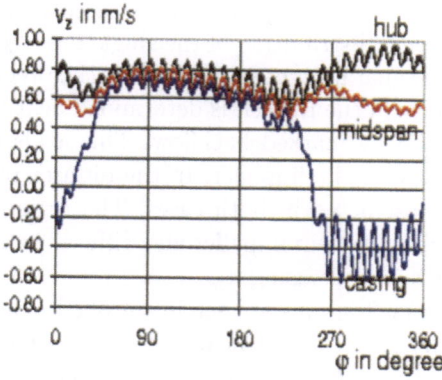

Fig. 9. Transport velocity in front of impeller

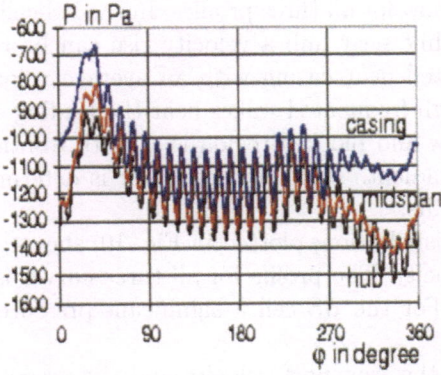

Fig. 10. Pressure in front of impeller

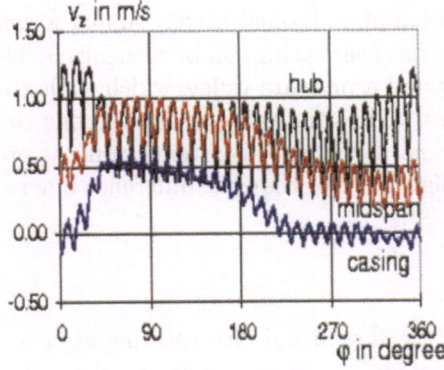

Fig. 11. Transport velocity in interspace

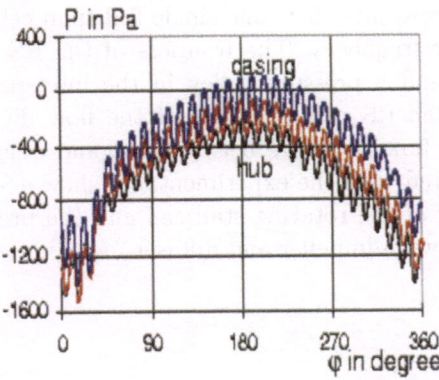

Fig. 12. Pressure in interspace

nearly identical inflow for all three profiles. In the RS-cell the velocity profiles are separating. While near hub a velocity rise can be recognized a marked backflow is predicted near casing with an average velocity of v_z=-0.4 m/s with significant high frequented wakes near the leading edge. The transition between sound flow and blocked region is at both frontiers characterized by a steep velocity gradient. The blockage factor is determined to λ=0.38 due to the velocity profile.

The pressure distribution, plotted in Fig. 10, shows, in tendency, similar pattern as the velocity. The profile for all three curves is nearly identical at the sound inflow. For the RS cell a significant pressure rise near casing is simulated.

The profiles of the transport velocity and the pressure in the interspace are shown in Figs. 11 and 12. Near casing v_z indicates the separation between sound flow and RS-cell. The blocked region reaches to the interface due to the negative transport component. Near hub increasing velocity in the RS-cell can be detected.

The pressure distribution is qualitatively similar for all three cuts, whereby a constant pressure rise near casing can be recognized. The RS cell is bounded by a pressure peak and a pressure valley, which leads to a blockage factor of λ=0.45. This agrees well to the value calculated due to the temporal curve. The pressure difference between the extreme values is predicted to ΔP=1600 Pa. This value is slightly lower than the difference due to the temporal curve.

5   Summary

The simulation of the flow instability rotating stall in a single-stage axial compressor is presented. The interest is focussed to structure and motion of the fully developed RS cell, whereas the inception of RS is not studied. The domain of calculation is splitted into 60 subdomains and simulated parallely. The interaction between impeller and diffuser is taken into account by a coupling algorithm with overlapping grids.

The simulation results show one single full span cell, which rotates with 51 % of the rotor frequency. The frontiers of the RS cell are bounded by a pressure peak and a pressure valley in the interspace between impeller and diffuser. In the RS cell a change of the flow direction with strongly three-dimensional flow structure between hub and casing is simulated. The results are compared with the experiment and show good agreement for the angular frequency of the rotating stall cell and the pressure distribution in the interspace between impeller and diffuser.

References

Day, I. J., Cumpsty, N. A.: The Measurement and Interpretation of Flow within Rotating Stall Cells in Axial Compressors. Journal of Mechanical Engineering Sciences, Vol. 20 (1978)

Day, I. J., Greitzer, E. M., Cumpsty, N. A.: Prediction of Compressor Performance in Rotating Stall. Journal of Engineering for Power, Vol.100 (1978)

Ginter, F.: Berechnung der instationären, turbulenten Strömung in hydraulischen Strömungsmaschinen. Ph. D., Institute for Fluid Mechanics and Hydraulic Machinery, University of Stuttgart, Germany, 1997

Gyarmathy, G.: Impeller-Diffuser Momentum Exchange during Rotating Stall. ASME-Paper 96-WA/PID-6, ASME Winter Annual Meeting, Atlanta, (1996)

Harada, H.: Performance Characteristics of Shrouded and Unshrouded Impellers of a Centrifugal Compressor. Journal of Engineering for Gas Turbines and Power, Vol. 107, 1985

He, L.: Computational Study of Rotating Stall Inception in Axial Compressors. Journal of Propulsion and Power, Vol. 13, 1997

Heitele, M., Helmrich, T., Maihöfer, M., Ruprecht, A.: New Insight into an Old Product by Parallel Computing. Proceedings of the Fifth European SGI/Cray MPP Workshop, Bologna, 1999

Hoying, D. A., Tan, C. S., Huu Doc Vo, Greitzer, E. M.: Role of Blade Passage Flow Structures in Axial Compressor Rotating Stall Inception. Journal of Turbomachinery, Vol. 121, 1999

Ismail, J. O., He, L.: Three Dimensional Computation of Rotating Stall Inception. Proceedings of the 2nd European Turbomachinery Conference, Antwerpen, 1997

Nishizawa, T., Takata, H.: Numerical Study on Rotating Stall in Finite Pitch Cascades. ASME 94-GT-258

Outa, E., Dai, K., Chiba, K.: An N-S Simulation of Stall Cell Behaviour in a 2-D Compressor Rotor-Stator System at Various Loads. ASME 94-GT-257

Pampreen, R. C.: Compressor Surge and Stall. Concepts ETI, Norwich, 1993

Ruprecht, A.: Finite Elemente zur Berechnung dreidimensionaler, turbulenter Strömungen in komplexer Geometrien. Ph. D., Institute for Fluid Mechanics and Hydraulic Machinery, University of Stuttgart, Germany, 1989

Ruprecht, A., Bauer, C., Gentner, C., Lein, G.: Parallel Computation of Stator-Rotor Interaction in an Axial Turbine. ASME PVP Conference, CFD Symposium, Boston, 1999

Saxer A. P., Saxer-Felici, H. M., Ginter, F., Interbitzin, A., Gyarmathy, G.: Structure and Propagation of Rotating Stall Cells in an Axial Compressor. 3rd European Conference on Turbomachinery - Fluid Dynamics and Thermodynamics, London, 1999.

Euler and Navier-Stokes Solutions for Flapping Wing Propulsion

M.F. Neef and D. Hummel

Institut für Strömungsmechanik, Technische Universität Braunschweig,
Bienroder Weg 3, 38106 Braunschweig, Germany

Abstract. The flapping flight of birds is investigated with the help of a finite-volume method for numerical computation of the governing conservation laws. Sufficient spatial and temporal discretization of the computational domain necessitates the use of high performance computing. Thrust is calculated from unsteady pressure distributions for 2D and 3D cases without and with account for viscous forces. The results for velocities near the wing surface and in the whole computational domain allows studying of time-dependent circulation and vortices.

1 Introduction

With the help of todays resources in high performance computing, the calculation of the unsteady flow field around moving surfaces has become feasible, as results for the flow around helicopter rotor blades have already shown. However, while most of the aerodynamic research in Computational Fluid Dynamics (CFD) is aimed at a more sophisticated understanding and use of conventional lifting surfaces and propulsive systems, the flapping flight of birds still remains comparatively unexplored.

Analytical results for an oscillating flat plate are available since the 30'ies of last century [1], still providing an easy means for quick comparison with more sophisticated approaches. While experimental results for thrust generation are rather rare, recent years have spawned a number of publications where CFD was used to incorporate all physical effects which govern the flow field around the flapping wing. This includes solution of the Navier- Stokes equations for oscillating airfoils in forward flapping flight (2D) [2,3] and for insect wings in hovering flight (3D) [4]. For the cruising flight of birds, computational results for finite span wings have been obtained with Panel-Methods [5] or Euler-Codes [6], not counting other results based on classical airfoil theory.

The reason for the high computational cost of studying flapping wings becomes apparent from the complexity of the involved flow phenomena. Due to the wing motion, the flow is unsteady and thus requires sufficient resolution in time for iterative solution procedures. The same requirement holds for discretization in space, if a 3D domain around the flapping wing is to be modeled with appropriate resolution of the boundary layer.

2   Wing Motion

The primary movement in flapping flight is a plunging motion vertical to the direction of the freestream, see Fig. 1a. The oscillation is assumed to be first order harmonic, thus the plunging coordinate, related to the chord length c of the airfoil, can be written as

$$\frac{z}{c}(t) = \frac{z_1}{c}\,\cos(2\pi f t), \tag{1}$$

with plunging amplitude $\frac{z_1}{c}$. The flapping frequency f is often expressed in nondimensional form as the reduced frequency k:

$$k = \frac{\pi f c}{U_\infty}. \tag{2}$$

Here, U_∞ denotes the freestream velocity. Typical values for k in flight of large birds are in the order of $k \approx 0.1$, with higher k for smaller birds and insects. Although pure plunging may produce a thrust force, the performance of the system can be improved by a superimposed pitching motion according to Fig. 1b, which can be defined as

$$\alpha(t) = \alpha_0 + \alpha_1 \cos(2\pi f t + \Phi). \tag{3}$$

The pitching with amplitude α_1 can experience a phase shift Φ relative to the plunging, while a mean angle of attack $\alpha_0 > 0$ will provide a mean lift.

If a bird's wing of finite span is considered, the plunging may be expressed as a linear function of the spanwise position of an airfoil section or, in other terms, can be written as an angular motion with flapping amplitude Ψ_1, see Fig. 1c. The corresponding 3D pitching becomes a twisting of the wing with

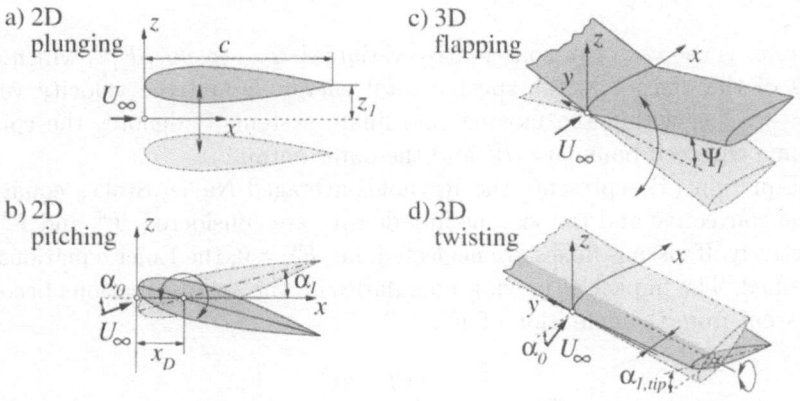

Fig. 1. Definition of wing motion

varying twist along span with an amplitude of $\alpha_{1,\text{tip}}$ at the tip, as depicted in Fig. 1d.

The actual motion of a birds wing may also include a forward and backward motion, a dynamic flexing and an oscillating fuselage position, which will be neglected here. These additional modes could be included in the present model, but they make parameter studies too complex. Despite, the flow around the wing is assumed to be fully attached throughout the flapping cycle, which imposes moderate reduced frequencies and amplitudes, which are experienced in the cruising flight of large birds.

3   Numerical Method

The presented results have been obtained using the CFD Code FLOWer, which was compiled for the NEC-SX5 architecture in sequential mode. The software has been developed by the Institute of Design Aerodynamics at the DLR Braunschweig, Germany. For the flapping wing problem, the conservation laws for mass, momentum and energy in integral form are solved on structured meshes. For discretization in space, a second-order cell-vertex method for finite volumes is employed. Integration with respect to time utilizes a five stage Runge-Kutta scheme to converge the solution, which can be accelerated by using implicid residual smoothing and a multigrid scheme [7]. A considerable cut in computational cost for calculating unsteady flows is achieved with an implicit dual time stepping scheme.

The flapping wing problem requires the possibility to account for the motion of the wall boundaries, i.e. at least some of the grid points are exposed to a transient motion and the flow becomes unsteady. For this purpose, the governing equations for a moving coordinate system are written as follows [8]:

$$\frac{\partial}{\partial t}\int_V w\, dV + \int_{\partial V}(F^c - F^v)\,n\, dS + \int_V g\, dV = 0, \tag{4}$$

Here, w is the vector of conservative variables, $w = [\rho, \rho q, \rho E]^T$, which consists of the density ρ, the specific total energy E and the velocity vector $q = [u, v, w]$ within the moving coordinate system. V denotes the control volume with cell boundary ∂V and the outer normal n.

Equation (4) represents the Reynolds-averaged Navier-Stokes equations if the convective and the viscous flux-density are considered, F^c and F^v respectively. If viscous fluxes are neglected, i.e. $F^v = 0$, the Euler equations are obtained. The impact of moving boundaries on the set of equations becomes apparent from the definition of F^c,

$$F^c = \begin{bmatrix} \rho(q - q_b) \\ \rho q \bullet (q - q_b) + pI \\ \rho E(q - q_b) + pq \end{bmatrix}, \tag{5}$$

where q_b acconts for the boundary velocity for each cell face of the computational grid. Pressure p in the momentum law is multiplied with a 3×3 identity matrix I for the three coordinate directions. For predefined motions of the wing, the q_b's are known in advance of the calculation and can be provided for the solution of (4). This is achieved in two different ways. Either, the motion of the wing can be regarded as a rigid body motion, e.g. pure plunging and/or pitching, which allows the use of a single grid and the analytical description of the motion. In this case, source term g in (4) accounts for the time derivatives of the unity direction vectors, if a rotating coordinate system is used for the description of pitching. The second way for implementation of wing motions is the use of a time-history of predefined meshes, with the wing placed at different positions in each grid. For an oscillating airfoil, the provision of seven meshes per period is sufficeint to calculate the desired q_b's for each time step by means of discrete Fourier analysis. In this case, the wing may also be deformed during the motion, e.g. wing twisting can be considered, which affords flexible cell volumes in the surrounding grid. A geometric conservation law accounts for the time-dependent change of local cell volumes.

Computational cost is relatively low, if the oscillating wing is regarded as two dimensional (see Fig. 1a,b). The use of high performance computing becomes helpful in the solution of the Euler equations for a 3D wing with aspect ratio $A = 8$ undergoing a flapping and twisting motion. Due to the symmetry of the problem, only one half of the wing needs to be modeled (see

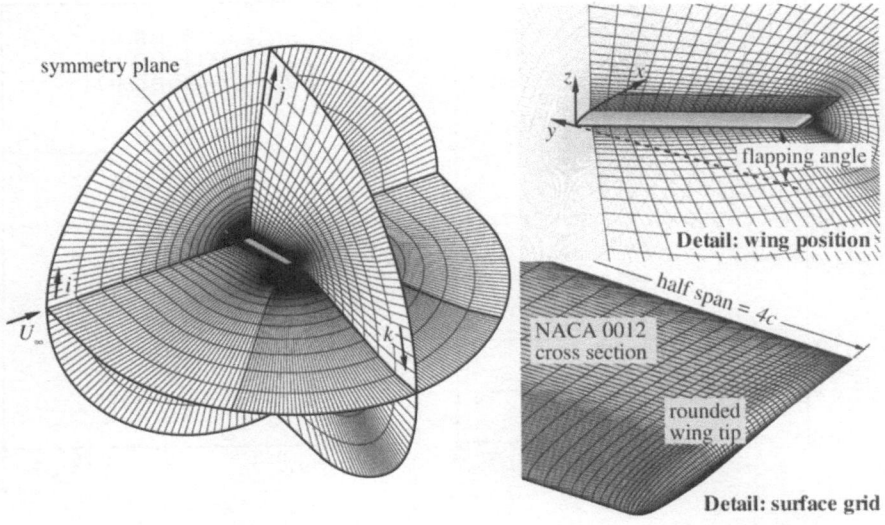

Fig. 2. Flexible mesh for 3D Euler calculation using OO-topology and 204 800 gridpoints

Fig. 2), with 200 000 grid points giving satisfactory results. For the solution of viscous flow, sufficient resolution of the boundary layer requires up to 1.0 Mio. overall gridpoints. Concerning temporal resolution, the dual time stepping approach allows relatively large outer time steps, usually 40 to 80 per period. Together with 50 inner iterations in each time step and two motion cycles to calculate for each flapping wing case, about 4 000 to 8 000 iterations for the whole grid are needed for solution of the five governing equations. If extensive turbulence modeling is employed, up to two more equations are to be solved. Accounting for dynamic stall effects or better resolution of the wake behind the wing may afford additional expenses in temporal and spatial discretization.

Since the governing equations are solved in compressible form, the code encounters stability and convergence problems for very low Mach numbers. Thus, all calculations were performed for a Mach number of $M = 0.3$, where compressiblity effects are still negligible.

4   Results

The exemplary result for a time-accurate calculation of the flow around a flapping airfoil is shown in Fig. 3. The calculation first of all yields the unsteady pressure distribution and – for viscous forces – the shear stresses at

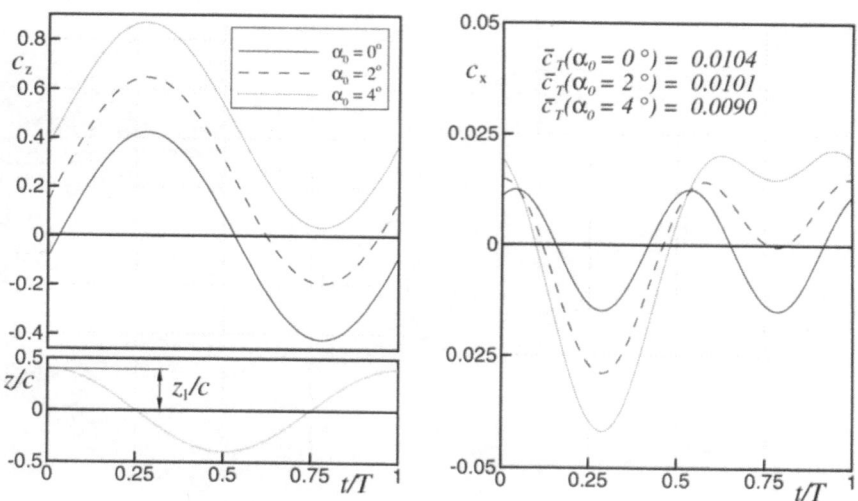

Fig. 3. Vertical and horizontal force coefficient of a pure plunging motion. (2D Navier-Stokes calculation, NACA 0014 airfoil, $k = 0.1$, $z_1/c = 0.4$, $Re = 1.0 \times 10^6$, Baldwin-Lomax turbulence model)

the wing surface. Integration over the whole surface results in the corresponding aerodynamic force, which can be split up in directions perpendicular and
vertical to the freestream. In dimensionless form, a horizontal and a vertical
force coefficient is obtained, c_x and c_z respectively. Non-dimensionalization
is achieved by referring the aerodynamic force to the product of dynamic
pressure in the freestream and the reference area, e.g. $c_x = F_x/(\frac{1}{2}\rho U_\infty^2 S_{ref})$.
The time-history of motion and force coefficients in Fig. 3 shows the effect of
a pure plunging motion on an airfoil in 2D flow. The vertical force oscillates
with time, but its mean value is only larger than zero, if a mean angle of attack $\alpha_0 > 0$ is employed. Regarding the horizontal force coefficient, a positive
value denotes drag, whereas negative values in Fig. 3b show the appearance of
instantaneous thrust. In Navier-Stokes calculations the contribution of pressure (Index p) and wall friction (Index f) are calculated indepentently, thus
the horizontal force coefficients can be expressed as $c_x = c_{x,p} + c_{x,f}$. The
friction induced part $c_{x,f}$ is always positive and almost independent of time,
hence only $c_{x,p}$ is of interest for the thrust generation. Therefore, the mean
thrust per period $\bar{c}_T$ is defined as:

$$\bar{c}_T = -\frac{1}{T} \int_0^T c_{x,p} \, dt. \tag{6}$$

For increasing mean angles of attack α_0 shown in Fig. 3, the mean thrust
output diminishes sightly for increasing α_0 due to growing airfoil drag, which
shows its effect on the pressure distribution.

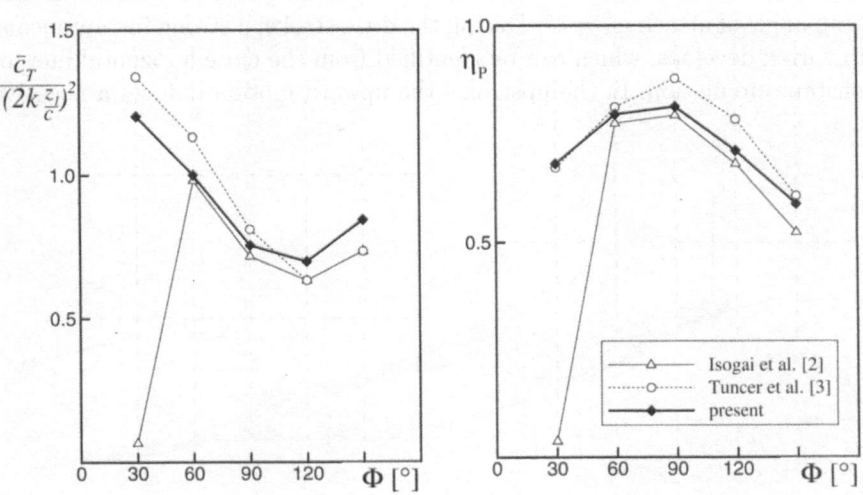

Fig. 4. Mean thrust coefficient and propulsive efficiency of a combined plung-
ing/pitching motion. (2D Navier-Stokes calculation, NACA 0012 airfoil, $k = 0.15$,
$z_1/c = 1.0$, $\alpha_0 = 0°$, $\alpha_1 = 10°$, $Re = 1.0 \times 10^5$, Baldwin-Lomax turbulence model)

Thrust output is of major concern for the performance of the propulsive system. It can be studied by varying the motion paramters such as frequency and amplitude, as well as phase shift Φ, if a pitching motion is superimposed. This has been investigated for a NACA 0012 airfoil in [2,3], thus providing a good basis for comparison with present results. Thrust $\bar{c}_T$ and efficiency η_P are calculated, the latter being defined as mean pressure-based power output divided by input power due to the plunging and pitching motion. $\bar{c}_T$ in [2] is defined in accordance with (6), but the result is divided by an additional factor $(2kz_1/c)^2$. Respecting these definitions, the good agreement with results from the present code can be seen in Fig. 4. Note the general result for highest efficiencies near phase shifts of $\Phi = 90°$. In this case, the pitching advances the plunging by a quarter of a period. This characteristic motion can also be observed in the flight of birds.

For the 3D wing, the computational method offers a wide range of possibilities to evaluate the obtained unsteady flow field data, e.g. the instantaneous circulation distribution along the flight path of the wing. For this purpose, the difference between the surface velocities of upper and lower side of the wing in an Euler calculation is used to calculate the instantaneous vortex density distribution of the wing. Integration of this distribution in chordwise direction yields the circulation Γ as a function of wing span for a time t during the flapping cycle. The circulation $\Gamma(y)$ can be obtained for distinct instants or – identically – for distinct positions along the flight path in x-direction as shown in Fig. 5. Plotting the distribution of $\Gamma(x,y)$ in a top view of Fig. 5 produces a contour plot with lines of constant circulation. A result is shown in Fig. 6, where two different wing motions are considered.

In Fig. 6a, the wing is undergoing a pure plunging motion with a constant mean angle of attack $\alpha_0 = 4°$. During the downstroke, a strong instantaneous tip vortex develops, which can be identified from the close horizontal lines of constant circulation. In the upstroke, the upward motion induces a negative

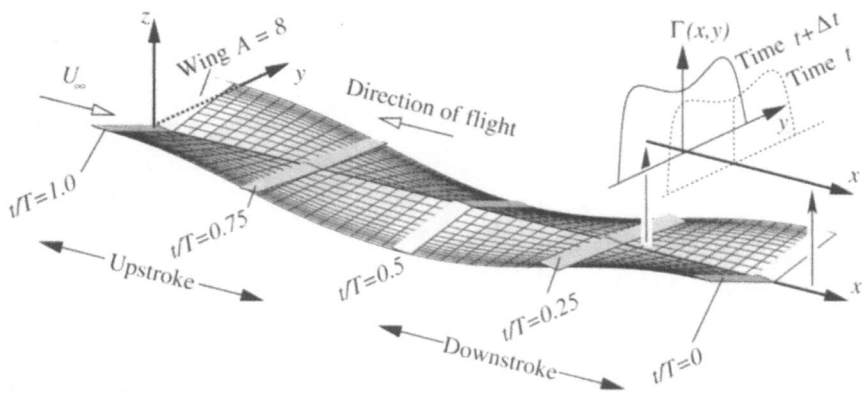

Fig. 5. Flight path of the flapping and twisting wing, $\alpha_0 = 4°$

instantaneous angle of attack that counterweights the positive α_0 thus suppressing the formation of a tip vortex. Vertical isolines indicate the existence of transverse vortices due to aceleration and deceleration of the moving wing. These vortices are especially strong when the wing changes its direction of motion, e.g from down- to upstroke. For a more complicated motion such as bird-like flapping and twisting of the wing, the circulation distribution becomes more complex, see Fig. 6b. While the formation of a tip vortex will follow the same principle as in case (a), starting and stopping vortices are particularily strong at outboard wing locations. This is indicated by the close isolines near the tip, where vertical motion of the wing is at a maximum.

The formation of the tip vortices can also be followed by plotting the velocities in a plane behind the trailing edge as shown in Fig. 7. Here, the flapping and twisting wing is in the middle of a downstroke ($t/T = 0.25$), with a vortex rotating counterclockwise due to the induced angle of attack, while the mean angle of attack is zero. The contours show the streamwise vorticity ω_x in the plane 0.5 chord lengths downstream of the trailing edge. Additionally, the instantaneous pressure distribution is plotted for four different spanwise positions of the wing. Since the vertical motion of the flapping wing is largest at the tip, the pressure peaks are highest for the outboard position.

Solving the Navier-Stokes equations for an oscillating wing in 3D flow allows to account for both induced and viscous drag, the former being absent

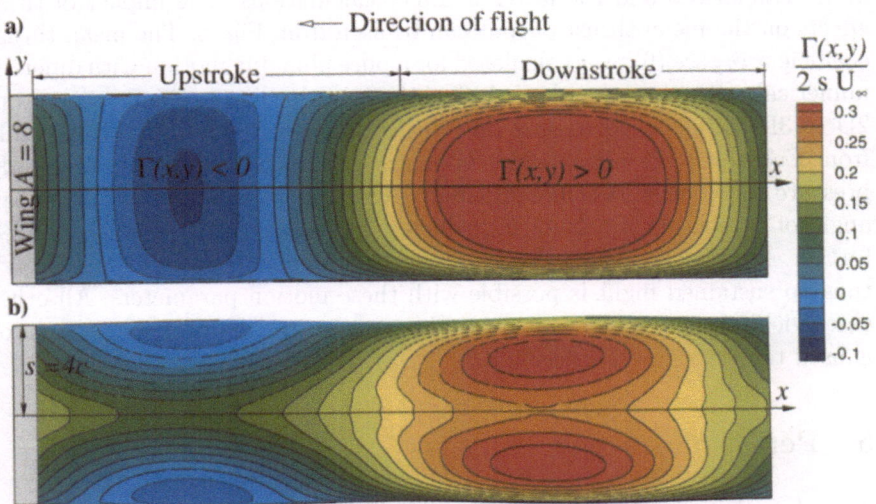

Fig. 6. Circulation along the flight path for a plunging (a) and a flapping and twisting (b) wing at a mean angle of attack of $\alpha_0 = 4°$. (3D Euler-Calulation for $A = 8$ wing, NACA 0012 airfoil, $k = 0.1$, (a) $z_1/c = 0.524$, (b) $\Psi_1 = 15°$, $\alpha_{1,\text{tip}} = 4°$, $\Phi = 90°$)

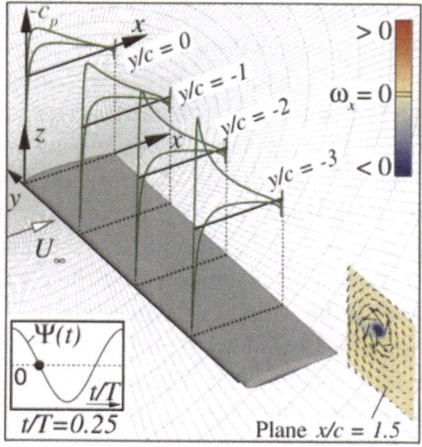

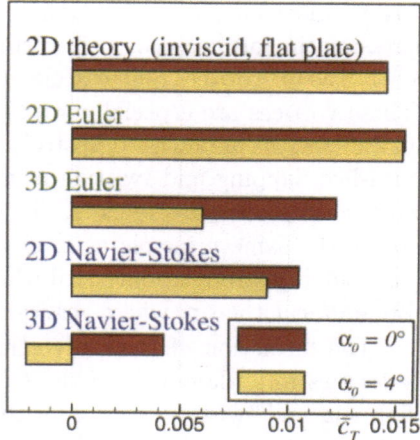

Fig. 7. Instantaneous pressure distribution c_p and tip vortex during downstroke of a flapping and twisting wing, (2D Navier-Stokes calculation for $A = 8$ wing, NACA 0012 airfoil, $k = 0.1$, $\Psi_1 = 15°$, $\alpha_{1,tip} = 4°$, $\Phi = 90°$, $\alpha_0 = 0°$, $Re = 2 \times 10^5$, k-ω Turbulence- Model)

Fig. 8. Mean thrust coefficient for a pure plunging motion for different approaches, (NACA 0014 airfoil, $k = 0.1$, $z_1/c = 0.4$, $A = 8$ for 3D, $Re = 10^6$, Balwin-Lomax Turbulence-Model for Navier-Stokes)

in 2D calulations and the latter in Euler calculations. The impact of these effects on the mean thrust output can be seen from Fig. 8. The mean thrust coefficient $\bar{c}_T$, see (6), was calculated for a pure plunging motion with different numerical methods and with analytical theory according to Küssner [1]. From 2D to 3D, thrust is decreasing due to the influence of the tip vortex, while from Euler to Navier-Stokes the influence of skin friction on the unsteady pressure distribution becomes apparent. Furthermore, the influence of a mean angle of attack can be studied, reducing the mean thrust coefficient even further. For the 3D Navier-Stokes case, $\bar{c}_T$ becomes negative for $\alpha_0 = 4°$, thus no sustained flight is possible with these motion parameters. All other cases yield an excess thrust which either accelerates the plunging system or is used to overcome the friction drag.

5   Performance

The above results were obtained running the FLOWer code in sequential mode on one of the NEC-SX5 processors. In all cases a performance of about 2 GFLOP/sec was obtained which amounts half of the possible peak performance of one processor. The vector operation ratio was always more than 99%, with vector lengths between 190 and 209 for cases with and without turbulence modelling respectively. The memory requirement and run time is mainly dependent on grid size and temporal resolution. For the 3D Euler cases with about 200 000 gridpoints and about 5 000 overall iterations, the CPU time spent was just more than 1h with 1GB memory requirement. For the 3D Navier-Stokes solutions, including turbulence modeling on a mesh with more than 1 Mio. gridpoints and using 8000 overall iterations (calculation for result in Fig. 7), the CPU time increases up to 16h and requires up to 2.4GB of main memory.

6   Conclusions

The NEC-SX5 platform has been used sucessfully to calculate the unsteady viscous flow around a flapping wing. Besides the obtained thrust output of the motion, the result provides an insight into the time dependent formation of tip vortices and the drag effects. Although lift and drag is obtained with the same airfoil, the interaction of motion, vortex formation and viscous drag is rather weak for the investigated parameter range. More complex results are expected for cases with additional flow separation, which would require further grid refinement and higher temporal resolution.

Notes and Comments. This work was funded by the Deutsche Forschungs-gemeinschaft (DFG) as part of the Graduate College "Structure-Fluid Interaction" at the Technical University of Braunschweig.

References

1. Küssner, H. G.: Zusammenfassender Bericht über den instationären Auftrieb von Flügeln. Luftfahrtforschung **30** (1936) 410–424
2. Isogai, K., Shinmoto, Y., Watanabe, Y.: Effects of dynamic stall on propulsive efficiency and thrust of flapping airfoil. AIAA J. **37** (1999) 1145–1151
3. Tuncer, I. H., Walz, R., Platzer, M. F.: A computational study on the dynamic stall of a flapping airfoil. AIAA Paper 98-2519 (1998)
4. Liu, H., Kawachi, K.: A numerical study of insect flight. J. Comp. Phys. **146** (1998) 124–156
5. Smith, M. J. C.: Simulating moth wing aerodynamics: Towards the development of flapping-wing technology. AIAA J. **34** (1996) 1348–1355
6. Neef, M. F., Hummel, D.: Euler solutions for a finite-span flapping wing. Proceedings of the Conference: "Fixed, Flapping and Rotary Wing Vehicles at Very Low Reynolds Numbers", Notre Dame, 5–7 June 2000, 75–99
7. Radespiel, R., Rossow, C., Swanson, R. C.: Efficient cell vertex multigrid scheme for the three-dimensional Navier-Stokes equations. AIAA J. **28** (1990) 1464–1472
8. Heinrich, R., Bleecke, H.: Simulation of unsteady, three-dimensional viscous flows using a dual-time stepping method. "Notes on Numerical Fluid Mechanics", Vol. 60, Vieweg Verlag, Braunschweig 1997, 173–180

Hindcasting the Uptake of Anthropogenic Trace Gases with an Eddy-Permitting Model of the Atlantic Ocean

René Redler[2], Jens-Olaf Beismann[1], Lars Czeschel[1], Christoph Völker[3], Joachim Dengg[1], and Claus W. Böning[1]

[1] Institut für Meereskunde an der Universität Kiel, Düsternbrooker Weg 20, 24105 Kiel, Germany
[2] C&C Research Laboratories, NEC Europe Ltd., Sankt Augustin, Germany
[3] Alfred-Wegener Institut für Polar- und Meeresforschung, Bremerhaven, Germany

Abstract. The ocean takes up a large fraction of the pertubation CO_2 that enters the atmosphere by human activity. A realistic representation of this uptake in numerical models is essential for future climate studies. Uptake of CO_2 or other atmospheric trace gases is strongly influenced by oceanic physical variability at spatial scales between 20 and 100 km. Our main goal is to study the effect of this mesoscale variability on the cumulative uptake of anthropogenic CO_2 and chlorofluorocarbons using an existing model of the ocean circulation in the Atlantic that resolves a significant part of that variability explicitly because of its grid spacing of about 20 km. Results are compared with simulated trace gas distribution obtained from a model with coarser resolution.

1 Introduction

The increasing atmospheric partial pressure of carbon dioxide during the last two centuries has caused an increase in the air-sea flux and oceanic storage of carbon. Ocean carbon models are essential tools in predicting the future storage of this so-called anthropogenic carbon in the world oceans and in assessing how this storage will be influenced by a possible global warming. A recent comparison of results obtained with four ocean carbon models [7] shows that estimates of the mean uptake of carbon by the ocean in the decade 1980-1989 have a considerable spread from 1.5 to 2.2 Gigatons carbon per year (GtC/y, $1 Gt = 10^9 t = 10^{12}$ kg), with strong regional differences between the models, especially in the Southern Ocean.

A realistic representation of the uptake of anthropogenic CO_2 in numerical models is essential for future climate studies. The processes that are mainly responsible for the uptake, deep water formation in polar regions and ventilation of water masses between the wind-mixed layer and the pycnocline, are influenced by strong oceanic physical variability at spatial scales between 20 and 100 km. These processes have to be parameterised in current ocean carbon cycle models. However, it has been shown that details of the parameterization can strongly influence trace gas uptake in the models [11]. Furthermore, the simulated volume of subtropical mode water increases in models of the North Atlantic Ocean when the model grid spacing allows a part of the mesoscale eddy spectrum to be resolved [3].

At the present stage, global simulations of the oceanic circulation at eddy-permitting or even eddy-resolving spatial resolution are restricted to just a few years by limitations on computing time. However, modeling the uptake of transient atmospheric gases such as chlorofluorocarbons (CFCs) or anthropogenic CO_2 requires model integrations over timescales of a century, while modeling the full oceanic carbon or nutrient cycle requires even longer integrations of thousands of years, dictated by the slow ventilation of the abyssal ocean. A way out of this dilemma is offered by the use of regional (basin-scale) models that represent only a part of the world oceans and prescribe the exchange with adjacent ocean basins in some reasonable way. Such models have the potential to be run over longer timescales at eddy-permitting resolution, thus allowing to study the effects of the internal oceanic variability on ventilation processes and therefore on transient tracer uptake.

Our main aim is to study the effect of this mesoscale variability on the cumulative uptake of anthropogenic CO_2 and of CFCs using an existing model of the ocean circulation in the Atlantic that resolves a significant part of that variability explicitly because of its grid spacing of about 20 km.

The specific aims of this project are

- to study the effects of mesoscale variability on ocean uptake of anthropogenic tracers directly, i.e. not only by comparing different parameterizations in coarse-resolution models. The focus is here on the ventilation of intermediate and mode waters that will be the main sink for anthropogenic CO_2 during the next century. The timescale of this ventilation (more than a decade) requires model integrations over the time span from 1900 to today.
- to validate the ventilation of deep and intermediate water masses in the model by direct comparison with oceanic tracer observations. It is for this purpose that we include CFCs in the model simulation.
- to test parameterizations of subgrid-scale processes in coarse-resolution models with a view on the effect on tracer uptake. This can be done by comparing the simulation with the existing model simulations at coarser resolution.

2   Description of the numerical model

The FLAME (*Family of Linked Atlantic Model Experiments* [1]) code is based on the Modular Ocean Model (MOM, [8]) from the Geophysical Fluid Dynamics Laboratory in Princeton.

---

[1] http://www.ifm.uni-kiel.de/to/FLAME

Technical details addressing the model configuration and parallelization strategy have already been published at the beginning of this project [10]. For further details the reader is kindly referred to this document accessible via http://www.rzg.mpg.de/mpp-workshop/proceedings.html.

Our model area covers the region of the Atlantic Ocean and is situated between 70°N and 70°S with open boundaries across the Antarctic Circumpolar Current in the Drake Passage (70°W) and south of Africa at 30°E. The horizontal resolution of the locally rectangular grid is 1/3° in longitude and $1/3° \times \cos(\phi)$ in latitude (ϕ) resulting in a mesh size of 37 km at the Equator decreasing to 12.5 km at the subpolar boundaries. The vertical is discretised in 45 levels.

2.1 Software and Physics

MOM is a Fortran code based on the so called "primitive equations" [6] which are derived from the Reynolds-averaged Navier-Stokes equations under various assumptions, like the Boussinesq-, spherical and hydrostatic approximation. From these equations MOM calculates prognostic values of the horizontal velocity, potential temperature (Θ) and salinity (S) distribution. Density and pressure are constructed from the Θ and S fields. In addition to that we calculate the distribution of anthropogenic CO_2 and a species of chlorofluorocarbons (CFC-11) as supplementary tracers. The equations are discretised on a regular three-dimensional grid with depth as vertical coordinate.

For solving the momentum equations, the horizontal velocities are split up into a vertical average and its deviations. The vertically averaged transport can be expressed by a 2-dimensional stream function which is described by a Poisson equation. The latter is solved using an iterative conjugate gradient method. Additional constraints are needed to keep the values around the islands constant during every iteration step.

Due to the hydrostatic approximation the vertical component of the momentum equation degenerates and vertical transport due to hydrostatic instabilities (convection) has to be parameterised, vertical velocities are constructed from the divergent horizontal velocity field.

The remaining 2nd order non-linear partial differential equations for Θ, S and the vertical shear of horizontal momentum are solved on regular 3-D grids using the finite volume/difference approach. Explicit time stepping is used. Central differences are applied in space. Time discretisation is done with central differences as well but replaced at regular intervalls by a backward Euler timestep in order to damp the computational mode.

For horizontal diffusion of tracers, e.g. Θ, S, CFC-11 and anthropogenic CO_2 we use mixing along isopycnal surfaces [9,2]. This implicit approach requires the inversion of a tridiagonal matrix accompanied by a rotation of the mixing tensor into the direction of surfaces of constant density.

Simulating the oceanic uptake and spreading of anthropogenic tracers requires an appropriate representation of the flow of dense water masses across shallow sills which in our model is achieved by a bottom boundary layer parameterization [1].

2.2 Parallelization

The original version of MOM2.1 was designed for vector computers. Running a single experiment of a high-resolution model over long simulation periods would require almost a year on a single CPU on dedicated modern vector architectures. Therefore it is tempting to take advantage of the parallelization approach to run such type of experiments on parallel machines such as the Cray T3E or the NEC SX5.

A parallel version of MOM2.1 has been created using 2-dimensional grid partitioning. To run the code most efficiently on both types of platforms work sharing and data distribution has been implemented, and data exchange between CPUs is done either via Cray's explicit shared memory software (SHMEM library) on the T3E or with MPI on the SX5.

In ocean and atmosphere models the number of grid points in each horizontal direction exceeds the number of grid points applied in the vertical direction by an order of magnitude. The code was originally developed for vector architectures like the Cray Y-MP or the Cray T90. There it is advantageous due to the need for long vector loops to calculate one of the horizontal dimensions in an inner loop, in our case the i-direction (see Fig. 1). Note that auxiliary arrays e.g. for advection and diffusion are not dimensioned over the whole computational domain. To save memory they are dimensioned over a memory window (jmw) only, which is usually two orders of magnitude smaller that the actual range of j-dimension. This technique (original to MOM) is advantageous for running the code on systems with distributed memory and

```
integer, parameter :: km  = 45    ! vertical dimension
integer, parameter :: imt = 308   ! length in i-direction
integer, parameter :: jmt = 600   ! length in j-direction
integer, parameter :: jmw = 10    ! width of memory window

real, dimension(imt,km,jmt) :: t
real, dimension(imt,km,jmw) :: adv, diff

Do j = 1, jmt
  jptr = ...
  Do k = 1, km
    Do i = 1, imt
      t(i,k,j) = adv(i,k,jptr) + diff(i,k,jptr)
    End Do
  End Do
End Do
```

Fig. 1. Simplified but typical loop in MOM within a timestep.

```
real, allocatable, dimension(:,:,:) :: t, adv, diff

allocate (  t (is_pe:ie_pe,km,js_pe:je_pe))
allocate (adv (is_pe:ie_pe,km,jmw))
allocate (diff(is_pe:ie_pe,km,jmw))

Do j = js_pe, je_pe
  jptr = ...
  Do k = 1, km
    Do i = is_pe, ie_pe
      t(i,k,j) = adv(i,k,jptr) + diff(i,k,jptr)
    End Do
  End Do
End Do
```

Fig. 2. Parallelised loop.

only small memory on each processor like T3E, but at the cost of CPU time: This technique requires internal data copying in order to perform a correct calculation of advection and diffusion terms at the borders of the memory window. For details the reader is referred to the MOM manual. To achieve an optimal performance, the memory window should be as large as possible and ideally should encompass the whole computational domain of each processor.

The outer j-loop is a natural first choice for 1-dimensional grid partitioning. Here each CPU calculates a part of the grid ranging from j-indices js_pe to je_pe (Fig.2) which are calculated at run time based on the number of processors that are requested. With respect to the underlying data structure it would have been optimal to use the k-direction for grid partitioning in the second dimension. Due to strong physical dependencies in this direction and the resulting amount of message passing this strategy is unfortunately prohibitive. Thus, for use on massively parallel architectures the code was additionally parallelised in i-direction. For reasons that cannot be discussed in detail in this paper, parallelization in i-direction does not scale very well, but it nevertheless allows us to double the number of CPUs on a system like T3E with out losing much of the performance. Since on SX5 it is mandatory to maintain sufficiently long vector lengths we did not make use of the parallelization in i-direction there.

3   Optimization of the code

The optimization strategy for vector computers is to vectorise as many loops as possible accomplished with long vector length. RISC processors like the DEC Alpha chips of a T3E require a different strategy that is to reduce traffic to the central memory by reusing data whenever it is possible. To achieve a better performance on RISC systems the most CPU time demanding subroutines and loops have been optimised the following way:

- Memory traffic is reduced by collapsing the outer two loops and passing data in one dimensional arrays or scalars. This kind of data stays in cache.
- Reordering of loops or array indices is done in order to access as many arrays as possible with stride one.

After treating all major subroutines, especially the solving of the tendency equation for the 3-D fields and the inversion of the tridiagonal matrix, a speedup of 2 has been achieved compared with the original MOM 2.1. For all optimizations steps addressing an increased performance on RISC systems the vector code has been maintained: In a preprocessing step one or the other piece of code is chosen for compiling.

In a second step, extensions of the code addressing the computation of CFCs and anthropogenic CO_2 as well as some physical parameterizations had been added. These modifications developed and tested on RISC processors turned out to be highly inefficient on SX5 mainly due to extensive use of Fortran functions. To achieve maximum performance on this machine, Fortran functions have been changed to subroutines that are called over the whole range of the inner loops. Due to the larger memory available, a reformulation of the BBL parameterization [1] helped to save additional CPU time.

4 Performance

Performance of the code has been measured on a Cray T3E (Konrad-Zuse Zentrum für Informationstechnik Berlin), NEC SX5e (Höchstleistungs-Rechenzentrum Universität Stuttgart) and a NEC SX5/B8 (Nationaal Luchtvaart Laboratorium Amsterdam).

The **Cray T3E LC 384** is a heterogeneous system based on 408 DEC Alpha EV5.6 processors in the following configuration:

    48 PEs with 600 MHz and 512 MB
    88 PEs with 600 MHz and 128 MB
  264 PEs with 450 MHz and 128 MB
    8 PEs with 450 MHz and 256 MB

The **NEC SX5/B8** has 8 processors with a peak performance of 8 GFlop/s each and 64 Gbyte total memory. The **NEC SX5Be-32M2** consists of 2×16 processors with a total peak performance of 128 GFlop/s with 80 Gbyte total memory.

All performance data presented here have been measured on non-dedicated platforms. On the T3E memory and processors have been used exclusively by our application while the interconnects have been shared with other applications. Since on the SX5 we also had to share memory and processes (especially on the NEC SX5Be-32M2) those data represent only a lower performance limit. On the T3E the Cray performance analysis tool PAT has been used, on the SX5 measurements have been carried out using Flow Trace.

For each benchmark run we calculated 216 timesteps (3 days of model simulation) encompassing two additional tracer (anthropogenic CO_2 and CFC-

11) with its complete physics. For the parallel benchmarks the numbers given represent average values over the whole set of processors.

Due to the limited memory on a single EV5.6 processor we are not able to compare sequential performance for our production code between the different architectures. Instead we concentrate here on the scalability of the code on the different platforms.

Figure 3 shows the parallel efficiency as a function of processors. Clearly visible is the loss of performance for higher number of processors. This deviation from a linear speedup can be almost exclusively addressed to the conjugate gradient algorithm: three global operations within the iteration loop are the major impediments on our way to a linear speedup. Almost all other subroutines that have to deal with the vertical shear of the velocities or the advection and diffusion of tracers show an almost linear speed up (with more than 99% efficiency).

With the MOM configuration we have chosen for the production runs we achieve 26 to 41 % of the theoretical peak performance on the SX5, depending on the number of processors and the specific architecture, leading to performance rates of 8.46 GFlop/s (4 CPUs, SX5/B8) and 10.6 GFlop/s (8 CPUs, SX5Be). To achieve a similar performance on the T3E-900 approx-

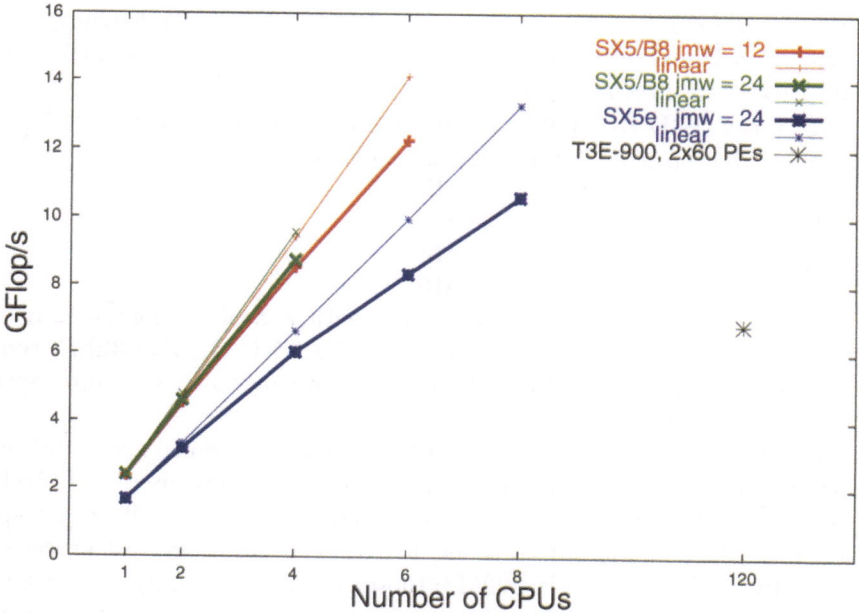

Fig. 3. Measured Gflop/s on the SX5 and T3E with MOM. The x-axes has linear spacing from 1 to 8. The value for the T3E is obtained using 120 processing elements.

imately 185 processors (450 MHz EV5.6) would be needed. In our testcase with 120 processing elements we achieve 6.8 GFlop/s.

5 Simulation results

As an example of the results of the model configuration run at HLRS, Fig. 4 shows a comparison of CFC concentrations on a zonal section in the North Atlantic from an early stage of the integration of the eddy-permitting model and an experiment with a non-eddy resolving configuration. Although absolute tracer concentrations and penetration depths are similar in these experi-

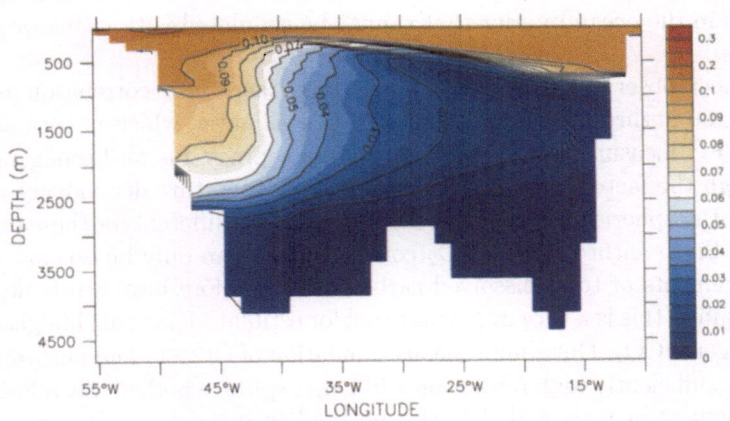

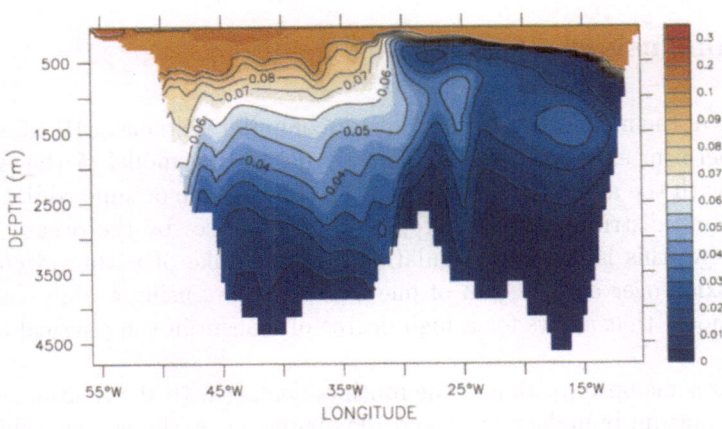

Fig. 4. Concentration of CFC-11 (in *pmol/l*) on a vertical section at 50°N for the simulated month September 1961. Upper panel: result from a 4/3° model, lower panel: result from the 1/3° model run at HLRS.

ments, there are a number of differences that emphasise the role of mesoscale eddies for water mass formation and tracer redistribution.

The coarse-resolution model (upper panel) shows a rather smooth tracer field at this latitude. The penetration of the CFC tongue from the Deep Western Boundary Current to the east at depths of around 1500 m is hampered by the fact that the deep circulation in this model is sluggish because of the comparatively high diffusivity that is required for numerical stability. The eddy-permitting model (lower panel) allows for a more vigorous circulation and shows clearly the effect of explicitly simulating mesoscale eddies. It is interesting to note that an interpolation of the solution from the high-resolution configuration onto the coarse grid would look very different from the field shown in the upper panel. This indicates that there is a rectifying effect of mesoscale eddies on the distribution of anthropogenic tracers in the ocean interior that cannot be simulated with coarse-resolution models.

Tracer observations by [5] suggest that there is a correlation between CFCs and anthropogenic CO_2 in the North Atlantic, which in turn suggests that CFC measurements could be used as a proxy for anthropogenic CO_2 although the factors governing the uptake (temperature dependence of solubility, atmospheric concentration history) are very different for these two tracers [4]. Since anthropogenic CO_2 concentrations can only be estimated from measurements of total dissolved carbon and therefore have a non-negligible uncertainty, this is a very important tool for estimating oceanic budgets of anthropogenic CO_2. Our simultaneous simulation of CFC-11 and anthropogenic CO_2 at sufficiently high resolution will help explore whether this relationship is stationary in time and if it can be used in other parts of the ocean as well.

6   Conclusions

The use of the moderately parallel vector computer systems at HLRS enables us to perform experiments with an eddy-permitting model of the Atlantic Ocean with the aim to quantify the effect of mesoscale oceanic eddies on the uptake and distribution of anthropogenic trace gases by the ocean. To our knowledge, this is the first simulation of the uptake of anthropogenic carbon dioxide over a time span of one hundred years using a high-resolution ocean model that allows for a high degree of realism in the physical circulation.

After some optimization in the routines dealing with the treatment of the oceanic bottom boundary layer and the air-sea gas exchange, we achieve 26 to 31 % of the theoretical peak performance of the SX5. Results from an early stage of the integration show important differences in the tracer distribution when compared to experiments at coarse resolution, but the experiment has not yet come to a stage where quantitative conclusions can be drawn.

Acknowledgements

The authors gratefully acknowledge the work of Klaus Ketelsen, SGI Munich for his work on optimising the code for the T3E. The work of the FLAME group is acknowledged for its contribution at various stages of the project. Model results have been analysed using the Ferret plotting package, kindly provided by Steve Hankin and his team from NOAA/PMEL.

References

1. A. Beckmann and R. Döscher. A method for improved representation of dense water spreading over topography in geopotential–coordinate models. *J. Phys. Oceanogr.*, 27:581–591, 1997.
2. M. D. Cox. Isopycnal diffusion in a z–coordinate model. *Ocean Modelling*, 74:1–5, 1987. unpubl. manuscr.
3. L. Czeschel. Modelluntersuchungen zur Ventilation der Hauptsprungschicht im subtropischen Nordatlantik. Master's thesis, Institut für Meereskunde an der Universität Kiel, 2000.
4. M.H. England and E. Maier-Reimer. Using chemical tracers to assess ocean models. *Rev. Geophys.*, 39:29–70, 2001.
5. A. Körtzinger, M. Rhein, and L. Mintrop. Anthropogenic CO_2 and CFCs in the North Atlantic ocean – a comparison of man-made tracers. *Geophys. Res. Lett.*, 26:2065–2068, 1999.
6. P. Müller and J. Willebrand. Equations for oceanic motions. In *Landolt-Börnstein, Group V, Oceanography, Volume 3b*, pages 1–14. J. Sündermann, Ed., Springer Verlag, Berlin, 1989.
7. J.C. Orr, E. Maier-Reimer, U. Mikolajewicz, P. Monfray, J.L. Sarmiento, J.R. Toggweiler, N.K. Taylor, J. Palmer, N. Gruber, C.L. Sabine, C. Le Quéré, R.M. Key, and J. Boutin. Estimates of anthropogenic carbon uptake from four three-dimensional global ocean models. *Global Biogeochem. Cycles*, 15(1):43–60, 2001.
8. R. C. Pacanowski. MOM 2 Documentation, User's Guide and Reference Manual. Technical Report 3, GFDL Ocean Group, 1995.
9. M. H. Redi. Oceanic isopycnal mixing by coordinate rotation. *J. Phys. Oceanogr.*, 12:1154–1158, 1982.
10. R. Redler, K. Ketelsen, J. Dengg, and C.W. Böning. A high-resolution numerical model for the circulation of the Atlantic ocean. In H. Lederer and Friedrich Hertweck, editors, *Proceedings of the Fourth European SGI/CRAY MPP Workshop*, pages 95 – 108. Max-Planck-Institut für Plasmaphysik, 1998.
11. C. Völker, J. Willebrand, and J. Dengg. Anthropogenic CO_2 uptake in the Atlantic studied with a basin-scale ocean general circulation model. submitted manuscript, 2001.

Flow with Chemical Reactions

Prof. Dr. Dietmar Kröner

Institut für Angewandte Mathematik, Universität Freiburg
Hermann-Herder-Str. 10, D-79104 Freiburg

The simulation of multicomponent reactive dynamical flows turns out to be a key technology for the improvement of the development of industrial combustion plants, for their design and for the reduction of pollution. The present standard of the modelling is based on the equations for conservation of mass, momentum, energy and the conservation of mass for the chemical species and on a detailed or reduced resolution of the chemical kinetics. Viscous effects are taken into account. Different turbulence models are used to resolve the different scales in space and time.

The computational power of the available parallel computers allows the performance of direct numerical simulation of reactive flows based on a detailed chemical reaction mechanism, at least in small domains in two space dimensions. It is still a challenging task to use a detailed mechanism for the chemical kinetics together with a direct numerical simulation in three spatial dimensions. Since the turbulence is a three dimensional phenomenon many effects cannot be treated with such simulations in 2D. On the other hand it is known that many other effects of turbulent combustions cannot be obtained by reduced models for the chemical reactions. Therefore many authors believe that the modelling of turbulent flames on the basis of detailed chemical reaction schemes is at least as important as an extension to three dimensions. Therefore turbulent modelling of reactive flows in three spatial dimensions is still based on reduced mechanism for the chemical kinetics.

After a period, which was mainly characterized by the development of efficient numerical codes, the implementation on parallel computers and the performance of test problems and benchmarks, we have now reached the status that the CPU time for typical industrial applications with acceptable accuracy can be considerably reduced. This implies that the numerical simulation will become more and more important for the industrial design process.

This was proved in the contribution of Berreth et al. They report on the simulation of advanced reburning in the in-furnace produced nitrogen oxides in three space dimensions based on a reduced mechanism for the chemical kinetics and an eddy dissipation concept for the turbulence. They got a very good computational efficiency for the parallelization and could strongly diminish the CPU time for a computation with more than one million gridpoints on the CRAY T3E with 128 processors.

The contributions of Lange and Tsai et al. concern the direct numerical simulation of flame kernels in two spatial dimensions. In order to resolve the typical lengthscales Tsai et al. have used a grid with more than three million gridpoints in 2D together with a detailed mechanism for the reaction. They investigate the influence of the turbulent combustion on the growth of the flame kernel and can describe the instationary ignition process and the flame propagation. Similar results are obtained in the paper of Lange. He also tries to optimize the load balancing while trying to minimize the length of the subdomain boundaries and therefore the amount of communication.

Furthermore Lange has done an extensive comparison between the performance using PVM and MPI and between the CRAY T3E and PC clusters, based on the code for the direct numerical simulations for flame kernels. It turns out that there is no significant difference between PVM and MPI during the normal time integration. But for I/O activities the MPI version outperforms the PVM version. The comparison between clusters with different processors and interconnect technologies with the CRAY T3E clearly shows and confirms the high potential of clustered hardware systems for high-performance computing. He also reports on some first experience with this code in metacomputing. His has run a numerical test on a system consisting of 16 processor of the T3E combined with 16 processors of the HITACHI SR8000.

Although we have obtained a high level of the quality and efficiency of numerical simulation there is still a large potential for further improvements. Since for the evolution of flame fronts most of the dynamics take place in a small region, dynamical adaptive grid refinement, combined with dynamical load balancing, will lead to a higher efficiency. Adaptive timestep control can be used for further improvements.

Implementation of Complex Chemical Reaction Mechanisms Into a 3D Furnace Simulation Code

Alexander Berreth[2], Frank Rückert[1], Dieter Förtsch[1], Benedetto Risio[2], Uwe Schnell[1], and Klaus R.G. Hein[1]

[1] Institute of Process Engineering and Power Plant Technology (IVD), University of Stuttgart, Pfaffenwaldring 23, 70569 Stuttgart (Germany)
[2] Recom Services, REaction and COMbustion Modeling, Bahnhofstrasse 9, D-71106 Magstadt (Germany)

Abstract. Numerical simulation of chemically reacting flows in technical furnaces is a very complex task because of the involved basic processes of fluid flow, turbulence, heat transfer by radiation, and reaction kinetics, as well as their interaction. In this paper a method of treating complex chemical reaction mechanisms in a Computational Fluid Dynamics (CFD) code is described and the computational efficiency of the implementation is assessed on a technical test case.

1 Introduction

In recent years, the application of computational fluid dynamics (CFD) has become increasingly attractive to study and optimize technical furnaces [1]. This is on account of the more powerful computer platforms now available for numerically simulating, with a reasonable response time, the complex processes comprising fluid flow, turbulence, heat transfer by radiation, and chemical reaction kinetics taking place in furnaces of typically $20 \times 20 \times 10\,\mathrm{m}$ in size. Each of the submodels describing the main processes may be described by models of varying accuracy and validity.

Reduced kinetic mechanisms that can represent the important features of the original detailed mechanisms greatly increase the possibility of the kinetic modelling of practical combustion problems and offer potential improvement of modelling accuracy in comparison to using global mechanisms. The present work applies a mechanism reduction technique in conjunction with the CHEMKIN library.

The simulation program AIOLOS has been developed for the numerical calculation of three-dimensional, stationary and dynamic, turbulent reactive flows in pulverised coal-fired utility boilers. AIOLOS contains submodels treating fluid flow, turbulence, homogeneous and hetereogeneous combustion, and heat transfer. In these submodels equations for calculating the conservation of mass, momentum, and energy are solved. A detailed description of the code can be found in [2], [3].

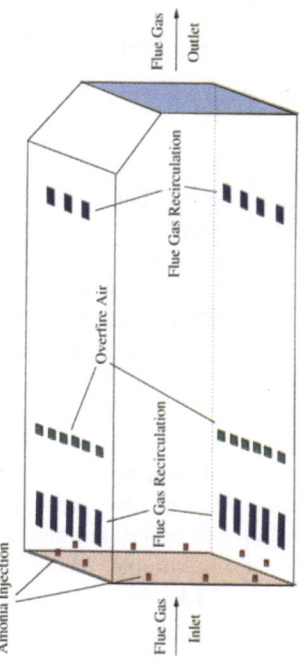

Fig. 1. Scheme of the Furnace

A scheme of the calculated test case is shown in Fig. 1. It is the upper part of a furnace, of physical size $10 \times 13.8 \times 24.8$ m, operated with advanced reburning, an efficient method to reduce the in-furnace produced nitrogen oxides (NO_x). The computational grid used is shown in Fig. 2. The grid dimensions are $70 \times 47 \times 71$ cells (total: $\approx 233{,}000$ cells; calculated: $\approx 200{,}000$ cells).

2 Basic Equations

The basic equations to be solved for obtaining the distribution of flow field, species concentrations, and temperature can be written in a common form given by the Generalized Transport Equation [4], [5] for a variable, ϕ:

$$\nabla \cdot (\rho \, v \, \phi) \quad = \quad \nabla \cdot (\Gamma_\phi \, \nabla \phi) + S_\phi \tag{1}$$

where: ρ is the fluid density, v is the velocity vector, Γ_ϕ is the generalized diffusion coefficient related to ϕ, and S_ϕ is the corresponding source term of the variable ϕ, that represents the quantities for momentum, energy and species, e.g. the velocity if Eq. 1 represents the Navier-Stokes equations.

In order to calculate the distribution of species mass fractions, w_j, of a species j, Eq. 1 is solved for $\phi = w_j$ with the species source term, S_j, defined

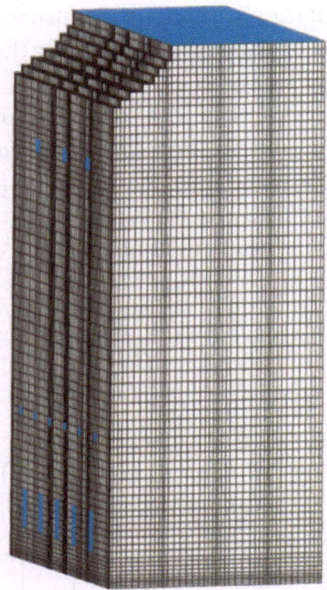

Fig. 2. Numerical Grid

by the rates of chemical reactions taking place. For a considered control volume this is given by the integrated rate of production/destruction, ω_j, of species j in the considered finite volume:

$$S_j = \frac{1}{V} \cdot \int_V \omega_j \, dV = \tilde{\omega}_j \qquad (2)$$

Due to the nonlinear dependence of the rate of production/destruction on local concentrations and temperature, i.e. , $\omega_j = f(w_1, w_2, ..., p, T)$, simple averaging is not appropriate: $\tilde{\omega}_j \neq \omega(\tilde{w}_1, \tilde{w}_2, ..., \tilde{p}, \tilde{T})$. Specifically, for very fast reactions, as in combustion, the overall reaction rates are often limited by the rate of mixing on a molecular level determined by the molecular viscosity and the local turbulence properties. Consequently, the effect of turbulence on the overall rate of homogeneous reactions has to be accounted for. Various models have been proposed to model this interaction, as reviewed in several books [6], [7].

One general method for modelling this interaction is the Eddy Dissipation Concept proposed by Magnussen [8] as an extension of earlier models [9]. The EDC has been successfully applied to gas combustion using detailed reaction mechanisms [10], [11] and to coal combustion using global reaction mechanisms [11], [12]. Nevertheless, when using detailed mechanisms the computational effort is enormous, as will be discussed below, so that these applications were limited to the simulation of small-scale furnaces.

The conceptual basis of the EDC is the fact that complete mixing on a molecular scale is achieved only in certain regions, and reactions can occur only there. Thus, in the EDC the total space is subdivided into a reaction space, the *fine structures*, and the *surrounding fluid*. All homogeneous reactions are assumed to take place only within these fine structures which are locally treated as a well stirred reactor transferring mass and energy only to the surrounding fluid. The mean residence time, τ^*, of fluid within the fine structures is modeled by [8]:

$$\tau^* = C_\tau \cdot \sqrt{\frac{\nu}{\epsilon}} \tag{3}$$

with: $C_\tau = 0.41$; the dissipation of turbulent kinetic energy, ϵ, and the kinematic viscosity, $\nu = 1.178 \cdot 10^{-4} \, m^2/s \, (T/1000 \, K)^{1.65}$. The mass fraction, γ^*, occupied by the fine structures is modeled by [8]:

$$\gamma^* = \left[C_\lambda \cdot \left(\frac{\nu \cdot \epsilon}{k^2} \right)^{0.25} \right]^\kappa \tag{4}$$

where: $C_\lambda = 2.13$; and κ takes values ranging from 2 to 3.

The reaction rates of all species are calculated on a mass balance for the fine structure reactor. Denoting quantities in the fine structures with an asterisk, the conservation equation of a species j can be written as:

$$\frac{\rho^* \cdot (w_j^* - \tilde{w}_j)}{\tau^* \cdot (1 - \gamma^*)} = M_j \cdot \omega_j^* \tag{5}$$

with the mass fraction inside the fine structures, w_j^*, the density-weighted mean with respect to time, is referred to as the "Favre average", is indicated by $\tilde{w}_j$. M_j is the molecular weight of the species j. The rate of production of the species j is given by ω_j^*.

So in Eq. 5 the new weight fractions of a species in a cell is composed by the weight fraction in the surrounding fluid and the weight fractions in the fine structures, which depend on the rate of production ω_j^* that is calculated by CARM on the basis of w_j^* of the former iteration and the temperature T^* in the fine structures. For the calculations of the rates of production in the CARM mechanism (see Fig. 3 and Fig. 4) refere [15]. ρ^* stands for the density of the fluid in the fine structures.

Therefore, Eq. 5 defines a system of non-linear differential equations (DE) which must be solved numerically in each cell of the computational grid.

The source term for a species j is then obtained from:

$$S_j = \frac{\gamma^* \cdot (w_j^* - \tilde{w}_j)}{\tau^* \cdot (1 - \gamma^*)} \cdot \tilde{\rho} V \tag{6}$$

$$= \quad \gamma^* \cdot \frac{M_j\, \omega_j^*}{\rho^*} \cdot \tilde{\rho} V \qquad (7)$$

Depending on the chemical reaction mechanism solving this DE system can be very time consuming, as discussed below.

3   Integration of Chemical Reaction Kinetics

The chemical kinetics used in this study is a reduced mechanism which is optimised to describe the selective non-catalytic reduction (SNCR) of nitrogen oxides, as it is relevant for the studied test case. It is based on the kinetic rates of Miller and Bowman [13] with recent literature modifications [14]. This detailed mechanism was reduced through sensitivity analysis, on plug flow reactor (PFR) and perfectly stirred reactor (PSR) calculations, and curve fitting using the Computer Assisted Reduction Method (CARM) developed by Chen [15]. The reduction method is briefly described in Fig. 3. This program automatically generates a FORTRAN computer code that can be used as a sub-routine in the 3D furnace simulation code AIOLOS.

Based on the above mentioned detailed reaction mechanism the global reactions given in Table 1 have been derived.

The kinetic expression of each single reaction is a complex function which includes prediction of additional species which are assumed to be in steady-state.

Conservation equations must be solved for each main species concentration. This leads to the introduction of the steady state conditions, i.e. the rate of consumption equals the rate of production. The reaction source terms of these steady-state species are thus zero and the concentrations in the fine

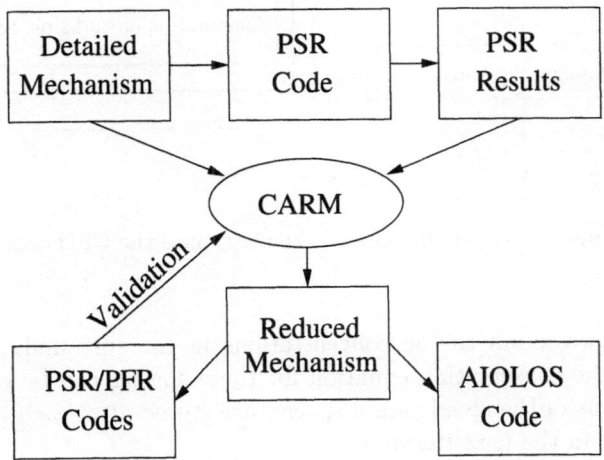

Fig. 3. Reduction of detailed reaction mechanisms using CARM [15]

Table 1. Stoichiometric relations considered for main species

1	$H_2O + 0.5\ O_2 \longleftrightarrow 2\ OH$
2	$0.5\ O_2 + CO \longleftrightarrow CO_2$
3	$0.5\ N_2 + 0.5\ O_2 \longleftrightarrow NO$
4	$HNCO + 0.5\ O_2 \longleftrightarrow 0.5\ N_2 + OH + CO$
5	$NO + 0.5\ N_2 \longleftrightarrow N_2O$
6	$NH_3 + NO + 0.5\ O_2 + CO \longleftrightarrow HNCO + 0.5\ N_2 + 2\ OH$

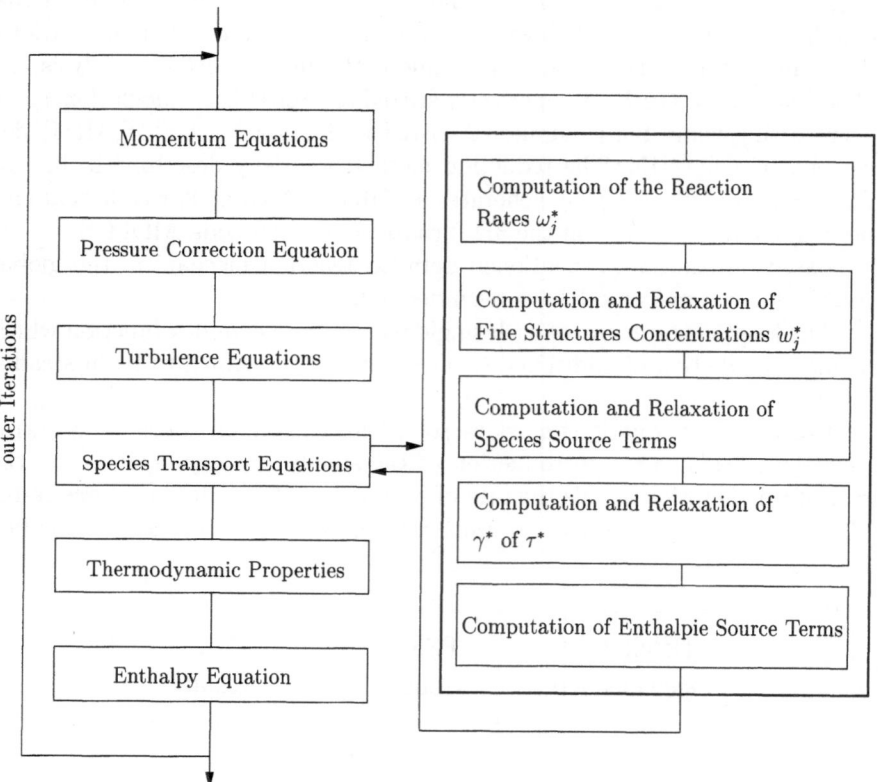

Fig. 4. Implementation of complex chemistry into the CFD code AIOLOS

structures correspond to the concentrations in the surrounding fluid. The solution of the conservation equation for these species can be omitted. The concentrations of the steady state species are stored after each iteration and are available in the next iteration.

The integration of the CARM routine for calculating the species and enthalpy source terms to be used in the CFD program is sketched in Fig. 4.

4   Results

4.1   Concentration Profiles

Typical simulation results are shown in Figs. 5 and 6 for the profiles along the furnace height: CO and O_2 in Fig. 5, and NO and NH_3 in Fig. 6.

Unfortunately, no experimental data are available for comparison with the predictions.

4.2   Performance

The efficiency of the implementation is assessed by the speed-up, defined as the ratio of the execution time on the actual number of processors to the execution time on one single processor, on the CRAY T3E using different numbers of processors, as given in Table 2.

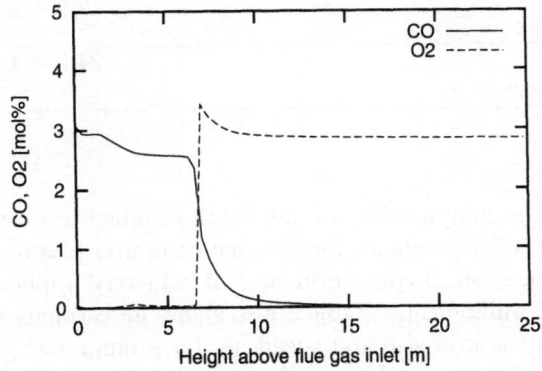

Fig. 5. CO and O_2 profiles along the furnace height

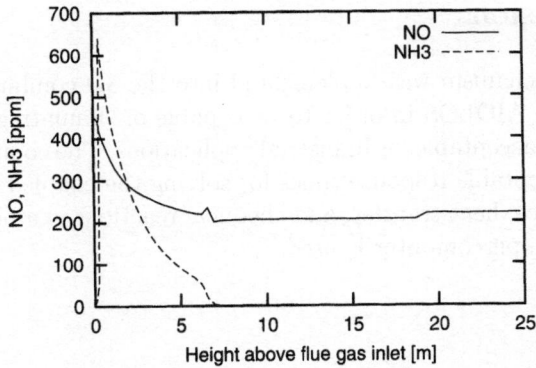

Fig. 6. NO and NH_3 profiles along the furnace height

Table 2. Computational efficiency of implementation

number of processors	speed-up
1	1
4	3.96
8	7.68
16	14.6
32	27.2

Table 3. Computational time on different computer platforms

platform	current test case with 200,000 cells (10,000 iterations)	typical industrial application with 1,500,000 cells (10,000 iterations)
workstation (XP1000 500 MHz)	67 h	500 h $\approx$ 20.8 days
Cray T3E 32 processors	10 h	70 h $\approx$ 2.9 days
Cray T3E 128 processors	–	24 h $\approx$ 1.0 days

In Table 3 the computational times for the studied test case (with 200,000 computational cells) are shown for different computer platforms and different numbers of processors. Typical grid sizes of industrial applications are in the range of 1 to 2 million cells. Table 3 also shows an estimated computational time, based on the grid size and speed-up, for a numerical grid of 1,500,000 cells. These numbers show the need for using high performance computers to allow response times acceptable for industrial purposes.

5   Conclusions

A reduced mechanism was implemented into the 3D combustion simulation CFD-program AIOLOS in order to be capable of simulating practical problems in a time acceptable for industrial applications. The computational times show that acceptable response times for solving the coupled problem of fluid flow, turbulence, heat transfer, and chemical reactions can be achieved when a high-speed supercomputer is used.

References

1. B. Risio, U. Schnell, and K.R.G. Hein. Towards a reliable and efficient furnace simulation tool for coal fired utility boilers. In E. Krause and W. Jäger, editors,

High-Performance-Computing in Science and Engineering '98 (Transactions of the High Performance Computing Center Stuttgart), pages 353–374. Springer-Verlag Berlin Heidelberg, 1999.

2. U. Schnell, R. Schneider, H.C. Magel, B. Risio, and J. Lepper. Numerical Simulation of Advanced Coal-Fired Combustion Systems with In-Furnace NOx Control Technologies. In *Third International Conference on Technologies and Combustion for a Clean Environment*, July 3-6, Lisbon (Portugal), 1995.

3. R. Schneider. *Beitrag zur numerischen Berechnung dreidimensional reagierender Strömungen in industriellen Brennkammern*. PhD thesis, University of Stuttgart, 1998.

4. S.V. Patankar. *Numerical Heat Transfer and Fluid Flow*. Hemisphere Publishing Corporation, 1980.

5. C. Hirsch. *Numerical Computation of Internal and External Flow. Vol. 1. Fundamentals of Numerical Discretization*. John Wiley and Sons, 1988.

6. P.A. Libby and F.A. Williams, editors. *Turbulent Reacting Flows*. Academic Press, San Diego, CA, second edition, 1994.

7. J. Warnatz, U. Maas, and R.W. Dibble. *Combustion*. Springer Verlag, 1996.

8. B.F. Magnussen. The Eddy Dissipation Concept. In *XI Task Leaders Meeting - Energy Conversion in Combustion, IEA*, 1989.

9. B.F. Magnussen, B.H. Hjertager, J.G. Olsen, and D. Bhaduri. Effects of Turbulent Structure and Local Concentrations on Soot Formation and Combustion in C_2H_2 Diffusion Flames. *Proceedings of the Combustion Institute*, 17:1383–1393, 1978.

10. I.R. Gran. *Mathematical Modeling and Numerical Simulation of Chemical Kinetics in Turbulent Combustion*. PhD thesis, University of Trondheim, 1994.

11. H.-C. Magel. *Simulation chemischer Reaktionskinetik in turbulenten Flammen mit detaillierten und globalen Mechanismen*. Ph.D. thesis, University of Stuttgart, 1997.

12. D. Förtsch, F. Kluger, U. Schnell, H. Spliethoff, and K.R.G Hein. A Kinetic Model for the Prediction of *NO* Emissions from Staged Combustion of Pulverized Coal. *Proceedings of the Combustion Institute*, 27:3037–3044, 1998.

13. J.A. Miller and C.T. Bowman. Mechanism and Modeling of Nitrogen Chemistry in Combustion. *Progress in Energy and Combustion Science*, 15:287–338, 1989.

14. A.J. Dean, R.K. Hanson, and C.T. Bowman. *Journal of Physical Chemistry*, 95:3180–3189, 1991.

15. J.-Y. Chen. Development of reduced mechanisms for numerical modelling of turbulent combustion. *Workshop on "Numerical Aspects of Reduction in Chemical Knetics"*, 1997.

Direct Numerical Simulation of Turbulent Flame Kernels Using HPC

Marc Lange

High-Performance Computing Center Stuttgart (HLRS), Stuttgart University
Allmandring 30, D-70550, Germany
E-mail: lange@hlrs.de

Abstract. A parallel code for the direct numerical simulation (DNS) of reactive flows with detailed models for chemical reactions and molecular transport is presented. This code is used as a benchmark for clustered systems with different CPU- and network-technologies. The performance of these clusters is compared with Cray T3E systems on which the code is known to perform very well. Another part of the paper deals with the application of DNS to study premixed flames emerging after induced ignition of turbulent mixtures.

1 Introduction

Energy conversion in numerous industrial power devices like automotive engines or gas turbines is still based on the combustion of fossil fuels. In most applications, the reactive system is turbulent and the reaction progress is influenced by turbulent fluctuations and mixing in the flow. The understanding and modeling of turbulent combustion is thus vital in the conception and optimization of these systems in order to achieve higher performance levels while decreasing the amount of pollutant emission. However, the interaction between chemical kinetics and fluid turbulence is very difficult to study experimentally. Often, terms which are fundamental in turbulent combustion modeling can not be obtained directly from experimental data. Analytical treatments of turbulent flames with complex chemistry encounter even higher difficulties, especially when there is an overlap of characteristic time-scales of fluid turbulence and chemical reactions which is typically the case in turbulent hydrocarbon-air flames.

During the last few years, direct numerical simulations (DNS), i.e. the computation of time-dependent solutions of the Navier-Stokes equations given in Sect. 2, have become one of the most important tools to study turbulent combustion. As DNS does not make use of any turbulence or turbulent combustion models, this technique may be interpreted as high-resolution (numerical) experiments, enabling new investigations on fundamental mechanisms in turbulence-chemistry-interaction and aiding in the refinement of turbulent combustion models. However, many of the DNS carried out so far have used simple one-step chemistry, although important effects cannot be captured by simulations with such oversimplified chemistry models [1,2]. By making efficient use of the computational power provided by massively parallel computers, it is possible to perform DNS of reactive flows using detailed chemical reaction mechanisms at least in two spatial dimensions. Nevertheless, computation time is still the main limiting factor for the DNS of reacting flows, especially when applying detailed chemical schemes.

2   Governing Equations for Detailed Chemistry DNS

Multicomponent reacting ideal-gas mixtures can be described by a set of coupled partial differential equations expressing the conservation of total mass

$$\frac{\partial \varrho}{\partial t} + \mathrm{div}(\varrho \boldsymbol{u}) = 0 \,, \tag{1}$$

momentum

$$\frac{\partial (\varrho \boldsymbol{u})}{\partial t} + \mathrm{div}(\varrho \boldsymbol{u} \otimes \boldsymbol{u}) = -\mathrm{grad}\, p + \mathrm{div}\, \mathsf{T} \,, \tag{2}$$

energy

$$\frac{\partial e_{\mathrm{t}}}{\partial t} + \mathrm{div}((e_{\mathrm{t}} + p)\boldsymbol{u}) = \mathrm{div}(\mathsf{T}\,\boldsymbol{u}) - \mathrm{div}\,\boldsymbol{q} \,, \tag{3}$$

and the masses of the N_{S} chemical species

$$\frac{\partial (\varrho Y_\alpha)}{\partial t} + \mathrm{div}(\varrho Y_\alpha \boldsymbol{u}) = M_\alpha \dot{c}_\alpha - \mathrm{div} \boldsymbol{j}_\alpha \,, \quad \alpha = 1, \ldots, N_{\mathrm{S}} \,. \tag{4}$$

Herein, ϱ denotes the density and $\boldsymbol{u}$ the velocity, Y_α, $\boldsymbol{j}_\alpha$, and M_α are the mass fraction, diffusion flux, and molar mass of the chemical species α respectively, T denotes the viscous stress tensor and p the pressure, $\boldsymbol{q}$ is the heat flux and e_{t} is the total energy given by

$$e_{\mathrm{t}} = \varrho \left(\frac{\boldsymbol{u}^2}{2} + \sum_{\alpha=1}^{N_{\mathrm{s}}} h_\alpha Y_\alpha \right) - p \,, \tag{5}$$

where h_α is the specific enthalpy of the species α.

The computation of the chemical source terms on the right-hand-sides of the species mass equations (4) is one of the most time-consuming parts in such DNS. The production rate $\dot{c}_\alpha$ of the chemical species α is given as the sum over the formation rate equations for all N_{R} elementary reactions,

$$\dot{c}_\alpha = \sum_{\lambda=1}^{N_{\mathrm{R}}} k_\lambda (\nu_{\alpha\lambda}^{(\mathrm{p})} - \nu_{\alpha\lambda}^{(\mathrm{r})}) \prod_{\alpha=1}^{N_{\mathrm{S}}} c_\alpha^{\nu_{\alpha\lambda}^{(\mathrm{r})}} \,, \tag{6}$$

where $\nu_{\alpha\lambda}^{(\mathrm{r})}$ and $\nu_{\alpha\lambda}^{(\mathrm{p})}$ denote the stoichiometric coefficients of reactants and products respectively, and c_α is the concentration of the species α. The rate coefficient k_λ of an elementary reaction is given by a modified Arrhenius law

$$k_\lambda = A_\lambda T^{b_\lambda} \exp\left(-\frac{E_{\mathrm{a}\lambda}}{RT} \right) \,. \tag{7}$$

The chemical reaction mechanism for the $H_2/O_2/N_2$ system which has been used in the simulations presented in Sect. 4 contains $N_S = 9$ species and $N_R = 37$ elementary reactions [3].

This system of equations is closed by the state equation of an ideal gas

$$p = \frac{\varrho}{\overline{M}} RT \qquad (8)$$

with R being the gas constant and $\overline{M}$ the mean molar mass of the mixture.

3   Structure and Performance of the Parallel DNS-Code

3.1   Basic Description of the DNS-Code

A code has been developed for the direct numerical simulation (DNS) of reactive flows on parallel computers with distributed memory using message-passing communication [4–6]. Detailed models are utilized for the computation of chemical reaction kinetics, of the thermodynamical properties, the viscosity, and the molecular and thermal diffusion velocities. At the moment we restrict ourselves to the simulation of two-dimensional flow fields. As turbulence is an inherently three-dimensional phenomenon, some of it's aspects cannot be reproduced precisely with such simulations. On the other hand, it has been shown that many aspects of turbulent combustion processes cannot be adequately represented if oversimplified models for chemistry or molecular diffusion are used [1,2]. We consider the inclusion of detailed chemical reaction schemes to be at least as important for the investigation of turbulent flames as an extension to three dimensions.

The spatial discretization in this code is performed using a finite-difference scheme with sixth-order central-dervatives, avoiding numerical dissipation and leading to very high accuracy. Depending on the boundary conditions, lower order schemes are used at the outermost grid points. The integration in time is carried out using a fourth-order fully explicit Runge-Kutta method. The time step of the integration is controlled by three different limiting conditions. A Courant-Friedrichs-Lewy (CFL) criterion and a Fourier criterion for the diffusion terms are checked to ensure the stability of the integration. An additional accuracy control of the result is obtained through time-step-doubling.

The fully explicit formulation leads to a parallelization strategy, which is based on a regular two-dimensional domain decomposition of the physical space, projected onto a corresponding two-dimensional processor topology. For a given computational grid and number of processors it is tried to minimize the length of the subdomain boundaries and thus the amount of communication. After this initial decomposition, each processor node controls a rectangular subdomain of the global computational domain. In addition to the grid points belonging to a node's subdomain, the values of the physical variables at the grid points of a three points wide peripheral surrounding region are stored on each node. Using these locally stored values, an integration step on the subdomain is carried out independently from the other nodes. After each integration step, the new inner boundary values of the subdomain are sent to and the new values of the surrounding region are received from the neighbouring nodes via message-passing communication.

The code has been successfully used on a variety of parallel platforms and message-passing libraries including Intel Paragon with NX, Parsytec GC/PowerPlus with Parix [5], Cray T3E systems using PVM and MPI (cf. Sect. 3.2), clusters of workstations or PCs (cf. Sect. 3.3), and metacomputing environments (cf. Sect. 3.4).

3.2 Performance on Cray T3E systems

Our main production platform is the Cray T3E on which versions of the code using PVM and MPI for the communication have been tested. During the normal integration in time, the performance difference between both versions was less than 1% CPU-time, whereas for the parts of the simulation in which values of the output variables from all subdomains are gathered for I/O, the MPI-version clearly outperformed the PVM-version [2]. In these parts, messages are sent with sizes scaling with the number of grid points per subdomain, whereas during the rest of the temporal integration the message-sizes scale with the number of grid-points along the sub-domain boundaries. Due to the fact that MPI delivers a higher communication bandwidth than PVM on the Cray T3E, the MPI-version performed better with increasing message-sizes in this comparison. As the latency for MPI_Send and MPI_Recv operations on the T3E has been reduced by a factor of two by making use of the E-register hardware message-queues [7] in newer versions of the Message Passing Toolkit (MPT 1.3.0.5, MPT 1.4), the MPI-version should also be superior in the computational part now, in which smaller messages are sent.

Recently, our main platform for production runs have therefore been Cray T3E systems using MPI. Using 64 PEs of a Cray T3E-900, a speedup of 57.9 has been achieved for a simulation of the $H_2/O_2/N_2$ system (9 species, 37 elementary reactions) on a 544^2 points grid which corresponds to small production runs. The average performance per PE within this simulation (including I/O) was 86.3 MFLOP/s. Using 256 processors to compute the same problem, a speedup of 189 was gained corresponding to a parallel efficiency of 73.8 % [8]. A tabulation of binary diffusion coefficients $D_{\alpha,\beta}$ leads to a reduction of the time needed for the computation (although FLOP rates are obviously also decreased). Table 1 lists the relative performance, i. e. the inverse of the ratio of execution time to the execution time on the T3E-1200, of one PE of a T3E-900 and of a T3E-1200 respectively for a test-case with a 50×50 points grid with and without tabulation of the binary diffusion coefficients. The performance of the 600 MHz PE (T3E-1200) is about 22 % higher than

Table 1. Performance comparison between Cray T3E-900 and Cray T3E-1200.

System	Tab. of $D_{\alpha,\beta}$	Rel. Performance
T3E-900	no	0.725
T3E-900	yes	0.820
T3E-1200	no	0.892
T3E-1200	yes	1.000

that of one 450 MHz PE (T3E-900) in the case with tabulation and about 23 % higher in the case without tabulation of the $D_{\alpha,\beta}$ in this test.

3.3   Performance on clusters of PCs

A big trend in parallel computing today is the use of clustered systems. Networks of workstations or clusters of PCs provide an excellent performance/price ratio – at least for problems which they are well suited for. Of course there are many applications in HPC that would mainly benefit from an increased memory bandwidth. For such applications one would preferably use vector computers such as the NEC SX-5.

As our code runs efficiently on the Alpha 21164 (EV5) found in the Cray T3E, it should be possible to achieve a good performance on other microprocessors, too. Computations have been performed on clusters with different processors and network hardware. In the following, results are presented of the clusters described below:

Alpha-Myrinet (Alpha-M):

> Nodes: 64 Compaq DS10 workstations, each having one 466 MHz Alpha 21264 (EV6) processor (2 MB cache) and 256 MB RAM
> Interconnect: Myrinet
> Software: SuSE Linux 6.3; Compaq Fortran CFAL 1.0, command: `fort` (via `mpif90`), uses `-O4` as default

Dual-Pentium-Myrinet (DP3-M):

> Nodes: 8 IBM dual-processor Netfinity servers, each having two 600 MHz Pentium III processors and 896 MB RAM
> Interconnect: Myrinet
> Software: RedHat Linux 6.2; PGI compiler version 3.1-3, command: `pgf90 -fast` / `pgf90 -o` (via `mpif90`)

Pentium-Ethernet (P3-E):

> Nodes: 16 PCs with 660 MHz Pentium III processors and 512 MB RAM
> Interconnect: 100 MBit Ethernet
> Software: RedHat Linux 6.2; PGI compiler version 3.1-3, command: `pgf90 -o` (via `mpif90`)

Table 2 lists the relative performance of one processor of the systems described above for different problem sizes. Again, all simulations were done with the mechanism describing the $H_2/O_2/N_2$ system (9 species, 37 elementary reactions). The performance is normalized with the problem size and the values are given relative to the performance of a T3E-1200 PE (using CF90 3.5, f90 -O3) on the problem with 50^2 gridpoints.

Table 2. Monoprocessor performance for different problem sizes related to the performance of the Cray T3E-1200 solving the 50 × 50 problem.

Problem Size	50 × 50	100 × 100	200 × 200	400 × 400
T3E-1200	1.00	1.11	1.19	1.23
Alpha-M	2.39	2.40	2.33	(not enough memory)
DP3-M (-fast)	1.36	1.41	1.35	1.32
DP3-M (-o)	1.21	.1.24	1.20	1.19
P3-E	1.28	1.25	1.22	1.26

The PE of the T3E performs better for the larger grids, whereas the differences of the other systems with respect to the differently sized problems remain within a few percent. The Pentium III systems are slightly faster in these tests than the Alpha 21164, and the Alpha 21264 performs about 75% better than the Pentium III with the PGI compiler at the higher optimization level. Although this is far from being an extreme example, the performance differences resulting from the use of different compiler options in the case of the Dual-Pentium-Myrinet cluster clearly show the impact of the compiler on the application performance.

The classical measure for the scaling behaviour in parallel computing is the speedup defined as

$$S(N, X) = \frac{t(1, X)}{t(N, X)}, \tag{9}$$

where $t(N, X)$ is the computation time for solving a problem X using N processors. In an ideal sense, $t(1, X)$ should be the time needed by one processor for solving the same problem using the optimal sequential algorithm (in an optimal implementation, ...), a concept which is typically not applicable in real-world applications. To match the classical definition of speedup, $t(1, X)$ should at least not include any unnecessary overhead dealing with the parallelization. The parallel efficiency is then defined as

$$E(N, X) = \frac{S(N, X)}{N} = \frac{t(1, X)}{N \cdot t(N, X)}. \tag{10}$$

Another approach is to measure the speedup for a scaled problem. In this case, the parallel efficiency

$$E_\mathrm{s}(N, X) = \frac{t(1, X)}{t(N, (N \cdot X))} \qquad (11)$$

shows more directly the performance loss due to the necessary communication. In addition, if one is more interested in using a parallel computer to solve a bigger problem than in solving the same problem in less time, this situation is better reflected in such a benchmark in which the load per processor is kept constant.

Table 3 shows the parallel efficiencies for two fixed problem-sizes achieved with 16 processors of each of the systems described above. Table 4 lists the scaled efficiencies for 16 processors and two different loads per processor. (For the parallelization on the DP3-M cluster, at least from the application programmers point of view, no differences between intra-node communication and inter-node communication have been made.)

Table 3. Parallel efficiency (in percent) using 16 processors for two problems of fixed size.

	$E(16, 400^2)/\%$	$E(16, 200^2)/\%$
T3E-1200	83.1	76.3
Alpha-M	–	82.6
DP3-M (-fast)	83.6	83.3
DP3-M (-o)	84.5	83.5
P3-E	84.6	81.4

Table 4. Scaled parallel efficiency (in percent) using 16 processors for two different loads per processor.

	$E_\mathrm{s}(16, 100^2)/\%$	$E_\mathrm{s}(16, 50^2)/\%$
T3E-1200	92.0	91.0
Alpha-M	90.4	80.5
DP3-M (-fast)	78.5	82.3
DP3-M (-o)	81.1	83.0
P3-E	85.7	77.5

One result is, that all the systems tested deliver reasonable parallel performance – at least up to 16 processors. But these numbers show also some problems of these classical performance measures. E. g., $E(16, 200^2)$ looks quite poor for the T3E, it is even worse than the efficiency of the Ethernet cluster. But from Table 2 it can be seen, that this is mostly due to the better relative performance of one PE for the 200×200 problem compared to that for 50×50 points, which is the load per PE when the 200×200 problem is solved using 16 processors. Taking this difference of 19% into account, one gets of course the same purged efficiency of 91% as is listed in Table 4 for the scaled efficiency $E_{\mathrm{s}}(16, 50^2)$. It should be noted, that although it is often a better indicator than $E(N, X)$, the scaled parallel efficiency $E_{\mathrm{s}}(N, X)$ also is not a measurement for the absolute network performance. In fact

$$E_{\mathrm{s}}(N, X) = \frac{t_{\mathrm{a}}(1, X)}{t_{\mathrm{a}}(N, (N \cdot X)) + t_{\mathrm{c}}(N, (N \cdot X))} , \qquad (12)$$

where t_{a} denotes the time spent for the computational part and t_{c} the time for communication and synchronization. In our code (at least in these tests without adaptivity) and several other parallel applications, except for the communication, the same operations are performed on each of N processors which solve a problem of size $(N \cdot X)$ as are performed on one processor which solves a problem of size X, and thus

$$t_{\mathrm{a}}(N, (N \cdot X)) = t_{\mathrm{a}}(1, X) . \qquad (13)$$

(In fact, there are some variations depending on the types of boundary conditions applied, but these are negligible for the benchmarks presented here.) The parallel efficiency E_{s} then becomes a function of the ratio of communication to computation performance

$$E_{\mathrm{s}}(N, X) = \frac{1}{1 + \frac{t_{\mathrm{c}}(N, (N \cdot X))}{t_{\mathrm{a}}(1, X)}} \qquad (14)$$

and decreases if the computational part is carried out in shorter time and t_{c} remains constant, i.e. the use of faster processors in combination with the same communication hardware leads to a decreased parallel efficiency E_{s}. In the same way, the better code optimization causes the decrease of the parallel efficiency in the case of the Dual-Pentium-Myrinet cluster.

Because of (13) we can easily calculate $t_{\mathrm{c}}(N, (N \cdot X))$, the reverse of which is a measure for the network performance (for a specific problem). Table 5 lists the times spent for communication and synchronization $t_{\mathrm{c}}(N, (N \cdot X))$ for our application. The problem-size per processor in these benchmarks is also typical for real production runs, the number of messages in both tests is the same, and the length of the messages between neighbouring nodes scales with the length of the domain-boundaries. From the times achieved on the Alpha-Myrinet cluster it can be seen, that for our application and 16 nodes the Myrinet is only slightly slower than the T3E-network in the case with the smaller messages and 76% faster in the case with the larger messages. As expected, the network performance of the Ethernet cluster is clearly lower than that of these two systems. The Dual-Pentium-Myrinet cluster suffers

Table 5. Time for communication and synchronization for 20 timesteps using 16 processors.

	$t_c(16, (16 \cdot 100^2))/s$	$t_c(16, (16 \cdot 50^2))/s$
T3E-1200	2,518	0.799
Alpha-M	1,427	0.813
DP3-M (-fast)	6.255	1.268
DP3-M (-o)	6.024	1.363
P3-E	4.295	1.816

from the very low parallel efficiency inside the nodes due to memory conflicts. The bus-based shared memory architecture also causes the performance of the dual-nodes to be more sensitive to the problem-size. A better systematics for comparisons like the presented ones may be the classification of one dual-node (using both processors) as the basic element of this system, to which the results of the scalability tests are related. One minor drawback in this case would be the need for different sets of input files for the different systems due to the number of nodes not being identical to the number of MPI-processes for all systems. However, the main reason to take one processor of the dual-board as the basis here, was the availability of only eight nodes. At least nine nodes are needed to get a topology, in which at least one node has four neighbours with respect to the two-dimensional domain-decomposition applied. Therefore, for our application the communication patterns of large configurations are much better reflected in a 16 node configuration (four "inner" nodes) than in a configuration with only eight nodes (no "inner" nodes).

3.4   Metacomputing

Besides clusters of commodity hardware like discussed in the last section, there is also a trend to clustered systems at the high-end of supercomputer systems in the form of clusters of SMPs (ASCI White, Hitachi SR8000) or clusters of PVPs (NEC SX-5M). It is also possible to combine two or more supercomputer systems into one cluster, then termed a metacomputer.

At least in the case of a heterogeneous metacomputer one has to deal with the problem that manufacturers typically do not care for interoperability with other vendor's systems. One of the design goals of PVM was this type of interoperability. Applications on different platforms can exchange messages with each other using PVM. As already said, there is a PVM-version of our DNS code. Unfortunately, the PVM implementation on the Cray T3E lacks this kind of interoperability. Over the last few years, several efforts have been made to extend MPI for heterogeneous distributed computing. One such effort is PACX-MPI [9], a library developed at the HLRS which allows to run MPI applications in metacomputing environments.

Several tests have been performed with our application using PACX-MPI in different metacomputing configurations inside the HLRS. In simulations of vortex-pair flame interactions using the Cray T3E-900 coupled with a SGI Origin 2000 we focussed on the efficiency of the implemented dynamic load-balancer in this environment. The load-imbalances in these computations stem from adaptive gridding as well as from the higher performance of the processors of the Origin 2000. Although one has to deal with the very different communication speeds in such metacomputing environments, promising performance levels have been achieved with very little modifications. A full production run, the DNS of autoignition in a turbulent mixing layer, has been performed using 16 PEs of the T3E and 16 processors (2 nodes) of a Hitachi SR8000 for about 8 hours. The results of our investigations regarding metacomputing are described in detail in [10].

4   Turbulent Flame Kernels

Investigations of flame-front turbulence interactions often employ synthetic turbulent flames, which are generated by superimposing an initially planar laminar premixed or non-premixed flame with turbulence [11–13]. Induced ignition is another phenomenon of practical importance, e. g. in Otto engine combustion and safety considerations. DNS studies of this process have been performed using simple one-step chemistry in a model configuration of an initially uniform premixed gas under turbulent conditions which is ignited by an energy source in a small region at the center of the domain [14]. Above a minimum ignition energy an expanding flame kernel is observed. An advantage of this configuration is that detailed experimental results are available now [15] for which DNS with conditions matching those of the experiments are possible.

In the simulations presented in this chapter, a cold ($T = 298\,\mathrm{K}$) hydrogen-air mixture is superimposed with a turbulent flow field computed by inverse FFT from a von-Kármán-Pao-spectrum with randomly chosen phases. The first phase of such a simulation is shown in Fig. 1. The energy-source is active in the first $15\,\mu\mathrm{s}$. During this time the mixture at the center of the domain heats up to about $3100\,\mathrm{K}$, radicals are formed, and the mixture ignites. A shockwave is observed which propagates outwards towards the boundaries of the domain. Non-reflecting outflow conditions based on characteristic wave relations [16] are imposed on all boundaries to allow the shock wave to leave the domain without disturbing the solution as can be seen from Fig. 1.

Figure 2 shows the temporal evolution of such a flame kernel in a turbulent flow field after the ignition. (A smaller energy source has been used for the ignition compared to the simulation shown in Fig. 1.) The first row of images shows the temperature at $t = 0.2\,\mathrm{ms}$, $t = 0.4\,\mathrm{ms}$, and $t = 0.6\,\mathrm{ms}$, respectively. In the middle row, the mass fraction of H_2O_2, a radical which is confined to a very thin layer in the flame, is shown at the same points in time. Below, the temporal evolution of vorticity is shown. The lack of a vortex-stretching

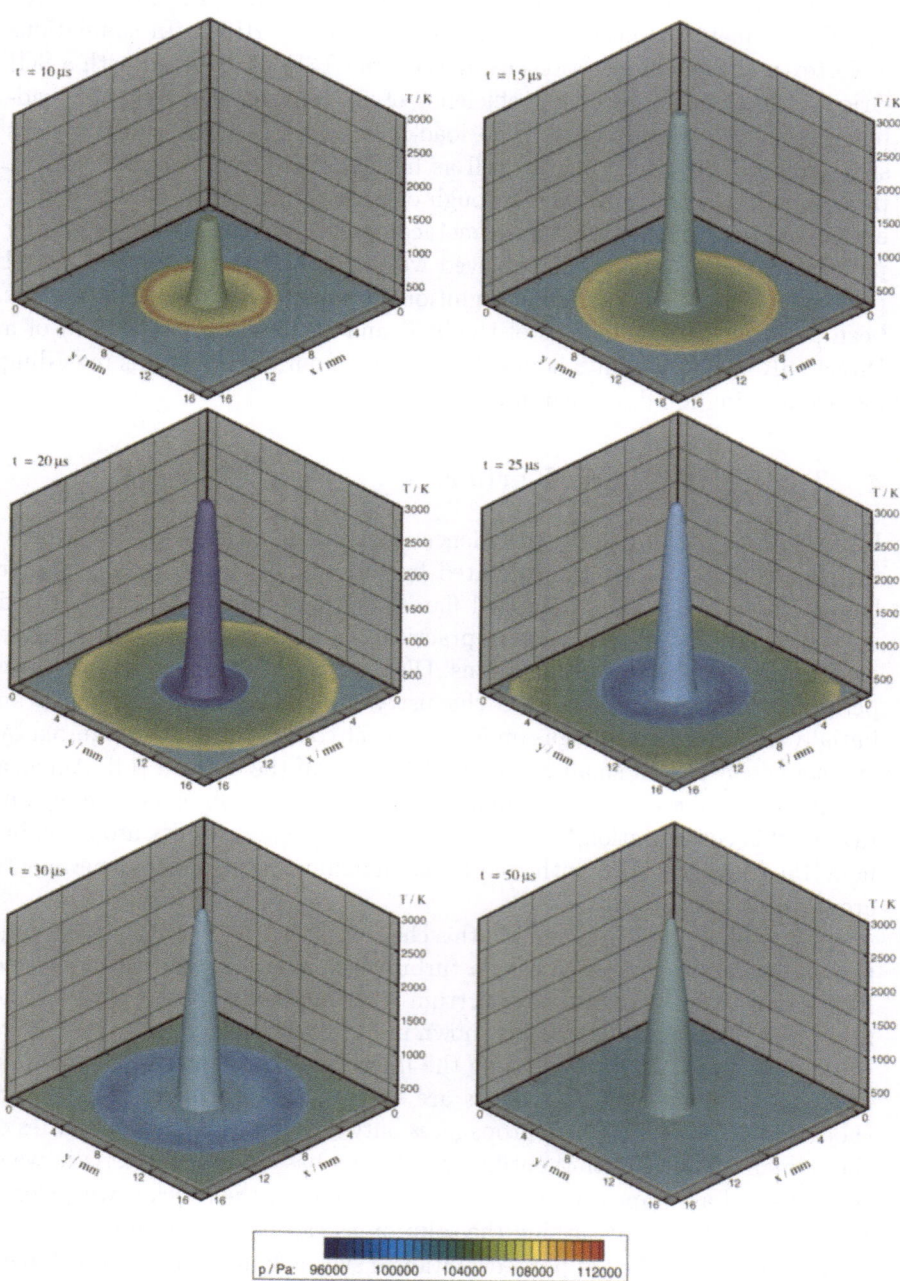

Fig. 1. Temporal evolution of pressure and temperature during the ignition of a turbulent hydrogen-air mixture.

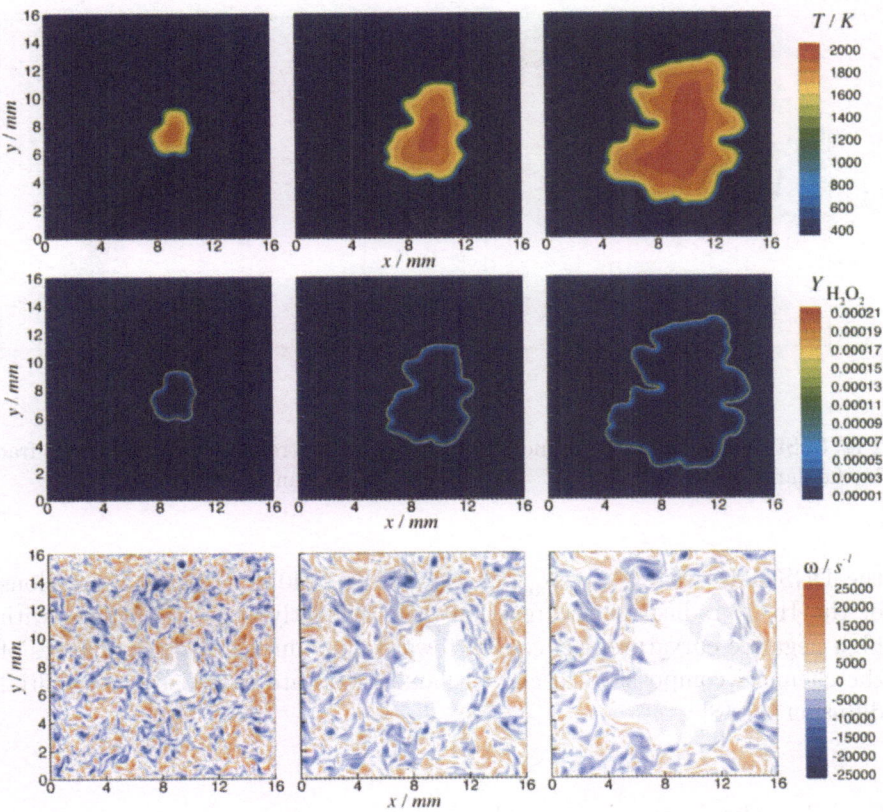

Fig. 2. Temperature, H_2O_2 mass fraction, and vorticity (top to bottom) at $t = 0.2\,ms$, $t = 0.4\,ms$, and $t = 0.6\,ms$ (left to right) in a turbulent premixed flame.

mechanism in two-dimensional simulations of decaying turbulence leads to an inverse cascade with growing structures. There is a very strong damping of the turbulence in the hot region of the burnt gas due to the high viscosity. A comparison shows a good qualitative agreement of the DNS presented with the results of an experimental investigation of turblent flames performed under similar conditions [15].

A quantitative analysis requires specialized tools for the postprocessing of the large and complex datasets generated by DNS. An extensible tool has been developed which allows to extract several features from these datasets. This is illustrated in Figs. 3 and 4 which exemplify results gained from the simulation of the turbulent flame kernel shown in Fig. 2. In Fig. 3 the strain rate and the flame thickness (based on the local temperature gradient) along the flame front at $t = 0.6\,ms$ are shown. The correlation of the H_2O_2 mass fraction and the curvature of the flame is shown in Fig. 4 for subsets of

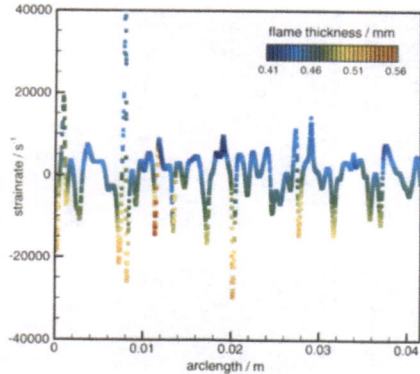
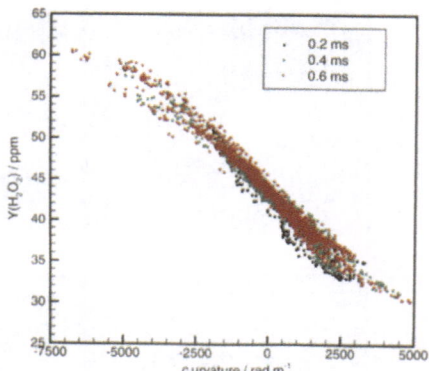

Fig. 3. Strain rate and flame thickness in the flame shown in Fig. 2.

Fig. 4. Correlation of H_2O_2 mass fraction and flame front curvature.

the DNS data with $T = (T_{max} + T_{min})/2 = 1140\,K$. High concentrations of the H_2O_2 radical in the reaction zone evidently occur in regions with high negative curvature, i. e. convex towards the burnt gas. Such changes of the chemical composition in curved flame fronts are caused by preferential diffusion [17,18].

5   Conclusions and Outlook

The results of using a parallel detailed-chemistry DNS-code as a benchmark on several clusters with different processors and interconnect technologies have been presented and compared with the Cray T3E. They clearly show the potential of clustered commodity hardware for high-performance computing. Scalability tests with at least 64 processors are in preparation. Heterogeneous metacomputing environments are a valuable facility for solving very large problems. However, efficient solutions for metacomputing typically require additional efforts compared to classical supercomputers. To be able to perform DNS with detailed models for chemistry and transport in three dimensions, it is crucial to make efficient use of the next generation supercomputing hardware. This implies making use of possible adaptivity in the discretization and in the physical models [19], employing efficient numerical schemes, and a thoughtful implementation with respect to the properties of the hardware used.

In the second part of the paper, results have been shown of the DNS of induced ignition of a turbulent mixture, which is followed by the propagation of a premixed flame. Postprocessing tools have been developed which allow the extraction of important flame characteristics like strain rate and curvature from the DNS-data. As an examle, the influence of the flame front curvature on the chemical composition in the reaction zone has been briefly discussed.

Acknowledgement

The author thanks the John von Neumann Institute for Computing at Jülich (NIC) and his home institution, the High Performance Computing Center at Stuttgart (HLRS), for providing him the opportunity to use the supercomputer systems installed at these sites. For the benchmarks presented in Sect. 3.3 access to their Alpha-Linux-Cluster-Engine (ALiCE) has been granted by Wuppertal University. The benchmarks on the Pentium-clusters have been performed in collaboration with IBM, Poughkeepsie.

References

1. Mantel, T. and Samaniego, J.-M., "Fundamental Mechanisms in Premixed Turbulent Flame Propagation via Vortex-Flame Interactions, Part II: Numerical Simulation," *Combustion and Flame*, Vol. 118, 1999, pp. 557–582.
2. Lange, M. and Warnatz, J., "Investigation of Chemistry-Turbulence Interactions Using DNS on the Cray T3E," *High Performance Computing in Science and Engineering '99*, edited by E. Krause and W. Jäger, Springer, Berlin, Heidelberg, New York, 2000, pp. 333–343.
3. Maas, U. and Warnatz, J., "Ignition Processes in Hydrogen-Oxygen Mixtures," *Combustion and Flame*, Vol. 74, 1988, pp. 53–69.
4. Thévenin, D., Behrendt, F., Maas, U., Przywara, B., and Warnatz, J., "Development of a Parallel Direct Simulation Code to Investigate Reactive Flows," *Computers and Fluids*, Vol. 25, No. 5, 1996, pp. 485–496.
5. Lange, M., Thévenin, D., Riedel, U., and Warnatz, J., "Direct Numerical Simulation of Turbulent Reactive Flows Using Massively Parallel Computers," *Parallel Computing: Fundamentals, Applications and New Directions*, edited by E. D'Hollander, G. Joubert, F. Peters, and U. Trottenberg, No. 12 in Advances in Parallel Computing, Elsevier Science, Amsterdam, 1998, pp. 287–296.
6. Lange, M. and Warnatz, J., "Detailed Simulations of Turbulent Flames Using Parallel Supercomputers," *High Performance Computing in Science and Engineering '98*, edited by E. Krause and W. Jäger, Springer, Berlin, Heidelberg, New York, 1999, pp. 343–352.
7. *Message Passing Toolkit: Release Notes*, No. 004-3689-001 in Online Software Publications, Cray Inc., 2000.
8. Lange, M. and Warnatz, J., "Direct Simulation of Turbulent Reacting Flows on the Cray T3E," *Proceedings of the 14th Supercomputer Conference '99 in Mannheim*, edited by H.-W. Meuer, 1999.
9. Gabriel, E., Resch, M., and Rühle, R., "Implementing MPI with Optimized Algorithms for Metacomputing," *Proceedings of the Third MPI Developer's and User's Conference*, edited by A. Skjellum, P. V. Bangalore, and Y. S. Dandass, 1999.
10. Gabriel, E., Lange, M., and Rühle, R., "Direct Numerical Simulation of Turbulent Reactive Flows in a Metacomputing Environment," *Proceedings of the International Conference on Parallel Processing*, 2001.

11. Poinsot, T. J., Candel, S., and Trouvé, A., "Applications of Direct Numerical Simulation to Premixed Turbulent Combustion," *Progress in Energy and Combustion Science*, Vol. 21, 1996, pp. 531–576.
12. Lange, M., Riedel, U., and Warnatz, J., "Parallel DNS of Turbulent Flames with Detailed Reaction Schemes," AIAA Paper 98-2979, 1998.
13. Thévenin, D. and Baron, R., "Investigation of Turbulent Non-Premixed Flames using DNS with Detailed Chemistry," *Direct and Large Eddy Simulation III*, edited by P. Voke, N. Sandham, and L. Kleiser, Kluwer Academic Publishers, 1999, pp. 323–334.
14. Echekki, T., Poinsot, T. J., Baritaud, T. A., and Baum, M., "Modeling and Simulation of Turbulent Flame Kernel Evolution," *Transport Phenomena in Combustion*, edited by S. H. Chan, Vol. 2, Taylor & Francis, 1995, pp. 951–962.
15. Renou, B., Boukhalfa, A., Puechberty, D., and Trinité, M., "Local Scalar Flame Properties of Freely Propagating Premixed Turbulent Flames at Various Lewis Numbers," *Combustion and Flame*, Vol. 123, 2000, pp. 507–521.
16. Baum, M., Poinsot, T. J., and Thévenin, D., "Accurate Boundary Conditions for Multicomponent Reactive Flows," *Journal of Computational Physics*, Vol. 116, 1995, pp. 247–261.
17. Law, C. K., "Dynamics of Stretched Flames," *Proceedings of the Combustion Institute*, Vol. 22, The Combustion Institute, Pittsburgh, 1988, pp. 1381–1402.
18. Lange, M. and Warnatz, J., "A DNS Study of Curvature Effects in Turbulent Premixed Flames," *Scientific Computing in Chemical Engineering II*, edited by F. Keil, W. Mackens, H. Voß, and J. Werther, Vol. 2. Simulation, Image Processing, Optimization, and Control, Springer, Berlin, Heidelberg, New York, 1999, pp. 126–133.
19. Lange, M., "Adaptive Chemistry Computation to Accelerate Parallel DNS of Turbulent Combustion," *High Performance Computing in Science and Engineering 2000*, edited by E. Krause and W. Jäger, Springer, Berlin, Heidelberg, New York, 2001, pp. 412–424.

Direct Numerical Simulations of Spark Ignition of H_2/Air-Mixture in a Turbulent Flow

Wilhelmina Tsai, Dietmar Schmidt and Ulrich Maas[1]

[1]Universität Stuttgart, Institut für Technische Verbrennung, Pfaffenwaldring 12, D-70569 Stuttgart, Germany

Abstract. In this work the spark ignition of a H_2/air mixture in a turbulent flow field is investigated by means of direct numerical simulations using both, detailed chemical kinetics and detailed transport models. Parameters like ignition energy, turbulent velocity, etc., are adapted to data given by a similar experiment [1]. The results are discussed in the context of the interaction of the turbulent flow field with chemical kinetics.

1 Introduction

Spark ignition plays an important role in many technical applications, like e.g. in IC engines and in safty devices. Therefore, much research has been carried out in the last years in order to understand the fundamentals of the ignition processes by means of experiments and numerical simulations. Using sophisticated experimental methods like Laser Induced Fluorescence (LIF), Raman scattering, Laser Doppler Anemometry (LDA), or Particle Image Velocimetry (PIV) high spatial and time resolutions of spark ignited system are investigated [1–3]. On the other hand the numerical calculations considering simplified [4] or detailed [5] chemical reaction mechanisms and transport models were performed recently using Direct Numerical Simulations (DNS) in order to investigate the interaction of turbulence, molecular transport and chemical reaction in spark ignited systems.

Because DNS requires intense computational efforts and costs the simulations are constrained only on very small areas (e.g. $1.0\,cm \times 1.0\,cm$) and low Reynolds-numbers. Nevertheless informations about the effect of the turbulence on the growth of a flame kernel can be extracted by the DNS-data allowing the description of the instationary ignition process and the following flame propagation.

In this work the code NSCORE developed by Poinsot and Baum [6,7] in the framework of their research on DNS of turbulent flames and spark ignited systems was implemented on the Cray T3E at the High Performance Computing Center Stuttgart (HLRS). In the following section the configurations of the experiment which is simulated are described. In section 3 the results of DNS are shown and discussed.

2   DNS of the Spark Ignition Processes

In NSCORE the governing conservation equation system of mass, momentum, energy and species masses is solved in two dimensions by a spatial discretization and time integration method of high order. For details of the numerical scheme please refer to [8,9]. As in [10] the detailed chemical scheme of Kee for combustion of H_2 in air (including 9 species and 20 reactions) is used. The NSCBC Boundary Conditions [11,12] avoiding reflection of acoustic waves into the calculation domains are used in all directions. As initial condition the turbulent flow field is initiated by an energy spectrum of the van Kármán-Pau typ [13]. To perform the DNS the conditions of experiments carried out in the framework of the experiments [1]-[3] are matched as close as possible. The gas mixture is 20vol% H_2 in air. It corresponds to a mass equivalent ratio of 0.68. The initial pressure is 1.01bar and initial unburnt gas temperature $T = 313K$. The laminar flame speed $s_l = 0.67m/s$ is calculated using PRE-MIX [14] and the flame thickness δ_l, defined by $\delta_l = (T_b - T_u)/(dT/dx)_{max}$, is 0.24mm with T_b and T_u temperatures in the burnt and the unburnt gas, $(dT/dx)_{max}$ being the maximum of temperature gradients.

The energy deposit of the spark ignition is performed by introducing heat power Q_{heat} into the gas phase by an exponential distribution

$$Q_{heat} = Q_{ign} \cdot \exp - \frac{(x - x_{ign})^2 + (y - y_{ign})^2}{2r_{ign}^2},$$

with x_{ign}, y_{ign} being the middle of the calculation domain, r_{ign} the extension of the energy deposit and Q_{ign} a constant parameter. In order to choose values for Q_{ign} and r_{ign} it's essential to understand the temperature evolution of the spark ignition process [3]. The first phase is a short breakdown phase when a conductive plasma channel between the electrodes takes place and the temperature reaches up to 60,000K. After the electrical resistance is dropped due to the plasma channel the second phase is initiated which is characterised by a maximal temperature limited to 6,000K. When the cathode fall increases the third phase begins and the temperature drops down to 3,000K. In contrast to these experimental data we switch off the energy deposit soon after the gas mixture is ignited and the temperature reaches approximately 3,000K because of the resulting small time steps. The time steps are controlled by the chemical numbers in order to avoid numerical oscillations due to large chemical rates. If temperature now exceeds 3,000K the time steps will be too small and the resulting CPU time gets prohibitively large. Therefore a total energy amount of 0.3mJ is deposited in the calculation domain over 36μs in contrast to 5.5mJ during 200μs given by the experiment. Nevertheless preparatory studies show that for the same chemical, thermo- and hydrodynamical conditions the evolution of temperature and species mass fractions of the free propagating flame after the ignition is not sensitive to the total amount of the energy introduced into the gas phase. As an example figure 1 shows the temperature and OH mass fraction of two test cases at a calculation time of

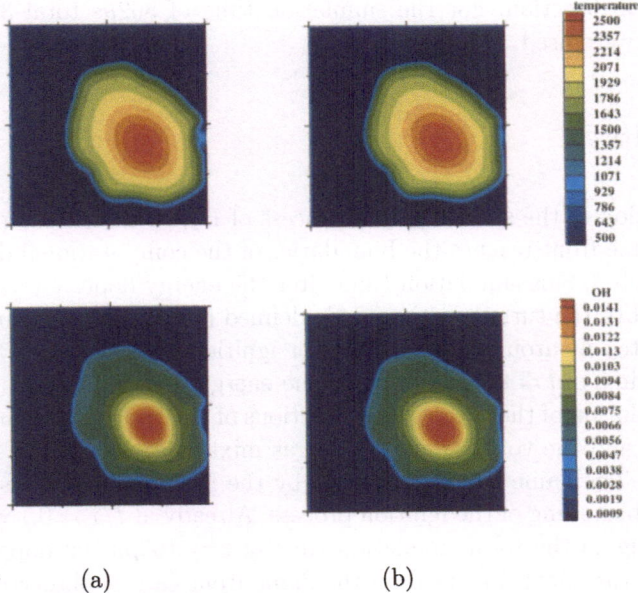

Fig. 1. Contour lines of temperature (upper row) and OH (lower row) of the simulated spark ignition system of a H_2/air mixture at $343\mu s$ after ignition and (a)=0.3mJ, (b)=0.8mJ total energy introduced.

approximately $343\mu s$ after the ignition. The simulation domain was $0.9\,cm$ $\times$ $0.9\,cm$ and $r_{ign} = 0.3mm$ for both cases. The energy input were 0.3mJ and 0.8mJ respectively. The discrepancis between both cases are negligible. Hence we state that the dynamical behaviour does not depend on the amount of spark energy when all other conditions are the same.

To introduce turbulence the experimental measured rms-velocity $u' = 0.75m/s$ was chosen. Using DNS the smallest length scale of the turbulent structure (the Kolmogorov length scale $l_k = 0.064mm$) has to be resolved. With this resolution the realization of the measured integral length scale ($l_0 = 4.3mm$) in the experiment as the greatest length scale of the turbulence will require a huge amount of grid points in our calculations. This is not possible because of the memory constraint on the CRAY T3E which limits the grid number to approximate 120 per processor in this study. The turbulent Reynolds-number defined by $l_0/l_k = Re^{3/4}$ remains consequently low. Nevertheless the laminar flame thickness of 0.26mm takes a value between the integral length scale of 4.3mm and the Kolmogorov length scale of 0.064mm so that the structure distorting the flame front is included in the turbulence of the DNS. The turbulent Reynolds-number is 46 in our simulations.

The DNS of spark ignition was carried out on the CRAY T3E of HLRS using 256 processors. The calculation domain was $1.8\,cm \times 1.8\,cm$ with 1808 grids in each direction. For the simulation time of $862\mu s$ total 372×256 CPU-hours were used.

3   Results

The simulation of the spark ignition process of H_2/air mixture is performed until the flame front reaches the boundaries of the computational domain. It corresponds to 0.8ms simulation time after the energy deposit was switched off. Referred to the turbulent time scale defined as $\tau = l_0/u'$ the simulations are terminated at around $t/\tau = 0.6$ after ignition corresponding to a total simulation time of $t = 862\mu s$ including the energy deposit. Figures 2-7 show the time evolution of the species mass fractions of O_2, HO_2, OH, temperature, heat release and the vorticity field. The gas mixture ignites in the middle of the domain. The flame kernel, indicated by the heat release, grows cylindrically at the beginning of the ignition process. Already at $t/\tau = 0.1$ which was corresponding to the total simulation time of $t = 195\mu s$ the impact of the surrounding turbulent flow field on the flame front can be observed and the flame front is wrinkled by the eddies of the flow. At the end of the simulation the flame surface is significantly enlarged compared with the initial laminar flame kernel.

Up to $t/\tau = 0.3$ after the ignition (total simulation time $t = 404\mu s$) the distribution of the OH mass fraction corresponds to that of the temperature: in the center of the flame kernel the temperature is high as well as the OH concentration. Afterwards however, OH produced in the flame front is consumed rapidly due to chemical reactions accompanied by the production of H_2O in the hot regions. As a result the OH concentration in the flame kernel falls down along the time evolution and OH can be found afterwards to be an indicator of the flame front. The largest amount of the heat release is produced in the flame front, whereas the highest temperatures of about 3,000K is located in the center of the flame kernel. The low temperature on the flame front is due to the fact that the heat release in the flame front is balanced by the heat conduction to the cold unburnt gas.

The mass fraction of O_2 as a stable initial species is shown, as well as HO_2 as a radical in competition to OH. In the chain-branching step hydrogen atoms react with molecular oxygen by $H + O_2 \rightarrow OH + O$ and in the chain termination step the three-body reaction $H + O_2 + M \rightarrow HO_2 + M$ creates less reactive hydroperoxyl radical HO_2. If the mass fraction of HO_2 increases faster than OH the flame propagation will be terminated by the chain breaking process, otherwise as in this study it is sustained by the chain branching process and a normal flame propagation takes place.

The initial vorticity field vanishes in the course of the simulations because no turbulent kinetic energy is introduced in the computational domain. This should be taken into account when comparing the DNS-data with the experimental findings.

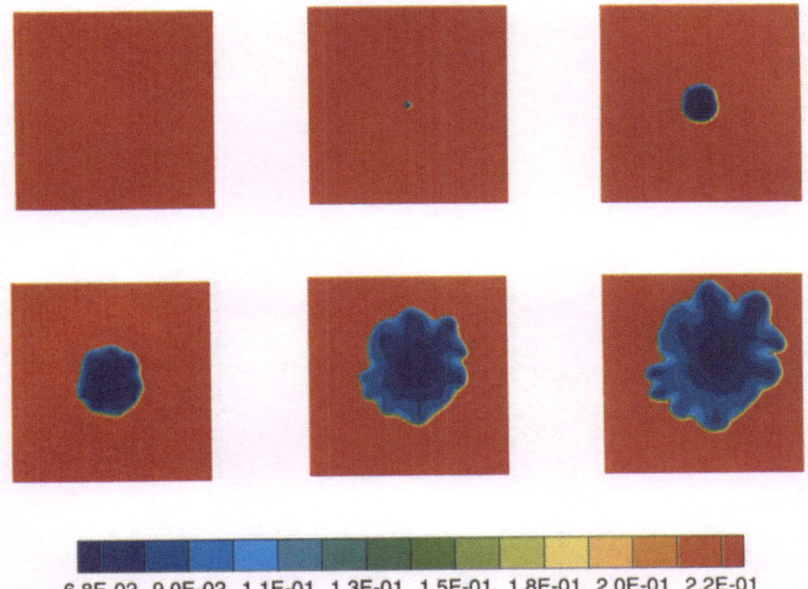

6.8E-02 9.0E-02 1.1E-01 1.3E-01 1.5E-01 1.8E-01 2.0E-01 2.2E-01

Fig. 2. Contour lines of O_2 of the simulated spark ignition system of a H_2/air mixture. Total simulation time from left to right, top to bottom are $t = 0\mu s$, $27\mu s$, $195\mu s$, $404\mu s$, $685\mu s$, $862\mu s$.

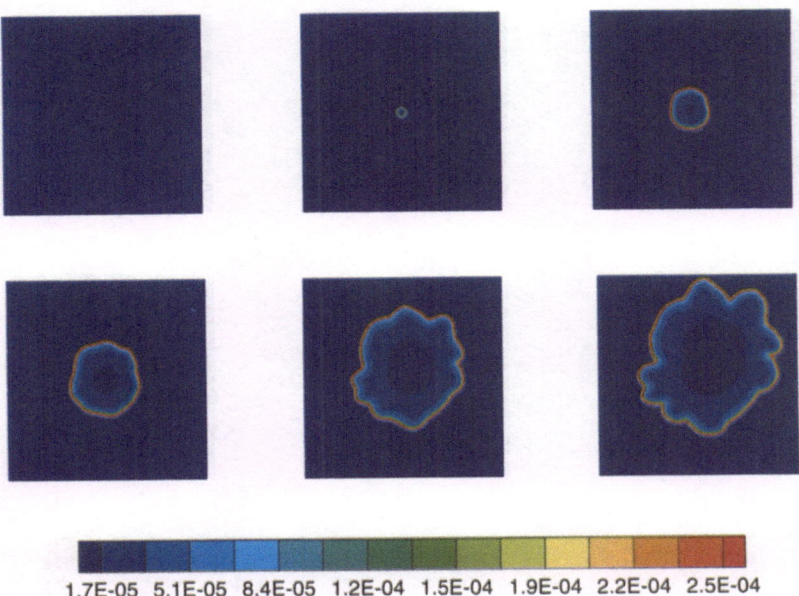

1.7E-05 5.1E-05 8.4E-05 1.2E-04 1.5E-04 1.9E-04 2.2E-04 2.5E-04

Fig. 3. Contour lines of HO_2 of the simulated spark ignition system of a H_2/air mixture. Total simulation time from left to right, top to bottom are $t = 0\mu s$, $27\mu s$, $195\mu s$, $404\mu s$, $685\mu s$, $862\mu s$.

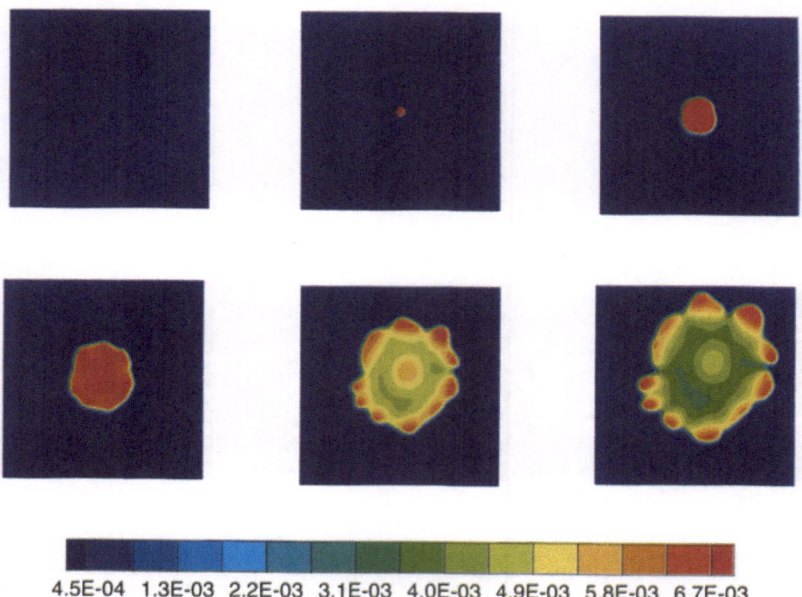

4.5E-04 1.3E-03 2.2E-03 3.1E-03 4.0E-03 4.9E-03 5.8E-03 6.7E-03

Fig. 4. Contour lines of OH of the simulated spark ignition system of a H_2/air mixture. Total simulation time from left to right, top to bottom are $t = 0\mu s$, $27\mu s$, $195\mu s$, $404\mu s$, $685\mu s$, $862\mu s$.

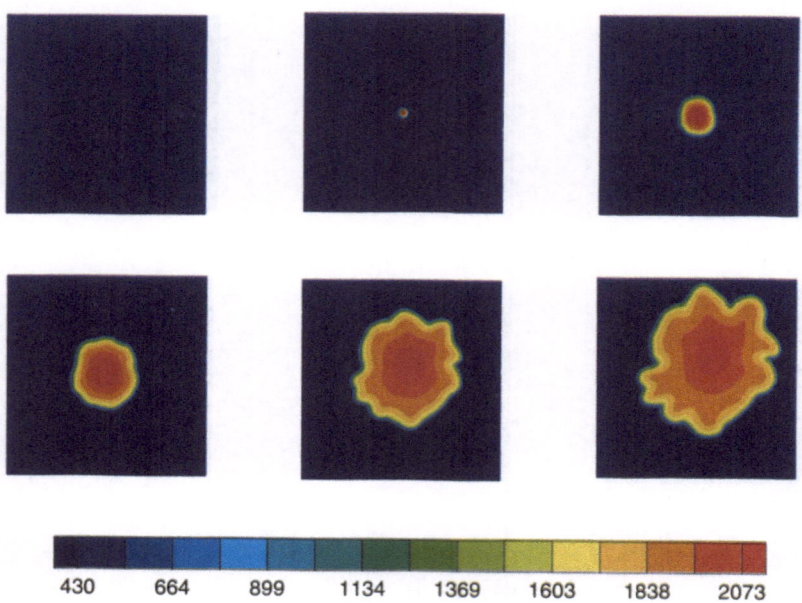

430   664   899   1134   1369   1603   1838   2073

Fig. 5. Contour lines of temperature of the simulated spark ignition system of a H_2/air mixture in Kelvin. Total simulation time from left to right, top to bottom are $t = 0\mu s$, $27\mu s$, $195\mu s$, $404\mu s$, $685\mu s$, $862\mu s$.

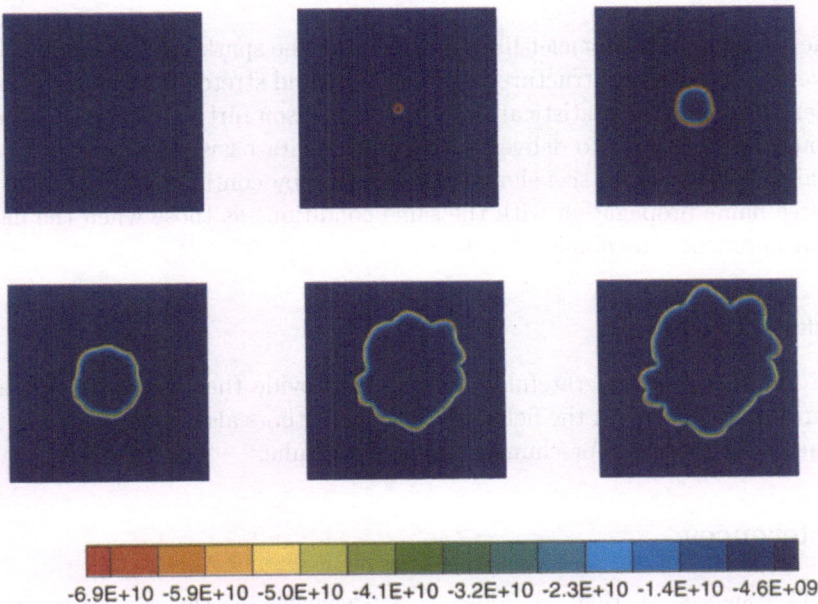

-6.9E+10 -5.9E+10 -5.0E+10 -4.1E+10 -3.2E+10 -2.3E+10 -1.4E+10 -4.6E+09

Fig. 6. Contour lines of heat release of the simulated spark ignition system of a H₂/air mixture in 1e-7W/cm³. Total simulation time from left to right, top to bottom are $t = 0\mu s$, $27\mu s$, $195\mu s$, $404\mu s$, $685\mu s$, $862\mu s$.

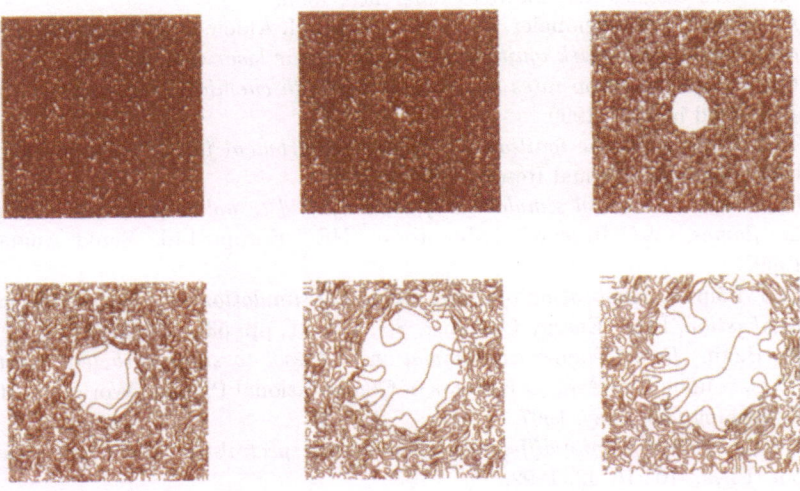

Fig. 7. Contour lines of normalized vorticity of the simulated spark ignition system of a H₂/air mixture. Total simulation time from left to right, top to bottom are $t = 0\mu s$, $27\mu s$, $195\mu s$, $404\mu s$, $685\mu s$, $862\mu s$.

4   Outlook

The analysis of the flamelet-like flame front in the spark ignition process will take account of local structures like curvature and stretch. This requires considerable amount of statistical data. For this reason further DNS calculations should be performed to deliver the necessary data basis. Moreover, the constraint of the domain size should be overcome by continuing the simulation of free flame propagation with the same conditions as those when the flame front approches the domain bords.

Acknowledgements

Dr. Markus Baum is gratefully thanked to provide the code NSCORE and share his experience in the field of DNS. The authors also acknowledge to the financial support by Forschungszentrum Karlsruhe.

References

1. C. Kaminski, J. Hult, M. Alden, S. Lindenmaier, A. Dreizler, U. Maas, M. Baum, *Spark ignition of turbulent methane/air mixtures revealed by time resolved laser induced fluorescence and direct numerical simulations*, Combust. Inst. Vol 28, 2000.
2. S. Lindenmaier, A. Dreizler, U. Maas, C. Kaminski, J. Hult, *Experimental Investigation and Mathematical Simulation of the Ignition of Methane/Air Mixtures*, 1. Dessauer Gasmotoren Conference, 1999.
3. A. Dreizler, S. Lindenmaier, U. Maas, J. Hult, M. Aldén, C. F. Kaminski, *Characterisation of a spark ignition system by planar laser induced fluorescence of OH at high repetition rates and comparison with chemical kinetic calculations*, submitted in JCP, 1999.
4. T. J. Poinsot, *Flame ignition in a premixed turbulent flow*, Center for Turbulence Research, Annual Research Briefs 1991.
5. M. Baum, *Numerical simulation of spark ignited turbulent premixed methane-air flames*, C&C Research Laboratories, NEC Europe Ltd., Sankt Augustin 1999.
6. T. J. Poinsot, *Application of direct numerical simulation to premixed turbulent combustion*, Prog. Energy Combust. Sci. Vol. 21, pp. 531–576, 1996.
7. M. Baum, *Direct Numerical Simulation – A tool to study turbulent reacting flows*, volume V of Annual Reviews of Computational Physics. World Scientific Publishing Company, 1997.
8. S. Lele, *Compact finite difference schemes with spectral-like resolution*. J. Comput. Phys., 103:16–42, 1992.
9. M.Baum, *Etude de l'allumage et de la structure des flammes turbulentes*. PhD thesis, Laboratoire d'Energetique Moleculaire et Macroscopique, Combustion (E.M2.C) du C.N.R.S. et de l'ECP, Paris, 1994.
10. W. Tsai, D. Schmidt, U. Maas, *Correlation Analysis of Premixed Turbulent Flames using Direct Numerical Simulations*, in: High Performance Computing in Science and Engineering 2000, Ed. E. Krause and W. Jäger, Springer, 2001.

11. T. J. Poinsot, S. Lele, *Boundary conditions for direct simulations of compressible viscous flows*, J. Comp. Phys., 101:104–129.

12. M. Baum, T. Poinsot, D. Thevenin, *Accurate Boundary Conditions for Multicomponent Reactive Flows*, Journal of computational physics 116, 247–261 (1994).

13. J. O. Hinze, *Turbulence*, 2nd Edition, McCraw Hill Book Company, 1975.

14. R. J. Kee, J. F. Crcar, M. Smooke, J. A. Miller, *A fortran program for modelling steady laminar one-dimensional premixed flames*, Technical Report SAND85-8240, Sandia Tech. Rep., 1985.

Detailed Simulation of Transport Processes in Reacting Multi-Species Flows Through Complex Geometries by Means of Lattice Boltzmann Methods

Thomas Zeiser[*,1], Hannsjörg Freund[2], Jörg Bernsdorf[3], Peter Lammers[1], Gunther Brenner[1], and Franz Durst[1]

[1] Lehrstuhl für Strömungsmechanik, Universität Erlangen-Nürnberg, Cauerstraße 4, D-91058 Erlangen, Germany
[2] Lehrstuhl für Technische Chemie I, Universität Erlangen-Nürnberg, Egerlandstraße 3, D-91058 Erlangen, Germany
[3] C&C Research Laboratories, NEC Europe Ltd., Rathausallee 10, D-53757 Sankt Augustin, Germany

Abstract. New numerical methods such as the lattice Boltzmann approach together with the increasing computational power can provide a detailed insight into the flow and transport processes in complex three dimensional geometries. The present paper demonstrates this approach by investigating inhomogeneities in the 3-D flow field of reacting species in a tubular fixed bed reactor of small tube-to-particle diameter ratio. For such reactors detailed 3-D simulations are of technical and economical interest as conventional modelling methods cannot correctly predict the local flow behaviour. However, these local inhomogeneities can significantly influence the reactor performance. Beside the discussion of the reaction engineering results, also the computational performance of the applied method is described.

1 Introduction

One fundamental problem in the area of computational fluid dynamics (CFD) is the discretization of the computational domain around complex geometries. The traditional mesh generation procedures are time-consuming and therefore expensive. Moreover, they can hardly be applied to complex structures without clear descriptions of the surface areas as, for example, for most kinds of porous media.

By the use of large regular and orthogonal lattices, arbitrary shaped structures can be discretized with the accuracy of the lattice spacing almost automatically. This so-called "marker-and-cell" description of the geometry can be generated from several sources: if available, CAD data can be discretized with appropriate sized voxels by commercially available software. In some cases, the marker-and-cell description can be generated synthetically, e.g. by analytic formulae (e.g. regular packing of spheres) or Monte Carlo methods (e.g.

---

[*] e-mail: thzeiser@lstm.uni-erlangen.de

randomly packed spheres). If both methods cannot be applied, computed tomography (3-D-CT), known from medical research and material testing, can
be used to get interference-free 3-D images of almost any geometry consisting
of a wide range of materials.

The discretization of complex geometrical structures on regular and orthogonal grids often results in very large computational domains because fine
grids must be used to resolve the geometric details. To handle these grids,
efficient numerical methods and data structures must be applied. With the
lattice Boltzmann method and recent vector-parallel computers, several tens
of millions of grid points can be handled in reasonable times. These methods
are very efficient and therefore can compete with other methods which use
smaller but more complex (e.g. body-adapted) grids.

We apply the lattice Boltzmann approach [5] together with different methods for the generation of the marker-and-cell representation of the geometry
to study in detail turbulence and reactive multi-species flow in complex geometries. The focus of the work in the research period 2000/2001 on the NEC
SX-5e supercomputers of the High-Performance Computing-Center Stuttgart
was to carry out first investigations of the inhomogeneities in tubular fixed
bed reactors of small tube-to-particle diameter ratios. Some results are summarised in this paper.

In Sec. 2, we first give a short description of the technical background
and the importance of the simulations we carried out. In the next section
(Sec. 3), we briefly describe the basic idea of the applied lattice Boltzmann
method for the simulation of the fluid flow and the coupled multi-species
transport. More details can be found in the cited references. In Sec. 4, the
generation of the geometric description of randomly packed tubular fixed bed
reactors by a Monte Carlo method imitating the technical filling process of
tubes with spherical particles is summarised. The following section (Sec. 5)
discusses the results and the technical relevance of the detailed simulations.
The last but one section (Sec. 6) provides information on the parallelisation
and vectorisation as well as the used computational resources and the efficiency of usage. Finally Sec. 7 draws some conclusions and points to areas of
further investigation.

2   Definition of the investigated problem: Tubular fixed bed reactors with small tube-to-particle diameter ratios

Tubular fixed bed reactors are of particular significance in the chemical process industries. They are, among others, used as catalytic reactors, adsorption
columns and separation towers. It is therefore not surprising, that tubular
fixed bed reactors and their mathematical description and modelling have
been a central topic of research for decades. Various models, usually based
on homogenisation approaches and semi-empirical correlations were derived

[1, 2, 13]. The drawback of many commonly used models is that they rely on symmetry assumptions in one or more directions. For tube-to-particle diameter ratios (aspect ratios) lower than approximately ten, circumferential symmetry (Θ-symmetry) is not fulfilled. Therefore, even sophisticated "non-parallel flow" models which take into account axial and radial variations of the porosity and consequently of the flow velocity and transport parameters are likely to predict wrong results. The complete 3-D structure of the randomly packed tubular fixed bed reactor has to be accounted for in order to predict the correct behaviour and to detect local inhomogeneities which may cause channelling and hot spots. Channelling occurs especially near the wall and for even aspect ratios also in the centre. Hot spots are of large technical and economical interest as they reduce the conversion rate and cause local catalyst deactivation leading to reduced selectivities and sometimes also to a permanent destruction of the catalyst. This demonstrates the importance of detailed information for the efficient and safe operation of these reactors. With new numerical methods using the increased computational power, the necessary details can be obtained nowadays.

3   Brief description of the lattice Boltzmann approach

The fundamental idea of the lattice Boltzmann methods [5] is to look at the physical background of fluid mechanics from a *statistical point of view*. Within typical Navier-Stokes CFD solvers the conservation equations for mass and momentum (Navier-Stokes equation), based on the homogenised quantities (pressure, density, flow velocity), are solved. In contrast, the underlying equation for the lattice Boltzmann technique is a space, momentum and time discrete formulation of the Boltzmann equation, describing the interaction of single fluid particles or an ensemble averaged particle density distribution function $N_i(t_*, r_*)$.

The current version of our lattice Boltzmann code is based on the 3-D 19-speed (D3Q19) lattice Boltzmann automata model with the so called single time Bhatnagar-Gross-Krook relaxation collision operator $\triangle_i^{Boltz}$ proposed by Qian [12]:

$$N_i(t_* + 1, r_* + c_i) = N_i(t_*, r_*) + \triangle_i^{Boltz}, \qquad i = 0, \dots, 18 \qquad (1)$$
$$\triangle_i^{Boltz} = \omega \left(N_i^{eq} - N_i \right), \qquad (2)$$

The first equation describes the advection. The density distribution function $N_i(t_*, r_*)$ with discrete momentum indirection c_i located at the discrete position r_* at the discrete time t_* is shifted within one iteration $(t_* + 1)$ in direction c_i to its next neighbouring lattice node located at $r_* + c_i$.

The second equation describes the "collision" of the particles, realized by a local redistribution of the density distribution function N_i while locally conserving mass and momentum. This is implemented as a single time relaxation approach with a local equilibrium distribution function N_i^{eq} and the relaxation parameter ω which determines the viscosity of the fluid (see eq. 7). The local equilibrium distribution function N_i^{eq} (eq. 3) is chosen in such a way, as to recover the time-dependent Navier-Stokes equation [12] in the low Mach number limit. It has to be calculated at each cell and at each time step from the local density ρ and the local macroscopic flow velocity $\boldsymbol{u}$. The factor t_p is a constant which only depends on the lattice direction.

$$N_i^{eq} = \rho t_p \left\{ 1 + 3\boldsymbol{c}_i\boldsymbol{u} + \frac{9}{2}(\boldsymbol{c}_i\boldsymbol{u})^2 - \frac{3}{2}u^2 \right\} \tag{3}$$

All macroscopic flow quantities can simply be calculated from the local particle distribution function $N_i(t_*, \boldsymbol{r}_*)$ by the following relations:

$$\text{density } \rho = \sum_{i=0}^{18} N_i \,, \tag{4}$$

$$\text{flow velocity } \boldsymbol{u} = \sum_{i=0}^{18} \boldsymbol{c}_i\, N_i / \varrho \,, \tag{5}$$

$$\text{pressure } p = c_s^2\, \varrho, \text{ with: } c_s = 1/\sqrt{3} \,, \tag{6}$$

$$\text{viscosity } \nu = \frac{1}{6} \left\{ \frac{2}{\omega} - 1 \right\} \,. \tag{7}$$

Wall boundary conditions can be integrated easily within the lattice Boltzmann framework by the so-called *bounce-back* rule which basically means that density distributions which hit a solid wall during the advection step are simply put back to the original cell but with opposite momentum (i.e. in the corresponding direction $-\boldsymbol{c}_i$). This results in a non-slip boundary condition at the wall and allows an easy and efficient handling[3, 9] of arbitrary complex geometries on the marker-and-cell mesh.

For the multi-species transport a similar approach can be applied for each species [6, 8]. Equations, analogous to 1 and 2, are solved for the additional density distributions functions $N_{i,s}$ of each species s. For the species, the relaxation parameter ω_s (which can be chosen differently for each species) determines the molecular diffusion coefficient. The equilibrium distribution function is chosen in such a way that the convection-diffusion equation is obeyed. The flow velocity, which is also required for the species equilibrium distribution function, is obtained from the carrier flow field. As only passive scalar transport is considered, the species do not influence the carrier flow.

4   Generation of the geometrical structure by a Monte Carlo method

The required detailed description of the geometric structure can be obtained in several ways. We apply a Monte Carlo method to synthetically generate realistic packings of spherical particles by imitating the technical filling

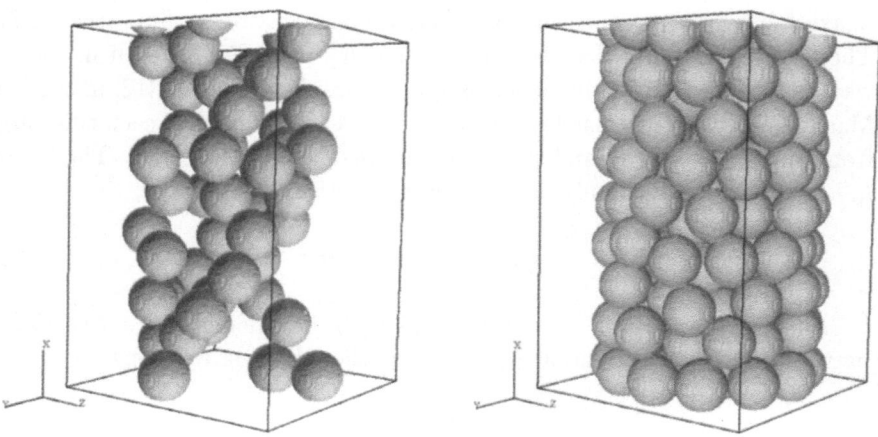

Fig. 1. Generation of randomly packed beds by a Monte Carlo simulation of the packing process. Loose packing generated by the raining step (left) and resulting dense packing after the compression step (right) [11].

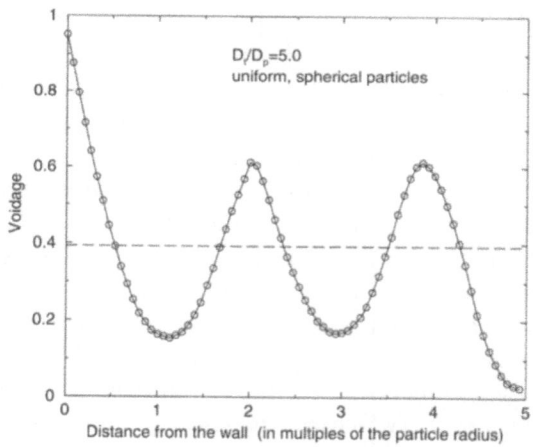

Fig. 2. Mean radial voidage in the randomly packed bed.

process consisting of two steps: first, a random placement of the particles (raining) and second, a compression step which leads to the close random packing [10]. Both steps are shown in Fig. 1 for mono-disperse spheres with a tube-to-particle diameter ratio of 5. The corresponding radial voidage profiles (Fig. 2) compare very well with integral experimental values [10].

5   Results of the detailed simulations

A common assumption of conventional models for tubular fixed bed reactors is Θ-symmetry. In the case of low aspect ratios this assumption is not valid. This can clearly be seen in Fig. 3 and 4 which shows the results of our detailed simulations of a tube-to-particle diameter ratio of 5 and a packing length of 7.5 particle diameters.

These figures show surface plots of the velocity magnitude in different cross-sections, coloured with the velocity and the concentration distribution of the reactant and the product respectively. In these surface plots, two different information of a two dimensional plane are depicted in the three dimensional space with one scalar value (velocity magnitude) as third dimension and the second scalar value (axial velocity or concentration) as colour on this generated surface.

The first plane at position -0.1 is located at a tenth part of the particle diameter in front of the packing. The next cross-sections are shown at the 1.8, 3.75 and 5.7 particle diameters apart from the beginning of the packed structure. The last two cross-sections show the distribution at the corresponding multiples of the particle diameter in the empty tube behind the packing.

The local flow velocity and consequentially also the reactant and product concentrations oscillate according to the geometric structure of the packed bed in the different cross sections. Large variations occur in both, the radial and the circumferential direction.

Due to the large changes in the flow patterns behind the packing, it seems to be questionable if experimental data measured e.g. with hot wire anemometry behind the packing can be extrapolated to the inside of the packing [1, 2, 4]. The flow at a tenth part of the particle diameter behind the packing (position $+0.1$) has already changed remarkably. This effect can especially be observed in the concentration distributions.

The increased voidage near the wall leads to generally higher velocities in this region. As a result, the reactant concentration is rarely consumed in the outer region. Therefore, the reactant concentrations in the wall region is still high at the end of the catalytic packing. This behaviour is commonly called "breakthrough". Adsorption processes are one example in which this effect can be especially crucial. Generally, in the case of chemical processes, local maxima of the product concentration or minima of the reactant indicate zones of high reactive activity. For exothermic reactions, this may result in hot spots and consequentially to the deactivation of the catalyst or reduced selectivity.

A quantitative analysis of the effects of these inhomogeneities is currently under way. But before systematical parameter studies can be performed, some enhancements of the algorithm have to be implemented to further reduce computing times.

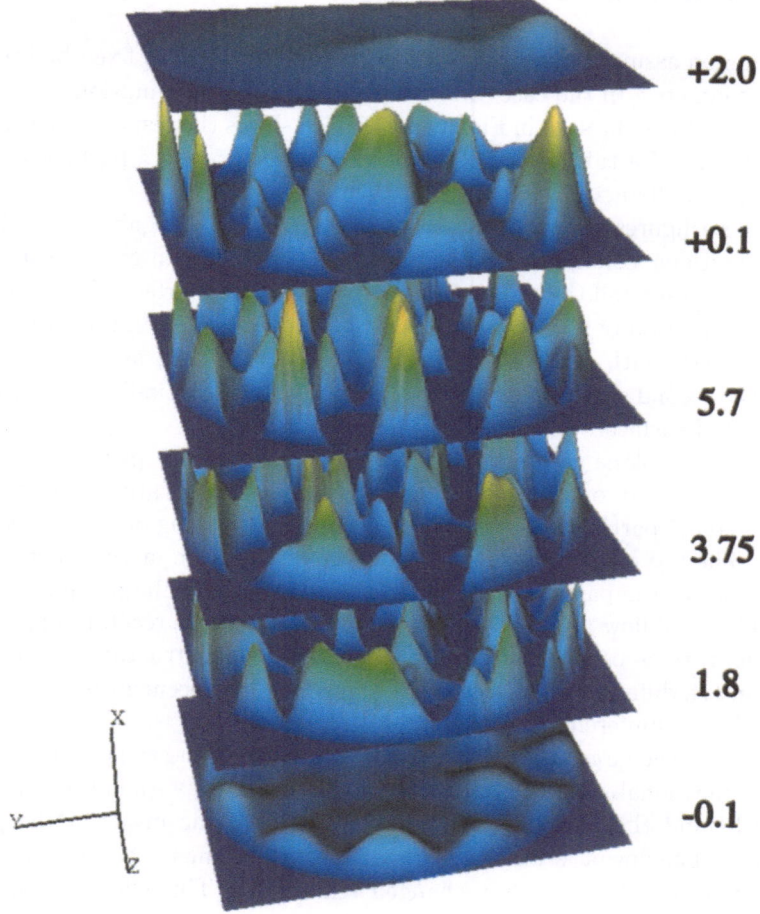

Fig. 3. Elevated surfaces of the velocity magnitude in different cross-sections of the tubular reactor. The axial positions in front ($-$), inside (unsigned) and behind ($+$) the packing are given in multiples of the particle diameter. The height of the surface represents the velocity magnitude, the colour scale (ranging from blue to red) shows the axial velocity [7].

6   Vectorisation, parallelisation, performance and resource usage

The used lattice Boltzmann code is vectorised throughout all major parts. In typical production runs, about 2.55 GFlops are achieved per CPU on a NEC SX5e (i.e. almost 65% of the peak performance) if only the flow field is calculated. In a combined calculation of the flow field, the mass transport

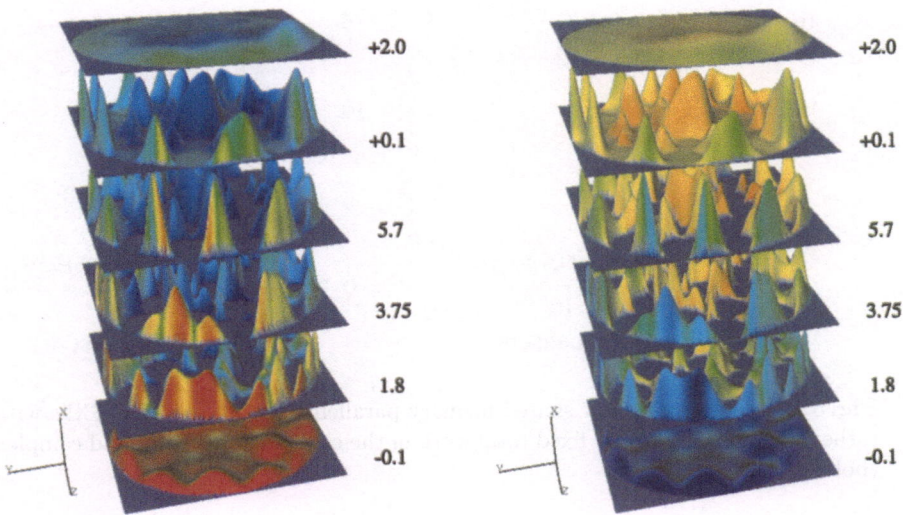

Fig. 4. Surface plots in different cross-sections of the tubular reactor as above. The colour represents the reactant concentration (left) and the product concentration (right) [7]. The height of the elevated surfaces corresponds again to the velocity magnitude.

and chemical surface reaction, still more than 2.0 GFlops (i.e. 50% of the peak performance) are typically sustained.

Due to the local character of the lattice Boltzmann algorithm, parallelisation is rather simple and, more over, likely to be efficient. For the parallelisation, simple domain decomposition methods together with message passing are straight forward. Within the current project, we did not use message passing as parallelisation paradigm but shared memory parallelisation. The project was started using NEC's proprietary micro-tasking instruction set, in the mean time we gradually shifted to the general OpenMP standard. It is often claimed, that the efficiency of shared memory parallelisation is rather low. Fig. 5 shows the results of extensive tests [15] with rather small test-cases (about 10% of typical production runs) on the NEC SX5e where the number of processors was gradually increased keeping the total work (i.e. the total grid size) constant. Up to 15 processors, almost linear speedups and efficiencies higher than 90% are achieved. Less systematical tests with larger test-cases prove this behaviour. The concurrent MFlops rate is about 15 GFlops on 6 processors (calculation of the flow field only) and about 16 GFlops on 8 processors (calculation of both the flow field and the mass transport with chemical surface reactions) for production runs.

Typical production runs were performed with micro-tasking parallelisation on 6–8 processors using up to 10 GB of memory. During the start-up,

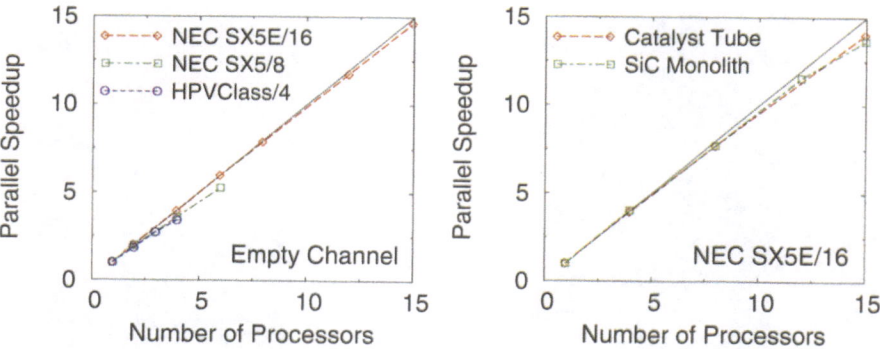

Fig. 5. Parallel speedup for shared memory parallelisation on the NEC SX5e with rather small test-cases and fixed total work in the case of simple (top) and complex (bottom) geometries [15].

several files with about 3 GB in total had to be transferred from the file server to the scratch area and read-in to feed the solver with the geometrical structure and some initial results. After about 50 000–100 000 time steps (approximately 8 hours elapsed time), the results were dumped back to disk (usable for restart) together with data files for the visualisation during the post-processing step (up to 5 GB in total). As a last step concerning the NQS jobs, all results were transfered back to the file server. From there, the data was copied by `ftp` to our local workstations for the post-processing. The overhead of `ssh` (even with simple cipher algorithms) compared to `ftp` proved to be much too high to efficiently apply this secure transfer method for large data files.

7   Conclusions

The lattice Boltzmann approach in combination with advanced techniques for the generation of the marker-and-cell representation of the geometry can successfully be applied to detailed simulations of the flow and coupled transport of reacting species in tubular fixed-bed reactors and other complex geometries. The simulations thus can supplement experimental investigations. At the moment, the required resolutions are too high to be applied in the daily design process of new chemical reactors. But the big advantage of the described detailed simulations is that only physical material parameters such as the molecular diffusion coefficient are necessary. Dispersion coefficients which are required for normal modelling approaches and which have to be determined from semi-empirical correlations or experiments, are not an input parameter but a result which can be extracted from the simulation results [14, 15] for further use with conventional modelling methods. The detailed simulations therefore might become an important tool during the design process in the future.

The effects of the observed local inhomogeneities on the integral reactor performance must still be examined in more detail. For this purpose, different operating conditions (e.g. Damköhler and Reynolds numbers) and geometrical properties (e.g. tube-to-particle diameter ratio) have to be considered in the following projects.

Vector-parallel supercomputers can be utilised efficiently with the lattice Boltzmann method. Between 50% and almost 65% of the peak performance can be obtained in production runs. The shared-memory parallelisation proved to be an appropriate and easy to handle parallelisation scheme on the NEC SX-5e series with efficiencies of 90-95% even when using up to 15 processors.

Acknowledgements

This work is supported by the German Research Foundation (DFG) under grant For262/2. The Monte Carlo simulations of the randomly packed beds were performed with a program developed by Yong-Wang Li in the framework of an Alexander von Humboldt (AvH) foundation fellowship. Additional support by Rudolf Fischer, NEC ESS, with some optimisation problems of the lattice Boltzmann solver is gratefully acknowledged.

References

1. R. Adler. Stand der Simulation von heterogen-gaskatalytischen Reaktionsabläufen in Festbettreaktoren – Teil 1. *Chem.-Ing.-Technik*, 72(6):555–564, 2000.
2. R. Adler. Stand der Simulation von heterogen-gaskatalytischen Reaktionsabläufen in Festbettreaktoren – Teil 2. *Chem.-Ing.-Technik*, 72(7):688–699, 2000.
3. J. Bernsdorf, F. Durst, and M. Schäfer. Comparison of cellular automata and finite volume techniques for simulation of incompressible flows in complex geometries. *Int. J. Numer. Met. Fluids*, 29:251–264, 1999.
4. O. Bey and G. Eigenberger. Fluid flow through catalyst filled tubes. *Chem. Engng. Sci.*, 57(8):1365–1376, 1997.
5. S. Chen and G. D. Doolen. Lattice Boltzmann method for fluid flows. *Annu. Rev. Fluid Mech.*, 30:329–364, 1998.
6. E. G. Flekkøy. Lattice Bhatnagar-Gross-Krook models for miscible fluids. *Phys. Rev. E*, 47(6):4247–4257, 1993.
7. H. Freund, E. Klemm, G. Emig, T. Zeiser, G. Brenner, and F. Durst. Detailed 3D-simulations of single phase reacting flow in randomly packed beds with low aspect ratios. In *Proceedings of 3rd European Congress of Chemical Engineering, Nuremberg, 26-28 June 2001*, 2001. published on conference CD.
8. X. He, N. Li, and B. Goldstein. Lattice Boltzmann simulation of diffusion-convection systems with surface chemical reaction. Molecular Simulation, Internet Conference, Apr. 1999.

9. T. Inamuro, M. Yoshino, and F. Ogino. A non-slip boundary condition for lattice Boltzmann simulations. *Phys. Fluids*, 7(12):2928–2930, 1995.

10. Y.-W. Li, T. Zeiser, P. Lammers, G. Brenner, E. Klemm, G. Emig, and F. Durst. Direct simulation of the structure and consequential flow field in a packed bed. *AIChE Journal*, submitted, 2000.

11. Y.-W. Li, T. Zeiser, P. Lammers, G. Brenner, E. Klemm, G. Emig, and F. Durst. Direct numerical simulations of the sphere structure in packed beds and the flow with chemical reactions. submitted to NATURE.

12. Y. H. Qian, D. d'Humières, and P. Lallemand. Lattice BGK models for Navier-Stokes equation. *Europhys. Lett.*, 17(6):479–484, Jan. 1992.

13. E. Tsotsas. Entwicklungsstand und Perspektiven der Modellierung von Transportvorgängen in durchströmten Festbetten. *Chem.-Ing.-Technik*, 72(4):313–321, 2000.

14. T. Zeiser. Untersuchung von Diffusionsvorgängen in porösen Medien mit dem Lattice-Boltzmann Verfahren. Diplomarbeit, Lehrstuhl für Strömungsmechanik, Universität Erlangen-Nürnberg, 2000.

15. T. Zeiser, G. Brenner, P. Lammers, and J. Berndsorf. Performance aspects of lattice Boltzmann methods for application in chemical engineering. In C. Jenssen, T. Kvamdal, H. Andersson, B. Pettersen, A. Ecer, J. Periaux, N. Satofuka, and P. Fox, editors, *Parallel Computational Fluid Dynamics 2000, Proceedings of the Parallel CFD 2000 Conference, May 22-25, Trondheim, Norway*, pages 407–414. Elsevier, 2001.

Structural Mechanics

Prof. Dr. Peter Wriggers

Institut für Mechanik und Numerische Mechanik, Universität Hannover
Appelstr. 9A, D-30167 Hannover, Germany

Computational methods have a long history in Structural Mechanics. They are used for the design of beams, shells and solids which are load carrying parts in general structures like high rise towers or bridges. Lately the computational tools are also applied to other engineering areas which include geotechnical problems. In this problem class one is immediately confronted with the fact that only three-dimensional analysis can be used for a good prediction of the real behaviour of the structure and soil. Furthermore often different length scales need high-performance computing as an inevitable tool for such large and complex applications. This is especially the case for computations in which multi-field analysis is needed to model the physics correctly.

All contributions which are contained in this section are concerned with geotechnical applications or with materials which are present in soils. The first contribution discusses nonlinear boundary value problems which occur in geotechnics and their numerical solutions. Due to the nature of the problem different length-scales have to be considered in such analysis which very fine finite element meshes for their resolution. Three different topics are investigated in the contribution. These are the bearing capacities of pile foundations, deformation of soils due to waste landfill and a soil freezing processes. All problems include material nonlinear behaviour. The results depict that, even when using high computing power with standard software packages, only relatively small sized problems can be tackled which shows the need for further developments in this area.

The second contribution in this section tackles the problem of propagating waves in heterogeneous media. The first part is interesting from the point of view that often times for travelling waves can be measured in solids and then it is possible to estimate the amount of cracks inside the medium. A finite difference technique is applied for the large scale numerical simulations which are then compared to known theoretical results for different crack distributions. In the second part seismic waves due to scattering are considered. Again a model for theoretical predictions is compared with the numerical analysis on a parallel computer.

All contributions underline the need of high performance computing tools when real three-dimensional solid mechanics problems have to be solved numerically. This is especially true for heterogeneous media where different length scales are present and have to be resolved using very fine grids.

Numerical Modelling of Geotechnical Boundary Value Problems

Hans Hügel

Universität Kaiserslautern, Fachgebiet Bodenmechanik und Grundbau

Abstract. The nature of geotechnical boundary value problems is characterized by large subsurface body, nonlinear mechanical behaviour of soils/rocks, soil/structure interaction and the fact that most boundary value problems are coupled, e.g. water or heat transport in saturated or unsaturated soils. In this report numerical solutions of different categories of geotechnical boundary value problems are discussed regarding their physical nature and the numerical expense to solve the problems.

1 Introduction

Nature of Geotechnical Problems

Geotechnical boundary value problems enclose the following fields: static/dynamic stress-displacement analysis (structural engineering), water and moisture transport in soils as well as pollutant migration and heat transfer in soils. Almost all geotechnical problems are so-called coupled problems, i.e. different fields have to be solved simultaneously, e.g. consolidation of cohesive soils requires simultaneous solution of static stress-displacement analysis and groundwater flow in a saturated medium. Furthermore soil/structure interaction is present in most geotechnical problems, i.e. contact conditions between soil and structures on ground surface or in subsoil.

Numerical Modelling of Geotechnical Problems

Non-linear geotechnical boundary value problems are solved using the Finite Element Method. Regarding the numerical expense of solving such problems, the following topics have to be taken into account:

- The dimensions of the discretized section of subsoil are generally very large compared to characteristic length of structures in soil or on ground surface, e.g. foundations, tunnels, walls etc. Example: in open pit mining FE-Models with a depth of 1000 m are common, but structures with a characteristic length of 0.1 m can occur as well.
- The mechanical behaviour of soils as a representative of granular materials is non-linear. The stiffness and strength of soils depend on several field variables, e.g. stress level, direction of deformation, void ratio, degree of saturation etc. If the material models include such effects, the initial fields of these variables have to be known, simulated or approximated based on sound engineering judgement.

- Geometrical non-linearity is often given, e.g. for problems accompanied by occurrence of shear zones and problems which are close to critical states. A mechanical modelling of shear zones occurrences in soils is still under development and is not treated in this report.
- Non-linear boundary conditions are often present as a result of contact between soil and structures like foundations, piles, tunnels, sheet piles etc. If several contacts have to be taken into account, the numerical expense can increase very fast, especially for problems with finite deformations.

The numerical cost of solving geotechnical boundary value problems is usually characterized by simultaneous occurrence of all three non-linearities and the disadvantageous ratio of the characteristic length of the soil body to the characteristic length of structures in soil and on the ground surface.

Simulations

The numerical simulations presented in the following sections were carried out at the Scientific Supercomputing Center (SSC) at University of Karlsruhe applying the implicit FE-code *Abaqus 5.8.*[1] All FE-jobs were carried out under IBM SP-256. Simulations on machine SP-SMP just began and are not presented in this report.

2   Simulation of Subsequently Piled Foundations

2.1   Situation

Existing foundations are piled subsequently in order to increase their bearing capacity. Therefore, small diameter concrete piles are installed under the already loaded foundation block. Large scale model tests are carried out to study the problem in sandy soils [6], see Fig. 1. Here a foundation block with dimensions $0.8 \times 0.8 \times 0.2$ m is subjected to a centric vertical load F in a box with dimensions $4.5 \times 4.5 \times 5$ m which is filled with medium sand. Four piles with a length of 1.2 m and a diameter of 0.06 m are installed symmetrically.

2.2   Simulation

The nature of the boundary value problem is characterized by path dependent material behaviour of sand, by simulation of the pile installation process as well as contact conditions along the surfaces sand/foundation and sand/pile. Although, from geometrical point of view, the problem has two symmetric axes, the contact conditions between the load piston and the foundation block allows the block to rotate. The FE-simulations include single-symmetric as well as double-symmetric models. The FE-mesh shown in Fig. 2 is single-symmetric in order to allow imperfections due to a small eccentricity of load piston as well as non-homogeneous distribution of initial void ratio e_0 in the soil body.

---

[1] *Abaqus* is a trademark of HKS Inc., Rhode Island, USA

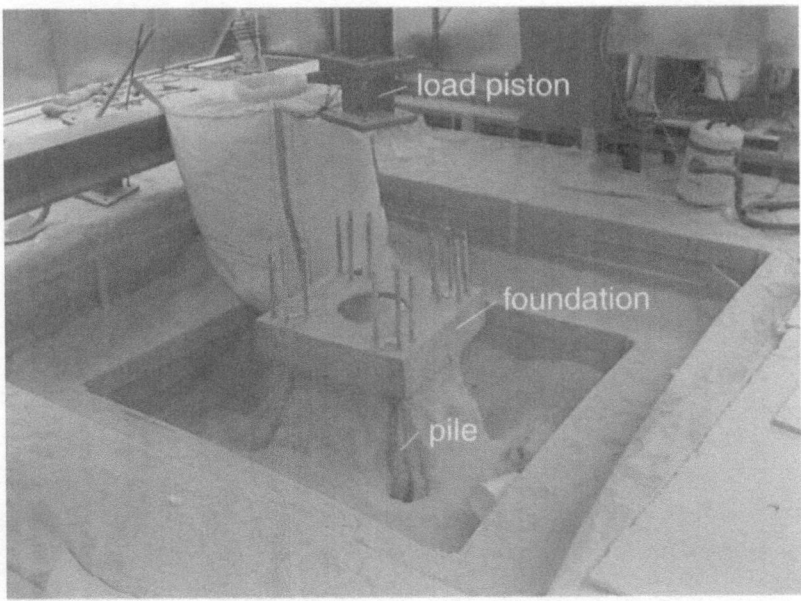

Fig. 1. Model test on subsequently piled foundation – photo taken after test

The mechanical behaviour of the sand is described with a hypoplastic constitutive model of the rate type, see [4,5]:

$$\mathring{T} = f_b f_e \frac{1}{\mathrm{tr}(\hat{T} \cdot \hat{T})} \left[F^2 D + a^2 \hat{T} \, \mathrm{tr}(\hat{T} \cdot D) + f_d a F(\hat{T} + \hat{T}^*) \|D\| \right] \qquad (1)$$

with *Cauchy* stress tensor T, deviatoric stress tensor T^*, normalized *Cauchy* stress tensor $\hat{T} = T/\mathrm{tr}\, T$, *Jaumann* stress rate tensor $\mathring{T}$ and rate of deformation tensor D. Equation 1 includes stiffness and strength of sand depending among other quantities on mean stress level p (so-called barotropy) and on void ratio e (so-called pyknotropy). f_b, f_e and f_d are scalar functions introducing barotropy and pyknotropy, a and F are scalar functions depending on stresses only [4]. The constitutive model was implemented in *Abaqus* using the user subroutine *umat*.

The mechanical behaviour of the concrete block was modelled elastic, the stiffness of installed piles (concrete) was modelled, depending on the phase change from liquid to solid state, i.e. Young's modulus E and Poisson's ratio ν depend on time [3].

Numerical simulations were carried out with the following variations:

- single- and double symmetric 3D-problem with/without imperfections in terms of exccentricity and initial void ratio,
- simulation of unpiled and piled foundation,
- variation of initial void ratio $e_0 = $ const.

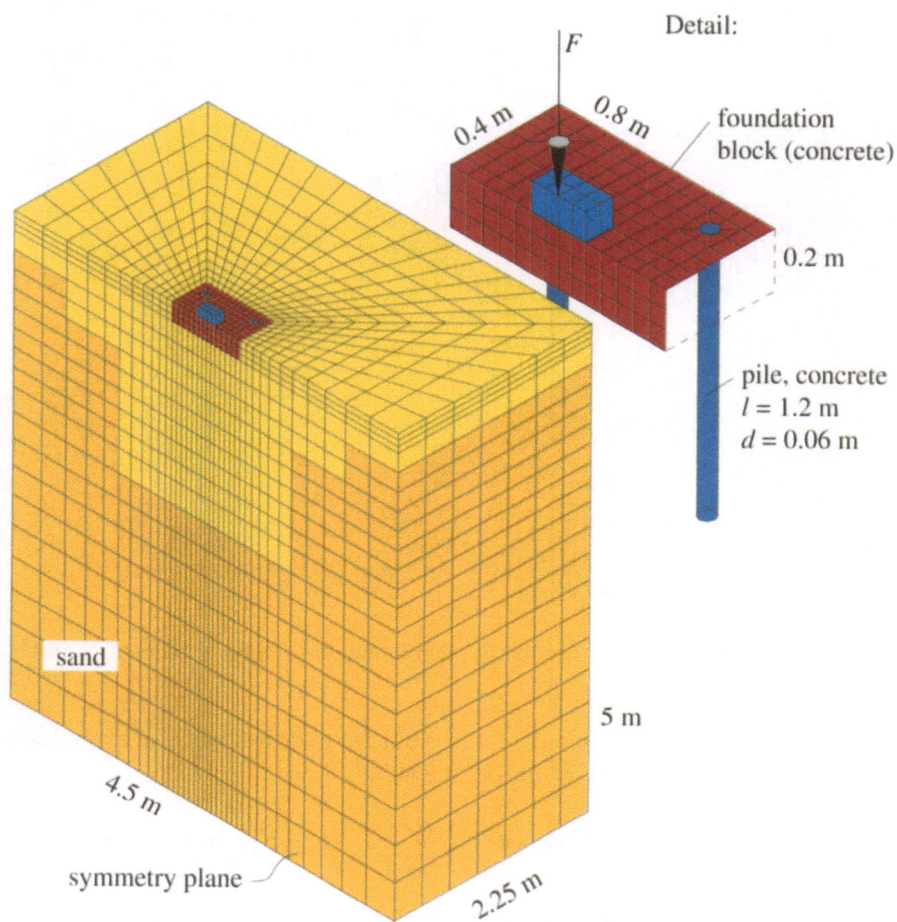

Fig. 2. FE-discretization of subsequently piled foundation – here single-symmetric 3D-problem

2.3   Results

The main interest of the calculations was focused on the effect of piles on the bearing capacity as well as the influence of the piling process. Figure 3 shows calculated load-displacement curves for original and piled foundation block for varying initial void ratios e_0. Although the global behaviour is quite linear, large strain gradients occur at all corners and at the pile tip especially for increasing vertical displacement of the foundation block.

Figure 4 shows a contour plot of stresses σ_{33} in the soil. The stress concentration under the pile tip shows that the pile force at the top is composed

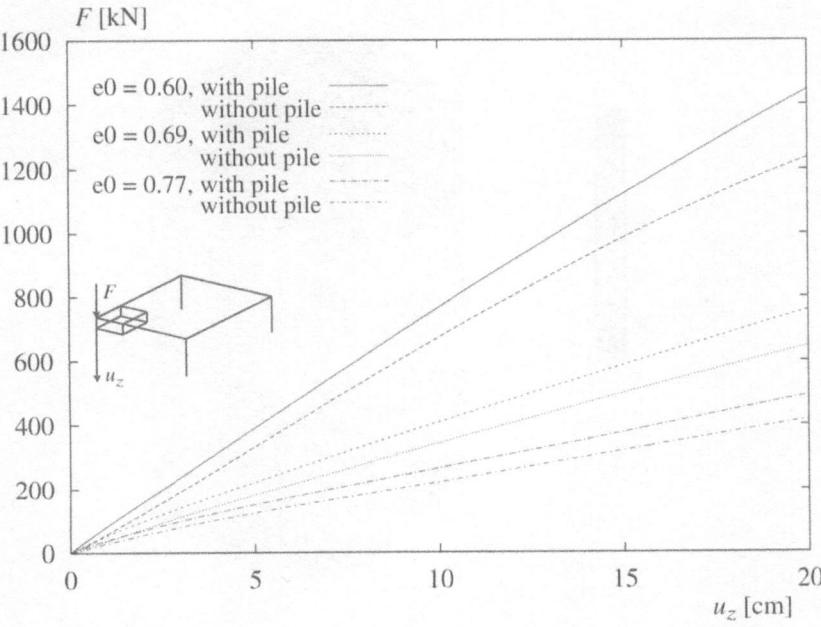

Fig. 3. Calculated load-displacement behaviour for original and piled foundation block for varying initial void ratio e_0

of friction and tip forces. The effective pile length cannot be predicted for this relatively poor discretized pile body.

2.4 Computational Aspects

Hypoplasticity is not working with yield functions as in elasto-plasticity. As a result corrections of stresses to a yield surface during integration of constitutive equation is not possible. Therefore the load increments have to be kept very small in comparison to elasto-plasticity and consequently the job run-time can increase dramatically. Furthermore, hypoplasticity is describing the realistic vanishing stiffness for $p \to 0$. This leads to a very small stiffness near ground surface compared to high stiffness of the concrete foundation block and, therefore, to ill-conditioned global jacobian matrices. Furthermore five discretized contact surfaces in the FE-model make the problem very difficult to solve.

Data of the presented FE-job's are:

– Total No. of variables (d.o.f.): 30075 for single-symmetric 3D-problem. Increasing the number of variables is necessary to model the pile more realistic.

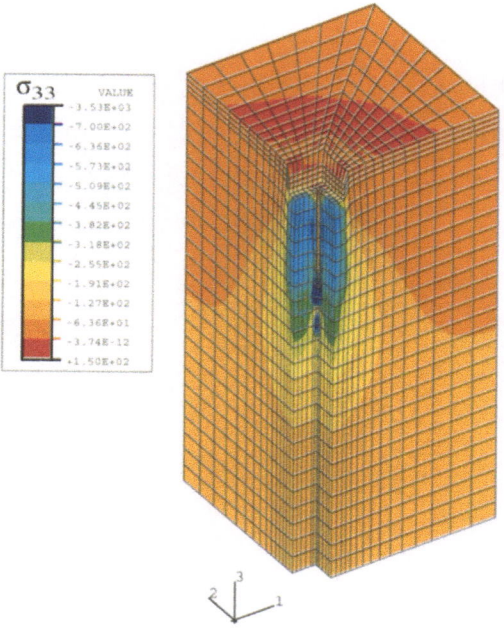

Fig. 4. Calculated distribution of stresses σ_{33} for double-symmetric 3D-problem (centric load and uniformly distributed initial void ratio)

- Data size: for single-symmetric 3D-model with quadratic continuum elements about 1.9 GByte. Models with more variables couldn't be implemented on SP-256.
- Job-runtime: 86 hours for single-symmetric 3D-model with linear continuum elements.
- Restart file size: 116 MByte per increment.
- Solver: sparse solver, unsymmetric jacobian matrix due to material model and nonlinear boundary conditions.

3   Simulation of Deformations in Waste Landfills

3.1   Situation

Waste landfills are investigated with respect to the deformation of the waste body due to biological degradation. The waste body in the landfill is protected by a multi-barrier liner system consisting of mineral layers and geosynthetics, see Fig. 5. The liner system consists of drainage layer, geotextile, geomembrane and a compacted clay layer from top to bottom.

The waste is undergoing a biological degradation process which leads to time-dependent subsidences of the waste body (so-called bio-consolidation).

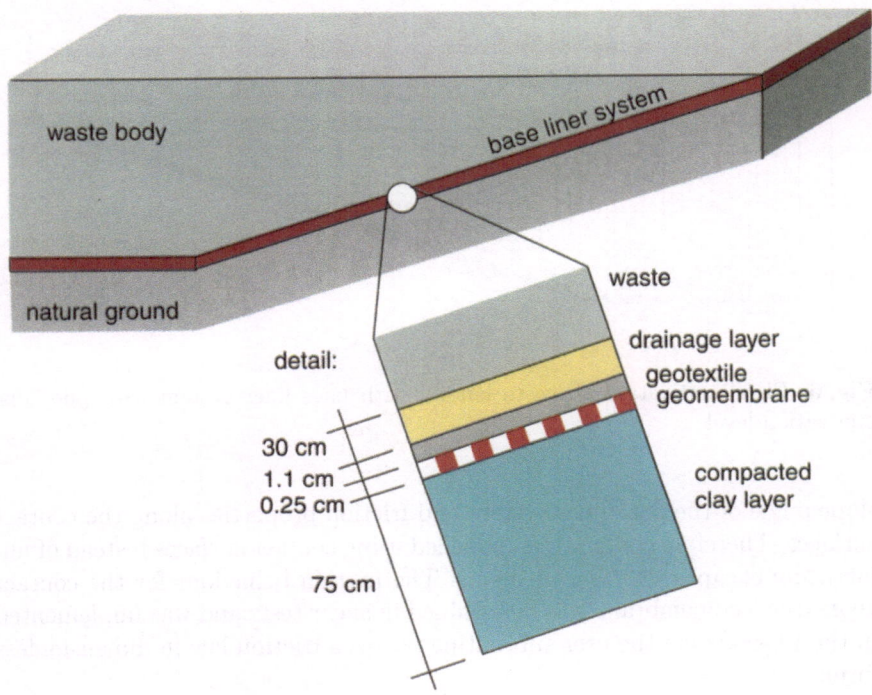

Fig. 5. Section of the investigated waste landfill with base liner system in detail

The corresponding deformations leads to stress changes in the base liner system. Especially the corresponding loading of the geomembrane as an important sealing element of the base liner system is investigated in this project [1,7].

3.2 Simulation

The problem is modelled with a plane strain stress-displacement analysis with time-dependent material behaviour, see Fig. 6 for the FE-discretization [1,7].

The following models are used to describe the mechanical behaviour of the materials:

- Waste: linear elastic, perfectly plastic model with time-dependent volumetric shrinking strain rate
- Drainage layer: linear elastic, perfectly plastic model
- Geotextile: orthotrop elastic
- Geomembrane: nonlinear elastic
- Clay layer: critical state model (modified Cam Clay model)

The contact geotextile/geomembrane and geomembrane/clay was modelled with contact surfaces. Numerical simulations [7,1] show maximal relative tangential displacements of up to 2 m depending on the degradation process,

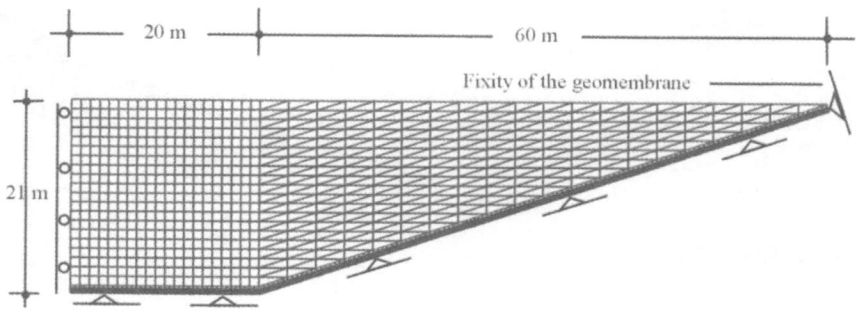

Fig. 6. FE-discretization of waste landfill with base liner system with poor discretization level

slope angle of the base liner system and friction properties along the contact surfaces. Therefore contact was modelled using contact surfaces instead of numerically cheaper interface elements. The friction behaviour for the contact geotextile/geomembrane was determined in shear tests and was implemented in the FE-code via the user subroutine *fric* as a friction law in dimensionless form:

$$\frac{\tau}{\sigma \tan \varphi_{\text{peak}} + a_{\text{peak}}} = \gamma \left[2.8 \exp\left(-1.25\gamma\right) + 0.2\right] \tag{2}$$

with contact normal stress σ, contact shear stress τ, peak value of friction angle φ_{peak}, peak value of adhesion a_{peak}, standardized displacement $\gamma = u/u_{\text{peak}}$ and relative tangential displacement u. $u_{\text{peak}} = 1$ cm for geotextile/geomembrane and $u_{\text{peak}} = 0.1$ cm for geomembrane/clay. The contact geomembrane/clay was described with a Coulomb friction model:

$$\frac{\tau}{\sigma + a} = \tan \delta \tag{3}$$

with the surface friction angle δ and adhesion a.

3.3   Results

In Fig. 7 elastic settlements occurring immediately after installation of waste in the landfill and subsidences due to biological degradation are distinguished.

The resulting deformation of the waste body causes a loading of the geomembrane. In Fig. 8 normalized tension stresses of the geomembrane are shown for varying friction parameters in geotextile/geomembrane (top) and geomembrane/clay contact surfaces (bottom) for varying friction parameters along the contact surfaces. This results are used to design the geomembrane considering the long-term behaviour of waste, e.g. municipal solid waste.

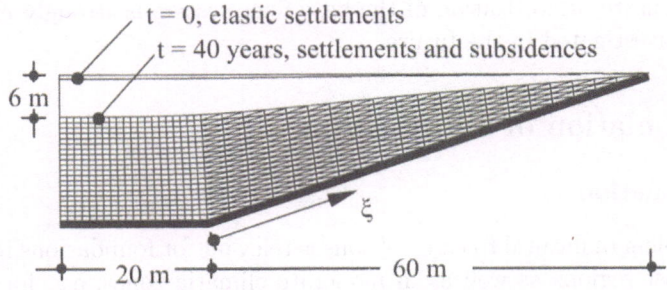

Fig. 7. FE-result: deformations of waste body due to biological degradation

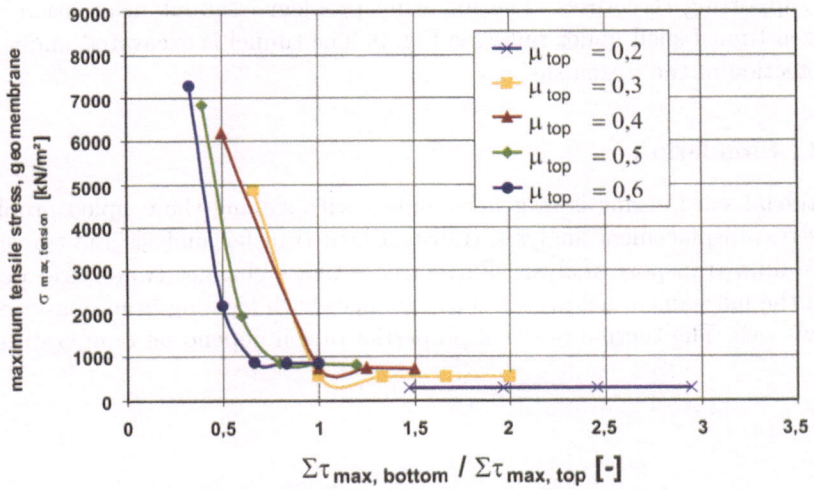

Fig. 8. Calculated normalized tension stresses in geomembrane for varying friction parameters along geotextile/geomembrane and geomembrane/clay contact surface

3.4   Computational Aspects

The numerical solution of the problem is dominated by the contact surfaces which show instability due to sudden switch from sticking to sliding.

Data for presented FE-jobs are:

- Total No. of variables (d.o.f.): 5530 for shown poor discretized plane strain FE-model.
- Data size: 186 MByte
- Job run time: 2 hours for presented FE-model for small slope angle.
- Restart file size: 15 MByte per increment.
- Solver: sparse solver, unsymmetric jacobian matrix due to nonlinear boundary conditions.

More accurate discretization of the base liner system is strongly needed and will be investigated in the future.

4   Simulation of Soil Freezing

4.1   Situation

The problem of natural freezing of soils is relevant for foundations in so-called permafrost regions as well as in moderate climatic zones, e.g. for frost suspectible mineral materials in road design and landfill design. Artificial soil freezing is applied to produce an artificial material with a stiffness comparable to concrete. Such frozen soil bodies are used in shaft and tunnel design as supporting structures. The following problem is about installation of a frozen tunnel shell under rails, see Fig. 9. The tunnel is excavated under the protection of the frozen shells.

4.2   Simulation

Artificial soil freezing is in general linked with solving the coupled problem of stress-displacement analysis, transient heat transfer analysis and transient pore fluid transport analysis. Furthermore phase changes (water/ice) occur and the mechanical behaviour of soil changes with freezing from non-viscous to viscous. The thermo-physical properties of soil depend on temperature θ,

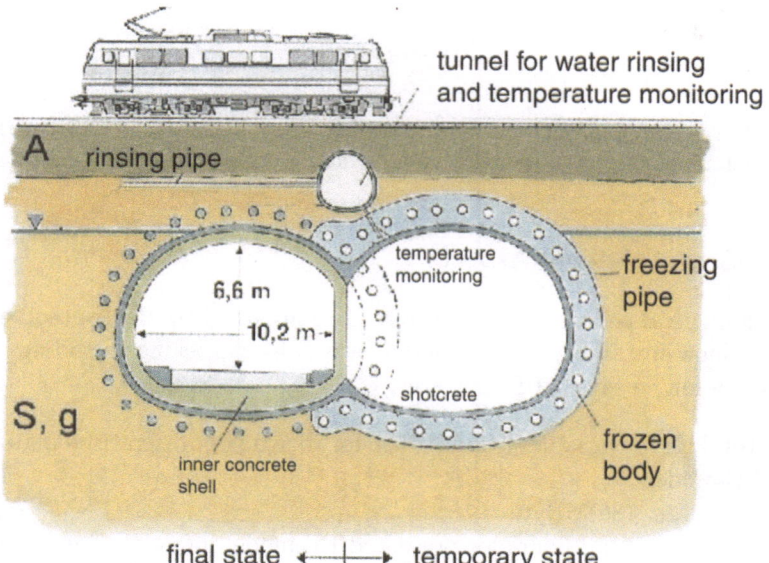

Fig. 9. Ground freezing as protection for tunnel excavation (here temporary state)

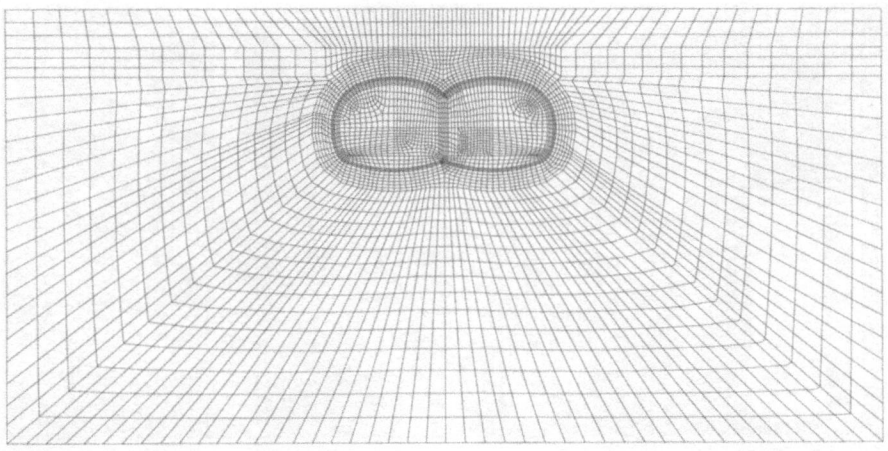

Fig. 10. FE-discretization of the plane strain problem

void ratio e and degree of saturation S_r. The mechanical parameters depend among other quantities on the mean stress level p, direction of deformation and temperature θ.

The boundary value problem described in the previous section is modelled as a coupled plane strain static stress-displacement and transient heat transfer analysis, see Fig. 10 for the FE-discretization of the problem.

The simulation includes the sequence of the following processes:

– initial stress state of ground,
– freezing of left shell,
– excavation of left tunnel and installation of shotcrete shell,
– installation of left inner concrete shell,
– freezing of right shell,
– excavation of right tunnel and installation of shotcrete shell,
– installation of right inner concrete shell,
– thawing.

4.3 Results

Figures 11 and 12 show the distribution of temperature in the subsoil after a period of $t = 32.4$ days of artificial freezing. The soil was frozen using calcium chloride circulating in pipes installed in the soil, see also Fig. 12 where the pipes can be seen as points with a temperature of $\theta = -35°$ C. The temperature distribution shows that after this period, a closed frost shell had developed so that the soil in the tunnel can be excavated without disturbing deformations of the surrounding soil and especially of the rails on the ground surface.

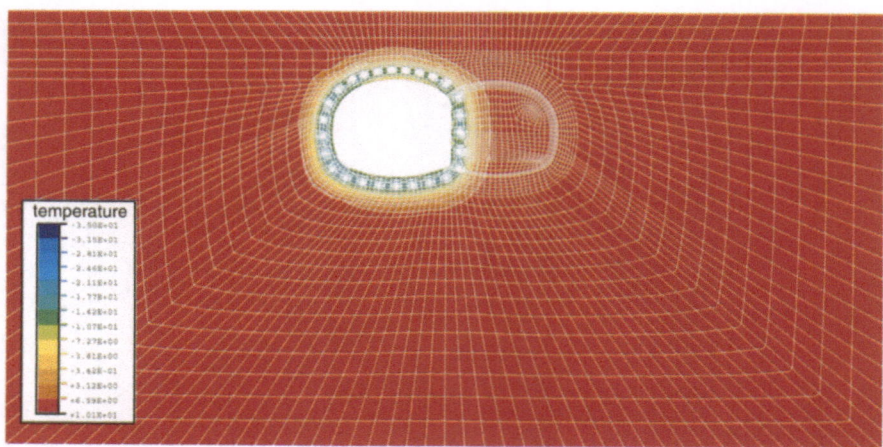

Fig. 11. Calculated distribution of temperature in subsoil after a period of 32.4 days of soil freezing

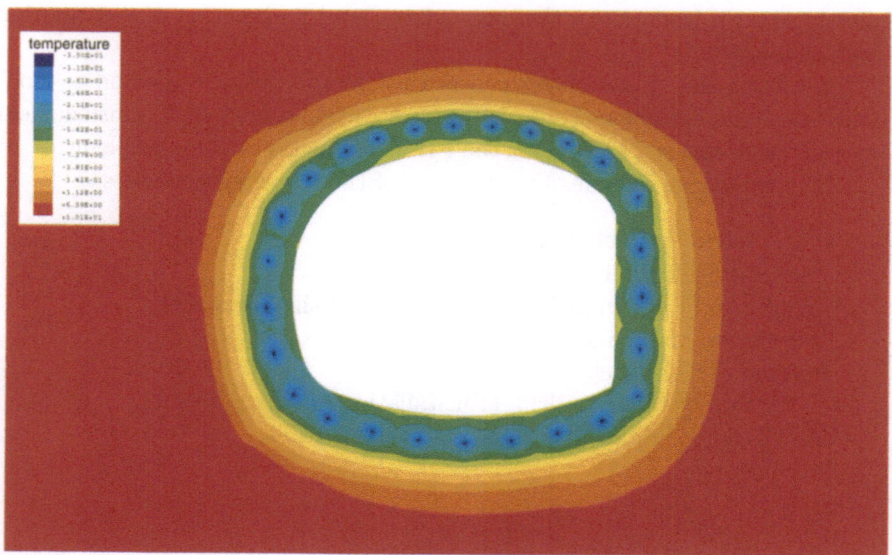

Fig. 12. Calculated distribution of temperature in subsoil after a period of 32.4 days of soil freezing (detail)

4.4   Computational Aspects

Although the amount of d.o.f. of this two-dimensional problem is small compared to 3D-problems presented, the CPU-time was relatively high due to sufficiently small time increment to simulate the phase changes accurately.

Data for presented FE-job's are:

- Total No. of variables (d.o.f.): 300045
- Data size: 115 MByte
- Job run time: 75 hours for one freezing process. Total time for two freezing periods, one thawing period and all structural processes: 233 hours
- Solver: sparse solver, unsymmetric jacobian matrix

In order to consider the three-dimensional bearing behaviour of the frozen shell during tunnel excavation a 3D-model is desirable but could not be realized with the single-CPU FE-code.

5   Summary and Outlook

The projects presented above showed clearly the difficulty in dealing with large scale geotechnical boundary value problems. There is a clear need to use a parallelized version of the FE-code *Abaqus* to overcome the shortcomings presented in the simulations in terms of discretization and job run time.

Acknowledgement

The author and his colleagues from University of Kaiserslautern want to thank the Scientific Supercomputing Center (SSC) of the University of Karlsruhe for supporting their work in the field of numerical methods in geotechnics. We would like to emphasize especially the cooperativeness of Dr. Paul Weber and Clemens Howar from SSC.

References

1. Abel K., Meißner H.: Numerical parameter studies for inclined landfill liners. Green 3 – 3rd International Symposium on Geotechnics Related to the European Environment, Berlin, Germany, 2000
2. Biehl R.: Tunnelsicherung durch Bodenvereisung. Diplomarbeit am Fachgebiet Bodenmechanik und Grundbau der Universität Kaiserslautern, 2001
3. Giese S.: Dreidimensionale Simulation des nichtlinearen Randwertproblems *nachträgliche Verdübelung von Fundamenten*. Studienarbeit am Fachgebiet Bodenmechanik und Grundbau der Universität Kaiserslautern, 2000
4. Hügel H.: Prognose von Bodenverformungen. Veröffentlichungen des Instituts für bodenmechanik und Felsmechanik der Universität Karlsruhe, Heft **136**, 1996
5. Hügel H.: Numerik in der Geotechnik 2. Skriptum zur gleichnamigen Vorlesung am Fachgebiet Bodenmechanik und Grundbau der Universität Kaiserslautern, 2001

6. Meißner H., Shen Y.L.: Foundations, improved by piles. 4th German-Polish Symposium Recent Developments in Civil Engineering and Environmental Engineering, Eds. Meißner H., Hügel H.M., Kaiserslautern, Germany, *pp. 11–17*, 1999
7. Meißner H., Abel K.: Numerical investigations of inclined landfill liner systems. GeoEng 2000 – International Conference on Geotechnical and Geological Engineering. Melbourne, Australia, 2000 (published on CD-ROM)
8. Meißner H., Hügel H.: Frost insulation of cover and base liner of waste landfills. Proceedings of 3rd International Congress on Environmental Geotechnics, Lisboa, Portugal, *pp. 133–138*, 1998

Wave Propagation in Heterogeneous Media.
Part 1: Effective Velocities in Fractured Media

Erik H. Saenger[2], Heiko Priller[1], Christian Grosse[3], Peter Hubral[1], and
Serge A. Shapiro[2]

[1] Geophysical Institute, Karlsruhe University, 76187 Karlsruhe, Germany
[2] Fachrichtung Geophysik, Freie Universität Berlin, 12249 Berlin, Germany
[3] Institut für Werkstoffkunde im Bauwesen, University of Stuttgart, 70550
   Stuttgart, Germany

Abstract. This paper is concerned with a numerical study of effective velocities
in two types of fractured media. We apply the so-called rotated staggered finite
difference grid technique. Using this modified grid it is possible to simulate the
propagation of elastic waves in a 2D or 3D medium containing cracks, pores or free
surfaces without hard-coded boundary conditions. Therefore it allows an efficient
and precise numerical study of effective velocities in fractured structures. We model
the propagation of plane waves through a set of different randomly cracked media.
In these numerical experiments we vary the crack density. The synthetic results are
compared with several theories that predict the effective P- and S-wave velocities
in fractured materials. For randomly distributed and randomly oriented rectilinear
intersecting thin dry cracks the numerical simulations of velocities of P-, SV- and
SH-waves are in excellent agreement with the results of a new critical crack density
(CCD) formulation. On the other hand for randomly distributed rectilinear parallel
thin dry cracks three different classical theories are compared with our numerical
results.

1   Introduction

The problem of effective elastic properties of fractured solids is of considerable interest for geophysics, for material science, and for solid mechanics.
In this paper we consider the problem of a fractured medium in two dimensions. With this work some broad generalizations can be elucidated that will
help solving problems with more complicated geometries. Strong scattering
caused by many dry cracks can be treated only by numerical techniques because an analytical solution of the wave equation is not available. So-called
boundary integral methods are well suited to handle such discrete scatterers in a homogeneous embedding. They allow the study of SV-waves [3,9],
SH-waves [2] and P-waves [6] in multiple fractured media, but they are restricted to non-intersecting cracks. Finite difference (FD) methods discretize
the wave equation on a grid. They replace spatial derivatives by FD operators using neighboring points. The wave field is also discretized in time,
and the wave field for the next time step is generally calculated by using
a Taylor expansion. Since the FD approach is based on the wave equation

without physical approximations, the method accounts not only for direct waves, primary reflected waves, and multiply reflected waves, but also for surface waves, head waves, converted reflected waves, and waves observed in ray-theoretical shadow zones [5]. Additionally, it automatically accounts for the proper relative amplitudes. Consequently, FD solutions of the wave equation are widely used to study scattering of waves by heterogeneities (e.g. [7,1]). In this paper we present a numerical study of effective velocities of two types of fractured 2D-media using the rotated staggered grid technique (see Fig. 1). We model the propagation of plane waves through well defined fractured regions. The numerical setup is described in section 2. Our numerical results for intersecting and parallel cracks are discussed in section 3 and 4.

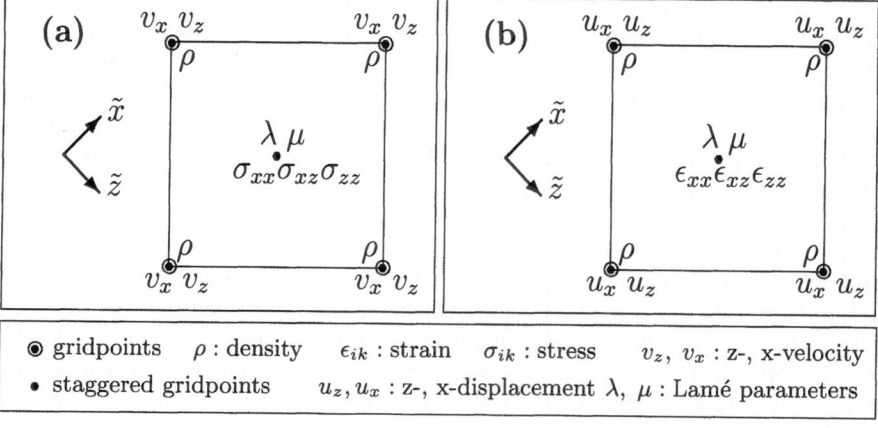

Fig. 1. Elementary cells of the rotated staggered grid technique. Locations where strains, stresses, displacement (velocity) and elastic parameters are defined. (a) velocity-stress FD technique. (a) time domain FD technique.

2   Experimental Setup

As mentioned above, the rotated staggered FD scheme [11] is a powerful tool for testing theories about fractured media. In order to test these formalisms we design some numerical elastic models which include a region with a well known crack density. The cracked region was filled randomly with rectilinear dry cracks. A typical model contains 1000×1910 grid points with an interval of 0.0001m. In the homogeneous region we set $v_p = 5100\ m/s$, $v_s = 2944\ m/s$ and $\rho_g = 2700\ kg/m^3$. Table 1 summarizes the relevant parameters of all the models we use for our experiments. For the dry cracks we set $v_p = 0\ m/s$, $v_s = 0\ m/s$ and $\rho_g = 0.0001\ kg/m^3$ which approximates vacuum. To obtain effective velocities in different models we apply a body force plane source at the top of the model. The plane wave generated in this way propagates through the fractured medium. With two horizontal lines of 1000 geophones at the top and at the bottom, it is possible to measure the time-delay of the mean peak amplitude of the plane wave caused by the inhomogeneous region. With the time-delay one can calculate the effective velocity. The direction of the body force and the source wavelet (i.e. source time function) can vary to generate two types of shear (SH- and SV-) waves and one compressional (P-) wave. The source wavelet in our experiments is always the first derivative of a Gaussian with a dominant frequency of 50kHz and with a time increment of $\Delta t = 5 * 10^{-9}s$. From the modeling point of view it is important to note that all computations are performed with second order spatial FD operators and with a second order time update. To obtain accurate modeling results with up to 40000 time steps we have to use the large scale parallel computer Cray T3E. Computing time was approximately 4 hours for 20000 time-steps on 64 CPUs.

Table 1. Crack models for numerical calculations. The models with an x attached to its number have intersection of cracks. The models with a p attached to its number have parallel cracks (see Fig. 2). Note that 0.0001m is the size of grid spacing and the size of the crack region is always 1000*1000 grid points.

No.	crack density ρ	length of cracks [0.0001m]	number of cracks	porosity ϕ of the crack region	number of model realizations
1x	0.050	56	63	0.0045	1
2x	0.100	56	126	0.0091	1
3x.1-3x.6	0.200	56	252	0.0181	6
4x	0.300	56	378	0.0270	1
5x.1-5x.6	0.401	56	504	0.0360	6
6x	0.601	56	756	0.0539	1
7x	0.801	56	1007	0.0720	1
8p	0.025	56	32	0.0018	1
9p	0.050	56	64	0.0036	1
10p	0.100	56	128	0.0073	1
11p	0.200	56	255	0.0145	1

3   Intersecting Cracks

In this section we consider randomly distributed and randomly oriented rectilinear intersecting thin dry cracks in 2D-media (see Fig. 2). Our goal is to compare the numerical results of the present study with the predicted effective velocities of a new critical crack density (CCD) formalism. The CCD formulation is valid for 2D (i.e. 3D transversely isotropic) fracturing configurations with intersecting cracks. We introduce an additional factor into

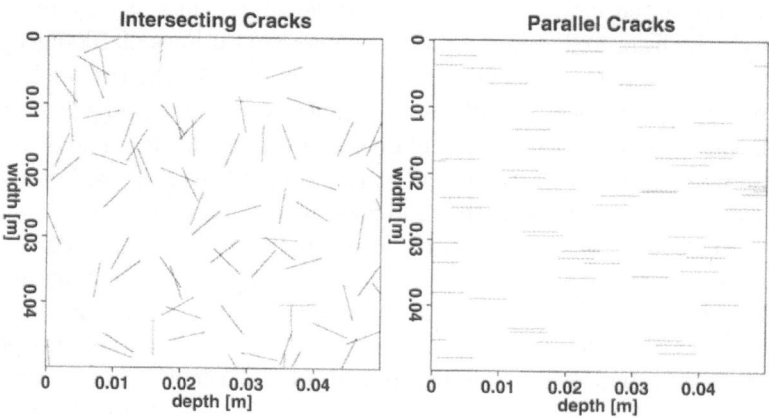

Fig. 2. Left-hand side: Randomly distributed and randomly oriented rectilinear intersecting thin dry cracks in homogeneous 2D-media. A part of Model 3x.1 is shown. Right-hand side: Randomly distributed rectilinear parallel non-intersecting thin dry cracks in homogeneous 2D-media. A part of Model 11p is shown.

results of a modified self-consistent theory (described e.g. in [3]) to include the physical behavior at the critical crack density (similar to the critical porosity described in [8]). We propose the following formulae for the effective elastic moduli $< E >$ and $< \mu >$:

$$< \mu_{CCD} >= \mu_0 e^{-\pi(\rho/2)\left(\frac{\rho_c}{\rho_c-\rho}\right)^{1/2}}, < E_{CCD} >= E_0 e^{-\pi\rho\left(\frac{\rho_c}{\rho_c-\rho}\right)^{1/2}},$$
$$\text{with: } \rho = \frac{1}{A}\sum_{k=1}^{n} l_k^2,$$

where μ_0 is the shear modulus, E_0 is the Young's modulus, ν_0 is the Poisson's ratio of the unfractured medium, ρ is the crack density (as in [4]), ρ_c is the critical crack density (can be found in [10] for our models), $2l_k$ is the rectilinear length of a crack and A is the representative area. Now, we discuss the numerical results on effective wave velocities. They are depicted with dots in Fig. 3. For comparison, the predictions of the CCD formulation described above are shown in the same figure with lines. We show the normalized effective velocities for three types of waves. The relative decrease of the effective velocity for one given crack density is in the following succession: For SH-waves we obtain the smallest decrease followed by SV-waves. For P-waves it is largest. For each wave type we perform numerical FD-calculations with different crack densities to obtain the effective velocity. For these measurements depicted in Fig. 3 we use the models No. 1x,2x,3x.1-3x.6,4x,5x.1-5x.6,6x,7x (see Table 1). A final result is that our numerical simulations of P-, SV- and SH-wave velocities are in excellent agreement with the predictions of the CCD formulation.

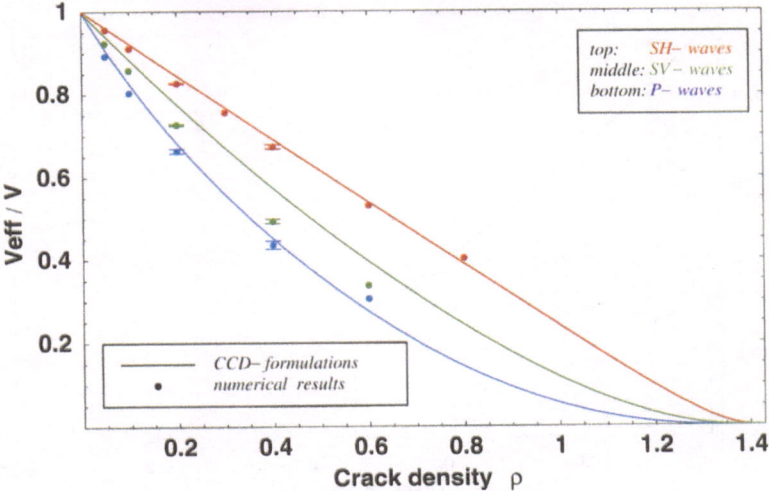

Fig. 3. Normalized effective velocity versus crack density. Dots: Numerical results of this study, Lines: Theoretical predictions of the CCD-formulations. The error bar denotes the standard deviation for different model realizations.

4   Parallel Cracks

This section considers randomly distributed rectilinear parallel non-intersecting thin dry cracks in 2D-media (see Fig. 2). The paper of Kachanov (1992) discusses three different theoretical descriptions of effective velocities in such a case. Namely, they are the "theory for non-interacting cracks (NIC)", the "modified (or differential) self-consistent theory (MSC)" and a extension of the modified self-consistent theory (EMSC) by Sayers and Kachanov (1991). In order to give an overview we present here the effective Young's modulus $< E >$ for the three theories:

$$< E_{NIC} >= E_0 \; \tfrac{1}{1+2\pi\rho}, \quad < E_{MSC} >= E_0 \; e^{-2\pi\rho},$$
$$< E_{EMSC} >= E_0 \; \tfrac{1}{1+2\pi\rho*e^{\pi\rho}},$$

where E_0 is the Young's modulus of the unfractured medium and ρ is the crack density.The effective velocities of SV- and P- waves predicted by this three theories for parallel cracks are plotted in Fig. 4 using lines. Our numerical results for parallel cracks can be seen in the same figure using dots. Details of the models No. 8p-11p used for the experiments are displayed in Table 1. We show the calculations for two types of waves. The relative decrease of the effective velocity for one given crack density is in the following order: For SV-waves we obtain the smallest decrease followed by P-waves. The main result of the investigations in this section is the fact, that the three

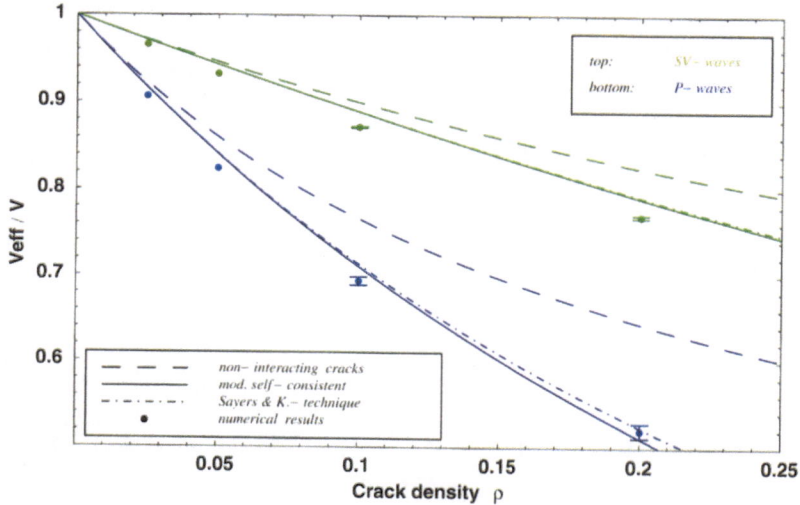

Fig. 4. Normalized effective velocity for parallel cracks versus crack density. Dots: Numerical results, Lines: Different theoretical predictions.

theories for parallel cracks can only be applied successfully to a dilute crack density. For large crack densities further research is necessary.

5   Conclusions

We present a numerical tool to calculate effective velocities in fractured media. Finite-difference modeling of the elastodynamic wave equation is very fast and accurate. In contrast to a standard staggered grid, high-contrast inclusions do not cause instability difficulties for our rotated staggered grid. Thus, our numerical modeling of elastic properties of dry rock skeletons can be considered as an efficient and well controlled computer experiment. We propose a heuristic approach called the critical crack density (CCD) formulation. This formulation introduces the critical crack density into the modified (differential) self consistent media theory. The CCD formulation predicts effective velocities for SV-, SH- and P- waves in fractured 2D-media with intersecting rectilinear thin dry cracks. The numerical results support predictions of this new formulation. Moreover, we show that different theories for effective velocities for parallel cracks can only be applied successfully to a dilute crack density.

References

1. Andrews, D. J. and Ben-Zion, Y. (1997). Wrinkle-like slip pulse on a fault between different materials. *Journal of Geophysical Research*, 102, No. B1:553–571.
2. Dahm, T. and Becker, T. (1998). On the elastic and viscous properties of media containing strongly interacting in-plane cracks. *Pure Appl. Geophys.*, 151:1–16.
3. Davis, P. M. and Knopoff, L. (1995). The elastic modulus of media containing strongly interacting antiplane cracks. *J. Geophys. Res.*, 100:18.253–18.258.
4. Kachanov, M. (1992). Effective elastic properties of cracked solids: critical review of some basic concepts. *Appl. Mech. Rev.*, 45(8):304–335.
5. Kelly, K. R., Ward, R. W., Treitel, S., and Alford, R. M. (1976). Synthetic seismograms: A finite-difference approach. *Geophysics*, 41:2–27.
6. Kelner, S., Bouchon, M., and Coutant, O. (1999). Numerical simulation of the propagation of p waves in fractured media. *Geophys. J. Int.*, 137:197–206.
7. Kneib, G. and Kerner, C. (1993). Accurate and efficient seismic modelling in random media. *Geophysics*, 58:576–588.
8. Mukerji, T., Berryman, J., Mavko, G., and Berge, P. (1995). Differential effective medium modeling of rock elastic moduli with critical porosity constraints. *Geophysical Research Letters*, 22(5):555–558.
9. Murai, Y., Kawahara, J., and Yamashita, T. (1995). Multiple scaterring of sh waves in 2-d elastic media with distributed cracks. *Geophys. J. Int.*, 122:925–937.
10. Robinson, P. C. (1983). Connectivity of fracture systems – a percolation theory approach. *J. Phys. A: Math. Gen.*, 16:605–614.
11. Saenger, E. H., Gold, N., and Shapiro, S. A. (2000). Modeling the propagation of elastic waves using a modified finite-difference grid. *Wave Motion*, 31(1):77–92.

Wave Propagation in Heterogeneous Media. Part 2: Attenuation of Seismic Waves Due to Scattering

Tobias M. Müller[2], Christof Sick[2], and Serge A. Shapiro[2]

[1] Geophysical Institute, Karlsruhe University, 76187 Karlsruhe, Germany
[2] Fachrichtung Geophysik, Freie Universität Berlin, 12249 Berlin, Germany

Abstract. We present a scattering attenuation model based on the statistical wave propagation theory in random media. It is suitable for the weak wavefield fluctuation regime and has practically no restriction in the frequency domain. The presented formulas allow to quantify scattering attenuation in complex geological regions using simple statistical estimates from well-log data. This knowledge is important for further petrophysical interpretations of reservoir rocks. To test our theory we perform numerical simulations of seismic wave propagation in 3-D elastic random media using a finite-difference solution of the elastodynamic wave equation. From the synthetic seismograms we determine the scattering attenuation (Q^{-1}) with help of spectral decay methods. We find good agreement of the frequency-dependent Q values and the theoretical predictions.

1 Introduction

Fundamental signatures of seismic waves in rocks are the attenuation and the dispersion. It is of great importance for the interpretation of seismic data to quantify the magnitude as well as the frequency dependence of attenuation. Usually the attenuation of seismic wavefields are characterized with help of the quality factor Q. E.g., the knowledge of Q is needed for a correct estimation of the magnitude of reflection coefficients. From a practical point of view, there are however serious problems connected with the determination of attenuation [2].One of them is the fact that attenuation is not only caused by absorption, i.e. due to viscoelastic effects (anelasticity, presence of fluids) but also due to the heterogeneous composition of rocks and reservoirs on many length scales. A further difficulty is the frequency dependence of attenuation measurements. That is why laboratory experiments cannot easily be compared with field measurements.Deterministic approaches are not suitable to describe the complex structures of reservoirs. In contrast to this, stochastic models provide an interesting alternative and are ideally used complementary with deterministic models. Analytical results are obtained within the framework of wave propagation theory in random media. Theoretical methods developed in order to quantify scattering attenuation include the meanfield theory using the Born approximation or the traveltime-corrected meanfield formalism which is commonly used in seismological studies [8].The mean-field theory overestimates the scattering attenuation, whereas the traveltime-corrected meanfield excludes large wavenumbers so that scattering on large-scale heterogeneities is not taken into account. It requires a heuristically chosen cut-off wavenumber (or equivalently a scattering angle) which can be only determined by numerical tests. In addition, numerous numerical studies characterized the scattering attenuation in random media (e.g. [4,11,3,5]). It is the purpose of this study to derive a model of scattering attenuation that is applicable in connection with standard methods in reservoir geophysics. Within the framework of the extension of the O'Doherty-Anstey theory to 2-D and 3-D random media [6] we obtain tractable expressions for scattering Q. We discuss the relation between this model of scattering attenuation and the meanfield and traveltime-corrected meanfield approximations, respectively. To test this theory, we perform numerical simulations of wave propagation in 3-D elastic random media. With help of a spectral decay analysis of the recorded seismograms we determine frequency-dependent Q^{-1} values and find good agreement with theoretical predictions.

2   Theory

2.1   New model of scattering attenuation

Based on the Rytov approximation and the causality principle, we give a description of scattering attenuation for plane waves propagating in 2-D and 3-D weakly heterogeneous elastic solids. Shapiro et al. (1996) derived analytical expressions for the phase velocity in random media which is practically valid for all frequencies. Applying the Kramers-Kronig relationship (which follows from the causality, passivity and linearity of the medium) to the results for the phase velocity, we calculate the attenuation (for a detailed derivation see Müller et al., 2001). For plane wave propagation in 3-D we obtain

$$Q^{-1} = 4\pi^2 k \int_0^\infty d\kappa \, \kappa \, \Phi(\kappa) \left[H(\kappa - 2k) - \frac{\sin(\kappa^2 L/k)}{\kappa^2 L/k} \right], \qquad (1)$$

where $k = \frac{\omega}{c_0}$ denotes the wavenumber, c_0 is the constant background velocity, L the travel-distance. $\Phi(\kappa)$ is the fluctuation spectrum which contains the second-order statistics of the medium's fluctuations, i.e. the variance σ^2 and the correlation length a of the P-wave (S-wave) velocity in rocks. H denotes the Heaviside step function. Note that the corresponding results in 2-D can be obtained by skipping κ in the integral over κ and dividing by π. The validity range of equation (1) in terms of the wave parameter $D = 2L/(ka^2)$ is $\frac{1}{\pi L/\lambda} \leq D \leq \left(\frac{L}{a}\right)^2$ if $L > \max\{\lambda, a\}$ where λ denotes the wavelength. Note that equation (1) is also restricted to the weak wavefield fluctuation regime. We emphasize that the scattering attenuation estimate (1) is dependent on the travel-distance. Equation (1) describes the scattering attenuation of most probable, so-called typical, primary wavefields in single realizations of the random medium. This is because the wavefield description in the light of the 3-D O'Doherty-Anstey theory is based on the mean of log-amplitude and phase fluctuations (Müller and Shapiro, 2001). The use of an averaged phase implicitly assumes that traveltime fluctuations are neglected. Therefore, an averaged traveltime-corrected primary wavefield can be rigorously formulated by a similar approach. The corresponding scattering attenuation estimate $^{ttc}Q^{-1}$ is given by

$$^{ttc}Q^{-1} \approx \frac{1}{2} Q^{-1} , \tag{2}$$

as shown in Shapiro and Kneib (1993). In the section 'Numerical experiments' we discuss the data analysis of plane wave transmission experiments which leads to Q^{-1} and $^{ttc}Q^{-1}$, respectively.

2.2   Relations among scattering attenuation models

The importance of seismic wave scattering was early recognized as a vital aspect of seismic data analysis and rock characterization (for an overview we refer to Sato and Fehler, 1998, and the references therein). In this section we show how the above presented scattering attenuation model is related to scattering attenuation of the meanfield and traveltime-corrected meanfield obtained in the single scattering (Born) approximation. To do so, we examine the scattering attenuation estimate given in the traveltime-corrected Born approximation (see equation 5.20 in Sato and Fehler, 1998)

$$^{TSc}Q^{-1} = \frac{k^3}{2\pi} \int_{\Psi_c}^{\pi} \Phi \left(2k \sin \left[\frac{\Psi}{2} \right] \right) \sin(\Psi) \, d\Psi , \tag{3}$$

where Ψ_c denotes the scattering angle. Theoretical and numerical considerations suggest $\Psi_c \approx 29°$. Note that equation (3) refers to a traveltime-corrected averaged field in the sense that all waveforms have to be aligned to a certain reference time and then summed up. After some straightforward calculations we can express $^{TSc}Q^{-1}$ as

$$^{TSc}Q^{-1} = \frac{k}{2\pi} \int_0^{2k} d\kappa \, \kappa \Phi(\kappa) - \int_0^{2k \sin\left[\frac{\Psi_c}{2}\right]} d\kappa \, \kappa \Phi(\kappa) . \tag{4}$$

From the latter equation it is easy to see how $^{TSc}Q^{-1}$ is related to the meanfield attenuation: For $\Psi_c = 0$ the second integral vanishes and the first term corresponds to the meanfield attenuation (see e.g. equation (A2-17) in Shapiro and Kneib, 1993). Exactly the same expression can be observed in equation (1) by considering only the first term and reducing the integral due to the Heaviside step function. By comparing the remaining terms in equations (1) and (4), it is possible to derive in the long travel-distance limit (Fraunhofer approximation) a relationship between the scattering angle (needed for $^{TSc}Q^{-1}$) and the travel-distance which enters into equation (1):

$$\Psi_c \approx 2 \arcsin\left[\sqrt{\frac{\pi}{8Lk}}\right] .$$ (5)

Equations (3)-(5) show that both models of scattering attenuation based on the single scattering approximation can be derived from the more general equation (1).

3   Numerical experiments

3.1   FD simulations in 3-D elastic random media

In order to explore the accuracy of the different scattering theories we simulate a plane wave propagating in a single random medium realization characterized by an isotropic exponential autocorrelation function (correlation distance=45m). Similar transmission simulations in 2-D acoustic random media were performed by Frankel and Clayton (1986), Shapiro and Kneib (1993). 3-D acoustic finite-difference modeling was done by Frenje and Juhlin, 2000. Here, we present results from 3-D random media modeling based on the elastodynamic wave equation. The geometry as well as the medium parameters are of the order of reservoir scales and rocks in hydrocarbon exploration, respectively. The reservoir P-wave velocity fluctuations have an average velocity of 3km/s and a standard deviation of 5 percent. Density and S-velocities were derived from P-velocities using empirical relations for sandstones. The source wavelet is a 70 Hz Ricker wavelet. The size of the numerical grid is 480x480x1200 meter in x-,y-,z-direction, respectively. The spacing between gridpoints is 2m. A periodic boundary condition is applied at the sides of the numerical mesh. The modeling was performed on the massive parallel supercomputer CRAY T3E by applying a parallelized 3-D staggered grid FD program (Bohlen and Milkereit, 2001). The total memory requirement was approximately 4 Gbytes. Computing time was approximately 5 hours for 2000 time-steps on 128 CPUs.

3.2   Scattering attenuation estimates

In order to obtain scattering attenuation estimates we apply the so-called spectral decay method. More specifically, we consider the decay of the logarithm of amplitude spectrum with travel-distance. The slope of the regression line is then directly linked with a global scattering Q estimate by $Q^{-1} = \frac{2\alpha}{k}$ where $\alpha = -\frac{\ln(A(\omega))}{L}$ with the amplitude spectrum $A(\omega)$ of the wavefield. However, care should be taken how the analysis is done. Shapiro and Kneib (1993) investigated the influence of the different processing steps. Stacking

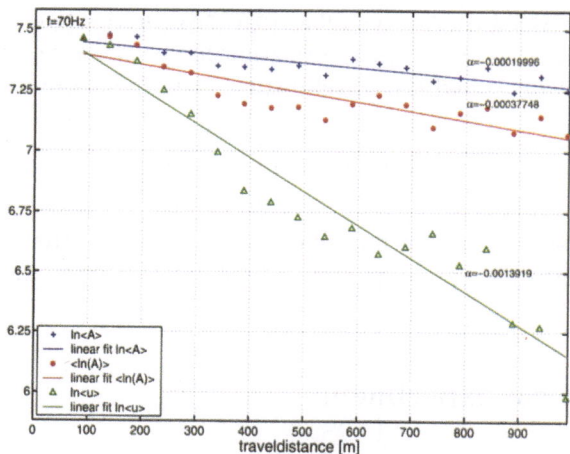

Fig. 1. Determination of Q^{-1} from the slopes the differently processed amplitude spectra for a single frequency plotted vs. travel-distance (dotted curves). We can observe that the largest estimates will be obtained for the meanfield $\langle u \rangle$ (green curve). Conversely, Q^{-1} corresponding to the traveltime-corrected averaged field yields the lowest estimates (blue curve). The linear fit to the mean log-amplitude values results in estimates within $^{ttc}Q^{-1} < Q^{-1} < {}^{mean}Q^{-1}$ (red curve).

all seismograms of the common-travel-distance gather in the time domain and analyzing the logarithm of the amplitude spectra vs. L, yields the the meanfield attenuation. Averaging the amplitude spectra and then analyzing its logarithm vs. L corresponds to the traveltime-corrected attenuation estimate $^{ttc}Q^{-1} = -\frac{2}{k}\frac{\ln(\langle A(\omega) \rangle)}{L}$, where the angular brackets denote the averaging operator. Changing the order of operations, we obtain the scattering attenuation estimate corresponding to the mean of the log-amplitude spectra $Q^{-1} = -\frac{2}{k}\frac{\langle \ln(A(\omega)) \rangle}{L}$. The effect of these 3 processing schemes on the determination of Q^{-1} values can be seen in Fig. 1. Note that strictly speaking it is not correct to apply a linear regression to obtain Q^{-1} and interpreting the result with equation (1) because of its travel-distance dependency. However, restricting the analysis to an intermediate depth range (i.e. respecting the validity range of eq. (1)), where the wavefield fluctuations are not saturated, we obtain reliable estimates. Moreover, the determination of Q^{-1} values from the first arrivals in seismograms is very sensitive to the applied window length around the primary arrivals (Frenje and Juhlin, 2000). We applied a window length corresponding to 1.0 times the source wavelength for $L = 0$ and that increases with L to account for the broadening and the traveltime fluctuations of the primary wavefield. The Q^{-1} estimates are shown in Fig. 2. Increasing the window length produces lower Q^{-1} since parts of the coda are included in the analysis. We note that within the investigated frequency range $5 \leq ka \leq 10$, the trend of increasing $^{ttc}Q^{-1}$ with increasing ka remains

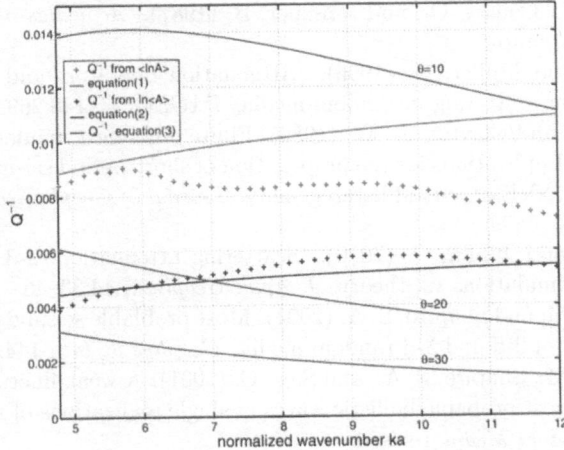

Fig. 2. Estimated Q^{-1} values as a function of ka from synthetic data vs. theoretical predictions. The green and blue curves correspond to equations (1) and (2), respectively. Scattering attenuation estimates using equation (3) are plotted for different scattering angles (red curves).

unaffected by the window length. The frequency dependence as well as the magnitude of Q^{-1} can be well explained using our scattering attenuation model (see the green and blue curve in Fig. 2 corresponding to equations (1) and (2), respectively) with $L/a \approx 2$. According to equation (5) this implies a scattering angle of 20° when using equation (4) (see also the red curves in Fig. 2 that suggest scattering angles $\leq 20°$).

4 Conclusions

In this paper we present a new model of scattering attenuation valid in weakly heterogeneous elastic media. We discuss its relation to scattering attenuation estimates based on the single scattering theory. We perform finite-difference simulations of seismic wave propagation in 3-D random media to check the analytical results. The application of the spectral decay method proves to be somewhat difficult for the inversion of Q^{-1} estimates because of the complex waveforms observed. Nevertheless, an accurate data processing yields Q^{-1} values whose magnitude and frequency dependence can be explained with the proposed scattering attenuation model. To study the frequency dependence of Q^{-1} in a broader frequency range, more simulations are required.

References

1. Bohlen, T. and Milkereit, B. (2001). Parallel 3-d viscoelastic finite difference seismic modelling. Amsterdam.

2. Bourbié, T., Coussy, O., and Zinszner, B. (1987). *Acoustics of porous media*. Editions Technip.

3. Fang, Y. and Müller, G. (1996). Attenuation operators and complex wave velocities for scattering in random media. *PAGEOPH*, 148:269–285.

4. Frankel, A. and Clayton, R. W. (1986). Finite difference simulations of seismic scattering: Implications for the propagation of short-period seismic waves in the crust and models of crustal heterogeneity. *Journal of Geophysical Research*, 91, No. B6:6465–6489.

5. Frenje, L. and Juhlin, C. (2000). Scattering attenuation: 2-d and 3-d finite difference simulations vs. theory. *J. Appl. Geophys.*, 44:33–46.

6. Müller, T. M. and Shapiro, S. A. (2001). Most problable seismic pulses in single realizations of 2-d and 3-d random media. *Geophys J. Int.*, 144:83–95.

7. Müller, T. M., Shapiro, S. A., and Sick, C. (2001). A weak fluctuation approximation of most probable ballistic waves in single realizations of random media. *Waves Random Media*, revised.

8. Sato, H. and Fehler, M. (1998). *Seismic wave propagation and scattering in the heterogenous earth*. Springer.

9. Sayers, C. and Kachanov, M. (1991). Single-scattering approximations for coefficients in biot's equations of poroelasticity. *Int. J. Solids & Struct.*, 7(6):671–680.

10. Shapiro, S. A. and Hubral, P. (1999). *Elastic waves in random media*. Springer, Heidelberg.

11. Shapiro, S. A. and Kneib, G. (1993). Seismic attenuation by scattering: Theory and numerical results. *Geophys J. Int.*, 114:373–391.

12. Shapiro, S. A., Schwarz, R., and Gold, N. (1996). The effect of random isotropic inhomogeneities on the phase velocities of seismic waves. *Geophys J. Int.*, 127:783–794.

Computer Science

Prof. Dr. Christoph Zenger

Institut für Informatik, Technische Universität München
D-80290 München

In this section four contributions are collected where Computer Science aspects play a dominant role.

In the article "Fast parallel particle simulations on distributed memory architectures" by Hipp et al. two particle methods (a smoothed particle hydrodynamics code and a one-dimensional Particle-in-Cell code with Monte-Carlo-collisions) are studied with respect to efficient implementations on various parallel computers including the Cray T3E of the HLRS.

The solution of PDEs based on the UG-code is investigated in "High-accuracy simulation of density driven flow in porous media" by Bastian et al. in a very complicated application problem using up to 16.7 million grid points where linear systems of equations up to more than 35 million unknowns have to be solved.

These two articles can be considered as prototypes of the state of the art in the simulation of complicated natural and technical processes.

Another area of technical relevance belonging to the growing field of image processing and image compression is addressed in "ParWave: Parallel Wavelet Video Coding" bei Feil et al. Compression rates for videos can be improved if more complex methods are used requiring high performance computer systems exploiting the inherent parallelism of the algorithms.

The last contribution "Compiler-Generated Vector-based Prefetching on Architectures with Distributed Memory" by Müller investigates compiler-based methods for overcoming delays caused by network latency. Vector pipelining strategies are used for regular and irregular data structures arising in typical applications.

Note that all contributions in this section come from "traditional" fields of applications. Data mining, large data base applications or related problems in bioinformatics are still missing.

Fast Parallel Particle Simulations on Distributed Memory Architectures *

M. Hipp[1], S. Kunze[2], M. Ritt[1], W. Rosenstiel[1], and H. Ruder[2]

[1] Wilhelm-Schickard-Institut für Informatik, Universität Tübingen
[2] Institut für Astronomie und Astrophysik, Universität Tübingen

Abstract. One of the major goals of the Sonderforschungsbereich (SFB) 382 is the development of parallelization strategies for physical applications. In particular, we focus on simulation methods based on particles.

In this paper, we present two different particle methods, a three-dimensional Smoothed particle hydrodynamics (SPH) code and a one-dimensional Particle-in-Cell (PiC) code with Monte-Carlo collisions (MCC). For both methods, a brief introduction to the physical model and its implementation is given. We discuss implementation and runtime aspects and detail the parallelization of the codes.

We talk about our experience porting the codes and running them on a couple of distributed memory machines, such as Cray T3E, Hitachi SR8000 and a Linux Cluster and present performance measurements of the codes. For the SPH method, some of the physical results of the simulation runs are explained.

1 Introduction

In our project the numerical method SPH is used to simulate astrophysical problems. In the last years we gained a lot of experience with several different parallel SPH codes in the programming languages Fortran, C and C++ using MPI and SHMEM on the Cray T3E [18]. We could show, that an efficient implementation of SPH is possible on machines with distributed memory [19]. In the last year the code was improved to simulate three dimensional problems and simulations with particle injection. Furthermore, the code was optimized to handle more particles with the same amount of memory allowing us to run bigger problems.

1.1 Motivation

To obtain a good spatial resolution for three dimensional problems it is necessary to increase the number of particles and the number of interactions per particle. Both result in an overall computation time and memory consumption which is too big for small desktop machines.

The overall performance for SPH simulations on vector SMP machines is not very good, because the SPH method does not provide long vectors. This

---

* This project is funded by the DFG within SFB 382: *Verfahren und Algorithmen zur Simulation physikalischer Prozesse auf Höchstleistungsrechnern* (Methods and algorithms to simulate physical processes on supercomputers).

results in a bad price-performance ratio for vector SMP machines like the NEC SX5 compared to machines with distributed memory (like Cray T3E, Hitachi SR8000 and becoming popular PC based Clusters) if it is possible to extract and group the communication parts of the code. So, we decided to do the parallelization using message passing with explicit communication.

2   Smoothed particle hydrodynamics

2.1   About SPH

Smoothed particle hydrodynamics (SPH) was introduced by Lucy [14] and Gingold & Monaghan [4]. It is a grid-less Lagrangian particle method for the solution of the hydrodynamic equations. Instead of solving the equations on a grid, the fluid is modelled by small interacting packets of matter that move along with the flow and carry mass and momentum. Hydrodynamic variables such as density, pressure, and temperature are assigned to each particle. The values of these quantities are determined by the interactions with the neighbour particles.

SPH is especially suited for the simulation of accretion disks because SPH can handle large density contrasts, open boundaries are easily implemented, and SPH posesses an adaptive resolution both by the variation of the inter-action range of each particle and also by the particle mass. So it is possible to resolve the most interesting regions very fine. Readers interested in the basic principles of SPH find detailed reviews of the SPH method in Benz [2] or Monaghan [15].

2.2   Physical Problem

Accretion discs are very common structures. They play an important role in galaxy formation and feed the central engines in the nuclei of active galaxies. All stars form via accretion, the left-over of this process provides the material for the formation of planets.

Most binary star systems form accretion disks at some evolutionary state. Here we are concerened with cataclysmic variables, which are good laboratories to study the physics of accretion disks.

Cataclysmic variables (CVs) are close binaries with mass transfer from the secondary to the primary. The donor, a low-mass, late type main sequence star, fills its Roche lobe and loses mass to the accretor, a white dwarf (WD). In many CVs the magnetic field strength of the WD is so small that it can be neglected. In this case the overflowing matter forms an accretion disk around the WD, and the accretion process is governed by the viscous evolution of the disk.

One aspect of the physics of accretion discs in CVs is the interaction of the in-falling gas stream with the rim of the accretion disk. As both flows, the stream and the disk, are highly supersonic, the development of shock fronts is expected at the impact zone.

This interaction region, the so-called "bright spot" or "hot spot", can be seen in many high-inclination CVs, e.g. U Gem [9], and Z Cha [20], as a hump in the orbital light curve shortly before the eclipse of the WD.

From the large range of different bright spot sizes, locations and intensities, and their variability, it is already clear that the underlying physics is rather complex and has to be approached by numerical simulations. Closely related to the bright spot is the question of how much of the in-falling gas stream is stopped at the edge of the disk and stored there, and how much of the stream can flow over and under the disk surface to inner parts of the disk.

By observation, there are several features seen in CVs and low-mass X-ray binaries (LMXBs) that can be explained by stream-disk overflow. In CVs the accretion stream reveals itself by its high velocity in Doppler maps and phased spectra [11,16,7]. Furthermore, many CVs and LMXBs show so-called absorption dips in X-ray and UV around orbital phase 0.7 to 0.8. Some systems also show a shallower dip at about phase 0.1. The absorption dips at phase 0.7 can be explained if stream material overflows the disk at several disk scale heights after being deflected by shock interaction at the bright spot region.

Still missing are numerical simulations of the stream-disk impact with high spatial resolution, and simulations that take into account the further fate of the overflowing matter. For this purpose it is necessary to use the full Roche potential. Also lacking are comparable simulations for different kinds of systems with differing orbital periods, mass ratios, and mass transfer rates. Our simulations of different systems cover a wide range of these parameters, and we hope to fill this gap to some extent.

2.3   Physical Results

Simulation Setup In order to achieve a reasonable spatial resolution the disk should contain at least 50 000 particles. Simply starting the simulation with an empty disk and waiting until some form of quasi-steady state is reached is far too time consuming to allow for parameter studies. This problem can be circumvented by making use of one of the more pleasant features of SPH. Since the particles are actually to be interpreted as integration points rather than fluid particles, one is free to substitute a given particle distribution with another, equivalent distribution that represents the same physical situation, within the accuracy of the method. The trick is to take a certain number of data sets of different time steps and simply concatenate them to get a new single data set. If we take, e.g., 10 data sets and concatenate them, the mass of each individual particle has to be divided by that factor.

This method allows for the construction of viscously evolved disks with almost arbitrary particle number in a short time.

The in-falling gas stream is set up in a way that gaussian shaped density distributions in the horizontal and vertical directions are fulfilled. The theoretical results of Lubow & Shu [12,13] are used to determine the vertical and horizontal scale heights of the stream. Particles are inserted a bit downstream of the inner Lagrangian point with appropriate velocity. The density in the stream is lower than in the disk. In order to reach comparable resolution, we used more but less massive particles for the stream than for the disk.

Results The astrophysical results of these simulations are discussed in detail in [10]. Here we can only show some exemplary results. In Fig. 1 the particle distribution of the simulation of the stream disk interaction and stream overflow of the dwarf nova IP Pegasi are shown. Details of the simulations are given in the figure captions. For clarity only the particles inserted during the last orbital period are shown. The most important results of this simulations are the rather massive stream overflow over the disk, resulting in the disposition of the in-falling material not only, and not even predominantly in the outer part of the disk, but rather at a radius close to the center. The other striking feature is the elevation of stream matter high above the disk plane, neatly explaining the x-ray dips that are often observed at this orbital phase in many medium to high inclination systems.

Apart from the exemplary results given here, the simulations span a large range of mass ratios, orbital periods, mass transfer rates and thermal states of the disk. Not only could we show that stream disk overflow always plays an important role in CV disks, our results also explain the UV and X-ray dips often observed in these sysytems. Under certain circumstances, namely a small disk or an enhanced mass transfer rate, a second absorption region occurs around orbital phase 0.2. This feature is observed in some systems and could not be explained before. Moreover, the re-impact of the overflowing gas onto the disk close to the white dwarf around orbital phase 0.5 can be seen as a second bright spot. Such a feature has been observed in the dwarf nova WZ Sagittae. Also this observation had no explanation yet.

To our knowledge these are the only high-resolution simulations of the stream-disk impact in CVs including the full Roche potential, which is necessary because the structure of the outer disk rim is heavily influenced by the tidal forces from the secondary star. Due to the high computational requirements these simulations profit from the use of parallel machines. Although it is in principle possible to perform such simulations on top-of-the-notch workstations, only parallel machines make it possible to cover such a large range of parameters in a reasonable time.

2.4   Parallel SPH Implementation

The parallel implementation of the used SPH code is written in C and parallelized using MPI. It is based on a code formerly written and optimized for the Cray T3E with the SHMEM communication library. The communication

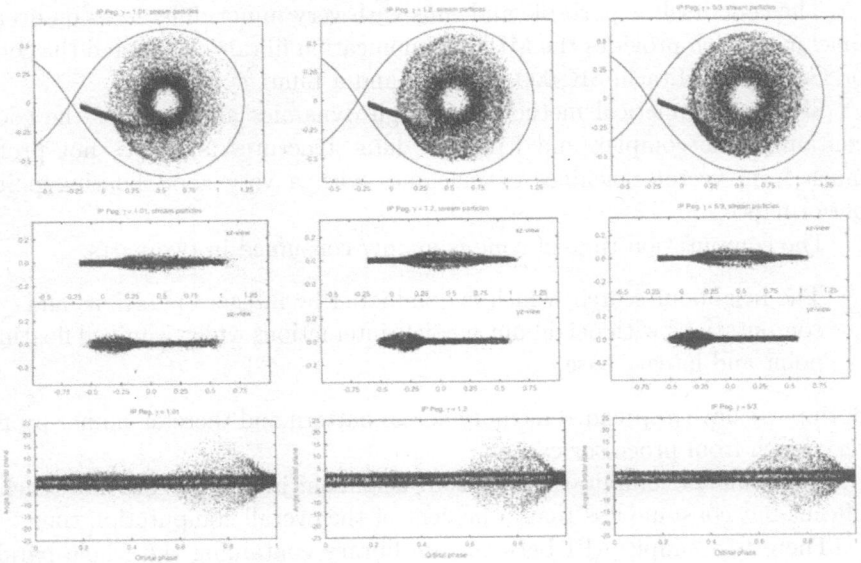

Fig. 1. Simulation of the dwarf nova IP Peg (primary mass: 1.15 $M_\odot$, secondary mass: 0.67 $M_\odot$, orbital period $3^h 48^m 20^s$, mass transfer rate: $10^{-10}\,M_\odot\,\mathrm{yr}^{-1}$) The left column shows particle distributions from a simulation with nearly isothermal equation of state, the right column is derived from a simulation with adiabatic equation of state, and in the simulation displayed in the middle column the polytropic coefficient was set to 1.1. Displayed are particles inserted during the last half orbital cycle. The upper row shows the distribution of the overflowing matter projected onto the orbital plane, the middle row shows edge-on views of the disks from two perspectives, namely perpendicular to and along the system axis. The lower row shows what the stream overflow would look like when seen from the white dwarf. Disk particles are not plotted. These simulations show that a substantial fraction of the accretion stream flows over the disk surface directly to inner parts of the disk even when the disk rim is geometrically thicker than the accretion stream at the impact region. The stream disk overflow can explain the X-ray and UV dips which are observed in many systems.

code is separated from the physical calculation in an own module allowing us to easily switch between different communication libraries such as MPI, PVM or SHMEM.

With older revisions of the MPI libraries on the Cray T3E we measured big performance differences between MPI and SHMEM for some operations, but after an update to new MPI libraries the difference between the SHMEM and MPI version was very small, with some performance advantages for the – slightly improved and optimized – MPI implementation.

The code itself is portable and runs with very minor differences on every machine, which provides the MPI communication library. We tested the code on Cray T3E, Hitachi SR8000, IBM SP and a Linux system.

SPH is a numerical method with high dynamics and therefor the code contains some complex and irregular data structures and does not profit much from vector machines or machines with a very good floating point performance.

The computation time of code is mainly consumed in two parts:

- The neighbour search, which is dominated by integer operations and
- computations with neighbour particle interactions which is mixed floating point and integer based.

Both parts have a spreaded memory access pattern and therefor cannot profit very much from processor caches.

Some smaller computations without neighbour interaction are more cache efficent but consume less than 5 percent of the overall computation time.

There is a simple API between the library containing the whole parallelization and the code containing the physics. The abstraction simplifies the extension of a simulation without knowing much about the parallelization. A user can add new physical quantities by allocating a new *parallel field*.

For the parallel computation the library provides an iterator concept to step through all particles and their neighbours and later communicates the new data. For all particle computations the user does not need to add explicit communication. But to prevent unnecessary communication the user explicitly has to register the parallel fields in the library, that are necessary for further computations. The library then ensures that for all particles returned by the iterator, that the necessary data is available on the node. If the field is no longer necessary for parallel computations, the user has to unregister the field.

On the former SHMEM based implementation there was a *load stealing* mechanism to optimize the rough static load–balancing *on-the-fly*. This part was removed in the MPI version, because it was heavily dependent on SHMEM one-sided communication operations. It showed that the MPI version of the parallel SPH code with its static load–balancing scales very well without the load stealing.

2.5   Performance results

We measured the performance of the improved SPH 3D Code with particle injection on three different machines. Cray T3E and Hitachi SR8000 at the HLRS in Stuttgart and on the Kepler Cluster, a Linux Cluster installed in Tübingen. The Kepler Cluster has Dual SMP Pentium III nodes with 650 MHz processor speed. For the communication the cluster has a Myrinet network with a peak MPI bandwidth of about 115 MByte/s and a one-way latency of about $7\mu s$.

We measured two different problems. A small problem with about 34 000 particles and a bigger problem with 360 000 particles.

One-node performance The small problem needs less than 100MB memory allowing us to compare the raw application performance of the three different machines, without communication. The code has fairly complex data structures and is therefor dominated by the integer and memory performance of the processors. So, using a Pentium III with its fairly bad floatingpoint performance is no disadvantage over the Alpha processors in the Cray T3E or the processors in the Hitachi SR8000. The application has a total floatingpoint performance of about 60 MFlop/s on one Pentium III processor.

Cray T3E	Hitachi SR8000	Kepler Cluster
903, 4 s	721, 3 s	467, 9 s

Parallel performance For better comparison of the three machines we decided to plot runtime information instead of speedup because the big problem fits only into at least 24 Cray T3E nodes. The big problem computes the *right hand side* 36 times and the small problem computes the *right hand side* 90 times. The three curves are all–over (wall clock) time, the time for a the nearest neighbour search and the parallel overhead including communication and additional work for load balancing (see Figs. 2, 3, 4) .

3   Particle-in-Cell with Monte-Carlo collisions

3.1   About PiC/MCC

The Particle-in-Cell method allows the simulation of large numbers of particles with short range interactions. In contrast to other particle methods such as SPH, the forces moving the particles are calculated using a fixed grid. Characteristic values of the particles are weighted on the grid using a *kernel function*. The momentum equations on the grid are solved using some standard method (for example finite differences). From the results the forces are calculated and interpolated back to the particles to solve the equations of motion. The basic algorithmic steps are summarized in Fig. 5.

The method reduces computational complexity from $O(n^2)$ to $O(n)$ for interpolating the n particles to the grid and the grid back to the particles. The complexity for solving the momentum equations on the grid is usually $O(g \log g)$ for g gridpoints.

A detailed description of the method can be found in [1]

3.2   Physical problem

In our application the PiC method is used to simulate the electrostatic plasma of a direct current glow discharge in a tube. The simulation is effectively one-dimensional, since the problem has a cylindrical symmetry and we are not

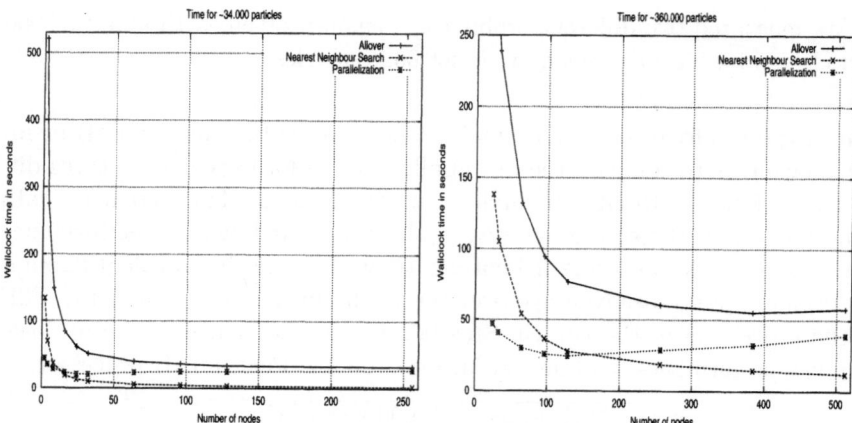

Fig. 2. The **Cray T3E** plot starts with 2 nodes for 34 000 particles and 24 nodes for 360 000 particles. The small problem does not scale very good for runs with more than 64 nodes. If we assume a speedup of 24 on 24 nodes for the big problem, we can achieve a maximum speedup of 128 for 384 processors. On 256 we have a speedup of about 124 and near 50 percent efficency. It showed, that a part with individual all–to–all communication becomes dominant for higher node numbers.

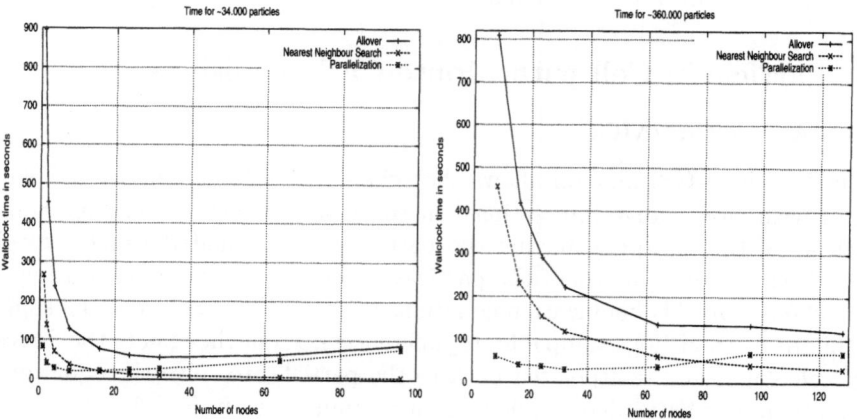

Fig. 3. We ran our tests on the **Hitachi SR8000** with pure MPI. Node numbers start at 2 nodes for the small problem and 8 nodes for the big problem. One can see a significant increase of the parallel overhead on more than 64 processors. Again the part with the individual all–to–all communication becomes very dominant. One should note, that we did not optimize the code for the Hitachi SR8000. Using threads instead of message passing for the inner node communication would probably increase the performance significantly. Therefor the maximum speedup compared with the one node run is limited to 53 on 128 nodes for the big problem.

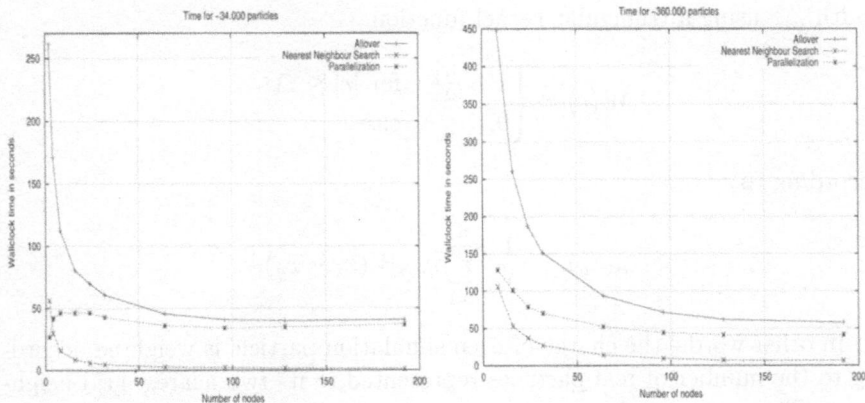

Fig. 4. On the **Kepler Linux Cluster** node numbers start at 2 nodes for the small problem and 8 nodes for the big problem. The Pentium III Processors have a fairly good integer performance and the contribution of the nearest neighbour search to the all-over time is very small. The Kepler cluster does not show the dominant all–to–all communication but has an expensive broadcast which is not this dominant on SR8000 and T3E. The big problem has a maximum speedup compared to the one node run of about 45 on 96 nodes using only one processor per node. This is worse compared to the other machines but the overall computation time using 64 nodes (128 processors) is about the same compared to 256 Cray T3E nodes.

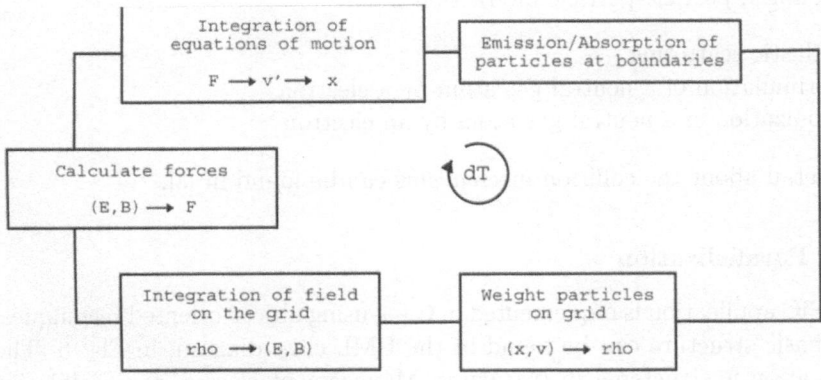

Fig. 5. Particle-in-Cell algorithm

interested in the radius components. Each simulation particle i represents a number η_i of physical particles (electrons or ions), where $\eta \approx 2 \cdot 10^9$. The charge q_i of the particles is weighted to a regular one-dimensional grid of width Δx using a triangular kernel function

$$W(r) = \begin{cases} 1 - \frac{|r|}{\Delta x} & \text{for } |r| \le \Delta x \\ 0 & \text{else} \end{cases}$$

according to

$$\rho(x_g) = \frac{1}{\Delta x} \sum_{i=1}^{n} \eta_i q_i W(x_i - x_g).$$

In other words, the charge of each simulation particle is weighted according to the number of real particles represented to its two nearest grid neighbours. The normalization by Δx results in the discretized charge density ρ. Next, the electrostatic potential Φ and the electric field E are computed on the grid using discretized versions of

$$\Delta \Phi(x) = -\frac{1}{\epsilon_0} \rho(x)$$

and

$$E(x) = -\nabla \Phi(x).$$

To model the particle-particle interactions in the plasma, the PiC method has been extended by Monte-Carlo collision processes. The last step computes randomized particle-particle interactions:

- Elastic scattering
- Stimulation of a neutral gas atom by a electron
- Ionization of a neutral gas atom by an electron

Detail about the collision mechanisms can be found in [8].

3.3 Parallelization

The PiC application is implemented in C++ using object-oriented techniques. The basic structure can be found in the UML class diagram in Fig. 6. The application is structered in two parts: Management classes, responsible for providing a flexible and easy to configure user interface. The execution concept is centered around the CTask object, responsible for executing some part of the physical simulation. CTask and derived objects can also be made persistent in a textual representation, which serves to configure the physical environment and the execution order of calculations. Simulations with different physical behaviour can be tailored using this single configuration file [3,8].

For the parallelization two aspects had to be considered: Data dependencies and load–balancing. Most of the steps in the PiC and MC collision code

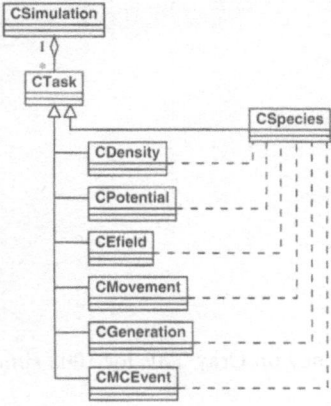

Fig. 6. Simplified class diagram of the PiC-MCC simulation

are independent calculations updating all particles. The only dependency lies in the global grid. Since, in our case the grid is much smaller than the number of particles ($g \ll n$), a natural approach is to parallelize over the number of particles and keep the grid redundant. After weighting the particles on the local grid, a global reduction operation provides each processor with a global grid. Next, each processor solves the grid equations in parallel. Measurements showed that this part of the calculation is lower than 1%, allowing for speedups ≥ 100.

Since the distribution of particles to processes is arbitrary, initially an optimal load-balance can be guaranteed. The particle-particle interactions calculated by Monte-Carlo collisions are divided between all processes and applied to the local subset of particles. Due to random variations of the number of particles created or destroyed, a variation of the load-balancing over the time can be expected. Since the MC processes are equally distributed we expected the resulting load imbalance to be small.

The parallelized code is targeted for distributed memory architectures and is parallelized using TPO++, an object-oriented message-passing library [6,5].

3.4   Performance measurements

The performance of the application has been measured on the Cray T3E and the Kepler cluster. Both are distributed memory architectures with different characteristics: Single nodes of the Cray T3E are less performant (450 Mhz) and equipped with less memory (128 MB) compared to the Kepler cluster (650 Mhz dual processors and 512 MB memory per processor). On the other hand, the network of the Cray T3E reaching 300 MB/s is more performant than the 115 MB/s of Kepler.

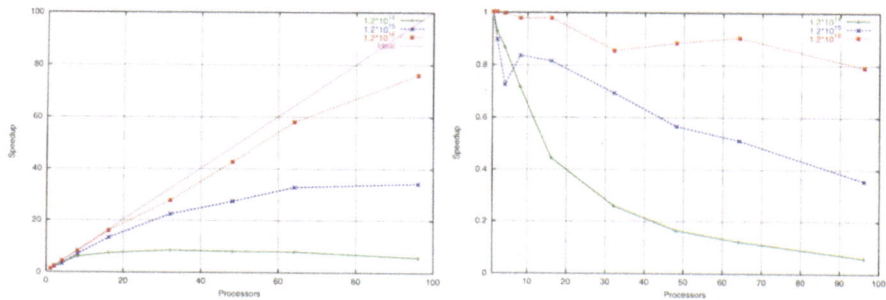

Fig. 7. Speedup and efficiency on Cray T3E for 1000 simulation steps and different number of particles.

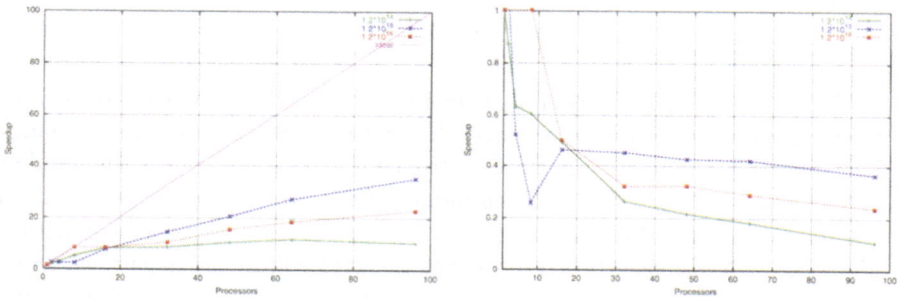

Fig. 8. Speedups and efficiency on Cray T3E for 5000 simulation steps and different number of particles.

The performance of the application is compared for three different numbers of real particles ($1.2 \cdot 10^{14}$, $1.2 \cdot 10^{15}$ and $1.2 \cdot 10^{16}$), different numbers of processors (1–96) and for different numbers of simulation steps (1000 and 5000) to analyze a possible load imbalance over the time.

The speedup and efficiency results for the Cray T3E can be found in Figs. 7 and 8, for the Kepler cluster in Figs. 9 and 10. Note that the speedup results on the Cray T3E are based on the 4-processor runs, since the problem is too large for 1 and 2 processors.

Both results show that the number of particles should be larger than $1.2 \cdot 10^{15}$ to obtain reasonable speedups. For $1.2 \cdot 10^{16}$ particles the application scales very good, reaching about 80% efficiency for 1000 simulation steps. The comparison of 1000 and 5000 simulation steps shows an unexpected decrease in performance. A detailed investigation reveals, that the load-imbalance due to the non-deterministic Monte-Carlo events gets worse with increasing number of steps, resulting in some processors which are re-

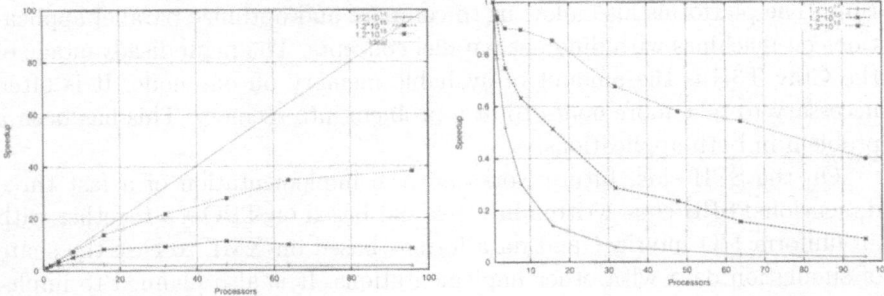

Fig. 9. Speedup and efficiency on Kepler for 1000 simulation steps and different number of particles.

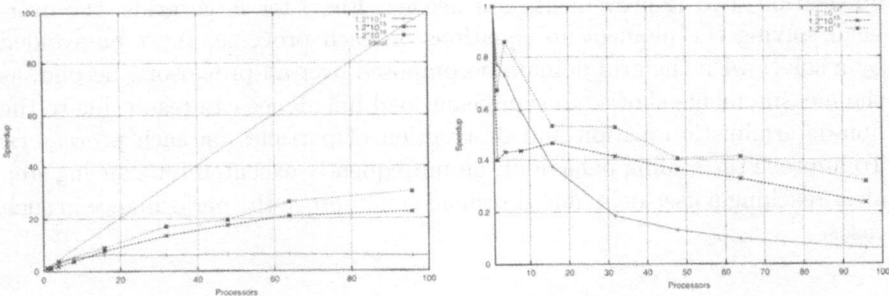

Fig. 10. Speedups and efficiency on Kepler for 5000 simulation steps and different number of particles.

sponsible for 3 times of the particles compared to the optimal load-balance, effectively reducing the speedup by the same factor.

Comparing both architectures, we find for all runs the speedups on Kepler being lesser than the speedups on Cray T3E. This is due to different ratio of CPU performance to network performance of the two architectures, resulting in lesser speedup for Kepler, which has faster nodes and a slower communication network. A comparison of the total (wall clock) execution time of both runs confirms this, with Kepler being about 50% faster on single-node runs. The runtime advantage of Kepler gets smaller with increasing number of processors.

4   Conclusions and future work

The parallel 3D SPH code allows us to run bigger simulations in a reasonable time. The code is portable and all platforms are suitable for SPH production runs. The platforms also allow us to compare and optimize parallel applications on machines with different parallel concepts. The main disadvantage of the Cray T3E is the amount of available memory on one node. It is often necessary to take more nodes to fit a problem into memory. This has been a problem in both applications.

On the SPH side, future works are the implementation of a fast three dimensional SPH code written in C++ and based on TPO++ together with an uniform I/O interface and data format based on XML to ease the share of simulation data with other implementations. It is also planned to implement a version with mixed MPI and thread-based parallelization. Especially machines with a mixed SMP/Message Passing architecture such as Hitachi SR8000 and the Kepler Cluster would profit from this optimization.

On the PiC side, we plan to extend the one-dimensional Particle-in-Cell code to solve two or three-dimensional problems. With respect to the parallelization, two improvements are needed: First, for large grids, the overhead solving the momentum equations on each processes must be avoided by a solver with the grid domain decomposed over all processors. Second, as the measurements showed, a significant load-imbalance can result due to the non-deterministic creation and destruction of particles on each processors. To improve the scaling behaviour, an unfrequently executed rebalancing step after reaching a user-definable threshold can improve the performance in such cases.

References

1. R.W. Hockney, J.W. Eastwood. *Computer simulation using particles*. Adam Hilger, Philadelphia, 1988.
2. W. Benz. Smooth Particle Hydrodynamics - a review. In J.R. Buchler, editor, *Numerical Modelling of Nonlinear Stellar Pulsations Problems and Prospects*, page 269, 1990.
3. Th. Daube and H. Schmitz. Opar: Open architecture c++ plasma simulation code. Ruhr-Universität Bonn, 1998.
4. R.A. Gingold and J.J. Monaghan. Smoothed particle hydrodynamics: theory and application to non-spherical stars. *Mon. Not. R. astr. Soc.*, 181:375–389, 1977.
5. Tobias Grundmann, Marcus Ritt, and Wolfgang Rosenstiel. Object-oriented message-passing with TPO++. In Arndt Bode, Thomas Ludwig, Wolfgang Karl, and Roland Wissmüller, editors, *Lecture notes in computer science*, pages xx–yy. Springer-Verlag, 2000.
6. Tobias Grundmann, Marcus Ritt, and Wolfgang Rosenstiel. TPO++: An object-oriented message-passing library in C++. pages 43–50. IEEE Computer society, 2000.
7. C. Hellier and E.L. Robinson. PX Andromedae and the SW Sextantis phenomenon. *apj*, 431:L107–L110, 1994.
8. A. Klaedtke. Particle in cell Simulationen mit Monte Carlo collisions. Master's thesis, Universität Stuttgart, Juli 1999.

9. W. Krzeminski. The eclipsing binary U Geminorum. *apj*, 142:1051, 1965.
10. S. Kunze, R. Speith, and F.V. Hessman. Substantial stream-disc overflow found in three-dimensional sph simulations of cataclysmic variables. *mn*, 322:499–514, 2001.
11. S.H. Lubow. On the dynamics of mass transfer over an accretion disk. *apj*, 340:1064–1069, 1989.
12. S.H. Lubow and F.H. Shu. Gas dynamics of semidetached binaries. *apj*, 198:383–405, 1975.
13. S.H. Lubow and F.H. Shu. Gas dynamics of semidetached binaries. ii - the vertical structure of the stream. *apj*, 207:L53–L55, 1976.
14. Leon B. Lucy. A Numerical Approach to Testing the Fission Hypothesis. *Astron. J.*, 82(12):1013–1924, December 1977.
15. J.J. Monaghan. Smoothed particle hydrodynamics. *Annual Review of Astronomy and Astrophysics*, 30:543–574, 1992.
16. A.W. Shafter, F.V. Hessman, and E.-H. Zhang. Photometric and spectroscopic observations of the eclipsing nova-like variable PG 1030 + 590 (DW Ursae Majoris). *apj*, 327:248–264, 1988.
17. S. Kunze, E. Schnetter, and R. Speith. Development and Astrophysical Applications of a Parallel Smoothed Particle Hydrodynamics Code with MPI. In W.Jäger E.Krause, editor, *High Performance Computing in Science and Engineering '99*, pages 52 – 61. Springer, 2000.
18. T. Bubeck, M. Hipp, S. Hüttemann, S. Kunze, M. Ritt, W. Rosenstiel, H. Ruder, and R. Speith. Parallel SPH on Cray T3E and NEC SX-4 using DTS. In W. Jäger E. Krause, editor, *High Performance Computing in Science and Engineering '98*, pages 396 – 410. Springer, 1999.
19. T. Bubeck, M. Hipp, S. Hüttemann, S. Kunze, M. Ritt, W. Rosenstiel, H. Ruder, and R. Speith. SPH test simulations on a portable parallel environment. In W. Kluge, editor, *Physics and Computer Science*, pages 139 – 155. Department of Computer Science, University of Kiel, 1999.
20. J. Wood, K. Horne, G. Berriman, R. Wade, D. O'Donoghue, and B. Warner. High-speed photometry of the dwarf nova Z Cha in quiescence. *mn*, 219:629–655, 1986.

High-accuracy Simulation of Density Driven Flow in Porous Media

Peter Bastian, Klaus Johannsen, Stefan Lang, Christian Wieners,
Volker Reichenberger, Gabriel Wittum

Interdisziplinäres Zentrum für Wissenschaftliches Rechnen
Technical Simulation Group
Universität Heidelberg
Im Neuenheimer Feld 368
69120 Heidelberg

Abstract. A three-dimensional test case for the density driven flow equations in porous media recently proposed by Oswald, Scheidegger and Kinzelbach is investigated numerically. It was shown in [14] that the results from the physical experiment can be reproduced if parameters are adjusted correctly and that a mathematical benchmark can be defined as an idealization of the physical experiment. Intensive numerical investigations were carried out, with calculations on up to 16.7 million grid points that required the efficient solution of linear systems with more than 35 million unknowns.

1 Introduction

Density driven flow is characterized by flow patterns that are influenced by density differences in the fluid system. This flow type appears in porous media in such highly relevant applications as sea water intrusion in coastal aquifers, upconing of saline waters from deep aquifers, flow around salt domes used as repositories, propagation of dense plumes emanating from land fills and geothermal heat production. If the effects of density driven flow in these situations are neglected, the results can be severely wrong even for small density differences. However, since the resulting equations are difficult to solve numerically because of their coupled nonlinear nature, too simplistic models are often employed. Accurate simulation of density driven flow still poses a considerable challenge even to combinations of todays most advanced software tools and computing resources.

In this paper a recently proposed benchmark problem is investigated. This test case has been studied experimentally and computationally, and only by employing massive parallel computing was it possible to simulate the fluid flow in a fully satisfactory way.

After introducing the experiment we describe the governing equations of density driven flow and explain how the mathematical benchmark problem was derived from the experiment. Then we describe the numerical methods employed in the solution process and present results from the simulation.

2   The saltpool experiment

The three-dimensional test cases proposed by Oswald, Scheidegger and Kinzelbach describe a series of laboratory experiments for a typical variable-density problem which involves stable layering of saltwater below freshwater and time-dependent upconing of the saltwater due to water discharge. If the density differences between saltwater and freshwater are varied, the flow pattern changes significantly due to gravity effects.

Figure 1 shows the setup of the experiment. In phase one, the cube of side length L filled with a homogeneous porous medium of porosity n and freshwater is recharged with saltwater from the hole in the lower center of the cube, causing a discharge of freshwater through the upper four holes. In the second phase all holes are closed and the saltwater layer equilibrates to an almost horizontal distribution with a thin mixing zone. In the third phase the freshwater is recharged through the inflow hole at a constant rate, with outflow only through the hole in the opposing upper corner. The governing parameters for this phase are shown in Table 1. The salt mass fraction varies over time at the outflow hole. Two initial maximum salt mass fractions are considered here, 1% and 10% which will be labeled *saltpool case 1* and *saltpool case 2*.

The parameters in Table 1, are:

V_s	Volume of saltwater recharged during phase 1.
D_m	Molecular diffusion coefficient.
K	Intrinsic permeability.
α_l, α_t	Longitudinal and transversal dispersion lengths.
ρ_f, ρ_s	Minimum and maximum densities.
μ_f	Dynamic viscosity of fresh water.

We refer to the values of the salt mass fraction at the corresponding points as the vectors $g_1^{exp} \in \mathbb{R}^{26}$ for saltpool case 1 and $g_2^{exp} \in \mathbb{R}^{31}$ for saltpool

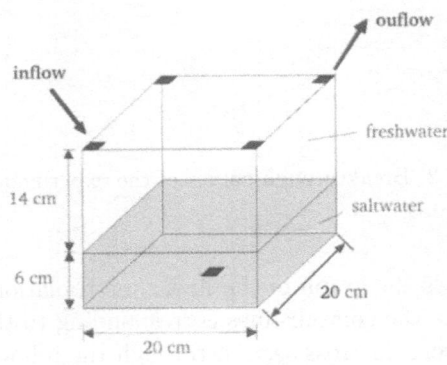

Fig. 1. The saltpool experiment

Table 1. Parameters of the experiment

	saltpool, case 1			saltpool, case 2			units	
	ref.	min.	max.	ref.	min.	max.		
c_{max}	1%			10%				$-$
V_s	8.64			8.9964			10^{-4}	m^3
T_3	8412	8410	8414	9594	9592	9596		s
L	200	199	201	ditto			10^{-3}	m
Q_3	1.89	1.872	1.908	1.83	1.802	1.848	10^{-6}	$m^3 s^{-1}$
n	0.372	0.370	0.375	ditto				$-$
D_m	10	7	12	ditto			10^{-10}	$m^2 s^{-1}$
K	10	8.9	11	ditto			10^{-10}	m^2
α_l	1.2	0.6	1.5	ditto			10^{-3}	m
α_t	0.12	0.03	0.25	ditto			10^{-3}	m
ρ_f	998.23			998.23				$kg\, m^{-3}$
$\rho_s/\rho_f - 1$	7.6	7.0	8.2	73.5	72.5	74.5	10^{-3}	$-$
μ_f	1.002	0.93	1.02	ditto			10^{-3}	$kg\, m^{-1}s^{-1}$

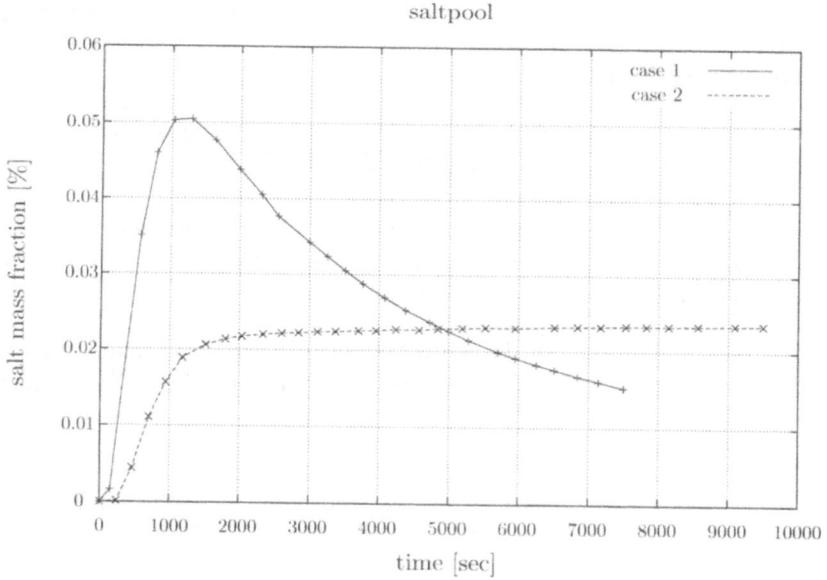

Fig. 2. Breakthrough curves of the experiment

case 2. Figure 2 shows the graph of the linear interpolation of these vectors vs. time. Fig. 4 shows the contour-lines corresponding to the values $0.1c_{max}$ and $0.5c_{max}$ in the vertical cross section through the inflow and the outflow hole at the end of phase 3 together with numerical results obtained there.

For a complete description of the experiments performed see [18].

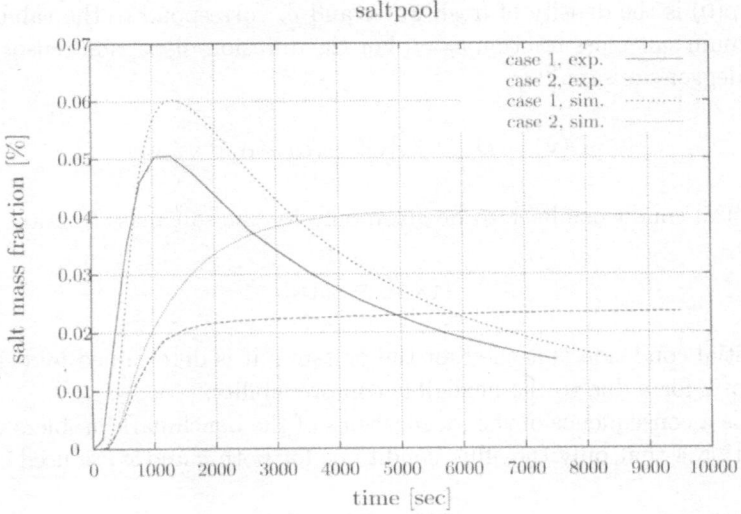

Fig. 3. Breakthrough curves obtained by d^3f for the parameters from Table 1.

3   Basic equations

The derivation of the governing equations based on mass conservation principles can be found in [4, 11, 15]; we only state the resulting system:

$$n\partial_t\rho(c) + \nabla \cdot (\rho(c)\mathbf{v}) = \rho(C)Q, \tag{1}$$

$$n\partial_t(\rho(c)c) + \nabla \cdot (\rho(c)(c\mathbf{v} - \mathbb{D}\nabla c)) = \rho(C)CQ, \tag{2}$$

with the source and sink terms $Q(x)$, the density $\rho(c)$, the diffusion-dispersion tensor $\mathbb{D}(\mathbf{v})$ and the the mass averaged velocity $\mathbf{v}$ given by Darcy's law

$$\mathbf{v} = -K/\mu(c)(\nabla p - \rho(c)\mathbf{g}), \tag{3}$$

where $\mu(c)$ is the dynamic viscosity and $\mathbf{g}$ the vector of gravity. Finally, C either denotes the salt mass fraction of the injected brine ($Q > 0$) or $C \equiv c$ in case of a sink ($Q < 0$). Inserting (3) into (1) and (2) leads to a nonlinear system of two partial differential equations for the unknown salt mass fraction c and the pressure p. To close the system, material laws for ρ and μ and an expression for the diffusion-dispersion tensor have to be provided. For ρ and μ different dependencies are used in the literature (see [10, 11]). We choose

$$\frac{1}{\rho} = \left(1 - \frac{c}{c_{max}}\right)\frac{1}{\rho_f} + \frac{c}{c_{max}}\frac{1}{\rho_s}, \tag{4}$$

$$\mu = \mu_f(1 + 1.85c - 4.1c^2 + 44.5c^3). \tag{5}$$

$\rho_f = \rho(0)$ is the density of fresh water and ρ_s correspond to the value at the maximum salt mass fraction c_{max}. For the diffusion-dispersion tensor we use Scheidegger's ansatz [20]

$$\mathbb{D}(\mathbf{v}) = D_m + \alpha_t |\mathbf{v}| \mathbb{I} + (\alpha_l - \alpha_t) \mathbf{v}\mathbf{v}/|\mathbf{v}|.$$

Initial conditions have to be given only for the salt mass fraction

$$c(\mathbf{x}, 0) = c_0(\mathbf{x}).$$

No initial condition is needed for the pressure; it is determined by (1) which is elliptic for p due to the negligible compressibility.

It is a consequence of the idealizations of the benchmark problem defined in section 4 that only (no-)flux conditions for both c and p are used,

$$\mathbf{n} \cdot (\mathbf{v}c - \mathbb{D}\nabla c) = 0, \qquad \mathbf{n} \cdot \mathbf{v} = 0. \tag{6}$$

A detailed discussion of boundary conditions can be found [15].

4   The benchmark problem

While physical benchmarks consist of precisely measured experiments under laboratory conditions, a mathematical benchmark defines a mathematical problem that is an approximation of the physical problem. The mathematical benchmark doesn't contain errors due to inappropriate physical modeling, the benchmark solution is compared to the exact solution of the mathematical problem. In most cases the exact solution of the problem does not exist in analytical form; in these cases highly accurate reference solutions have to be employed for comparisons.

Based on the experiments, two benchmark problems corresponding to the setup of saltpool case 1 and saltpool case 2 were defined in [14]. In order to derive well-defined mathematical problem descriptions some idealizations have to be made. Additionally, some parameters of the experiment have to be adapted if the solution of mathematical descriptions should reproduce the experimental results.

For the benchmark problem only the third phase of the experiment is considered. The definition of the mathematical problem consists of the equations to be solved, the material laws, the geometry and the initial and boundary conditions. The major part has already been defined in the previous sections. We approximate the geometry by a cube of side length $L = 0.2$m. The inflow and outflow holes are not modeled geometrically but by inserting point sinks and sources at the corresponding corners of the cube. The initial condition for the salt mass fraction c is the continuous, piecewise linear function defined by

$$c(\mathbf{x}, 0) = c_{max} \begin{cases} 1 & \text{if } x_3 \leq x_m - \omega/2 \\ 1/2 - (x_3 - x_m)/\omega & \text{if } x_m - \omega/2 < x_3 < x_m + \omega/2, \\ 0 & \text{if } x_m + \omega/2 \leq x_3 \end{cases}$$

(7)

where $x_m = V_s/nL^2$ denotes the vertical position of the initial mixing zone, assuming a perfect horizontal interface between saltwater and freshwater. ω is the width of the transition zone, and x_3 the z-component of $\mathbf{x}$. x_m depends on the parameters of the experiment while ω is a new parameter for the test cases. It has been derived from the experimental results and we choose the value $\omega = 8mm$ in both cases.

Fig. 3 shows the breakthrough curves from a numerical simulation using the values from Table 1 on a mesh with 262.144 grid points and a fixed time step size of $\Delta t = 70.1s$ for saltpool case 1 and $\Delta t = 39.975s$ for saltpool case 2. The computed breakthrough curves exhibit a correct qualitative behavior, but only a modestly correct approximation of the measured curves. Similar observations have already been made by [1, 17, 18]. In order to match the experimental results and to obtain a mathematical benchmark that models the experiment, a parameter fit was carried out by [14]. There details can be found on how the parameters K, n and α_t have to be modified and how the actual mathematical benchmark is defined.

5   Numerical simulation

The program package d^3f [5] is used for solving equations (1)–(3) from the benchmark problem. d^3f has been designed to simulate variable density flow in porous media in two- and three-dimensional complex geometries. It consists of three major parts plus various additional tools. The major parts are the *preprocessor*, the *simulator* and the *postprocessor*. The *preprocessor* supports the interactive design of the geometry and the specification of the physical parameters. The *simulator* is a tool to discretize and solve the density driven flow equations in porous media. It includes tools for the grid generation. Finally, the *postprocessor* is a highly efficient visualization and data extraction tool. It is based on the software package $GRAPE$ (see [16]).

The *simulator* is based on the software package UG [3], a toolbox for discretizing and solving partial differential equations on unstructured grids. *simulator* uses all the advanced features of UG ranging from handling complex geometries in two and three dimensions to adaptive multigrid methods. More precisely, *simulator* can handle polygonal shaped domains in 2d and polyhedral shaped domains in 3d. UG can handle hybrid meshes, i.e. triangles and quadrilaterals in 2d and tetrahedra, pyramids, prisms and hexahedra in 3d. In combination with error estimators the grids can be refined and de-refined locally and the time step size can also be variably adjusted. *simulator* uses a fully implicit/fully coupled solution technique for the vertex-centered finite volume discretization with a consistent velocity approximation. A monotonous first order and a non-monotonous second order spatial discretization are available. In the temporal discretization a first order implicit Euler or a second order diagonally implicit Runge-Kutta method can be applied. The solution strategy for the nonlinear algebraic equations is based on a Newton method using a multigrid iteration for the linear sub-problems. All parts of *UG* as well as *simulator* are parallelized so that *simulator* can be employed on a wide range of computers from PCs to massively parallel supercomputers.

The simulations in the next section use a hexahedral grid that is well suited for the cubic geometry. A hierarchy of grids with nine levels ranging from 0 to 8 is employed for the multigrid algorithm used for the solution of the linearized equations. The grids are created by successive uniform refinement. Table 2 shows the grid spacing, the number of hexahedra and the number of grid points n on the different levels. The spatial discretization of (1), (2) uses vertex-centered finite-volumes for a continuous, piecewise trilinear ansatz-space. No stabilization is used for the convective terms and a consistent velocity approximation [6] is applied. The discretization is locally mass-conserving and second order consistent for the unknowns c and p. The boundary conditions (6) are applied in a weak sense.

For the time discretization we use a second order diagonally implicit Runge-Kutta method with time step size Δt. The scheme is second order consistent and unconditionally stable both in the L_2 and the L_∞ norm [9,19]. The time discretization is described in detail in [14].

For the salt mass fraction the initial distribution $c(\mathbf{x}, 0)$ is given by (7). Since this function cannot be represented exactly on the grids a L_2-projection is used to establish the initial conditions. This results in a non-monotonous function with an exact total initial salt mass.

Table 2. The hierarchy of grids

Level i	$h_i[mm]$	# of hexahedra	# of grid points
0	200	1	8
1	100	8	27
2	50	64	125
3	25	512	729
4	12.5	4,096	4,913
5	6.25	32,768	35,937
6	3.125	262,144	274,625
7	1.5625	2,097,152	2,146,689
8	0.78125	16,777,216	16,974,593

For each time step a system of nonlinear algebraic equations has to be solved twice; here this is done by a Newton method with a nonlinear defect reduction by a factor of 10^{-6} in the Euclidean norm. The linear system is obtained by an exact calculation of the Jacobian—numerical differentiation introduces errors that destroy the superlinear convergence. Due to the non-monotonous spatial discretization and the full linearization including the dependence on e.g. the Darcy velocity within the transport equation, the Jacobians are not M-matrices. Standard smoothing procedures cannot be applied successfully. Therefore a modified point-block SSOR-iteration is used instead [12]. On the average 9 linear V-cycle multigrid iterations are required to solve a linear system. The multigrid method is accelerated using Bi-CGStab [2, 21].

6 Numerical Results

In order to provide reliable reference solutions for the benchmark in absence of analytical solutions the convergence order of the method of the preceding section was investigated numerically. The time step sizes for the two test cases are

$$\Delta t_1^i = (1/2)^i \, 70.1 \text{sec and } \Delta t_2^i = (1/2)^i \, 39.975 \text{ sec}, \quad i = 0, 1, 2, \qquad (8)$$

where i denotes the time grid level.

In Figure 4 we show the iso-lines of the salt mass fraction (experiment: 0.1%, 0.5%; simulation: 0.1%, 0.5%, 0.9%) for test case 1 and the salt mass fraction (experiment: 1%, 5%; simulation: 1%, 5%, 9%) for saltpool test case 2. For saltpool case 1 a time step width of $\delta t = 17.525$ sec was used and a time step width of $\delta t = 19.9875$ sec for saltpool case 2. Both solutions were computed on grid level 7. The salt mass fraction is evaluated in the diagonal cross section passing vertically through the inflow and outflow hole (see Fig. 2). The numerical results are in excellent agreement with the experimental data.

To illustrate the grid convergence we compare a sequence of breakthrough curves obtained on successively refined grids, both in space and time. We have chosen the following values $g_s^{i,j}$ for the linear interpolation of the breakthrough curves

$$
\begin{aligned}
&g_s^{-1,4} && \text{grid level 4, time step size } -1 \\
&g_s^{0,5} && \text{grid level 5, time step size } 0 \\
&g_s^{1,6} && \text{grid level 6, time step size } 1 \\
&g_s^{2,7} && \text{grid level 7, time step size } 2 \\
&g_1^{0,8} && \text{grid level 8, time step size } 0
\end{aligned}
$$

The time step size parameter is used according to (8) and s is either 1 for saltpool case 1 or 2 for saltpool case 2. For the coarsest spatial grid (level 4) we have chosen a time step size of $\Delta t_1^{-1} = 140.2s$ resp. $\Delta t_2^{-1} = 79.95s$. For

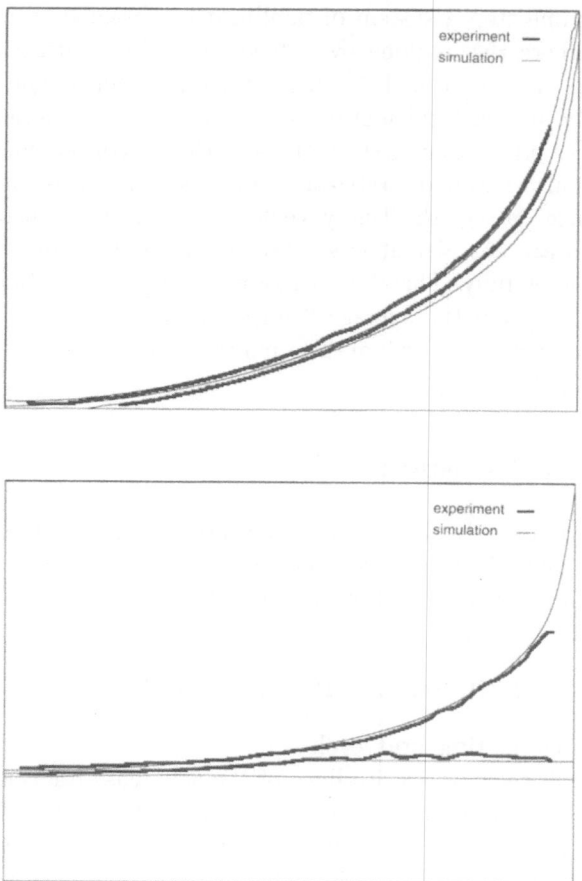

Fig. 4. Iso-lines of the salt mass fraction at $t = 8412\ sec$ for *saltpool, case 1*, Experiment: 0.1%, 0.5%; simulation: 0.1%, 0.5%, 0.9% (top) and iso-lines of the salt mass fraction at $t = 9594\ sec$ for *saltpool, case 2*, Experiment: 1%, 5%; simulation: 1%, 5%, 9%.

the analysis given above these time step sizes could not be adopted since they lead to convergence problems on finer spatial grids. A good approximation for *saltpool, case 1* is achieved already on the coarse grid levels (see Fig. 5). For *saltpool, case 2* a qualitatively correct solution is obtained only on finer grids, level ≥ 6 (also in Fig. 5). The size of the time step turned out to be of minor importance for the discretization scheme used.

The computations on grid level 8 with 16 million nodes represent the largest simulation up to date with software based on UG. The computation ran for 28 days on 128 processors of the IBM RS/6000 SP-256.

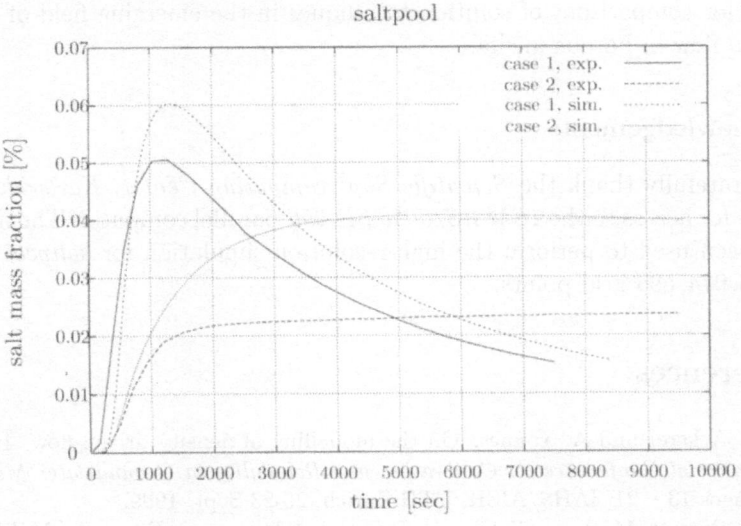

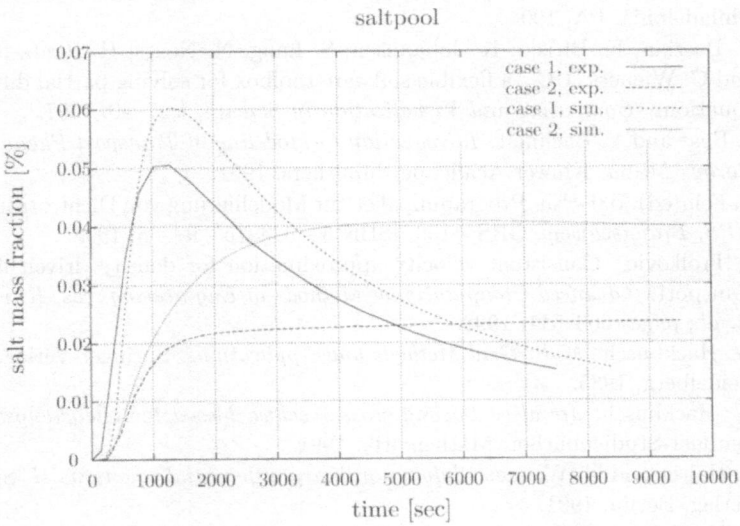

Fig. 5. Grid convergence of *saltpool, case 1* (top) and *saltpool, case 2* (bottom)

7   Conclusions

A careful analysis of experimental results and numerical simulations leads to
the definition of a mathematical benchmark that models a physical experiment. By employing advanced numerical simulation methods and parallel
computing it is possible to compute a reference solution that can serve as a

basis for comparisons of solution techniques in the emerging field of density driven flow in porous media.

Acknowledgements

We gratefully thank the *Scientific Supercomputing Center, Karlsruhe, Germany* for access to the *IBM RS/6000 SP-256* parallel computer. The machine has been used to perform the high-resolution simulation for *saltpool, case 2* on $16,974,593$ grid points.

References

1. P. Ackerer and A. Younes. On the modelling of density driven flow. In *International Conference on Calibration and Reliability in Groundwater Modelling*, pages 13 – 21. IAHS/AISH, ETH Zürich, 20-23 Sept. 1999.
2. R. Barrett, M. Berry, T. Chan J. Demmel, J. Donato, J. Dongarra, V. Eijkhout, R. Pozo, Ch. Romine, and H. van der Vorst. *Templates for the Solution of Linear Systems: Building Blocks for Iterative Methods, 2nd Edition*. SIAM, Philadelphia, PA, 1994.
3. P. Bastian, K. Birken, K. Johannsen, S. Lang, N. Neuss, H. Rentz-Reichert, and C. Wieners. UG - a flexible software toolbox for solving partial differential equations. *Computing and Visualization in Science*, 1:27–40, 1997.
4. J. Bear and Y. Bachmat. *Introduction to Modeling of Transport Phenomena in Porous Media*. Kluwer Academic Publishers, 1991.
5. E. Fein(ed). d3f - Ein Programmpaket zur Modellierung von Dichteströmungen. *GRS, Braunschweig*, GRS - 139, ISBN 3 - 923875 - 97 - 5, 1998.
6. P. Frolkovic. Consistent velocity approximation for density driven flow and transport. *Advanced Computational Methods in Engineering, eds. R. van Keer et. al.*, pages 603–611, 1998.
7. W. Hackbusch. *Multi-Grid Methods and Applications*. Springer-Verlag, Berlin, Heidelberg, 1985.
8. W. Hackbusch. *Iterative Lösung grosser schwachbesetzter Gleichungssysteme*. Teubner-Studienbücher: Mathematik, 1991.
9. E. Hairer and G. Wanner. *Solving ordinary differential equations II*. Springer-Verlag, Berlin, 1991.
10. A.W. Herbert, C.P. Jackson, and D.A. Lever. Coupled groundwater flow and solute transport with fluid density strongly dependent upon concentration. *Water Resources Research*, 24:1781–1795, 1988.
11. Ekkehard O. Holzbecher. *Modeling Density-Driven Flow in Porous Media*. Springer-Verlag, Berlin, Heidelberg, 1998.
12. K. Johannsen. Modified SSOR - a smoother for non M-matrices. *in preparation*.
13. K. Johannsen. On the stability of the Crank-Nicolson-scheme for density driven flow in porous media. unpublished.
14. K. Johannsen, W. Kinzelbach, S. Oswald, G. Wittum *Numerical Simulation of density driven flow in porous media* Advances in Water Resources, submitted.
15. Anton Leijnse. *Three-dimensional modeling of coupled flow and transport in porous media*. PhD thesis, University of Notre Dame, Indiana, 1992.

16. A. Wierse M. Rumpf. GRAPE, eine objektorientierte Visualisierungs- und Numerikplattform. *Informatik Forschung und Entwicklung*, 7:145–151, 1992.
17. S. Oswald. *Dichteströmungen in porösen Medien: Dreidimensionale Experimente und Modellierung*. PhD thesis, Institut für Hydromechanik und Wasserwirtschaft, ETH Zürich, 1998.
18. S.E. Oswald, M. Scheidegger, and W. Kinzelbach. A three-dimensional physical benchmark test for verification of variable-density driven flow in time. *Water Resources Research*, submitted.
19. R. Rannacher. Numerical analysis of nonstationary fluid flow. Technical Report SFB 123, Preprint 492, University of Heidelberg, Germany, November 1988.
20. A.E. Scheidegger. General theory of dispersion in porous media. *Journal of Geophysical Research*, 66:3273, 1961.
21. H. van der Vorst. Bi-CGSTAB: A fast and smoothly converging variant of bicg for the solution of nonsymmetric linear systems. *SIAM J. Sci. Stat. Comp.*, 13:631 – 644, 1992.

ParWave: Parallel Wavelet Video Coding*

M. Feil, R. Kutil, R. Norcen, and A. Uhl

RIST++ and Dept. of Scientific Computing, Salzburg University, Austria

Abstract. In this work, we discuss parallel algorithms for three distinct approaches for wavelet-based video coding and the performance of their corresponding MPI implementations on the HLRS Cray T3-E.

1 Introduction

In recent years there has been a tremendous increase in the demand for digital imagery. Applications include consumer electronics (Kodak's Photo-CD, HDTV, SHDTV, Video-on-Demand, and Sega's CD-ROM video game), medical imaging (digital radiography), video-conferencing and scientific visualization. The problem inherent to any digital image or digital video system is the large bandwidth required for transmission or storage.

Unfortunately, many compression techniques and applications demand execution times that are not possible using a single serial microprocessor [19], which leads to the use of general purpose high performance computers for such tasks (beside the use of DSP chips, FPGAs, or application specific VLSI designs). In this context, several papers have been published describing real-time image and video coding on general purpose parallel architectures – see for example JPEG [4], MPEG-1,2,4 [1], H.261 [10], vector quantization [12], and fractal compression [22].

Image and video coding methods that use wavelet transforms [21] have been successful in providing high rates of compression while maintaining good image quality and have generated much interest in the scientific community as competitors to DCT based compression schemes. With the finalization of the wavelet based JPEG2000 standard [2] and the inclusion of a wavelet algorithm for synthetic/natural hybrid coding in MPEG-4 [20] there is no doubt left that wavelet compression has to be considered state of the art nowadays. Therefore, a thourough investigation of parallel versions of these algorithms seems mandatory.

A significant amount of work has been devoted to wavelet/subband based video coding (see e.g. [8] for 3-D wavelet/subband coding and [13] for 2-D coding with motion estimation). Also, the matching pursuit algorithm has been applied successfully to video compression [17] – although this algorithm is fairly different as compared to wavelet codecs, the underlying basis functions and some fundamental ideas are closely related to wavelet-based coding.

---

* This work has been partially supported by the Austrian Science Fund (project 13903) and by the Austrian Academy of Sciences.

Within the project "ParWave", we have focused onto three areas related to wavelet-based video coding. In Section 2, we investigaste approaches to parallelize a 2-D wavelet packet video coding system with respect to different granularity levels. Section 3 covers parallel algorithms for 3-D SPIHT video coding. In Section 4, we discuss parallel algorithms for a Matching Pursuit intra frame codec based on image tiling.

2   Wavelet Packet Video Coding with Motion Compensation

Wavelet packets [23] represent a generalization of the method of multiresolution decomposition and comprise the entire family of subband coded (tree) decompositions. Whereas in the wavelet case the decomposition is applied recursively to the coarse scale approximations only (leading to the well known (pyramidal) wavelet decomposition tree), in the wavelet packet decomposition the recursive procedure is applied to all subbands, which leads to a complete wavelet packet tree (i.e. a quadtree in the 2D case) and more flexibility in frequency resolution.

The wavelet packet "best basis algorithm (bba)" [3] performs an adaptive optimization of the frequency resolution of a complete wavelet packet decomposition tree by selecting the most suitable frequency subbands for signal compression.

Recently, wavelet packet based compression methods have been developed [14] which outperform the most advanced wavelet coders (e.g. SPHIT [18]) significantly for textured images in terms of rate-distortion performance. Therefore, wavelet packet decomposition currently attracts much attention for 2-D and 3-D video coding as well.

This section deals with the parallelization of wavelet packet based video coder: the generation of an entire wavelet packet tree, the choice of the best representation of the input signal according to a predefined cost measure (e.g. the entropy) and the coding of the generated wavelet coefficients by using a WP zerotree coder.

In the following sections we suppose to have a video input stream of f_size frames ($0 \leq f < f\_size$). These frames should be processed by assorting them into GOPs of gop_size ($0 \leq g < gop\_size$) which are composed of 1 I-frame and $gop\_size - 1$ P-frames (see Fig.2). As motion compensation algorithm we

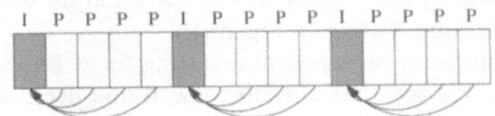

Fig. 1. A sequence of GOPs

employ the arithmetic difference between I- and P-frame but all results are also extendable to more sophisticated MC solutions.

2.1 Parallelization methods

Examing the sequence of frames in the encoding video stream (see Fig.2) we find three levels for parallelization, which are ordered by their granularity: intra-frame parallelization, frame-to-frame parallelization and group-of-picture parallelization.

Intra-frame parallelization The intra-frame parallelization of zerotree coding is described in all details in Section 3, so the following descriptions will focus on the decomposition step of a WP video coder. For the intra-frame parallelization of a WP transform we find two possible partitioning approaches:

- subband based (SB) partitioning. On a architecture with a number of processor elements equal to a power of 4 this kind of partitioning is straight forward: each PE p (out of a pool of 4^a PEs) gets input data corresponding to a rectangle area of size 2^{Xmax-a} by 2^{Ymax-a} (2^{Xmax} by 2^{Ymax} is the size of an input frame) and calculates the corresponding subbands situated in this area. During the first a levels each subband is shared by more than one PE. This implies additional communication between neighbouring PE after the calculation step to redistribute the results. During level a and higher, subbands and also the four children of each of them reside on each PE and no additional communication is needed. The generalization for an architecture with an arbitrary number of PEs implies a sophisticated variant of the host-node principle and was presented in all details in [6,7].
- stripe (ST) partitioning. Due to the fact that the parallel SPIHT algorithm is based on the locality of zerotrees (as seen in Fig.4(c) and explained in Section 3), this is the optimal partioning scheme for subsequent parallel zerotree coding. The input data is splitted into stripes of equal sizes and distributed among the PEs which perform the filtering in parallel on local data. Communication is needed in every step in order to provide the required border data to each PE.

Frame-to-frame parallelization In the next coarser level of parallelization we assign a single frame to each PE in a round-robin fashion. Because each P-frame is referring to its preceding I-frame, we find dependence between the PEs processing the I-frames (grey in Fig.2(a)) and all other PEs which are competent for P-frames (white in Fig.2(a)).

Figure 2(a) shows an experimental setting for 6 PEs, 100 frames and a GOP size of 10. We see that the first few frame cycles are needed as an initialization phase to organize the arrangement of I- and P-frames but after approx. 3 GOP the frame distribution is organized in an efficient manner and only a small parallelization overhead is detectable.

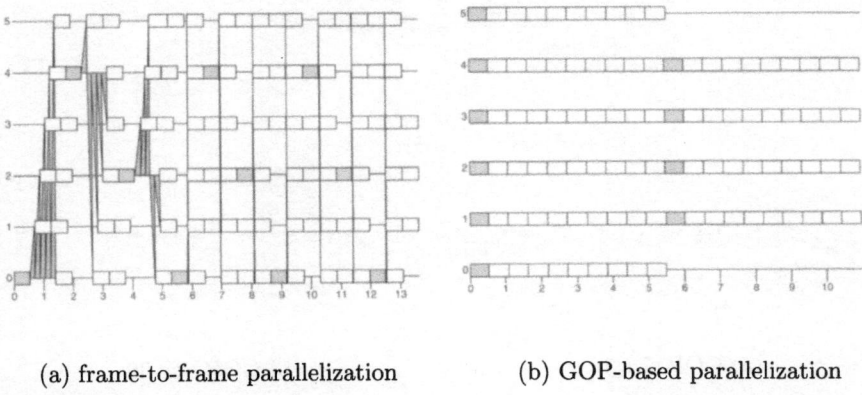

(a) frame-to-frame parallelization   (b) GOP-based parallelization

Fig. 2. Coarse grained parallelization methods

Group-of-picture parallelization In group-of-picture based parallelization we assign independent GOPs to each PE. Obviously, we don't have any additional communication between PEs (Fig.2(b)), but the memory demand is rather high (one PE which takes over a synchronising role has to cache the incoming video frames and rearrange the outgoing bitstream) and the processing of a GOP is delayed in relation to the number of PEs. Therefore, this kind of parallelization is more suitable for off-line video processing than real-time video processing.

Additionally, we see in Fig. 2(b) that the GOP based algorithm has a strong tendency to load unbalancing if there are not enough GOPs (which is equal to not enough input frames) to fill up all PEs evenly. In the worst case we find the same execution time for $p_{size} * n + 1$ GOPs ($n \in \mathbf{N}$) as for $p_{size} * (n + 1)$ GOPs. On the other hand, this algorithm is the most efficient of all presented parallelization strategies if the input stream is not bounded or very large (as often found in practice): there is no parallelization overhead at all, so the theoretical speedup would be linear.

2.2   Experimental Results

In this section we show timing results for the encoding of a 200 frame video. Each GOP (of size 10 resp. 30) starts with one I-frame, followed by 9 resp. 29 P-frames which are computed by a basic arithmetic difference between the reconstructed I-frame and the actual frame and afterwards decomposed. As input video stream we used the "coastguard" sequence, with 176 by 144 pixels and 4 levels of decomposition (using the "Daubechies 6" Wavelet).

In Fig. 3 we find speedup results for experiments on the HLRS Cray T3E for GOP-sizes of 10 resp. 30 frames.

For a GOP-size of 10 frames (Fig. 3(a)) we find both GOP-based and frame-based algorithms equally performing very well on both architectures.

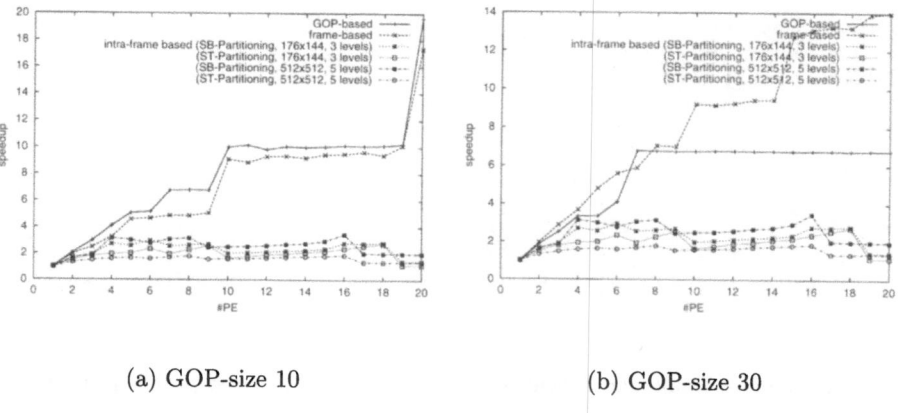

(a) GOP-size 10   (b) GOP-size 30

Fig. 3. Cray T3E

The plateau of the GOP-based algorithm after 10 PEs is due to the non-uniform distribution of 20 GOPS on 11...19 PEs (as seen in Fig.2(b)), but interestingly also the frame-based algorithms exhibits this effect. This results from dependencies between I- and P-frames which for certain settings are solved very efficiently in the initialization phase, but remain decisive for other settings.

When the GOP-size is raised to 30 frames (Fig.3(b)), the plateau mentioned above is already detectable after 7 PEs for the GOP-based algorithm whereas the frame-based algorithm still scales quite effectively because the delaying I-frame appears less frequent. Comparing this behaviour of both coarse grained parallelization strategies we see that the frame based algorithms performs the better the *larger* a GOP is, but is quite independent from the overall number of input frames. The GOP based algorithm instead is sensitive to the *number* of GOPs (and therefore tightly coupled to the number of input frames): if there are enough GOPs available, the GOP based parallelization performs optimal in terms of speedup.

The intra-frame based parallelizations show a rather poor performance in both settings, because the additional communication needed to organize the distribution of a single frame between all PEs cancelles the calculation power of numerous PEs.

Consequently, the algorithm of choice is the GOP-based approach if the available memory is large enough to buffer the incoming video stream and if real time processing is not required. Frame based parallelization works very well likewise if the group-of-pictures size is large enough to bypass dependecies of I- and P-frames. Of course it is also possible to mix these two strategies: use the GOP based algorithm as long as enough frames for a balanced load are available; after that continue with frame based processing. The intra-frame based parallelizations do not scale very well for large processor pools and are therefore not recommended.

3   Video Coding using 3-D SPIHT

Most video compression algorithms rely on 2-D based schemes employing motion compensation techniques. On the other hand, rate-distortion efficient 3-D algorithms exist which are able to capture temporal redundancies in a more natural way. Unfortunately, these 3-D algorithms often show prohibitive computational and memory demands (especially for real-time applications). Therefore, MIMD architectures seem to be an interesting choice for such an algorithm.

A significant amount of work has already been done on parallel wavelet transform [9,15]. Here we concentrate on the parallelisation of the coding part. As opposed to [5] we will produce a bit-stream that is compatible to the sequential 3-D variant [8] of the SPIHT algorithm [18].

3.1   Parallel Wavelet Transform and Zero-Trees

The wavelet filtering is performed in parallel on local data (stripe partitioning in the time domain is used). Before each decomposition step, border data has to be exchanged between neighbouring PEs due to the filter length. After that, transformed data are found distributed as shown in Fig. 4.

Zero-tree based algorithms arrange the coefficients of a wavelet transform in a tree-like manner, i.e. each coefficient has a certain number of child coefficients in another sub-band (mostly 4 in the 2-D, 8 in the 3-D case, see Fig. 4(c)).

Furthermore, a zero-tree is a sub-tree which entirely consists of insignificant coefficients. The significance of a coefficient is relative to a threshold which plays an imported role in the SPIHT algorithm ($\mathrm{sig}(c) \Leftrightarrow |c| \geq$ threshold). The statistical properties of transformed image or video data (self-similarity) ensures the existence of many zero-trees. With the help of these zero-trees, sets of insignificant coefficients can be encoded efficiently. We will see that sometimes the root coefficient of the subtree (or even its direct offspring) does not have to be insignificant.

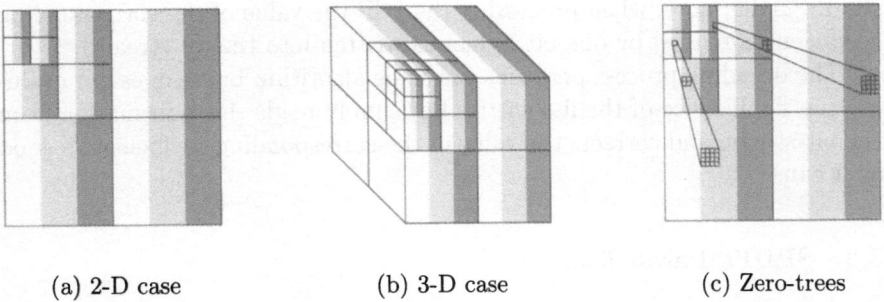

(a) 2-D case   (b) 3-D case   (c) Zero-trees

Fig. 4. Distribution of coefficients or list entries. Different colours indicate different PEs. (c) shows that zero-trees are local objects.

Zero-trees can be viewed as a collection of coefficients with approximately equal spacial position. While this fact implies that the coefficients significances are statistically related which is exploited by the SPIHT algorithm, this also means that zero-trees are local objects corresponding to the data distribution produced by the parallel wavelet transform (see Fig. 4(c)). This can be exploited by the parallelisation of the zerotree algorithms.

3.2   The SPIHT Algorithm

Although the SPIHT algorithm is sufficiently explained in the original paper [18] it is helpful in this context to reformulate the algorithm.

In the beginning the threshold is bigger than all the coefficients. Thus, all coefficients are insignificant. Then the threshold is repeatedly divided by 2 and the changes in significance have to be coded. Before and after each step (refinement step) the significance of the coefficients is represented by three lists:

LIS List of insignificant set of pixels. This list includes all roots of zero-trees. An entry in this list can be of two types: Type A – all descendants are insignificant, Type B – all descendants except the direct offspring are insignificant.
LIP List of insignificant pixels. This list includes coefficients that are insignificant but not part of a zero-tree corresponding to an LIS entry.
LSP List of significant pixels.

When processing a refinement step each entry of each list has to be tested for a change of significance and possibly be moved to another list. All entries inserted at the end of a list are also processed in the same refinement step until no more new entries are left.

Possible list entry transitions are LIS → LIS, LIS → LIP, LIS → LSP, LIS → BS, LIP → LSP, LIP → BS, LSP → BS where BS stands for "bit-stream" which means that each evaluation of a list entry has to be coded into the bit-stream (so the decoding process is able to reproduce the decision). Furthermore, when processing the LSP the value of the corresponding coefficient is refined by one bit (which is written into the bit-stream).

The decoding process performs the same algorithm but it does not evaluate the significance of the list entries but simply reads this information from the bit-stream and corrects the value of the corresponding coefficient as good as it can.

3.3   SPIHT Parallelisation

When parallelising the SPIHT algorithm we have to face the problem that it uses lists of coefficient positions and is therefore inherently sequential. The basic operations of the algorithm are: Moving an iterator all through a list, deleting elements at iterator position and appending elements at the end of a list. So the aim is to distribute the list so that each PE-local entry corresponds to a local coefficient where coefficients are distributed among the PEs as shown in Fig. 4.

For initial distribution this is a simple task but as coefficients are appended to the end of lists one has to provide a mechanism to indicate which parts of a list belong to which PE – or from a PEs view: where a sequence of local coefficients ends and parts of another PEs list should be inserted. This work is done by separators (see Fig. 5).

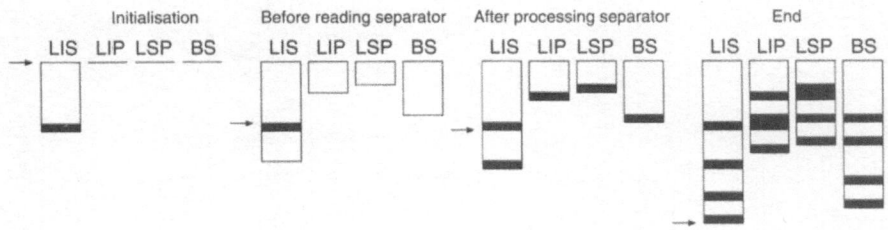

Fig. 5. Functionality of separators. Four states of the three lists and the bit-stream while processing the LIS.

The idea is to insert a separator at the end of each part of the list which entirely belongs to a single PE. So the initial distribution is to split the approximation sub-band into equal parts, assigning each part to a list on a single PE and appending a separator to the end of the list. From here on the sequential algorithm is performed locally with one exception: Each time the iterator meets a separator the separator is copied to the end of each destination list. A destination list is a list into which an entry could potentially have been inserted while processing previous iterator positions. Applying this principle the lists L_i on PE_i are split by separators into parts L_{ij} such that the assembled list $L_{11}L_{21}L_{31} \ldots L_{12}L_{22} \ldots$ is identical to the list the sequential algorithm would produce. The same is true for the bit-stream.

An important question is when the processing of a list is completed. Essentially, the procedure can stop if it has processed the last non-separator entry in the list. Unfortunately, this does not guarantee that each PE produces the same number of separators. But, this is a necessary condition for the correctness of the parallel algorithm because otherwise the correct order of the list-parts would be lost. Therefore, the global maximum number of separators has to be calculated (which unfortunately synchronises the PEs) and the lists have to be filled up with separators before continuing with the next list/refinement step.

The procedure of assembling the bit-stream (after collecting the PE-local bit-streams) is the only sequential part of the algorithm. Unfortunately, it gets more complicated and therefore consumes more calculation time when the number of PEs is increased.

3.4   Experimental Results

Video data size is always 864 frames with 88 by 72 pixels. The video sequence used here is the U-part of "grandma". The wavelet transform is performed up to a level of 3.

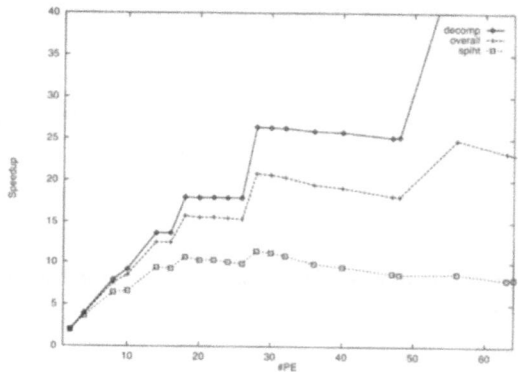

Fig. 6. Speedups for varying #PE and fixed compression rate (0.14 bpp).

Figure 6 shows speedups for a fixed compression rate: 0.14 bpp (bits per pixel, pixels in different frames are counted as different pixels). The speedup of the SPIHT parallelisation (without wavelet transform) seems to be limited by approximately 10. The reason for this is firstly the increased influence of the sequential part for higher #PE and secondly bigger load balancing problems for higher #PE due to the unevenly distributed complexity in different parts of the image.

Figure 7 shows speedup curves for fixed #PE and varying compression rate. Of course the wavelet decomposition is not dependent on the compression rate. Although the speedup of the coding part (SPIHT) increases with the bpp-value the overall speedup remains constant of drops slightly because the share in execution time of the coding part increases with the bpp-value.

4   Matching Pursuit Intra-frame Coding

Matching Pursuit Projection (MPP) is an approach to non-orthogonal transformation based compression. MPP or variants of it have been suggested

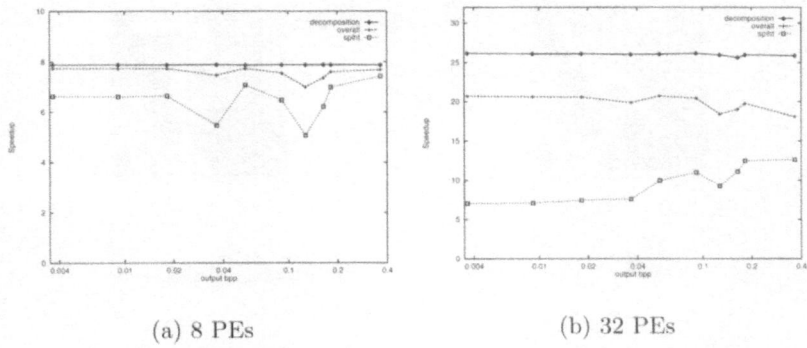

(a) 8 PEs   (b) 32 PEs

Fig. 7. Speedup for decomposition, SPIHT coding and overall speedup for varying compression rate.

for designing image compression and video compression algorithms and have been among the top-performing contributions to MPEG-4. However, the good rate-distortion performance of MPP has to be paid with an enormous computational complexity in the case of intra frame coding. Powerful sequential speedup techniques have been investigated in order to make the MPP approach useful to a wider range of applications (see next section). To further increase the performance of MPP image coding, we design several MIMD parallelizations of an MPP intra frame coder which is competitive in terms of rate-distortion but has a restricted usage due to its high computational demand. We present two coarse-grained parallelizations based on an image tiling and give experimental results for the HLRS CRAY T3E-900/512 and compare these with results on a Siemens HPC line cluster at PC^2, University of Paderborn.

4.1   MPP Image Coding

Let $D = \{g_\gamma\}_{\gamma \in \Gamma}$ be a dictionary consisting of arbitrary basis functions and f an image block. The *Matching Pursuit* algorithm [11] performs as follows:

1. $R^0 f = f$ and $n = 0$
2. $g_{\gamma_n} = \sup_{\gamma \in \Gamma} \langle R^n f, g_\gamma \rangle$
3. $R^{n+1} f = R^n f - \langle R^n f, g_{\gamma_n} \rangle g_{\gamma_n}$
4. If $|R^{n+1} f| < \varepsilon$ or $n > Max.Iterations$ then STOP
5. $n = n + 1$, goto 2

This algorithm selects iteratively the best dictionary function g_{γ_n}, that is, the function with the highest inner product of function and current residual $R^n f$, and subtracts it from the residual. The process stops when the norm of the residual is below a chosen threshold ε or a maximum number of functions (=iterations) $Max.Iterations$ is reached. The first iteration of a 2-dimensional MPP is illustrated in Fig. 8.

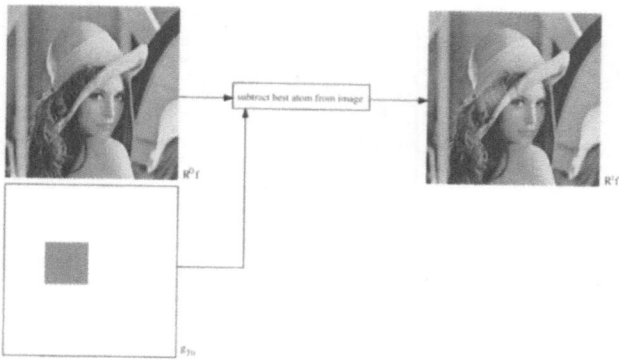

Fig. 8. Illustration of the first iteration: g_{γ_0} is the best matching atom (dictionary function) for residual $R^0 f$ (original image). $\langle R^0 f, g_{\gamma_0} \rangle\, g_{\gamma_0}$ is then subtracted from $R^0 f$ to retrieve residual $R^1 f$.

Schneider [16] introduced the Korn coder, a variant of MPP. The image compression quality of the Korn coder is comparable to state of the art wavelet image coders like the EZW or the SPIHT coder and clearly superior to JPEG (see Fig. 9).

The fundamental building blocks of the Korn coder are a set of 32×5 rotated Gabor functions. Every function can be translated in the xy-plane to be centered on an arbitrary position (u, v), all these possible translations form the dictionary D of the Korn coder [16]. All functions $g_\eta \in D$ of equal size build $Scale(s)$, $s \in \{0, 1, 2, 3, 4\}$ and have a support of 2^{s+2} pixels in both dimensions x and y.

Several speedup techniques are implemented in this coder to reduce the coding time for a 256×256 pixels image from a couple of days to only a few seconds:

Fig. 9. Compression ratio 35 achieved with JPEG (left: PSNR 24.11) and the Korn coder (right: PSNR 27.44): Image Lena 256^2.

Scale-by-Scale Speedup: In iteration m, the scale-by-scale-speedup always looks for a best matching atom $g_{\gamma_m} \in Scale(s)$ instead of investigating all functions $g_\eta \in D$. Encoding is done progressively from $Scale(4)$ atoms, describing the low-frequencies, to $Scale(0)$ atoms, describing the high-frequency information.

Block Decomposition with Overlap: The entire image is decomposed into so called logical blocks of size $p \times p$. To find a matching atom, only dictionary functions centered within one selected block (p_x, p_y) are investigated. In successive iterations, image blocks are covered consecutively row by row.

Updating Approach: Among consecutive iterations of a MPP, a lot of redundancy can be exploited: Inner products which are not influenced by the subtraction of g_{γ_n} can be reused in the next iteration $n + 1$. Even influenced inner products can be computed by one single product $(g_{\gamma_i} \in D)$:

$$\langle R^{n+1} f, g_{\gamma_i} \rangle = \langle R^n f, g_{\gamma_i} \rangle - \langle R^n f, g_{\gamma_n} \rangle \langle g_{\gamma_n}, g_{\gamma_i} \rangle .$$

Image and Basis Function Subsampling: A certain part of the image and dictionary functions can be subsampled by a factor *sub*. After looking for a best match in this subsampled space applying the same MPP techniques, the best match is subtracted from the original residual.

4.2 Parallel MPP Image Coding Based on Image Tiling

Let $P_1, \ldots, P_n$ be the set of available processors. The input image I is split recursively into n disjoint tiles (see Fig.10 left image) where every P_i gets one tile plus additional overlapping data (denoted local image data). According to the image tiling, P_i may have one or more neighbour processors. At this point we define 2 different types of parallel MPP iterations. To compute one parallel iteration for a $Scale(s)$ atom in *dependent-mode*, every P_i has to perform the following steps:

1. Find the best matching atom g_i of $Scale(s)$ applying the techniques of the Korn coder, and – if worth – subtract g_i from the local image data.
2. If the local image data of a neighbour processor of P_i is influenced by g_i, send update information to that processor.
3. Receive match information from neighbours to update the local image data.

To compute one parallel iteration in *independent-mode,* we skip steps 2 and 3 and restrict the subtraction of atom g_i to the image tile without overlapping data. Fig. 10 illustrates the image data after performing one iteration in independent-mode and one iteration in dependent-mode.

Offering these two types of parallel MPP iterations, we have a variety of possibilities to perform MPP in parallel. Independent-mode for atoms of all scales (denoted as brute-force method), though easy to implement, turns out

Fig. 10. Dependent- versus independent-mode with 3 processors: Residuals after performing the first parallel iteration (Left image: Initial encoding/region data; middle image: Dependent-Mode; right image: Independent-Mode).

to reduce the rate-distortion signifficantly for an increasing number of processors (Fig. 11(b)), since image data at the block-borders cannot be encoded efficiently thereby introducing severe blocking-artifacts. However, employing dependent-mode for atoms of all scales does not perform efficiently in parallel due to the high amount of synchronization especially for small scaled ($Scale(s)$, $s \in \{0, 1\}$) atoms. It turns out that regarding both, rate-distortion performance and parallel efficiency, it is best to encode scale 4,3, and 2 atoms in dependent-mode, while computing the two remaining update scales 0 and 1 (describing the details and high-frequencies) in independent-mode (denoted as best-dependent-mode). Even though the rate-distortion performance here is equal to the sequential case (Fig. 11(b)) the bitstreams are not identical. Figure 11(a) displays the speedup for the brute-force and best-dependent parallelization employing MPI on different architectures. We can observe that the high performance of the Cray's interconnection network is the obvious reason

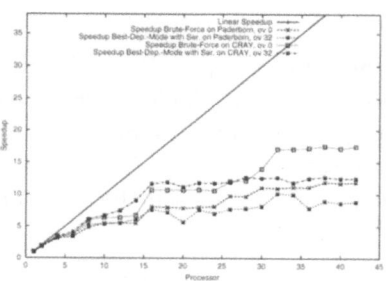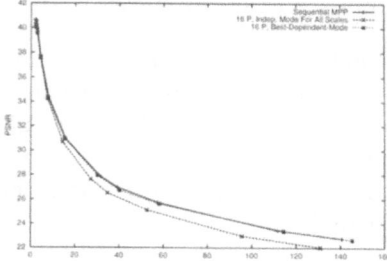

(a) Coarse-grained MPI on different architectures (*ov* stands for the overlap of an image tile).

(b) Rate-Distortion: The best-dependent-mode is able to keep the level of rate-distortion.

Fig. 11. Speedup and rate-distortion for parallel MPP image coding: Lena 256^2.

for its superior performance to the Siemens cluster. Although the brute-force parallelization is able to outperform the best-dependent-mode for a larger number of processors in terms of runtime efficiency, we must note that for these processor numbers the rate-disortion loss is already considerable and thus destroys the speedup advantage as compared with the best-dependent-mode. Anyhow, the scalability seems to be restricted for both presented parallel algorithms, which is due to the loss of compression quality on the one side (brute-force), and to the increasing amount of synchronization on the other side (best-dependent).

References

1. S.M. Akramullah, I. Ahmad, and M.L. Liou. Performance of software-based MPEG-2 video encoder on parallel and distributed systems. *IEEE Transactions on Circuits and Systems for Video Technology*, 7(4):687–695, 1997.
2. C. Christopoulos, A. Skodras, and T. Ebrahimi. The JPEG2000 still image coding system: an overwiew. *IEEE Transactions on Consumer Electronics*, 46(4):1103–1127, 2000.
3. R.R. Coifman and M.V. Wickerhauser. Entropy based methods for best basis selection. *IEEE Transactions on Information Theory*, 38(2):719–746, 1992.
4. G.W. Cook and E.J. Delp. An investigation of scalable SIMD I/O techniques with application to parallel JPEG compression. *Journal of Parallel and Distributed Computing*, 30:111–128, 1996.
5. C.D. Creusere. Image coding using parallel implementations of the embedded zerotree wavelet algorithm. In B. Vasudev, S. Frans, and S. Panchanathan, editors, *Digital Video Compression: Algorithms and Technologies 1996*, volume 2668 of *SPIE Proceedings*, pages 82–92, 1996.
6. M. Feil and A. Uhl. Algorithms and programming paradigms for 2-D wavelet packet decomposition on multicomputers and multiprocessors. In P. Zinterhof, M. Vajtersic, and A. Uhl, editors, *Parallel Computation. Proceedings of ACPC'99*, volume 1557 of *Lecture Notes on Computer Science*, pages 367–376. Springer-Verlag, 1999.
7. M. Feil and A. Uhl. Multicomputer algorithms for wavelet packet image decomposition. In *Proceedings of the International Parallel and Distributed Processing Symposium (IPDPS'2000)*, pages 793–798, Cancun, Mexico, 2000. IEEE Computer Society.
8. B.J. Kim, Z. Xiong, and W.A. Pearlman. 3-D set partitioning in hierarchical trees (3-D SPIHT). *IEEE Transactions on Circuits and Systems for Video Technology*, 8(10):1374–1387, 2000.
9. R. Kutil and A. Uhl. Optimization of 3-d wavelet decomposition on multiprocessors. *Journal of Computing and Information Technology (Special Issue on Parallel Numerics and Parallel Computing in Image Processing, Video Processing, and Multimedia)*, 8(1):31–40, 2000.
10. K.K. Leung, N.H.C. Yung, and P.Y.S. Cheung. Parallelization methodology for video coding – an implementation on the TMS320C80. *IEEE Transactions on Circuits and Systems for Video Technology*, 8(10):1413–1423, 2000.
11. S. Mallat and Z. Zhang. Matching pursuits with time-frequency dictionaries. *IEEE Trans. on Signal Process.*, 12(41):3397–3415, 1993.

12. M. Manohar and J.C. Tilton. Progressive vector quantization on a massively parallel SIMD machine with application to multispectral image data. *IEEE Trans. on Image Process.*, 5(1):142–146, 1996.
13. S.A. Martucci, I. Sodagar, T. Chiang, and Y.Q. Zhang. A zerotree wavelet video coder. *IEEE Transactions on Circuits and Systems for Video Technology*, 7(1):109–118, 1997.
14. F.G. Meyer, A.Z. Averbuch, and J.O. Strömberg. Fast wavelet packet image compression. *IEEE Trans. on Image Process.*, 9(5):792–800, May 2000.
15. O.M. Nielsen and M. Hegland. Parallel performance of fast wavelet transforms. *International Journal of High Speed Computing*, 11(1):55–74, 2000.
16. R. Norcen, P. Schneider, and A. Uhl. Matching pursuit projection — a parallelization. In G. Okša, R. Trobec, A. Uhl, M. Vajteršic, R. Wyrzykowski, and P. Zinterhof, editors, *Proceedings of the International Workshop on Parallel Numerics (Parnum 2000)*, pages 165–178, Bratislava, Slovakia, September 2000.
17. K. Osama, O. Al-Shaykh, E. Miloslavsky, T. Nomura, R. Neff, and A. Zhakhor. Video compression using matching pursuits. *IEEE Transactions on Circuits and Systems for Video Technology*, 9(1):123–143, 1999.
18. A. Said and W.A. Pearlman. A new, fast, and efficient image codec based on set partitioning in hierarchical trees. *IEEE Transactions on Circuits and Systems for Video Technology*, 6(3):243–249, 1996.
19. K. Shen, G.W. Cook, L.H. Jamieson, and E.J. Delp. An overview of parallel processing approaches to image and video compression. In M. Rabbani, editor, *Image and Video Compression*, volume 2186 of *SPIE Proceedings*, pages 197–208, 1994.
20. I. Sodagar, H.J Lee, P. Hatrack, and Y.Q. Zhang. Scalable wavelet coding for synthetic/natural hybrid coding. *IEEE Transactions on Circuits and Systems for Video Technology*, 9(2):244–254, 1999.
21. P.N. Topiwala, editor. *Wavelet Image and Video Compression*. Kluwer Academic Publishers Group, Boston, 1998.
22. A. Uhl and J. Hämmerle. Fractal image compression on MIMD architectures I: Basic algorithms. *Parallel Algorithms and Applications*, 11(3–4):187–204, 1997.
23. M.V. Wickerhauser. *Adapted wavelet analysis from theory to software*. A.K. Peters, Wellesley, Mass., 1994.

Compiler-Generated Vector-based Prefetching on Architectures with Distributed Memory

Matthias M. Müller

Computer Science Department
Universität Karlsruhe, Germany

Abstract. Network latency is the main hindrance for fast remote memory access in parallel computing. It prevents the fast execution of fine-grained data-parallel applications with a high amount of communication. This paper presents software controlled access pipelining with vector commands (VSCAP) to overcome this drawback in machines with distributed memory. VSCAP overlaps communication with both computation and communication by means of prefetching. VSCAP is implemented in the Karlsruhe HPF compiler KarHPFn. A comprehensive evaluation with 25 benchmarks from a wide range of data-parallel algorithm classes shows the efficiency of VSCAP on the Cray T3E. KarHPFn's VSCAP programs reach the theoretical peak communication performance on the T3E for regular communication. For dynamic communication patterns, the communication performance is unsurpassed by alternative approaches on this architecture.

1 Introduction

As microprocessors get faster and the gap between computation and remote memory access widens, network latency is the dominant factor of the execution time of fine-grained data-parallel programs. Instead of a single communication operation a processor can perform hundreds to thousand arithmetic operations. This situation is even worse by an order of magnitude once the software overhead of communication libraries is taken into account.

Software controlled access pipelining with vector commands (VSCAP) overcomes these shortcomings by means of prefetching. VSCAP does not solely rely on overlapping communication with computation, it also overlaps communication operations with other communication operations to hide even more network latency.

Research in prefetching can be divided into software [13], hardware [4,9,5], and hybrid prefetching [16,10,12]. VSCAP's prefetching approach is quite different as it does not address parallel architectures with cache coherent memory, it rather targets machines with distributed memory and explicit communication operations where data distribution is the responsibility of the programmer and not the system. With known data distribution, we can prefetch effectively and only what is needed and therefore keep the pressure on the network low.

VSCAP programs are generated automatically by the Karlsruhe proto-type HPF compiler KarHPFn.[1] The measurements of 25 benchmarks from a wide range of common algorithm classes show the efficiency of VSCAP compared to highly optimized shared-memory functions and to Portland Group's HPF compiler pghpf. For regular communication, KarHPFn generated VS-CAP programs are as fast as the shared-memory functions and up to 15 times faster than pghpf. For dynamic communication patterns, VSCAP is the technique with the best communication performance of the compared techniques.

VSCAP grounds on SCAP [17], the predecessor of VSCAP without vector operations. First experimental results of SCAP and VSCAP has been presented in [15,14]. These results show only practicability of SCAP and VSCAP in principle. Now, this paper presents a thorough evaluation of the VSCAP communication technique on a wide range of algorithm classes, and it explains the communication generation mechanism within KarHPFn.

2   VSCAP

This section explains the techniques used by VSCAP.

2.1   Overlapping Communication

The aim of VSCAP is to improve run-time of a program by overlapping several communication requests leading to a communication pipeline between prefetch and access instructions.

For a better understanding of the basic principle of VSCAP, it is first explained how communication is done usually. The processor issues a request to the network (downwards-arrow in Figure 1) and waits until the network replies (upwards-arrows). Only then, the processor continues its execution and issues a new request. This is done as long as the processor requires remote data-elements to perform its local part of computation. As the processor blocks after each data request, we call this *blocking communication*.

In *overlapping communication* the processor issues all its communication requests and the network is able to process them in an overlapped fashion. This leads to a shorter waiting period for the processor accessing the first and all other successive remote data-elements. Finally, communication is performed faster compared to the above mentioned blocking execution, see the lower half of Figure 1. To enable overlapping communication, the network interface has to provide a *prefetch buffer* that decouples the processor from the network execution.

VSCAP augments the overlapping communication with vector commands for prefetch and access. Instead of issuing a communication requests for each single non-local data-element, the processor can prefetch and access $L >$

---

[1] *Karpfen* ['karpfən] without 'h' is the german word for carp.

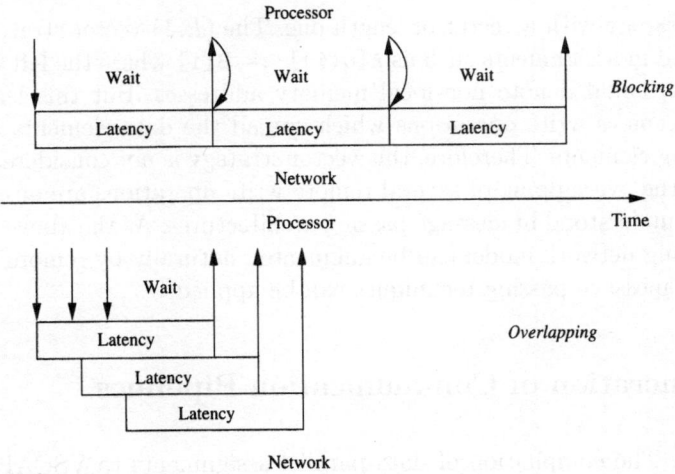

Fig. 1. Blocking vs. Overlapping Communication.

1 data-elements at once. L is the vector length of the vector commands. VSCAP's vector commands reduce prefetch and access overhead and improve communication time further.

2.2   Vector Strategies

In principal, vector commands for prefetching are only useful if displacements of the elements are equidistant and known at compile-time. Otherwise, if element addresses can be computed only at run-time as in dynamic communication patterns or if the distances of elements vary on a per-element basis, single-element prefetch instructions are used throughout the whole prefetch loop. For this reason we introduce the notion of a *vector strategy* that declares the usage of vector operations.

(p, a)-**vector strategy**

A (p, a)-*vector strategy* declares usage of vector operations for prefetch ($p \geq 1$) and access ($a \geq 1$) operations. Assuming vector lengths $p, a \in \{1, L\}$ there are four possible vector strategies:

Vector strategy	Explanation
(1,1)	Element wise prefetch and access operations
(1,L)	Element operations for prefetch but vector access
(L,L)	Vector operations both for prefetch and access
(L,1)	Vector prefetch but element wise access operation

This paper concentrates on the (L, L)- and $(1, L)$-vector strategies. The $(1, 1)$-vector strategy is not further mentioned as it is a special case of the

(L, L)-strategy with a vector of length one. The $(L, 1)$-vector strategy would be applied in assignments such as A[q(i)] := B[i] where the left hand side index $q(i)$ could denote non-local memory addresses. But this leads, however, to remote write operations which spread the data-elements across all processing elements. Therefore, this vector strategy is not considered further, because the overlapping of several remote write operations are already done and well understood in message passing architectures. As the above proposed overlapping network model can be augmented naturally by remote write operations, message passing techniques can be applied.

3   Generation of Communication Pipelines

pipelines. The compilation of data-parallel assignments to VSCAP involves two steps. The first step analyses communication primitives and maps them to transformation patterns. Transformation patterns describe the generation of program code using vector strategies as basic blocks. The second step generates the communication pipelines according to the selected transformation pattern and inserts them into the program code. An overview about KarH-PFn which is the implementation of the presented transformation for the Cray T3E finishes this section.

3.1   Mapping of Communication Primitives to Transformation Patterns

The starting point of the transformation is a data-parallel forall-statement. The first step maps the communication primitives of the data-parallel assignment to transformation patterns. The mapping uses the pattern matching technique used for communication detection [1]. The communication primitives are selected by an index pair. This pair is formed by the indexes of the left hand side and the right hand side array of a data-parallel assignment. Assume the following data-parallel assignment:

FORALL $i_1 = 0, \ldots i_n = 0$ **TO** $N_1 - 1, \ldots N_n - 1$ **DO**
  A$[i_1, \ldots i_n] :=$ B$[g_1(i_1), \ldots, g_n(i_n)]$;
END

The algorithm forms the index pairs $(i_j, g_j(i_j))$ $(j = 1, \ldots n)$ and selects the appropriate transformation pattern according to Table 1. The patterns are explained in Section 3.2.

Depending on the data-distribution of the left hand side array A and the index pair $(i_j, g_j(i_j))$ the table selects a transformation pattern. The selection assumes that the arrays A and B are distributed identically. If this assumption does not hold, the transformation pattern in the last row is selected. This pattern matching algorithm for communication detection is used in the first phase of the compilation.

Table 1. Mapping of communication primitives to transformation patterns. a, b are arbitrary integers, c an integer constant, f a functional symbol, and W an array symbol.

(lhs, rhs)	Comm. primitive	Transformation pattern BLOCK	CYCLIC
(i, b)	multicast	oneblock	oneblock
$(i, i + c)$	overlap shift	multiblock	oneblock
$(i, i + b)$	temporary shift	multiblock	oneblock
$(i, a * i + b)$	affine	multiblock	collect
$(i, f(i))$	precompute read	collect	collect
$(i, W[i])$	gather	collect	collect
other	precompute read	collect	

3.2 Transformation Patterns for Vector Strategies

Table 1 maps the communication primitives to three different transformation patterns: *oneblock*, *multiblock*, and *collect*. Altogether, there are four different transformation patterns: the three patterns above and another pattern for reductions called *reduction*. This last pattern is selected by calls to reduction functions within the source code. *Oneblock*, *multiblock*, and *reduction* generate code for the (L, L)-vector strategy. *collect* does this task for the $(1, L)$-vector strategy. The four patterns are sketched as follows.

Collect The *collect* transformation pattern is selected for dynamic communication patterns. It gathers data-elements spread across all processing elements. Element-wise prefetch instructions access memory locations on different processors.

Oneblock The *oneblock* transformation pattern assumes that all remote data-elements reside on one remote processor. Thus, the sole task of this transformation pattern is to emit the communication pipeline of the (L, L)-vector strategy.

Multiblock This transformation is used for regular data accesses spread across multiple processors $p_1 \ldots p_m$. As a single vector prefetch command can not access data from two different processors, communication is split into m phases. Each phase uses the (L, L)-vector strategy. The remote data accesses for each phase are emitted with the *oneblock* transformation.

Reduction Reductions are characterized by a computation of local results followed by a collection of the local results. To get enough remote memory accesses to hide network latency, VSCAP uses a reduction tree with a *fan-in* of $f > 2$. Thus, instead of overlapping only two remote memory accesses, f non-local data accesses can be used to hide the network latency. Finally, a reduction involves $log_f(P)$ collection steps to calculate the final result.

The generated pipelines are inserted into the program code. The next section shows the code for the *collect* and the *oneblock* transformation pattern.

3.3   Generation of Vector Strategies

The final step of the transformation generates the (L, L)- and $(1, L)$-vector strategies. The pipeline generation needs the description of the communication and iteration set. These sets are provided with regular section descriptors RSDs [7]. An RSD is a triple of the form $(low : high : inc)$ denoting all indexes from *low* to *high* in increments of *inc*. The RSDs are obtained using partitioning analysis techniques [11] and the owner-computes rule. The communication set $(low_B, high_B, inc_B)$ describes those indexes of B that require non-local data accesses of the local processor. The RSD $(low_A, high_A, inc_A)$ describes the local indexes of A that need the non-local data-elements from B. The calculation of these sets needs the data-distribution of A and B and the index function q. The implementation also needs the number of remote memory accesses $K = \frac{high_B - low_B}{inc_B} + 1 = \frac{high_A - low_A}{inc_A} + 1$.

The vector pipelines perform the copy operation $A[low_A + j * inc_A] = B[q(low_B + j * inc_B)]$ for regular accesses patterns $(q = Id)$ and for dynamic access patterns $(q = $ index function, $0 \leq j < K)$. Table 2 shows the conceptual implementation without possible optimizations. The left hand side shows the (L, L)- and the right hand side the $(1, L)$-vector strategy.

The implementation assumes $K \bmod L \equiv 0$, for brevity. Otherwise, a prefetch loop and an access loop with operations on a per-element basis would have to be inserted before loop 1 and loop 2, respectively. The first loop fills the prefetch buffer with capacity C_P avoiding an overflow of the prefetch buffer. The second loop alternates between vector accesses and successive prefetch instructions. The last loop accesses the remaining elements from the prefetch buffer.

3.4   Compilation within KarHPFn

Generation of VSCAP is incorporated into the Karlsruhe prototype HPF compiler. KarHPFn is a source-to-source compiler transforming a data-parallel HPF program into an executable Fortran 90 node program that uses the E-register operations of the T3E for communication.

The Cray T3E was used as target platform for the implementation since it is the only machine architecture providing overlapping communication by means of E-registers that can be used as prefetch buffers. Eight E-registers can be combined to a vector. Distance between successive vector elements have to be equidistant to ensure correct address translation. Thus, the T3E enables VSCAP with a vector length of $L = 8$.

Table 2. The (L,L)- and $(1,L)$-vector strategies. It is $K \bmod L \equiv 0$. vector_access(A[k], inc, addr) copies L entries from the prefetch buffer starting at address addr to the memory locations denoted by $A[k] + i * inc$, for $0 \leq i < L$. The macro RCLC is defined as RCLC ::= $low_B + (I_{Acc} - low_A)/inc_A * inc_B$.

(L, L)-vector strategy	$(1, L)$-vector strategy
Loop 1: Prefetch loop	

// Vector prefetch
$I_{Pre} := low_B$;
FOR i = 0 **TO** MIN(K, C_P)/L-1 **DO**
    addr := calc_addr(B[I_{Pre}]);
    vector_prefetch(addr,inc_B);
    $I_{Pre} := I_{Pre} + inc_B * L$;
END
	// Element prefetch
	$I_{Pre} := low_B$;
	FOR i = 0 **TO** MIN(K,C_P)-1 **DO**
	    addr := calc_addr(B[q(I_{Pre})]);
	    prefetch(addr);
	    $I_{Pre} := I_{Pre} + inc_B$;
	END

Loop 2: Combined access and prefetch loop	

$I_{Acc} := low_A$;
FOR i = 0 **TO** (K-C_P)/L-1 **DO**
    addr := calc_addr(B[RCLC]);
    vector_access(A[I_{Acc}],inc_A,addr);
    $I_{Acc} := I_{Acc} + inc_A * L$;
    addr := calc_addr(B[I_{Pre}]);
    vector_prefetch(addr,inc_B);
    $I_{Pre} := I_{Pre} + inc_B * L$;
END
	$I_{Acc} := low_A$;
	FOR i = 0 **TO** (K-C_P)/L-1 **DO**
	    addr := calc_addr(B[q(RCLC)]);
	    vector_access(A[I_{Acc}],inc_A,addr);
	    $I_{Acc} := I_{Acc} + inc_A * L$;
	    FOR j = 0 **TO** L-1 **DO**
	        addr := calc_addr(B[q(I_{Pre})]);
	        prefetch(addr);
	        $I_{Pre} := I_{Pre} + inc_B$;
	    END
	END

Loop 3: access loop	

FOR i = 0 **TO** MIN(K,C_P)/L-1 **DO**
    addr := calc_addr(B[RCLC]);
    vector_access(A[I_{Acc}],inc_A,addr);
    $I_{Acc} := I_{Acc} + inc_A * L$;
END
	FOR i = 0 **TO** MIN(K,C_P-L)/L-1 **DO**
	    addr := calc_addr(B[q(RCLC)]);
	    vector_access(A[I_{Acc}],inc_A,addr);
	    $I_{Acc} := I_{Acc} + inc_A * L$;
	END

4   Benchmarks and their Implementation

The set of benchmarks is categorized into algorithm classes AC. The distinctive properties are the relationship of communication T_{Comm} to computation time T_{Calc} and the data access pattern, see Table 3. The columns present the different amount of communication while the rows distinguish communication patterns. Each entry presents the algorithm class and the transformation pattern used by the compilation. The algorithm classes are characterized as follows.

AC1: Reductions are a common operation in parallel applications. The reduction in algorithm class 1 is structured in such a way that each processor performs a local reduction and then collects the results from other

Table 3. Data access patterns and the associated algorithm classes.

Access Pattern	$T_{\text{Comm}} < T_{\text{Calc}}$	$T_{\text{Comm}} \sim T_{\text{Calc}}$
Reduction	AC1, *reduction*	AC2, *multiblock*
Indexed Arrays	AC3, *multiblock*	AC4, *one/multiblock*
Indirect Indexed Arrays	AC5, empty	AC6, *collect*
nD-grid	AC7, *oneblock*	AC8, empty
Scatter	AC9, *oneblock*	AC10, *oneblock*

processors. Thus, the amount of communication is merely depending on the number of processors and small compared to computation. Benchmarks: LL3, LL4, LL24.

AC2: This reduction is not implemented as efficiently as the one of AC1. Rather than a local reduction each array element collects results from $log(N)$ neighbors, where N is the size of the array. Thus, communication grows with the size of the array. Benchmarks: LL2, LL5, LL11, LL13, LL14, LL19, LL23.

AC3: This algorithm class deals with indexed arrays. The index offsets are constants. Benchmarks: LL1, LL7, LL8, LL12, LL15, LL18.

AC4: The index offsets in these indexed arrays are arbitrary integer variables or the data-distribution involves a lot of communication. Benchmarks: LL6, Rotate.

AC6: This algorithm class represents dynamic communication caused by indexing. The indexes are computed dynamically from other data. Benchmarks: Indirect, Fire.

AC7: This algorithm class deals with blocked nD-grids in which communication is limited to the border of the blocks. Benchmarks: Jacobi, Laplace, PDE1.

AC9: In this kind of scatter operations, local data has to be spread to a subset of the processing elements. Benchmark: LL21.

AC10: These scatter operations spread data to *all* processing elements. Benchmark: Veltran.

The algorithm classes AC5 and AC8 are empty. For AC5, the author did not find an appropriate benchmark and for AC8, there exists no benchmark because the proportion of computation to communication is always high in nD-grid applications.

Most of the benchmarks are data-parallel versions of the Livermore loop kernels (*LL*) [6,17]. *Rotate* implements a cyclic shift of an array distributed cyclicly. *Indirect* performs indirect array access. *Fire* is a fluid dynamics package using the method of conjugate gradients on unstructured meshes [3]. *Jacobi* and *Laplace* perform successive over-relaxation on a 2D-grid. *PDE1* is a 3D-grid Poisson solver using red-black relaxation. *Veltran* is an application from geophysics that uses the method of conjugate gradients [8].

Most of the benchmarks contain only one of the access patterns of Table 3. The benchmarks containing more than one access pattern are classified into the algorithm class that dominates its run-time behavior. Thus, *LL13*, *LL14*, and *LL23* are classified into AC2; *LL6* into AC4; *Laplace* into AC7; and *Fire* into AC6. The benchmarks were compiled to four different versions.

BLOCK simulates blocking communication on the T3E.

VSCAP does prefetch and access with vectors of length $L = 8$.

SHMEM uses Cray's shared-memory system functions for communication. SHMEM delivers maximum communication performance on the T3E for regular communication patterns. SHMEM behaves like BLOCK in dynamic patterns because of lack of support by the system library.

PGHPF represents the executables of pghpf [2].

PDE1, *Fire*, and *Veltran* were measured with a fixed problem size and the number of processors was varied from 2 to 128. The remaining benchmarks were measured with varying problem sizes and a fixed number of 32 processors. *LL1*, *LL7*, *LL13*, *LL21*, *Jacobi* and *Laplace* were ran on 64 processors. All benchmark were compiled by KarHPFn except for *LL13* and *LL14*. These two kernels need the HPF SUM_SCATTER function which is not yet supported by KarHPFn. The author inserted this function manually.

5   Results

The run-times of the different versions of each benchmark are compared to the basic performance of BLOCK. Due to the large number of benchmarks, Table 4 summarizes the results of the measurements.

The columns of Table 4 present the different versions while the rows present the benchmarks. Each of the columns (VSCAP, SHMEM, and PGHPF) is divided into two sub columns *run-time improvement factor* (*Factor*) and *Virtualization* (*Virt*). Their entries show the maximum relative performance compared to BLOCK at the corresponding virtualization. For example, the *LL6* pair (22.3, 2048) for VSCAP denotes that VSCAP achieves the best result at a virtualization of 2048 local elements. In this case, it is 22.3 times faster then BLOCK.

First, have a look at the algorithm classes AC1 and AC2 that are dominated by a reduction. Three observations are supported by a comparison of the columns entitled *Factor*. First, except for *LL24* and *LL2*, VSCAP is faster than the shared-memory functions of SHMEM. Second, the performance of both VSCAP and SHMEM compared to BLOCK is higher for benchmarks in AC2 than in AC1. This is due to the different kind of reduction. As in AC1 the number of non-local data accesses is fixed to the number of processors and therefore limited to 32. The number of non-local data accesses in AC2 varies with virtualization. Thus, the benchmarks in AC2 can hide network latency better than BLOCK. And third, KarHPFn compiled VSCAP performs

Table 4. Run-time results.

Program	Number of PEs	VSCAP Factor	VSCAP Virt	SHMEM Factor	SHMEM Virt	PGHPF Factor	PGHPF Virt
Performance relative to BLOCK							
Algorithm Class AC1							
LL3	32	2.2	8	1.5	8	1.0	16384
LL4	32	2.2	128	1.4	128	0.9	16384
LL24	32	2.4	512	2.5	512	0.8	32768
Algorithm Class AC2							
LL2	32	30.0	32768	31.1	32768	1.7	32768
LL5	32	10.8	256	9.4	1024	1.3	64
LL11	32	15.9	1024	14.5	1024	1.2	1
LL13	64	6.7	4	6.4	4	4.6	256
LL14	32	7.9	4	6.5	4	0.2	64
LL19	32	9.4	512	8.0	512	2.1	32
LL23	32	7.1	512	6.6	512	0.6	2048
Algorithm Class AC3							
LL1	64	3.0	16	3.0	16	1.9	1
LL7	64	2.1	4	1.9	8	0.1	32768
LL8	32	1.0	64	1.0	64	0.8	1024
LL12	32	1.2	2048	1.2	2048	1.0	32768
LL15	32	2.0	1	2.0	1	1.2	8
LL18	32	2.0	4	2.0	4	1.0	32768
Algorithm Class AC4							
LL6	32	22.3	2048	20.0	2048	1.1	8
Rotate	32	32.9	512	31.4	512	1.1	512
Algorithm Class AC6							
Indirect	32	6.5	1024	0.9	1	1.1	64
Fire	128	3.2	2484	0.9	2484	1.1	2484
Algorithm Class AC7							
Jacobi	64	2.7	15^2	2.8	15^2	0.7	2047^2
Laplace	64	2.4	15^2	2.0	15^2	0.9	2047^2
PDE1	128	2.6	$16*32^2$	2.7	$16*32^2$	0.2	$16*32^2$
Algorithm Class AC9							
LL21	64	9.9	8	9.8	8	0.9	256
Algorithm Class AC10							
Veltran	128	4.6	8*34	4.7	8*34	0.9	8*34

better than PGHPF. For *LL2*, VSCAP is 15 times faster than PGHPF. This third observation is valid for all benchmarks.

Evaluation of AC3 and AC4 shows a similar result. AC3 and AC4 are characterized by indexed array accesses. For these classes, VSCAP is as fast as SHMEM for all benchmarks. The good performance of VSCAP and SHMEM for AC4 compared to BLOCK is striking. This behavior is due to the high amount of communication within these benchmarks.

The next algorithm class AC6 is dominated by a dynamic access pattern caused by indirect indexed array accesses. Now, VSCAP is more than 6 times faster than SHMEM for *Indirect* and more than 3 times faster for *Fire*. The last result was measured on 128 processors.

AC6 is also the only class in which PGHPF is faster than SHMEM. The benchmarks *LL13* and *LL14* of AC2 also have dynamic access patterns. But the behavior of these patterns are obliterated by the dominating indexed array access.

The next algorithm class AC7 shows performance of nD-grid applications. For *Jacobi* and *PDE1*, VSCAP is almost as fast as SHMEM. And both versions are more than two times faster than PGHPF for all benchmarks.

AC9 and AC10 show performance of scatter operations. *Veltran* shows performance on 128 processors. Both versions, VSCAP and SHMEM are more then 5 times faster than PGHPF and VSCAP is almost as fast as SHMEM.

6 Conclusions

This study presented software controlled access pipelining with vector commands (VSCAP) to hide network latency in architectures with distributed memory. The paper presented the compilation technique used to obtain VSCAP's communication pipelines and its implementation within the HPF compiler KarHPFn.

The executables of KarHPFn were compared to the shared-memory library of the T3E and to Portland Group's HPF compiler pghpf. On 25 benchmarks covering a wide range of parallel algorithm classes VSCAP is about as fast as T3E's system functions for regular communication patterns. But for dynamic communication patterns, VSCAP is more than 6 times faster. KarHPFn also outperforms the executables of pghpf in all tests; for benchmarks with a high proportion of communication KarHPFn code was up to 15 times faster.

The web page http://wwwipd.ira.uka.de/KarHPFn provides further information about KarHPFn and some additional optimization techniques for the T3E. There is also a short introduction into E-register programming.

Acknowledgments

I thank Thomas Warschko who gave me the possibility to develop VSCAP. I thank him, Walter Tichy and Michael Philippsen for their everlasting support, the fruitful discussions, and their thorough reviews. This work was sponsored by a scholarship of the Graduiertenkolleg Karlsruhe "Beherrschbarkeit komplexer Systeme".

References

1. Z. Bozkus. *Compiling Fortran 90D/HPF for Distributed Memory MIMD Computers*. PhD thesis, Syracuse University, June 1995.
2. Z. Bozkus, L. Meadows, S. Nakamoto, V. Schuster, and M. Young. PGHPF – An optimizing High Performance Fortran compiler for distributed memory machines. *Scientific Programming*, 6(1):29–40, Spring 1997.
3. P. Brezany, V. Sipkova, B. Chapman, and R. Greimel. Automatic parallelization of the AVL FIRE benchmark for a distributed-memory system. In J. Dongarra, K. Madsen, and J. Wasniewsky, editors, *Applied parallel computing: computations in physics, chemistry, and engineering science: second international workshop, PARA '95*, volume 1041 of *Lecture Notes in Computer Science*, pages 50–60, Lyngby, August 1996. Springer.
4. T. Chen and J. Baer. Reducing memory latency via non-blocking and prefetching caches. In *Fifth International Conference on Architectural Support for Programming Languages and Operating Systems*, pages 51–61, Boston, Massachusetts, October 1992. Also available as U. Washington CS TR 92-06-03.
5. C. Chi and C. Cheung. Hardware-driven prefetching for pointer data references. In *Twelfth International Conference on Supercomputing*, pages 384–395, Melbourne, July 1998.
6. J. Feo. An analysis of the computational and parallel complexity of the Livermore loops. *Parallel Computing*, 7(2):163–185, June 1988.
7. P. Havlak and K. Kennedy. An implementation of interprocedural bounded regular section analysis. *IEEE Transactions on Parallel and Distributed Systems*, 2(3):350–360, July 1991.
8. M. Jacob, M. Philippsen, and M. Karrenbach. Large-scale parallel geophysical algorithms in Java: a feasibility study. *Concurrency: Practice and Experience*, 10(11–13):1143–1153, September 1998. Special Issue: Java for High-performance Network Computing.
9. N. Jouppi. Improving direct-mapped cache performance by the addition of a small fully-associative cache and prefetch buffers. In *Seventeenth Annual International Symposium on Computer Architecture*, pages 364–373, Seattle, Washington, 1990.
10. A. Klaiber and H. Levy. An architecture for software-controlled data prefetching. In *Eighteenth Annual International Symposium on Computer Architecture*, pages 43–53, Toronto, May 1991.
11. C. Koelbel. Compile-time generation of regular communications patterns. In *Fourth Conference on Supercomputing*, pages 101–110, Alburquerque, New Mexiko, November 1991.
12. S. McKee, R. Klenke, K. Wright, W. Wulf, M. Salinas, J. Aylor, and A. Batson. Smarter memory: Improving bandwith for streamed references. *IEEE Computer*, 31(7):54–63, July 1998.
13. T. Mowry. *Tolerating Latency Through Software Controlled Data Prefetching*. PhD thesis, Department of Computer Science, Stanford University, March 1994.

14. M. Müller. KaHPF: Compiler generated data prefetching for HPF. In *High Performance Computing in Science and Engineering 1999*, pages 474–482. Springer, 2000.
15. M. Müller, T. Warschko, and W. Tichy. Prefetching on the Cray-T3E. In *Twelfth International Conference on Supercomputing*, pages 368–375, Melbourne, July 1998.
16. A. Rogers and K. Li. Software support for speculative loads. In *Fifth International Conference on Architectural Support for Programming Languages and Operating Systems*, pages 38–50, Boston, Massachusetts, October 1992.
17. T. Warschko. *Effiziente Kommunikation in Parallelrechnerarchitekturen.* PhD thesis, School of Computer Science, Universität Karlsruhe, December 1997.

Lecture Notes in Computational Science and Engineering

Vol. 1 D. Funaro, *Spectral Elements for Transport-Dominated Equations.* 1997. X, 211 pp. Softcover. ISBN 3-540-62649-2

Vol. 2 H. P. Langtangen, *Computational Partial Differential Equations.* Numerical Methods and Diffpack Programming. 1999. XXIII, 682 pp. Hardcover. ISBN 3-540-65274-4

Vol. 3 W. Hackbusch, G. Wittum (eds.), *Multigrid Methods V.* Proceedings of the Fifth European Multigrid Conference held in Stuttgart, Germany, October 1-4, 1996. 1998. VIII, 334 pp. Softcover. ISBN 3-540-63133-X

Vol. 4 P. Deuflhard, J. Hermans, B. Leimkuhler, A. E. Mark, S. Reich, R. D. Skeel (eds.), *Computational Molecular Dynamics: Challenges, Methods, Ideas.* Proceedings of the 2nd International Symposium on Algorithms for Macromolecular Modelling, Berlin, May 21-24, 1997. 1998. XI, 489 pp. Softcover. ISBN 3-540-63242-5

Vol. 5 D. Kröner, M. Ohlberger, C. Rohde (eds.), *An Introduction to Recent Developments in Theory and Numerics for Conservation Laws.* Proceedings of the International School on Theory and Numerics for Conservation Laws, Freiburg / Littenweiler, October 20-24, 1997. 1998. VII, 285 pp. Softcover. ISBN 3-540-65081-4

Vol. 6 S. Turek, *Efficient Solvers for Incompressible Flow Problems.* An Algorithmic and Computational Approach. 1999. XVII, 352 pp, with CD-ROM. Hardcover. ISBN 3-540-65433-X

Vol. 7 R. von Schwerin, *Multi Body System SIMulation.* Numerical Methods, Algorithms, and Software. 1999. XX, 338 pp. Softcover. ISBN 3-540-65662-6

Vol. 8 H.-J. Bungartz, F. Durst, C. Zenger (eds.), *High Performance Scientific and Engineering Computing.* Proceedings of the International FORTWIHR Conference on HPSEC, Munich, March 16-18, 1998. 1999. X, 471 pp. Softcover. 3-540-65730-4

Vol. 9 T. J. Barth, H. Deconinck (eds.), *High-Order Methods for Computational Physics.* 1999. VII, 582 pp. Hardcover. 3-540-65893-9

Vol. 10 H. P. Langtangen, A. M. Bruaset, E. Quak (eds.), *Advances in Software Tools for Scientific Computing.* 2000. X, 357 pp. Softcover. 3-540-66557-9

Vol. 11 B. Cockburn, G. E. Karniadakis, C.-W. Shu (eds.), *Discontinuous Galerkin Methods.* Theory, Computation and Applications. 2000. XI, 470 pp. Hardcover. 3-540-66787-3

Vol. 12   U. van Rienen, *Numerical Methods in Computational Electrodynamics. Linear Systems in Practical Applications.* 2000. XIII, 375 pp. Softcover. 3-540-67629-5

Vol. 13   B. Engquist, L. Johnsson, M. Hammill, F. Short (eds.), *Simulation and Visualization on the Grid.* Parallelldatorcentrum Seventh Annual Conference, Stockholm, December 1999, Proceedings. 2000. XIII, 301 pp. Softcover. 3-540-67264-8

Vol. 14   E. Dick, K. Riemslagh, J. Vierendeels (eds.), *Multigrid Methods VI.* Proceedings of the Sixth European Multigrid Conference Held in Gent, Belgium, September 27-30, 1999. 2000. IX, 293 pp. Softcover. 3-540-67157-9

Vol. 15   A. Frommer, T. Lippert, B. Medeke, K. Schilling (eds.), *Numerical Challenges in Lattice Quantum Chromodynamics.* Joint Interdisciplinary Workshop of John von Neumann Institute for Computing, Jülich and Institute of Applied Computer Science, Wuppertal University, August 1999. 2000. VIII, 184 pp. Softcover. 3-540-67732-1

Vol. 16   J. Lang, *Adaptive Multilevel Solution of Nonlinear Parabolic PDE Systems.* Theory, Algorithm, and Applications. 2001. XII, 157 pp. Softcover. 3-540-67900-6

Vol. 17   B. I. Wohlmuth, *Discretization Methods and Iterative Solvers Based on Domain Decomposition.* 2001. X, 197 pp. Softcover. 3-540-41083-X

Vol. 18   U. van Rienen, M. Günther, D. Hecht (eds.), *Scientific Computing in Electrical Engineering.* Proceedings of the 3rd International Workshop, August 20-23, 2000, Warnemünde, Germany. 2001. XII, 428 pp. Softcover. 3-540-42173-4

Vol. 19   I. Babuška, P. G. Ciarlet, T. Miyoshi (eds.), *Mathematical Modeling and Numerical Simulation in Continuum Mechanics.* Proceedings of the International Symposium on Mathematical Modeling and Numerical Simulation in Continuum Mechanics, September 29 - October 3, 2000, Yamaguchi, Japan. 2002. VIII, 301 pp. Softcover. 3-540-42399-0

Vol. 20   T. J. Barth, T. Chan, R. Haimes (eds.), *Multiscale and Multiresolution Methods.* Theory and Applications. 2002. X, 389 pp. Softcover. 3-540-42420-2

Vol. 21   M. Breuer, F. Durst, C. Zenger (eds.), *High Performance Scientific and Engineering Computing.* Proceedings of the 3rd International FORTWIHR Conference on HPSEC, Erlangen, March 12-14, 2001. 2002. XIII, 408 pp. Softcover. 3-540-42946-8

Vol. 22   K. Urban, *Wavelets in Numerical Simulation.* Problem Adapted Construction and Applications. 2002. XV, 181 pp. Softcover. 3-540-43055-5

For further information on these books please have a look at our mathematics catalogue at the following URL: http://www.springer.de/math/index.html